In this third volume of *The Quantum Theory of Fields*, Nobel Laureate Steven Weinberg continues his masterly exposition of quantum field theory. This volume presents a self-contained, up-to-date, and comprehensive introduction to supersymmetry, a highly active area of theoretical physics that is likely to be at the center of future progress in the physics of elementary particles and gravitation. The text introduces and explains a broad range of topics, including supersymmetric algebras, supersymmetric field theories, extended supersymmetry, supergraphs, non-perturbative results, theories of supersymmetry in higher dimensions, and supergravity. A thorough review is given of the phenomenological implications of supersymmetry, including theories of both gauge- and gravitationally mediated supersymmetry breaking. Also provided is an introduction to mathematical techniques, based on holomorphy and duality, that have proved so fruitful in recent developments. This book contains much material not found in other books on supersymmetry, including previously unpublished results. Problems are included.

From reviews of Volume I

'...an impressively lucid and thorough presentation of the subject... Weinberg manages to present difficult topics with richness of meaning and marvellous clarity. Full of valuable insights, his treatise is sure to become a classic, doing for quantum field theory what Dirac's *Quantum Mechanics* did for quantum mechanics. I eagerly await the publication of the second volume.'

S. S. Schweber, *Nature*

'For over twenty years there has been no good modern textbook on the subject. For all that time, Steven Weinberg has been promising to write one. That he has finally done it is cause for celebration among those who try to teach and try to learn the subject. Weinberg's book is for serious students of field theory... it is the first textbook to treat quantum field theory the way it is used by physicists today.'

Howard Georgi, *Science*

'Steven Weinberg, who contributed to the development of quantum chromodynamics and shared the Nobel Prize in Physics for his contributions to the electroweak theory, has written a definitive text on the physical foundations of quantum field theory. His book differs significantly from the long line of previous books on quantum field theory... To summarize, *Foundations* builds the structure of quantum field theory on the sure footing of physical insight. It is beautifully produced and meticulously edited... and it is a real bargain in price. If you want to learn quantum field theory, or have already learned it and want to have a definitive reference at hand, purchase this book.'

O. W. Greenberg, *Physics Today*

'In addition to a superb treatment of all the conventional topics there are numerous sections covering areas that are not normally emphasized, such as the subject of field redefinitions, higher-rank tensor fields and an unusually clear and thorough treatment of infrared effects... this latest book reinforces his high scholarly standards. It provides a unique exposition that will prove invaluable both to new research students as well as to experienced research workers. Together with Volume II, this will become a classic text on a subject of central importance to a wide area of theoretical physics.'

M. B. Green, *CERN Courier*

'I believe that what readers will find particularly helpful in this volume is the consistency of the whole approach, and the emphasis on quantities and properties that are directly useful to particle physicists. This is particularly true for those who are interested in the more phenomenological aspects. The reader only needs limited background knowledge, and a clear line is followed throughout the book, making it easy to follow. The author presents extremely thorough but elementary discusssions of important physical questions, some of which seem to be an original way of addressing the subject.'

J. Zinn-Justin, *Physics World*

'This is a well-written book by one of the masters of the subject... it is certainly destined to become a standard text book and should find its way to the shelves of every physics library.'

J. Madore, *Classical and Quantum Gravity*

'The book starts out with an excellent historical introduction, not found anywhere else, giving citations to many by now classic papers... a valuable reference work as well as a textbook for graduate students.'

G. Roepstorff, *Zentralblatt für Mathematik*

From reviews of Volume II

'It is a majestic exposition. The two volumes are structured in a logical way. Everything is explained with incisive clarity. Weinberg always goes to the heart of any argument, and includes many things that cannot be found elsewhere in the literature. Often I find myself thinking: "Ah! Now I understand that properly."... I find it hard to imagine a better treatment of quantum field theory than Weinberg's. All serious students and researchers will want to have these volumes on their shelves.'

John C. Taylor, *Nature*

'Weinberg's *Modern Applications* goes to the boundaries of our present understanding of field theory. It is unmatched by any other book on quantum field theory for its depth, generality and definitive character, and it will be an essential reference for serious students and researchers in elementary particle physics.'

O. W. Greenberg, *Physics Today*

'... Steven Weinberg is one of our most gifted makers of theoretical tools as well as a virtuoso in their use. His new book conveys both the satisfaction of understanding nature and the feel of the atelier, for the "modern applications" of its subtitle include both the derivation of physical consequences and the development of new tools for understanding and applying field theory itself... *Modern Applications* is a splendid book, with abundant useful references to the original literature. It is a very interesting read from cover to cover, for the wholeness Weinberg's personal perspective gives to quantum field theory and particle physics.'

Chris Quigg, *Science*

'Experienced researchers and beginning graduate students will delight in the gems of wisdom to be found in these pages. This book combines exposition of technical detail with physical insight in a unique manner that confirms the promise of Volume I and I have no doubt that these two volumes will rapidly constitute the classic treatment of this important subject.'

M. B. Green, *CERN Courier*

'...a valued reference and a mine of useful information for professional field theorists.'

Tom Kibble, *New Scientist*

'...a clear presentation of the subject, explaining the underlying concepts in much depth and in an accessible style. I expect that these volumes will become the first source we turn to when trying to answer the challenging questions asked by bright postgraduates when they first encounter quantum field theory...I have no doubt that *The Quantum Theory of Fields* will soon be found on the bookshelves of most particle theorists, and that it will be one of the main sources used in the preparation of lectures on the subject for postgraduate students.'

C. T. C. Sachrajda, *The Times Higher Education Supplement*

'...Weinberg has produced a masterpiece that will be a standard reference on the field for a long time to come.'

B. E. Y. Svensson, *Elementa*

From reviews of Volume III

'...has produced a treatise that many of us had long awaited, perhaps without fully realizing it...with the publication of *The Quantum Theory of Fields*, Vol. III, has performed an analogous service for supersymmetry...Although this volume is the third in a trilogy, it is quite different from its two predecessors, and it stands on its own...May a new generation of students imbibe its content and spirit.'

Physics Today

'The third volume of *The Quantum Theory of Fields* is a self-contained introduction to the world of supersymmetry and supergravity. It will be useful both for experienced researchers in the field and for students who want to take the first steps towards learning about supersymmetry. Unlike other books in this field, it covers the wide spectrum of possible applications of supersymmetry in physics.'

Hans Peter Nilles, *Nature*

'Weinberg is of course one of the creators of modern quantum field theory, as well as of its physical culmination, the standard model of all (nongravitational) interactions. It is...very timely that this latest part of his monograph, devoted to supersymmetry and supergravity, has just appeared. As a text, it has been pretested by Weinberg for a freestanding one-year graduate course; as a clear organizing reference to this extremely vast field, it will help the experts as well... Weinberg's style of presentation is as clear and meticulous as in his previous works.'

Stanley Deser, *Journal of General Relativity and Gravitation*

'Weinberg tries to be as elementary and clear as possible and steers clear of more sophisticated mathematical tools. Together with the previous volumes, this volume will serve as an invaluable reference to researchers and a textbook for graduate students.'

G. Roepstorff, *Zentralblatt MATH*

The Quantum Theory of Fields

Volume III
Supersymmetry

Steven Weinberg
University of Texas at Austin

CAMBRIDGE
UNIVERSITY PRESS

University Printing House, Cambridge CB2 8BS, United Kingdom

One Liberty Plaza, 20th Floor, New York, NY 10006, USA

477 Williamstown Road, Port Melbourne, VIC 3207, Australia

314-321, 3rd Floor, Plot 3, Splendor Forum, Jasola District Centre, New Delhi - 110025, India

79 Anson Road, #06-04/06, Singapore 079906

Cambridge University Press is part of the University of Cambridge.

It furthers the University's mission by disseminating knowledge in the pursuit of education, learning and research at the highest international levels of excellence.

www.cambridge.org
Information on this title: www.cambridge.org/9780521670555

First published 2000
Paperback edition published 2005
17th printing 2019

A catalogue record for this publication is available from the British Library

ISBN 978-0-521-67055-5 Paperback

Contents

Sections marked with an asterisk are somewhat out of the book's main line of development and may be omitted in a first reading.

Preface To Volume III

This volume deals with quantum field theories that are governed by supersymmetry, a symmetry that unites particles of integer and half-integer spin in common symmetry multiplets. These theories offer a possible way of solving the 'hierarchy problem,' the mystery of the enormous ratio of the Planck mass to the 300 GeV energy scale of electroweak symmetry breaking. Supersymmetry also has the quality of uniqueness that we search for in fundamental physical theories. There is an infinite number of Lie groups that can be used to combine particles of the same spin in ordinary symmetry multiplets, but there are only eight kinds of supersymmetry in four spacetime dimensions, of which only one, the simplest, could be directly relevant to observed particles.

These are reasons enough to devote this third volume of *The Quantum Theory of Fields* to supersymmetry. In addition, the quantum field theories based on supersymmetry have remarkable properties that are not found among other field theories: some supersymmetric theories have couplings that are not renormalized in any order of perturbation theory; other theories are finite; and some even allow exact solutions. Indeed, much of the most interesting work in quantum field theory over the past decade has been in the context of supersymmetry.

Unfortunately, after a quarter century there is no direct evidence for supersymmetry, as no pair of particles related by a supersymmetry transformation has yet been discovered. There is just one significant piece of indirect evidence for supersymmetry: the high-energy unification of the $SU(3)$, $SU(2)$, and $U(1)$ gauge couplings works better with the extra particles called for by supersymmetry than without them.

Nevertheless, because of the intrinsic attractiveness of supersymmetry and the possibility it offers of resolving the hierarchy problem, I and many other physicists are reasonably confident that supersymmetry will be found to be relevant to the real world, and perhaps soon. Supersymmetry is a primary target of experiments at high energy planned at existing accelerators, and at the Large Hadron Collider under construction at the CERN laboratory.

After a historical introduction in Chapter 24, Chapters 25–27 present the essential machinery of supersymmetric field theories: the structure of the supersymmetry algebra and supersymmetry multiplets and the construction of supersymmetric Lagrangians in general, and in particular for theories of chiral and gauge superfields. Chapter 28 then uses this machinery to incorporate supersymmetry in the standard model of electroweak and strong interactions, and reviews experimental difficulties and opportunities. Chapters 29–32 deal with topics that are mathematically more advanced: non-perturbative results, supergraphs, supergravity, and supersymmetry in higher dimensions.

I have made the treatment of supersymmetry here as clear and self-contained as I could. Wherever possible I take the reader through calculations, rather than just reporting results from the literature. Where calculations are too lengthy or complicated to be included in a book of this sort, especially in Chapter 28, I have tried to present simpler versions that give the reader an idea of the physical issues involved.

I have made a point of including topics here that have generally not been covered in earlier books, some because they are too new. These include: the use of holomorphy to study perturbative and non-perturbative radiative corrections; the calculation of central charges; gauge-mediated and anomaly-mediated supersymmetry breaking; the Witten index; duality; the Seiberg–Witten calculation of the effective Lagrangian in $N = 2$ supersymmetric gauge theories; supersymmetry breaking by modular fields; and a first look at the rapidly developing topic of supersymmetry in higher dimensions, including theories with p-branes.

On the other hand, I have shortened the usual treatment of two topics that seemed to me to have been well covered in earlier books. One of these is the use of supergraphs. Many of the previous applications of the supergraph formalism in studying the general structure of radiative corrections can now be handled more easily by using the arguments of holomorphy described in Sections 27.6 and 29.3. The other is supergravity. In Sections 31.1–31.5 I have given a detailed and self-contained treatment of supergravity in the weak-field limit, which makes it clear why the ingredients of supergravity theories — the graviton, gravitino, and auxiliary fields — are what they are, and which allows us to derive some of the most important results of supergravity theory, including the formula for the gravitino mass and for the gaugino masses produced by anomaly-mediated supersymmetry breaking. In Section 31.6 I have outlined the calculations that generalize supergravity theory to gravitational fields of arbitrary strength, but these calculations are so lengthy and unlovely that I was content to quote other sources for the results. However, in Section 31.7 I have given a fuller than usual treatment of gravitationally mediated supersymmetry breaking. I regret that I have not been able to include exciting work of the past decade on supersymmetry related to string theory, but string theory is beyond the scope of this book, and I did not want to report results for which I had not provided a basis of explanation.

I have given citations both to the classic papers on supersymmetry and to useful references on topics that are mentioned but not presented in detail in this book. I did not always know who was responsible for material presented here, and the mere absence of a citation should not be taken as a claim that the material presented here is original, but some of it is. I hope that I have improved on the original literature or standard textbook treatments in several places, as for instance in the proof of the Coleman–Mandula theorem; in the treatment of parity matrices in extended supersymmetry theories; in the inclusion of new soft supersymmetry-breaking terms in the minimum supersymmetric standard model; in the derivation of supercurrent sum rules; and in the proof of the uniqueness of the Seiberg–Witten solution.

I have also supplied problems for each chapter. Some of these problems aim simply at providing exercise in the use of techniques described in the chapter; others are intended to suggest extensions of the results of the chapter to a wider class of theories.

In teaching a course on supersymmetry, I have found that this book provides enough material for a one-year course for graduate students. I intended that this book should be accessible to students who are familiar with quantum field theory at the level it is presented in the first two volumes of this treatise. It is not assumed that the reader has gone through Volumes I and II, but for the convenience of those fortunate readers who have done so I use the same notation here, and give cross-references to material in Volumes I and II wherever appropriate.

* * *

I must acknowledge my special intellectual debt to colleagues at the University of Texas, notably Luis Boya, Phil Candelas, Bryce and Cecile De Witt, Willy Fischler, Daniel Freed, Joaquim Gomis, Vadim Kaplunovsky, and especially Jacques Distler. Also, Sally Dawson, Michael Dine, Michael Duff, Lawrence Hall, Hitoshi Murayama, Joe Polchinski, Edward Witten, and Bruno Zumino gave valuable help with special topics. Jonathan Evans read through the manuscript of this volume, and made many valuable suggestions. For pointing out various errors in the first printing of this book, I am greatly indebted to Stephen Adler, Jose Espinora, Tony Gherghetta, and San Fu Tuan. For corrections to the first printing of this volume I am indebted to several colleagues, especially Stephen Adler. Thanks are due to Alyce Wilson, who prepared the illustrations, to Terry

Riley for finding countless books and articles, and to Jan Duffy for many helps. I am grateful to Maureen Storey of Cambridge University Press for working to ready this book for publication, and especially to my editor, Rufus Neal, for his continued friendly good advice.

STEVEN WEINBERG

Austin, Texas
May, 1999

Notation

The big issue in choosing notation for a book on supersymmetry is whether to use a two-component or a four-component notation for spinors. The standard texts on supersymmetry have opted for the two-component Weyl notation. I have chosen instead to use the four-component Dirac notation except in the first stages of constructing the supersymmetry algebra and multiplets, because I think this will make the book more accessible to those physicists who work on particle phenomenology and model building. It would be a pity to see the growth of a separate enclave of supersymmetry specialists, who communicate well with each other but are cut off by their notation from the larger community of particle theorists.

There is no great difficulty anyway in converting expressions in four-component form into the two-component formalism. In the representation of the Dirac matrices used throughout this book, in which γ_5 is the diagonal matrix with elements +1, +1, −1, and −1 on the main diagonal, any four-component Majorana spinor ψ_α (such as the supersymmetry generator Q_α, the superspace coordinate θ_α, or the superderivative $\mathscr{D}_\alpha$) may be written in terms of a two-component spinor χ_a as

$$\psi = \begin{pmatrix} e\chi^* \\ \chi \end{pmatrix} ,$$

where e is the 2×2 antisymmetric matrix with $e_{12} = +1$. The two-component spinor χ_a is what in other books is often called $\psi^*_a = \bar{\psi}_{\dot{a}}$, while $(e\chi^*)_a$ would be called ψ^a. A summary of useful properties of four-component Majorana spinors is given in the appendix to Chapter 26.

Here are some other features of the notation used in this book:

Latin indices i, j, k, and so on generally run over the three spatial coordinate labels, usually taken as 1, 2, 3. Where specifically indicated, they run over values 1, 2, 3, 4, with $x^4 = it$.

Greek indices μ, ν, etc. from the middle of the Greek alphabet generally run over the four spacetime coordinate labels 1, 2, 3, 0, with x^0 the time coordinate. Where it is necessary to distinguish between spacetime coordinates in a general coordinate system and in a locally inertial system, indices μ, ν, etc. are used for the former and a, b, etc. for the latter.

Greek indices α, β, etc. from the beginning of the Greek alphabet generally (except in Chapter 24) run over the components of four-component spinors. To avoid confusion, I depart here from the conventions of Volume II, and use upper-case letters A, B, etc. to label the generators of a symmetry algebra. Components of two-component spinors are labelled with indices a, b, etc. In particular, four-component supersymmetry generators are denoted Q_α, while two-component generators (the bottom two components of Q_α) are called Q_a.

Repeated indices are generally summed, unless otherwise indicated.

The spacetime metric $\eta_{\mu\nu}$ is diagonal, with elements $\eta_{11} = \eta_{22} = \eta_{33} = 1$, $\eta_{00} = -1$.

The d'Alembertian is defined as $\Box \equiv \eta^{\mu\nu}\partial^2/\partial x^\mu \partial x^\nu = \nabla^2 - \partial^2/\partial t^2$, where ∇^2 is the Laplacian $\partial^2/\partial x^i \partial x^i$.

The 'Levi–Civita tensor' $\epsilon^{\mu\nu\rho\sigma}$ is defined as the totally antisymmetric quantity with $\epsilon^{0123} = +1$.

Dirac matrices γ_μ are defined so that $\gamma_\mu\gamma_\nu + \gamma_\nu\gamma_\mu = 2\eta_{\mu\nu}$. Also, $\gamma_5 = i\gamma_0\gamma_1\gamma_2\gamma_3$, and $\beta = i\gamma^0 = \gamma_4$. Where explicit matrices are needed, they are given by the block matrices

$$\gamma^0 = -i\begin{bmatrix} 0 & \mathbf{1} \\ \mathbf{1} & 0 \end{bmatrix}, \qquad \boldsymbol{\gamma} = -i\begin{bmatrix} 0 & \boldsymbol{\sigma} \\ -\boldsymbol{\sigma} & 0 \end{bmatrix},$$

where $\mathbf{1}$ is the unit 2×2 matrix, $\mathbf{0}$ is the 2×2 matrix with elements zero, and the components of $\boldsymbol{\sigma}$ are the usual Pauli matrices

$$\sigma_1 = \begin{pmatrix} 0 & 1 \\ 1 & 0 \end{pmatrix}, \quad \sigma_2 = \begin{pmatrix} 0 & -i \\ i & 0 \end{pmatrix}, \quad \sigma_3 = \begin{pmatrix} 1 & 0 \\ 0 & -1 \end{pmatrix}.$$

We also frequently make use of the 4×4 block matrices

$$\gamma_5 = \begin{bmatrix} \mathbf{1} & 0 \\ 0 & -\mathbf{1} \end{bmatrix}, \qquad \epsilon = \begin{bmatrix} e & 0 \\ 0 & e \end{bmatrix},$$

where e is again the antisymmetric 2×2 matrix $i\sigma_2$. For instance, our phase convention for four-component Majorana spinors s may be expressed as $s^* = -\beta\,\gamma_5\,\epsilon\, s$.

The step function $\theta(s)$ has the value $+1$ for $s > 0$ and 0 for $s < 0$.

The complex conjugate, transpose, and Hermitian adjoint of a matrix or vector A are denoted A^*, A^{T}, and $A^\dagger = A^{*\mathrm{T}}$, respectively. We use an asterisk * for the Hermitian adjoint of an operator or the complex conjugate of a number, except where a dagger † is used for the transpose of the matrix formed from the Hermitian adjoints of operators or complex conjugates of numbers. +H.c. or +c.c. at the end of an expression indicates the addition of the Hermitian adjoint or complex conjugate of the foregoing terms. A bar on a four-component spinor u is defined by $\bar{u} = u^\dagger \beta$.

Units are used with $\hbar$ and the speed of light taken to be unity. Throughout $-e$ is the rationalized charge of the electron, so that the fine structure constant is $\alpha = e^2/4\pi \simeq 1/137$. Temperatures are in energy units, with the Boltzmann constant taken equal to unity.

Numbers in parenthesis at the end of quoted numerical data give the uncertainty in the last digits of the quoted figure. Where not otherwise indicated, experimental data are taken from 'Review of Particle Physics,' The Particle Data Group, *European Physics Journal C* **3**, 1 (1998).

24
Historical Introduction

The history of supersymmetry is as peculiar as anything in the history of science. Suggested in the early 1970s, supersymmetry has been elaborated since then into a beautiful mathematical formalism that unites particles of different spin into symmetry multiplets and has profound implications for fundamental physics. Yet there is so far not a shred of direct experimental evidence and only a few bits of indirect evidence that supersymmetry has anything to do with the real world. If (as I expect) supersymmetry does turn out to be relevant to nature, it will represent a striking success of purely theoretical insight.

Chapter 25 will begin the construction of supersymmetry theories from first principles. In the present chapter we shall introduce supersymmetry along chronological rather than logical lines.

24.1 Unconventional Symmetries and 'No-Go' Theorems

In the early 1960s the symmetry $SU(3)$ of Gell-Mann and Ne'eman (discussed in Section 19.7) successfully explained the relations between various strongly interacting particles of different charge and strangeness but of the same spin. The idea then grew up that perhaps $SU(3)$ is part of a larger symmetry, which has the unconventional effect of uniting $SU(3)$ multiplets of different spin.[1] There is such an approximate symmetry in the non-relativistic quark model, under $SU(6)$ transformations on quark spins and flavors, analogous to an earlier $SU(4)$ symmetry of nuclear physics that had been introduced in 1937 by Wigner.[2] As described in detail in Appendix A of this chapter, this $SU(6)$ symmetry unites the pseudoscalar meson octet π, K, $\bar{K}$, and η, the vector meson octet ρ, K^*, $\bar{K}^*$, and ω, and the vector meson singlet ϕ in a single **35** multiplet, and also unites the spin 1/2 baryon octet N, Σ, Λ, and Ξ with the spin 3/2 baryon decuplet Δ, $\Sigma(1385)$, $\Xi(1530)$, and Ω in a single **56** multiplet. The $SU(6)$ symmetry scored a number of successes, but it is actually nothing but a consequence of the approximate spin and flavor independence of forces in the quark model; the $SU(6)$ symmetry is somewhat weaker than the assumption of spin and flavor independence, but, as shown in Appendix A, there is no evidence that the predictions of $SU(6)$ symmetry are any more accurate than those of complete spin and flavor independence.

Nevertheless, there were various attempts to generalize the $SU(6)$ symmetry of the non-relativistic quark model to a fully relativistic quantum theory.[3] These attempts all failed, and a number of authors showed under various restrictive assumptions that this is in fact impossible.[4] The most far-reaching theorem of this sort was proved in 1967 by Coleman and Mandula.[5] They adopted reasonable assumptions about the finiteness of the number of particle types below any given mass, the existence of scattering at almost all energies, and the analyticity of the S-matrix, and used them to show that the most general Lie algebra of symmetry operators that commute with the S-matrix, that take single-particle states into single-particle states, and that act on multiparticle states as the direct sum of their action on single-particle states consists of the generators P_μ and $J_{\mu\nu}$ of the Poincaré group, plus ordinary internal symmetry generators that act on one-particle states with matrices that are diagonal in and independent of both momentum and spin. We will use this theorem as an essential ingredient in our analysis of all possible supersymmetry algebras in four spacetime dimensions in Chapter 25, and in higher spacetime dimensions in Chapter 32. In Section 32.3 we will consider supersymmetry algebras in theories that involve extended objects, for which the Coleman–Mandula theorem does not apply.

Coleman and Mandula's proof is ingenious and complicated. A version is presented in Appendix B to this chapter. In the present section we shall give a very simple purely kinematic proof of one piece of this theorem, but this piece is enough to show clearly why an unconventional symmetry like $SU(6)$ is possible in non-relativistic but not in relativistic theories. We shall use Lorentz invariance to show that if the Lie algebra of all symmetry operators B_α that commute with the momentum generators P_μ consists of the P_μ themselves plus the Hermitian generators B_A of some finite-parameter semi-simple compact* Lie subalgebra $\mathscr{A}$, then the B_A must be the generators of an ordinary internal symmetry, in the sense that they act on single-particle states with matrices that are diagonal in and independent of both momentum and spin. No use is made in this theorem of the properties of the S-matrix, of the finiteness of the particle spectrum, or of assumptions about how the symmetry generators act on physical states. The Lie algebra of $SU(6)$ is of course both semi-simple

* For the definition of semi-simple and compact Lie algebras, see the footnote in Section 15.2.

and compact, so this theorem rules out the use of any such symmetry in relativistic theories to derive relations among particles of different spin.

Here is the proof. Let all symmetry generators that commute with the four-momentum P_μ form a Lie algebra spanned by the generators B_α. Consider the effect on these generators of a proper Lorentz transformation $x^\mu \to \Lambda^\mu{}_\nu x^\nu$, which is represented on Hilbert space by the unitary operator $U(\Lambda)$. It is easy to see that the operator $U(\Lambda)B_\alpha U^{-1}(\Lambda)$ is a Hermitian symmetry generator that commutes with $\Lambda_\mu{}^\nu P_\nu$, so since $\Lambda_\mu{}^\nu$ is non-singular, this operator must commute with P_μ, and therefore must be a linear combination of the B_α:

$$U(\Lambda)B_\alpha U^{-1}(\Lambda) = \sum_\beta D^\beta{}_\alpha(\Lambda)\, B_\beta \,, \tag{24.1.1}$$

with $D^\beta{}_\alpha(\Lambda)$ a set of real coefficients that furnish a representation of the homogeneous Lorentz group

$$D(\Lambda_1)\, D(\Lambda_2) = D(\Lambda_1\, \Lambda_2)\,. \tag{24.1.2}$$

Further, the $U(\Lambda)B_\alpha U^{-1}(\Lambda)$ satisfy the same commutation relations as the B_α, so the structure constants $C^\gamma_{\alpha\beta}$ of this Lie algebra are invariant tensors in the sense that

$$C^\gamma_{\alpha\beta} = \sum_{\alpha'\beta'\gamma'} D^{\alpha'}{}_\alpha(\Lambda)\, D^{\beta'}{}_\beta(\Lambda)\, D^\gamma{}_{\gamma'}(\Lambda^{-1})\, C^{\gamma'}_{\alpha'\beta'}\,. \tag{24.1.3}$$

Contracting this with the corresponding equation for $C^\alpha_{\gamma\delta}$, we find that

$$g_{\beta\delta} = \sum_{\beta'\delta'} D^{\beta'}{}_\beta(\Lambda)\, D^{\delta'}{}_\delta(\Lambda)\, g_{\beta'\delta'}\,, \tag{24.1.4}$$

where $g_{\beta\delta}$ is the Lie algebra metric

$$g_{\beta\delta} \equiv \sum_{\alpha\gamma} C^\gamma_{\alpha\beta}\, C^\alpha_{\gamma\delta}\,. \tag{24.1.5}$$

Because all of these generators commute with P_μ, we have $C^\alpha_{\mu\beta} = -C^\alpha_{\beta\mu} = 0$, so $g_{\mu\alpha} = g_{\alpha\mu} = 0$.

We will distinguish the symmetry generators other than the P_μ by using subscripts A, B, etc. in place of α, β, etc. Using the vanishing of $C^A_{\mu B} = -C^A_{B\mu}$ in Eq. (24.1.5) gives $g_{AB} = \sum_{CD} C^D_{AC}\, C^C_{BD}$. We assumed that the generators B_A span a compact semi-simple Lie algebra, so the matrix g_{AB} is positive-definite. Eqs. (24.1.4) and (24.1.2) show that the matrices $g^{1/2}D(\Lambda)g^{-1/2}$ furnish a real orthogonal and hence unitary finite-dimensional representation of the homogeneous Lorentz group. But because the Lorentz group is non-compact, *the only such representation is the trivial one*, for which $D(\Lambda) = 1$. (This is the place where relativity makes all the difference; the semi-simple part of the Galilean group is the *compact* group $SU(2)$, which of course has an infinite number of unitary finite-dimensional representations.) With $D(\Lambda) = 1$, the generators B_A commute with $U(\Lambda)$ for all Lorentz transformations $\Lambda^\mu{}_\nu$.

Acting on the state $|p, n\rangle$ of a single stable particle with momentum p^μ and spin and species labelled by a discrete index n, an operator like B_A that commutes with P_μ can only yield a linear combination of such states

$$B_A|p,n\rangle = \sum_m \Big(b_A(p)\Big)_{mn}|p,m\rangle \,. \tag{24.1.6}$$

The fact that the B_A commute with what we called 'boosts' in Section 2.5 implies that the $b_A(p)$ are independent of momentum, and the fact that the B_A commute with rotations implies that the $b_A(p)$ act as unit matrices on spin indices, so the B_A are the generators of an ordinary internal symmetry, as was to be proved.

24.2 The Birth of Supersymmetry

If theoretical physics followed logic in its evolution, then after the proof of the Coleman–Mandula theorem someone seeking exceptions to this theorem would have noticed that it deals only with transformations that take bosons into bosons and fermions into fermions and are therefore generated by operators that satisfy commutation relations rather than anti-commutation relations. This would have raised the question of whether a relativistic theory can have symmetries acting non-trivially on particle spins that take fermions and bosons into each other, and that therefore satisfy anticommutation relations rather than commutation relations. Exploring the possible structure of such a superalgebra along the lines described in the following chapter, supersymmetry would have emerged as the only possibility.

This is not what happened. Instead, supersymmetry arose in a series of articles on string theory and independently in a pair of little-noticed papers, about which more later, none of which show any sign that the authors were at all concerned with the Coleman–Mandula theorem.

Starting in the late 1960s, the effort to construct S-matrix elements for strong-interaction processes that would satisfy various theoretical requirements led to a new picture of various types of hadrons, as different modes of vibration of a string.[6] A point on a string labelled by a parameter σ will at time τ on some fixed clock have spacetime coordinates $X^\mu(\sigma, \tau)$, so the theory of a string's motion in d spacetime dimensions may be regarded as a two-dimensional field theory with d bosonic fields, with action

$$I[X] = \frac{T}{2}\int d\sigma \int d\tau \, \eta_{\mu\nu}\left[\frac{\partial X^\mu}{\partial \tau}\frac{\partial X^\nu}{\partial \tau} - \frac{\partial X^\mu}{\partial \sigma}\frac{\partial X^\nu}{\partial \sigma}\right]$$
$$= T\int d\sigma^+ \int d\sigma^- \, \eta_{\mu\nu}\frac{\partial X^\mu}{\partial \sigma^+}\frac{\partial X^\nu}{\partial \sigma^-}\,, \tag{24.2.1}$$

where T is a constant known as the string tension; $\mu = 0, 1, \ldots, d-1$; and $\sigma^\pm$ are the two-dimensional 'light-cone' coordinates $\sigma^\pm \equiv \tau \pm \sigma$. This action can be derived from a more general version, with complete invariance* under transformations of a pair of 'worldsheet coordinates' σ_k

$$I[X] = -\frac{T}{2}\int d^2\sigma \, \eta_{\mu\nu}\sqrt{\text{Det}\, g}\, g^{kl}\frac{\partial X^\mu}{\partial \sigma^k}\frac{\partial X^\nu}{\partial \sigma^l}\,, \tag{24.2.2}$$

by passing to a special coordinate system in which the worldsheet metric g_{kl} satisfies the condition

$$\sqrt{\text{Det}\, g}\, g^{kl} = \begin{pmatrix} 1 & 0 \\ 0 & -1 \end{pmatrix}. \tag{24.2.3}$$

In much the same way that in electrodynamics the problems introduced by the negative sign of the action for timelike photons are eliminated by the gauge invariance of the theory, here the problems introduced by the negative sign of $\eta_{\mu\nu}$ in Eqs. (24.2.1) and (24.2.2) for $\mu = \nu = 0$ are eliminated by the invariance of the action (24.2.2) (for appropriate boundary conditions) under general transformations of the worldsheet coordinates. In the special coordinate system in which the action takes the form (24.2.1), there is an important remnant of invariance under general worldsheet coordinate transformations: invariance under the global *conformal transformations*:

$$\sigma^\pm \to f^\pm(\sigma^\pm)\,, \tag{24.2.4}$$

with $f^\pm$ a pair of independent arbitrary functions.

The particles described by this string theory do not match those seen in the real world. Ramond[7] and Neveu and Schwarz[8] in 1971, aiming respectively to introduce particles with half-integer spin or with the quantum numbers of pions, suggested the addition of d fermionic field doublets $\psi_1^\mu(\sigma,\tau)$ and $\psi_2^\mu(\sigma,\tau)$. Shortly after, Gervais and Sakita[9] introduced an action for this theory:

$$I[X,\psi] = \int d\sigma^+ \int d\sigma^- \left[T\frac{\partial X^\mu}{\partial \sigma^+}\frac{\partial X_\mu}{\partial \sigma^-} + i\psi_2^\mu\frac{\partial}{\partial \sigma^+}\psi_{2\mu} + i\psi_1^\mu\frac{\partial}{\partial \sigma^-}\psi_{1\mu}\right], \tag{24.2.5}$$

* This symmetry is violated by quantum anomalies, like those discussed in Chapter 22, except in $d = 26$ spacetime dimensions for the purely bosonic theory, or $d = 10$ dimensions after the introduction of fermions.

and noted that conformal invariance could be maintained by extending the conformal transformations (24.2.4) to act also on the fermion fields

$$\psi_1^\mu \to \left(\frac{df^+}{d\sigma^+}\right)^{-1/2} \psi_1^\mu\,, \qquad \psi_2^\mu \to \left(\frac{df^-}{d\sigma^-}\right)^{-1/2} \psi_2^\mu\,. \tag{24.2.6}$$

Gervais and Sakita pointed out that, in addition to two-dimensional conformal invariance and d-dimensional Lorentz invariance, for appropriate boundary conditions this theory has a symmetry under infinitesimal transformations that interchange the bosonic field X^μ with the fermionic fields ψ_r^μ:

$$\begin{aligned}
\delta\psi_2^\mu(\sigma^+,\sigma^-) &= iT\,\alpha_2(\sigma^-)\,\frac{\partial}{\partial\sigma^-}X^\mu(\sigma^+,\sigma^-)\,,\\
\delta\psi_1^\mu(\sigma^+,\sigma^-) &= iT\,\alpha_1(\sigma^+)\,\frac{\partial}{\partial\sigma^+}X^\mu(\sigma^+,\sigma^-)\,,\\
\delta X^\mu(\sigma^+,\sigma^-) &= \alpha_2(\sigma^-)\,\psi_2^\mu(\sigma^+,\sigma^-) + \alpha_1(\sigma^+)\,\psi_1^\mu(\sigma^+,\sigma^-)\,,
\end{aligned} \tag{24.2.7}$$

where α_1 and α_2 are a pair of infinitesimal *fermionic* functions of σ^+ and σ^-, respectively, like the Grassmann variables introduced in Section 9.5. This was an example of what has subsequently come to be called supersymmetry, a symmetry connecting bosons and fermions, but thus far it was only a symmetry of a two-dimensional field theory, not of a physical theory in four spacetime dimensions.

A few years later Wess and Zumino[10] referred back to the example of supersymmetry that had been provided by References 7–9, and commented that it would be natural to try to extend the idea of supersymmetry to quantum field theories in *four* spacetime dimensions. They constructed several supersymmetric models. The simplest involved a single Majorana (self-charge-conjugate Dirac) field ψ, a pair of real scalar and pseudoscalar bosonic fields A and B, and a pair of real scalar and pseudoscalar bosonic auxiliary fields F and G, with invariance under the infinitesimal transformation**

$$\begin{aligned}
\delta A &= \left(\bar\alpha\,\psi\right)\,, \qquad \delta B = -i\left(\bar\alpha\,\gamma_5\,\psi\right)\,,\\
\delta\psi &= \partial_\mu(A + i\gamma_5 B)\gamma^\mu\alpha + (F - i\gamma_5 G)\alpha\,,\\
\delta F &= \left(\bar\alpha\,\gamma^\mu\,\partial_\mu\psi\right)\,, \qquad \delta G = -i\left(\bar\alpha\,\gamma_5\gamma^\mu\,\partial_\mu\psi\right)\,,
\end{aligned} \tag{24.2.8}$$

where α is an arbitrary constant infinitesimal Majorana fermion c-number

** The notation for Dirac matrices used here is explained in the Preface and in Section 5.4. The γ_5 used here (which satisfies $\gamma_5^2 = 1$) is a factor of i times that used by Wess and Zumino, and the covariant conjugate $\bar\psi$ of any spinor ψ is defined here as i times that of Wess and Zumino. For this reason, some of the phases in Eqs. (24.2.8)–(24.2.10) are different from those in Reference 10.

parameter. If we require invariance of the action under these transformations, then the most general real, Lorentz-invariant, parity-conserving, renormalizable Lagrangian density built out of these ingredients is

$$\begin{aligned}\mathscr{L} = & -\tfrac{1}{2}\partial_\mu A\, \partial^\mu A - \tfrac{1}{2}\partial_\mu B\, \partial^\mu B - \tfrac{1}{2}\bar{\psi}\gamma^\mu \partial_\mu \psi \\ & + \tfrac{1}{2}(F^2 + G^2) + m\left[FA + GB - \tfrac{1}{2}\bar{\psi}\psi\right] \\ & + g\Big[F(A^2 + B^2) + 2GAB - \bar{\psi}(A + i\gamma_5 B)\psi\Big] . \qquad (24.2.9)\end{aligned}$$

Since the auxiliary fields F and G enter quadratically, we can derive an equivalent Lagrangian by setting them equal to the values given by the field equations

$$F = -mA - g(A^2 + B^2)\,, \qquad G = -mB - 2gAB\,. \qquad (24.2.10)$$

The Lagrangian density then becomes

$$\begin{aligned}\mathscr{L} = & -\tfrac{1}{2}\partial_\mu A\, \partial^\mu A - \tfrac{1}{2}\partial_\mu B\, \partial^\mu B - \tfrac{1}{2}\bar{\psi}\gamma^\mu \partial_\mu \psi \\ & - \tfrac{1}{2}m^2\left[A^2 + B^2\right] - \tfrac{1}{2}m\,\bar{\psi}\psi \\ & - gmA(A^2 + B^2) - \tfrac{1}{2}g^2(A^2 + B^2)^2 - g\bar{\psi}(A + i\gamma_5 B)\psi\,. \qquad (24.2.11)\end{aligned}$$

This Lagrangian density exhibits relations not only between scalar and fermion masses, but also between Yukawa interactions and scalar self-couplings, which are characteristic of supersymmetric theories. Wess and Zumino also described supersymmetry transformations and gave a Lagrangian for a supermultiplet containing a vector field. (We shall go into all this in more detail in Chapter 26.) Finally, in a second paper, Wess and Zumino[11] recalled the Coleman–Mandula theorem and traced the apparent violation of this theorem to the fact that the symmetry generators here satisfy anticommutation rather than commutation relations. It was a few more years before Gliozzi, Scherk and Olive[11a] showed that it was possible to construct a superstring theory with spacetime as well as worldsheet supersymmetry by imposing suitable periodicity conditions on the fields of the Ramond–Neveu–Schwarz model.

Unknown to Wess and Zumino, at the time of their first papers on supersymmetry in four spacetime dimensions this symmetry had already appeared in a pair of papers published in the Soviet Union. In 1971 Gol'fand and Likhtman[12] had extended the algebra of the Poincaré group discussed in Section 2.4 to a superalgebra and used the requirement of invariance under this superalgebra to construct supersymmetric field theories in four spacetime dimensions. Their paper though prophetic gave few details, and was generally ignored until much later. Independently Volkov and Akulov[13] in 1973 discovered what today would be called spontaneously broken supersymmetry, but they used their formalism to identify the Goldstone fermion associated with supersymmetry breaking with the neutrino, an idea that met with no success. For most theorists, especially outside the Soviet Union, supersymmetry as a possible symmetry of nature in four spacetime dimensions began with the 1974 papers of Wess and Zumino.

Appendix A $SU(6)$ Symmetry of Non-Relativistic Quark Models

This appendix will describe the way that an $SU(6)$ symmetry that relates particles of different spin arises in non-relativistic quark models. This has nothing directly to do with supersymmetry, but it provides the historical background for the Coleman–Mandula theorem, which is an essential input to the construction of general supersymmetry algebras in Sections 25.1 and 31.1.

In general, the Hamiltonian of the non-relativistic quark model could depend not only on positions and momenta but also on the spin and flavor operators $\sigma_i^{(n)}$ and $\lambda_A^{(n)}$, where the $\sigma_i^{(n)}$ (with $i = 1, 2, 3$) act on the spin indices of the nth quark as the Pauli matrices σ_i defined by Eq. (5.4.18), while the $\lambda_A^{(n)}$ (with $A = 1, 2, \ldots, 8$) act on the flavor indices of the nth quark as the Gell-Mann $SU(3)$ matrices λ_A defined by Eq. (19.7.2). (Where n refers to an antiquark, $\sigma_i^{(n)}$ and $\lambda_A^{(n)}$ act as the matrices $-\sigma_i^T$ and $-\lambda_A^T$ of the contragredient representations.) If we were to assume only that there is no spin-orbit coupling, so that the total orbital angular momentum L_i is separately conserved, then we could conclude only that the Hamiltonian commutes with the total spin and unitary spin

$$S_i \equiv \tfrac{1}{2}\sum_n \sigma_i^{(n)}\,, \qquad T_A \equiv \tfrac{1}{2}\sum_n \lambda_A^{(n)}\,, \tag{24.A.1}$$

as well as L_i. On the other hand, if we were to suppose that the Hamiltonian depends only on quark positions and momenta, and not at all on spins or quark flavors, then such a Hamiltonian would commute not only with the total orbital angular momentum $\mathbf{L}$, but also with *each* of the operators $\sigma_i^{(n)}$ and $\lambda_A^{(n)}$. In between these two extremes there is the interesting possibility that in addition to commuting with the L_i, S_i, and T_A, the Hamiltonian also commutes with the operators

$$R_{iA} \equiv \tfrac{1}{2}\sum_n \pm\sigma_i^{(n)}\,\lambda_A^{(n)}\,, \tag{24.A.2}$$

where the sign is + or − for quarks and antiquarks, respectively.* The S_i, T_A, and R_{iA} form the Lie algebra of the group $SU(6)$, with commutation

* The minus sign for antiquarks arises because the terms in R_{iA} for antiquarks must act on spin and flavor indices as the matrix $-(\sigma_i\,\lambda_A)^T = -(-\sigma_i^T)(-\lambda_A^T)$.

relations

$$[S_i\,,\,S_j] = i\sum_k \epsilon_{ijk}S_k\,, \qquad [T_A\,,\,T_B] = i\sum_C f_{ABC}T_C\,, \qquad [S_i\,,\,T_A] = 0\,,$$
$$[S_i\,,\,R_{jA}] = i\sum_k \epsilon_{ijk}R_{kA}\,, \qquad [T_A\,,\,R_{iB}] = i\sum_C f_{ABC}R_{iC}\,, \tag{24.A.3}$$
$$[R_{Ai}\,,\,R_{Bj}] = i\delta_{ij}\sum_C f_{ABC}\,T_C + \tfrac{2}{3}i\delta_{AB}\sum_k \epsilon_{ijk}\,S_k + i\sum_{kC}\epsilon_{ijk}\,d_{ABC}R_{kC}\,.$$

Here f_{ABC} and d_{ABC} are respectively totally antisymmetric and totally symmetric numerical coefficients,[14] with independent non-vanishing values given by

$$\begin{aligned} &f_{123} = 1\,, \qquad f_{458} = f_{678} = \sqrt{3}/2\,, \\ &f_{147} = f_{165} = f_{246} = f_{257} = f_{345} = f_{376} = 1/2\,, \end{aligned} \tag{24.A.4}$$

and

$$\begin{aligned} &d_{146} = d_{157} = -d_{247} = d_{256} = d_{344} = d_{355} = -d_{366} = -d_{377} = 1/2\,, \\ &d_{118} = d_{228} = d_{338} = -d_{888} = 1/\sqrt{3}\,, \\ &d_{448} = d_{558} = d_{668} = d_{778} = -1/(2\sqrt{3})\,. \end{aligned} \tag{24.A.5}$$

This is the symmetry that remains if we include a spin- and flavor-dependent two-body interaction in the Hamiltonian that commutes with R_{iA} as well as with the S_i and T_A. There are such interactions, given by linear combinations of two-body operators of the form

$$H^{(nm)} \propto \left[1 \pm \sum_i \sigma_i^{(n)}\sigma_i^{(m)}\right]\left[\frac{2}{3} \pm \sum_A \lambda_A^{(n)}\,\lambda_A^{(m)}\right]\,, \tag{24.A.6}$$

where the sign $\pm$ is negative if one of the particles n, m is a quark and the other an antiquark and positive if they are both quarks or both antiquarks.

Of course, even in the non-relativistic quark model the $SU(6)$ symmetry is at best approximate. It is broken by spin–orbit and spin–spin forces, and also by the mass of the s quark, which reduces the flavor $SU(3)$ symmetry to the $SU(2)$ and $U(1)$ of isospin and hypercharge conservation. If we avoid the effects of this quark mass difference by restricting ourselves to hadrons built up from the light u and d quarks and antiquarks, then the only non-vanishing λ_A matrices are the λ_a with $a = 1, 2, 3$ (which for the u and d quarks are given by the Pauli matrices (5.4.18) (that in this context are conventionally called τ_a) and λ_8 (which is just the number $1/\sqrt{3}$ for the u and d quarks, and $-1/\sqrt{3}$ for the $\bar{u}$ and $\bar{d}$ antiquarks). The interaction (24.A.6) thus becomes

$$H^{(nm)} \propto \left[1 \pm \sum_i \sigma_i^{(n)} \sigma_i^{(m)}\right] \left[1 \pm \sum_A \tau_A^{(n)} \tau_A^{(m)}\right] . \tag{24.A.7}$$

Aside from quark number conservation, the remaining symmetry is then $SU(4)$, with generators S_i, T_a, and R_{ia} which commute with (24.A.7). This was proposed by Wigner[2] in 1937 as a symmetry of nuclear forces, though of course with protons and neutrons in place of u and d quarks. The interaction (24.A.7) is known in nuclear theory as a *Majorana potential* to distinguish it from an interaction that does not depend on spin or isospin, called a *Wigner potential*, or an interaction proportional to just the spin-dependent or just the isospin-dependent factor in (24.A.7), known respectively as a *Bartlett potential* and a *Heisenberg potential.*

It is amusing to note that, although in non-relativistic theories there is no theoretical barrier to symmetries like $SU(6)$ that act on spin as well as particle type, *there never was any experimental evidence for such an assumed* $SU(6)$ *symmetry of the non-relativistic quark model that is any better satisfied than the assumption of complete spin and flavor independence.* These assumptions are not the same; if the Hamiltonian of a system of N non-relativistic quarks and/or antiquarks is completely independent of spin and flavor, then its symmetry is $SU(6)^N$, not $SU(6)$. For instance, a two-particle interaction like (24.A.6) and various other multiparticle interactions break $SU(6)^N$ to $SU(6)$. Of course, all these symmetries are only approximate anyway. The question is whether $SU(6)$ is less badly broken than $SU(6)^N$?

This cannot be answered by studying the multiplet that contains the baryon octet, which consists of the nucleon and hyperons Λ, Σ, and Ξ. In the non-relativistic quark model these particles are interpreted as bound states of three quarks with zero orbital angular momentum. Because these states are color neutral, the wave function is completely antisymmetric in the suppressed color indices, and therefore it is completely symmetric under the combined interchange of spin and flavor. The baryon octet would therefore have to be put in the symmetric third-rank tensor representation **56** of $SU(6)$, which besides the baryon octet contains a spin 3/2 decuplet, which may be identified as the one consisting of the famous '3–3' resonance Δ and the $\Sigma(1385)$, $\Xi(1530)$, and Ω particles. (Numbers in parentheses give masses in MeV, where these are needed to distinguish the particles from others of the same isospin and strangeness but lower mass.) The $SU(6)$ symmetry leads to good predictions for the baryon magnetic moments: The quark charge operator is $q = e(\lambda_3/2 + \lambda_8/2\sqrt{3})$, so if quarks have the magnetic moments $3q/2m_N$ of Dirac particles of this charge and mass $m_N/3$, then the magnetic moment operator is

$$\mu_i = 3\mu_N \left[\frac{1}{2}R_{i3} + \frac{1}{2\sqrt{3}}R_{i8}\right] ,$$

where $\mu_N \equiv e/2m_N$ is the nuclear magneton, and R_{iA} is defined by Eq. (24.A.2). It is straightforward to calculate the matrix elements of this symmetry generator between the members of the **56** multiplet, with the result that the magnetic moments for p, n, Λ, Σ^+, Σ^-, Ξ^-, and Ξ^0 in units of μ_N are respectively +3, −2, −1, +3, −1, −1, and −2, which may be compared with the corresponding experimental values +2.79, −1.91, −0.61, +2.46, −1.16, −0.65, and −1.25. The agreement is fair, and somewhat better (except for Σ^-) if we take the quark magnetic moment to be a little less than $3\mu_N$. Because of the symmetry of the three-quark wave function, nothing new is learned here if we assume that the Hamiltonian is completely independent of spin and flavor; the states of zero angular momentum would still have to fall in a multiplet consisting of $6 \times 7 \times 8/6! = 56$ members. In particular, the operator (24.A.6) has the same value 4 for any state of two quarks that is symmetric under simultaneous interchange of their spins and flavors, whether it is symmetric both under interchange of spins and interchange of flavors, or antisymmetric under both interchanges.

In order to decide whether $SU(6)$ is any better than $SU(6)^N$, it is more useful to study the mesons, which in the non-relativistic quark model are interpreted as bound states of a quark and antiquark. If the Hamiltonian of these states is completely independent of spin and flavor then its symmetry is $SU(6)^2$, and the meson states fall into its 36-dimensional $(6, \bar{6})$ representation, while for $SU(6)$ symmetry we could only say that the mesons belong to either of the two representations of $SU(6)$ contained in $\mathbf{6} \times \bar{\mathbf{6}}$: the adjoint representation **35** or the singlet representation. To be more specific, the **35** consists of an $SU(3)$ singlet with spin $S = 1$, an $SU(3)$ octet with $S = 0$, and an $SU(3)$ octet with $S = 1$, corresponding to the $SU(6)$ generators S_i, T_A, and R_{iA}, which is split from the $SU(3)$ singlet state with $S = 0$ by the interaction (24.A.6). Since all these assumptions are approximate anyway, the question in deciding whether $SU(6)$ symmetry is more accurate than complete spin and flavor independence is whether the splitting of the $SU(3)$ singlet $S = 0$ state from the other 35 states of the same orbital angular momentum is any greater than the splittings within the **35** supermultiplet.

For orbital angular momentum $L = 0$ the quark–antiquark states have negative parity P, and positive or negative charge-conjugation quantum number C (for self-charge-conjugate states) according to whether the total spin S is zero or one, respectively. (For an explanation, see Section 5.5.) The **35** therefore consists of a singlet with $J^{PC} = 1^{--}$, a 0^{-+} octet, and a 1^{--} octet, which may be identified respectively as: the $\phi(1020)$; the pseudoscalar octet π, η, K, and $\bar{K}$; and the vector octet ρ, ω, K^*, and $\bar{K}^*$. There is also a 0^{-+} $SU(3)$ singlet η' at 958 MeV, which can be regarded as the $SU(6)$ singlet. The splitting of this singlet from the particles in the **35** multiplet is not distinctly greater than the splittings within the **35** multiplet.

It may be argued that the $L = 0$ mesons do not provide a good test of the symmetries of the non-relativistic quark model, because they include the Goldstone bosons π, η, K, and $\bar{K}$, which become massless for zero u and d quark masses and are therefore not well described by this model. Therefore let us consider the quark–antiquark states with $L = 1$. These states have P positive and C positive or negative according to whether $S = 1$ or $S = 0$, so the p-wave **35** consists of: $SU(3)$ singlets with $S = 1$ and $J^{PC} = 0^{++}$, 1^{++}, 2^{++}, which may be identified as the $f_0(1370)$, the $f_1(1285)$, and the $f_2(1270)$; an $S = 0$ 1^{+-} octet identified as $h_1(1170)$, $b_1(1235)$, $K_1(1400)$, and $\bar{K}_1(1400)$; and $S = 1$ octets: a 0^{++} octet consisting of $f_0(980)$, $a_0(980)$, $K_0^+(1950)$, and $\bar{K}_0^+(1950)$; a 1^{++} octet consisting of $f_1(1420)$, $a_1(1260)$, $K_1^+(1650)$, and $\bar{K}_1^+(1650)$; and a 2^{++} octet consisting of $f_2(1430)$, $a_2(1320)$, $K_2^+(1980)$, and $\bar{K}_2^+(1980)$. In addition to these 35×3 states, there is a another particle with the right quantum numbers to be the p-wave $SU(6)$ singlet: the 1^{+-} isoscalar $h_1(1380)$. Of course, we could interchange the identifications of $h_1(1170)$ and $h_1(1380)$, or identify the $SU(3)$ singlet and octet isoscalar 1^{+-} states as orthogonal linear combinations of $h_1(1170)$ and $h_1(1380)$. The important point is that there are *two* of these 1^{+-} isoscalars, with no indication that one of them, belonging to an $SU(6)$ singlet, is more strongly split from the particles of the **35** than the particles of the **35** are split from each other. Here too, then, there is no evidence that $SU(6)$ symmetry is any more accurate than the stronger assumption of complete spin and flavor independence.

Appendix B The Coleman–Mandula Theorem

This appendix provides a proof of the celebrated theorem of Coleman and Mandula[5] that the only possible Lie algebra (as opposed to super-algebra) of symmetry generators consists of the generators P_μ and $J_{\mu\nu}$ of translations and homogeneous Lorentz transformations, together with possible internal symmetry generators, which commute with P_μ and $J_{\mu\nu}$ and act on physical states by multiplying them with spin-independent, momentum-independent Hermitian matrices.* By 'symmetry generators' here is meant any Hermitian operators: that commute with the S-matrix; whose commutators are also symmetry generators; which take one-particle states into one-particle states; and whose action on multiparticle states is the direct sum of their action on one-particle states (as in Eq. (24.B.1)). A further technical requirement will be added when needed later. Apart from the general principles of relativistic quantum mechanics described in Chapters 2 and 3, the only other assumptions needed in this proof are:

Assumption 1 *For any M there are only a finite number of particle types with mass less than M.*

Assumption 2 *Any two-particle state undergoes some reaction at almost all energies (that is, at all energies except perhaps an isolated set).*

Assumption 3 *The amplitudes for elastic two-body scattering are analytic functions of the scattering angle at almost all energies and angles.***

It is not necessary to assume that the S-matrix is governed by a local quantum field theory. The proof presented here is somewhat rearranged and streamlined, and it spells out some steps that Coleman and Mandula left to the reader.

It is convenient to start by proving this theorem for the subalgebra consisting of those symmetry generators B_α that commute with the four-momentum operator P_μ. (This part of the theorem is of some interest in itself; it rules out symmetries in relativistic theories that act like the $SU(6)$ symmetry of the non-relativistic quark model.) The action of such

* As we shall see, in theories with only massless particles there is also the possibility that in addition to the generators P_μ and $J_{\mu\nu}$ there are additional generators D and K_μ that fill out the Lie algebra of the conformal group.[15]

** Strictly speaking, this assumption is not satisfied in theories with infrared divergences such as quantum electrodynamics, where, as shown in Section 13.3, the S-matrix element for any one scattering process involving charged particles actually vanishes, except for elastic forward scattering. In Abelian gauge theories like electrodynamics this problem can be avoided by applying the Coleman–Mandula theorem to the theory with a fictitious gauge boson mass, and then working only with 'infrared-safe' quantities like masses and suitably integrated cross-sections that are finite in the limit of zero gauge boson mass. There is no problem in non-Abelian gauge theories like quantum chromodynamics, in which all massless particles are trapped — symmetries if unbroken would only govern S-matrix elements for gauge-neutral bound states, like the mesons and baryons in quantum chromodynamics. As far as I know, the Coleman–Mandula theorem has not been proved for non-Abelian gauge theories with untrapped massless particles, like quantum chromodynamics with many quark flavors.

symmetry generators on multiparticle states is given by

$$B_\alpha |p\,m,\, q\,n,\, \ldots\rangle = \sum_{m'} \Big(b_\alpha(p)\Big)_{m'm} |p\,m',\, q\,n,\, \ldots\rangle$$
$$+\sum_{n'} \Big(b_\alpha(q)\Big)_{n'n} |p\,m,\, q\,n',\, \ldots\rangle + \cdots\,, \tag{24.B.1}$$

where m, n, etc. are discrete indices labelling spin z-components and particle type for particles of a definite mass $\sqrt{-p_\mu p^\mu}$, and the $b_\alpha(p)$ are finite Hermitian matrices, which define the action of the B_α on one-particle states.

Now, we can see from Eq. (24.B.1) that the mapping that takes the B_α into $b_\alpha(p)$ for some fixed p is a homomorphism in the sense that the commutation relations

$$[B_\alpha, B_\beta] = i\sum_\gamma C^\gamma_{\alpha\beta} B_\gamma \tag{24.B.2}$$

are also satisfied by the Hermitian matrices $b_\alpha(p)$:

$$[b_\alpha(p), b_\beta(p)] = i\sum_\gamma C^\gamma_{\alpha\beta}\, b_\gamma(p)\,. \tag{24.B.3}$$

A well-known theorem proved in Section 15.2 tells us that any Lie algebra of finite Hermitian matrices like $b_\alpha(p)$ must be a direct sum of a compact semi-simple Lie algebra and $U(1)$ algebras. However, we cannot immediately apply this result to the operator algebras B_α because the homomorphism between the operators B_α and matrices $b_\alpha(p)$ is not necessarily an isomorphism. For it to be an isomorphism would require also that whenever $\sum_\alpha c^\alpha b_\alpha(p) = 0$ for some coefficients c^α and momentum p, then $\sum_\alpha c^\alpha b_\alpha(k) = 0$ for all momenta k, which is equivalent to the condition $\sum_\alpha c^\alpha B_\alpha = 0$.

Instead of considering the homomorphism that maps the B_α into the one-particle matrices $b_\alpha(p)$, Coleman and Mandula considered the homomorphism that maps the B_α into the matrices that define the action of B_α on *two-particle states* with fixed four-momenta p and q:

$$\Big(b_\alpha(p,q)\Big)_{m'n',mn} = \Big(b_\alpha(p)\Big)_{m'm}\delta_{n'n} + \Big(b_\alpha(q)\Big)_{n'n}\delta_{m'm}\,. \tag{24.B.4}$$

The invariance of the S-matrix for the elastic or quasi-elastic scattering of two particles with four-momenta p and q into two particles with momenta p' and q', with masses $\sqrt{-p'_\mu p'^\mu} = \sqrt{-p_\mu p^\mu}$ and $\sqrt{-q'_\mu q'^\mu} = \sqrt{-q_\mu q^\mu}$, yields the condition

$$b_\alpha(p',q')\, S(p',q';p,q) = S(p',q';p,q)\, b_\alpha(p,q)\,. \tag{24.B.5}$$

Here $S(p',q';p,q)$ is a matrix of the same dimensionality as $b(p,q)$ and $b(p',q')$, defined in terms of the connected S-matrix elements $S(pm, qn \rightarrow$

$p'm', q'n')$ by

$$S(p\,m, q\,n \to p'm', q'n') \equiv \delta^4(p'+q'-p-q)\Big(S(p',q';p,q)\Big)_{m'n',mn} . \quad (24.B.6)$$

According to Assumption 2 and the optical theorem (see Section 3.6), for almost any choice of p and q the elastic scattering amplitude is non-vanishing in the forward direction, and Assumption 3 then tells us that the matrix $S(p',q';p,q)$ is non-singular for almost all p' and q' on the same mass shells and satisfying the conservation condition $p'+q'=p+q$, so for almost all such four-momenta Eq. (24.B.5) is a *similarity transformation.*

It follows then that if $\sum_\alpha c^\alpha b_\alpha(p,q) = 0$ for almost any fixed four-momenta p and q, then $\sum_\alpha c^\alpha b_\alpha(p',q') = 0$ for almost any four-momenta p' and q' on the same mass shells that satisfy $p'+q'=p+q$. Unfortunately, this does not tell us that $\sum_\alpha c^\alpha b_\alpha(p')$ and $\sum_\alpha c^\alpha b_\alpha(q')$ vanish, but only that these matrices are proportional to the unit matrix (with opposite coefficients). To do better, it is necessary to consider not the $b_\alpha(p)$ or $b_\alpha(p,q)$, but their traceless parts.

One immediate consequence of Eq. (24.B.5) is that

$$\mathrm{Tr}\, b_\alpha(p',q') = \mathrm{Tr}\, b_\alpha(p,q) . \quad (24.B.7)$$

With Eq. (24.B.4), this tells us that

$$\begin{aligned} &N(\sqrt{-q_\mu q^\mu})\,\mathrm{tr}\, b_\alpha(p') + N(\sqrt{-p_\mu p^\mu})\,\mathrm{tr}\, b_\alpha(q') \\ &\quad = N(\sqrt{-q_\mu q^\mu})\,\mathrm{tr}\, b_\alpha(p) + N(\sqrt{-p_\mu p^\mu})\,\mathrm{tr}\, b_\alpha(q) , \end{aligned} \quad (24.B.8)$$

where $N(m)$ is the multiplicity† of particle types with mass m, and the lower case t in 'tr' indicates a sum over one-particle rather than two-particle labels. In order for this to be satisfied for almost all mass-shell four-momenta for which $p'+q'=p+q$, it is necessary that the function $\mathrm{tr}\, b_\alpha(p)/N(\sqrt{-p_\mu p^\mu})$ be linear†† in p:

$$\frac{\mathrm{tr}\, b_\alpha(p)}{N(\sqrt{-p_\mu p^\mu})} = a_\alpha^\mu\, p_\mu , \quad (24.B.9)$$

with a_α^μ independent of p (and of everything else but the displayed indices.) We may define new symmetry generators by subtracting terms linear in

† These multiplicity factors were not shown explicitly by Coleman and Mandula. They are needed in justifying a step that Coleman and Mandula made without explanation, that of defining the symmetry generators $B_\alpha^\sharp$ with traceless kernels.

†† A constant term is easily seen to be ruled out in Eq. (24.B.9) by the existence of processes in which the number of particles is not conserved, processes which are inevitable in any relativistic quantum theory satisfying the cluster decomposition principle. Even if we considered only two-particle processes and did not use this argument, a constant term in Eq. (24.B.9) would only amount to a change in the action of internal symmetries on physical states.

the momentum operator P_μ:

$$B^\sharp_\alpha \equiv B_\alpha - a^\mu_\alpha P_\mu \,, \tag{24.B.10}$$

which according to Eq. (24.B.9) are represented on one-particle states by the traceless matrices

$$\left(b^\sharp_\alpha(p)\right)_{n'n} = \left(b_\alpha(p)\right)_{n'n} - \frac{\operatorname{tr} b_\alpha(p)}{N(\sqrt{-p_\mu p^\mu})}\,\delta_{n'n} \,. \tag{24.B.11}$$

Because P_μ commutes with B_α and the unit matrix commutes with everything, the commutators of the $B^\sharp_\alpha$ are the same as those of the B_α, and the commutators of the $b^\sharp_\alpha(p)$ are the same as those of the $b_\alpha(p)$:

$$[B^\sharp_\alpha, B^\sharp_\beta] = i\sum_\gamma C^\gamma_{\alpha\beta} B_\gamma = i\sum_\gamma C^\gamma_{\alpha\beta}\left[B^\sharp_\gamma + a^\mu_\gamma P_\mu\right] , \tag{24.B.12}$$

$$[b^\sharp_\alpha(p), b^\sharp_\beta(p)] = i\sum_\gamma C^\gamma_{\alpha\beta} b_\gamma(p) = i\sum_\gamma C^\gamma_{\alpha\beta}\left[b^\sharp_\gamma(p) + a^\mu_\gamma p_\mu\right] . \tag{24.B.13}$$

Also, Eq. (24.B.13) and the fact that commutators of the finite matrices $b^\sharp_\alpha(p)$ have zero trace[‡] imply that $\sum_\gamma C^\gamma_{\alpha\beta}\, a^\mu_\gamma = 0$, and using this in Eq. (24.B.12) then shows that the $B^\sharp_\alpha$ satisfy the same commutation relations as the B_α:

$$[B^\sharp_\alpha \,, B^\sharp_\beta] = i\sum_\gamma C^\gamma_{\alpha\beta}\, B^\sharp_\gamma \,. \tag{24.B.14}$$

Because $B^\sharp_\alpha$ is a symmetry generator, the scattering amplitude satisfies

$$b^\sharp_\alpha(p',q')\, S(p',q';p,q) = S(p',q';p,q)\, b^\sharp_\alpha(p,q) \,, \tag{24.B.15}$$

where the $b^\sharp_\alpha(p,q)$ are the matrices representing the $B^\sharp_\alpha$ on the two-particle states

$$\left(b^\sharp_\alpha(p,q)\right)_{m'n',mn} = \left(b^\sharp_\alpha(p)\right)_{m'm}\delta_{n'n} + \left(b^\sharp_\alpha(q)\right)_{n'n}\delta_{m'm} \tag{24.B.16}$$

and satisfy the same commutation relations as the $B^\sharp_\alpha$:

$$[b^\sharp_\alpha(p,q) \,, b^\sharp_\beta(p,q)] = i\sum_\gamma C^\gamma_{\alpha\beta}\, b^\sharp_\gamma(p,q) \,. \tag{24.B.17}$$

The advantage of dealing with these two-particle matrices is that, since $S(p',q';p,q)$ is a non-singular matrix, it follows that if $\sum_\alpha c^\alpha b^\sharp_\alpha(p,q) = 0$ for some fixed mass-shell four-momenta p and q, then $\sum_\alpha c^\alpha b^\sharp_\alpha(p',q') = 0$ for

‡ This is one of the places where we use Assumption 1, without which commutators need not have zero trace. Also, at this point it is crucial that we are dealing with commutation rather than anticommutation relations, since the unit matrix does not anticommute with other matrices, and anticommutators of finite matrices do not necessarily have zero trace.

almost all p' and q' on the same respective mass shells with $p'+q' = p+q$. Because we are dealing now with traceless matrices, this tells us that

$$\sum_\alpha c^\alpha\, b^\sharp_\alpha(p') = \sum_\alpha c^\alpha\, b^\sharp_\alpha(q') = 0\,. \tag{24.B.18}$$

We would like to conclude from this that $\sum_\alpha c^\alpha b^\sharp_\alpha(k) = 0$ for *all* mass-shell four-momenta k, but so far we have proved only that $\sum_\alpha c^\alpha\, b^\sharp_\alpha(p') = 0$ for almost all of those p' for which $q' = p+q-p'$ as well as p' is on the mass shell (and correspondingly for q'.) To get around this limitation, we can use a trick of Coleman and Mandula, noting that if $\sum_\alpha c^\alpha\, b^\sharp_\alpha(p,q) = 0$ then (24.B.18) and (24.B.16) together yield

$$\sum_\alpha c^\alpha\, b^\sharp_\alpha(p,q') = 0\,,$$

so that, according to Eq. (24.B.15),

$$\sum_\alpha c^\alpha\, b^\sharp_\alpha(k\,,\, p+q'-k) = 0\,,$$

and therefore

$$\sum_\alpha c^\alpha\, b^\sharp_\alpha(k) = 0\,, \tag{24.B.19}$$

for almost all mass-shell four-momenta k for which $p+q'-k$ is also on the mass shell. Now, the conditions that q' and $p+q-q'$ be on the mass shell leave two parameters free in q', so that we have enough freedom in choosing q' that the condition that $p+q'-k$ is on the mass shell leaves us free to choose $\mathbf{k}$ to be anything we like, at least within a finite volume of momentum space. This volume can be adjusted to be as large as we like by taking $\mathbf{p}$ and $\mathbf{q}$ sufficiently large, so if $\sum_\alpha c^\alpha b^\sharp_\alpha(p,q) = 0$ for some fixed mass-shell four-momenta p and q then $\sum_\alpha c^\alpha b^\sharp_\alpha(k) = 0$ for almost all mass-shell four-momenta k. But then if $\sum_\alpha c^\alpha b^\sharp_\alpha(k_0) \neq 0$ for some particular mass-shell four-momentum k_0, a scattering process in which particles with four-momenta k_0 and k scatter into particles with four-momenta k' and k'' will be forbidden by the symmetry generated by $\sum_\alpha c^\alpha B^\sharp_\alpha$ for almost all k, k', and k'', in contradiction with our assumption about the analyticity of the scattering amplitude. We conclude then that if $\sum_\alpha c^\alpha b^\sharp_\alpha(p,q) = 0$ for some fixed mass-shell four-momenta p and q then $\sum_\alpha c^\alpha b^\sharp_\alpha(k) = 0$ for *all* k, and therefore $\sum_\alpha c^\alpha B^\sharp_\alpha = 0$, so that the mapping that takes B_α into $b^\sharp_\alpha(p,q)$ is an isomorphism.

One immediate consequence is that, since the number of independent matrices $b^\sharp_\alpha(p,q)$ cannot exceed $N(\sqrt{-p_\mu p^\mu})\, N(\sqrt{-q_\mu q^\mu})$, there can be at most a finite number of independent symmetry generators B_α. As emphasized by Coleman and Mandula, in proving their theorem it is not necessary to make an independent assumption that the symmetry algebra is finite-dimensional.

The theorems of Section 15.2 tell us further that a Lie algebra of finite Hermitian matrices like $b^\sharp_\alpha(p,q)$ for fixed p and q is at most the direct sum of a semi-simple compact Lie algebra and some number of $U(1)$ Lie algebras. We have seen that this Lie algebra is isomorphic to that of the symmetry generators $B^\sharp_\alpha$, so the $B^\sharp_\alpha$ must also span the direct sum of at most a compact semi-simple Lie algebra and $U(1)$ Lie algebras.

Let us first dispose of the $U(1)$ Lie algebras. For any pair of mass-shell momenta p and q we can find a Lorentz generator J that leaves both p and q invariant. (If p and q are lightlike and parallel then take J to generate rotations around the common direction of $\mathbf{p}$ and $\mathbf{q}$. Otherwise, $p+q$ will be timelike, so we can take J to generate rotations around the common direction of $\mathbf{p}$ and $\mathbf{q}$ in the center-of-mass frame in which $\mathbf{p}=-\mathbf{q}$.) We can choose the basis of two-particle states to diagonalize J, so that

$$J|pm,qn\rangle = \sigma(m,n)|pm,qn\rangle\,. \tag{24.B.20}$$

Now, P_μ commutes with all $B^\sharp_\alpha$, and $[J,P_\mu]$ is a linear combination of components of P_μ, so P_μ commutes with all $[J,B^\sharp_\alpha]$, and therefore the symmetry generator $[J,B^\sharp_\alpha]$ must be a linear combination of the B_β, which by definition form a complete set of symmetry generators that commute with P_μ. More specifically, since the matrices representing a commutator of symmetry generators are necessarily traceless, $[J,B^\sharp_\alpha]$ has to be a linear combination of the $B^\sharp_\beta$. But any $U(1)$ generator $B^\sharp_i$ (taken Hermitian) in the algebra of the $B^\sharp_\beta$ would have to commute with all the $B^\sharp_\beta$, and hence in particular must commute with $[J,B^\sharp_i]$:

$$[B^\sharp_i,\,[J,\,B^\sharp_i]]=0\,.$$

Taking the expectation value of this double commutator in the two-particle basis in which J is diagonal, we have

$$0=\sum_{m',n'}\Big(\sigma(m',n')-\sigma(m,n)\Big)\,\left|\Big(b^\sharp_i(p,q)\Big)_{m'n',mn}\right|^2\,, \tag{24.B.21}$$

for any m and n. The indices run over a finite range, so if there were any σ for which there existed an m and n with $\sigma(m,n)=\sigma$ and an m' and n' with $\sigma(m',n')\neq\sigma$, with $(b^\sharp_i(p,q))_{m'n',mn}\neq 0$, then there would have to be a *smallest* such σ, in which case the right-hand side of Eq. (24.B.21) for this m and n would be positive-definite, contradicting Eq. (24.B.21). We conclude that $(b^\sharp_i(p,q))_{m'n',mn}$ must vanish for all m, n, m', and n' for which $\sigma(m',n')\neq\sigma(m,n)$. Because the algebra of the $b^\sharp_i(p,q)$ is isomorphic to that of the $B^\sharp_i$, this means that each of the $U(1)$ generators $B^\sharp_i$ commutes with J. Since we can choose $p+q$ to be in any timelike direction, it follows that each of the $U(1)$ generators $B_i^\sharp$ commutes with all the generators $J_{\mu\nu}$ of the homogeneous Lorentz group. The fact that they commute with what we called 'boosts' in Section 2.5 implies that the $(b_i^\sharp(p))_{n'n}$ are independent of three-momentum, and the fact that they commute with rotations implies that the $(b_i^\sharp(p))_{n'n}$ act as unit matrices on spin indices, so these generators are the generators of an ordinary internal symmetry.

This leaves the $B_\alpha^\sharp$ that generate a semi-simple compact Lie algebra. The argument of Section 24.1 (a somewhat more explicit version of the reasoning given by Coleman and Mandula) tells us that the generators of the semi-simple compact part of the Lie algebra commute with Lorentz transformations and, as shown for the $U(1)$ generators, this means that they too are the generators of internal symmetries. We have thus shown that the symmetry generators B_α that commute with P_μ are either internal symmetry generators or linear combinations of the components of P_μ itself.

Next, we must take up the possibility of symmetry generators that do not commute with the momentum operator. The action of a general symmetry generator A_α on a one-particle state $|p\,n\rangle$ of four-momentum p would be

$$A_\alpha\,|p,n\rangle = \sum_{n'}\int d^4p'\,\left(\mathscr{A}_\alpha(p',p)\right)_{n'n}|p',n'\rangle\,, \tag{24.B.22}$$

where n and n' are again discrete indices labelling both spin z-components and particle types. Of course, the kernel $\mathscr{A}_\alpha(p',p)$ must vanish unless both p and p' are on the mass shell. We shall show first that the $\mathscr{A}_\alpha(p',p)$ vanishes for any $p'\neq p$.

For this purpose, note that if A_α is a symmetry generator, then so is

$$A_\alpha^f \equiv \int d^4x\ \exp(\,iP\cdot x)\,A_\alpha\ \exp(-iP\cdot x)\,f(x)\,, \tag{24.B.23}$$

where P_μ is the four-momentum operator, and $f(x)$ is a function that can be chosen as we like. Acting on a one-particle state, this gives

$$A_\alpha^f\,|p,n\rangle = \sum_{n'}\int d^4p'\ \tilde{f}(p'-p)\left(\mathscr{A}_\alpha(p',p)\right)_{n'n}|p',n'\rangle\,, \tag{24.B.24}$$

where $\tilde{f}$ is the Fourier transform

$$\tilde{f}(k) \equiv \int d^4x\ \exp(ix\cdot k)\,f(x)\,. \tag{24.B.25}$$

Suppose that there is some pair of mass-shell four-momenta p and $p+\Delta$ with $\Delta\neq 0$ such that $\mathscr{A}(p+\Delta,p)\neq 0$. Generic mass-shell four-momenta q, p', and q' with $p'+q'=p+q$ will not have $q+\Delta$ or $p'+\Delta$ or $q'+\Delta$ on the mass shell. If we take $\tilde{f}(k)$ to vanish outside a sufficiently small region around Δ, then A^f_α will annihilate all one-particle states with four-momenta q, p', and q', but not the one-particle states with four-momentum p, so such a symmetry would forbid any scattering process in which any particles of momenta p and q go into any particles of momenta p' and q', in contradiction with the consequence of Assumptions 2 and 3 that there is some scattering at almost all energies and angles.

This result does not mean that any symmetry generator A_α must commute with P_μ, because the kernels $\mathscr{A}_\alpha(p', p)$ may include terms proportional to derivatives of $\delta^4(p' - p)$ as well as terms proportional to $\delta^4(p' - p)$ itself. To deal with this possibility, Coleman and Mandula made the 'ugly technical assumption' that the kernels $\mathscr{A}_\alpha(p', p)$ are *distributions*, which means that each can contain at most a finite number D_α of derivatives of $\delta^4(p' - p)$. To put this another way, each symmetry generator A_α is assumed to act on one-particle states as a polynomial of order D_α in the derivatives $\partial/\partial p_\mu$, with matrix coefficients that at this point are allowed to depend on momentum and spin. To use the above results for symmetry generators that commute with the momentum operators, Coleman and Mandula considered the D_α-fold commutator of momentum operators with A_α

$$B_\alpha^{\mu_1 \ldots \mu_{D_\alpha}} \equiv [P^{\mu_1}, [P^{\mu_2}, \ldots [P^{\mu_{D_\alpha}}, A_\alpha]\,] \ldots] \,. \tag{24.B.26}$$

The matrix elements of the commutators of $B_\alpha^{\mu_1 \cdots \mu_{D_\alpha}}$ with P^μ between states with four-momenta p' and p are proportional to $D_\alpha + 1$ factors of $p' - p$, times a polynomial of order D_α in momentum derivatives acting on $\delta^4(p' - p)$, and therefore vanish. Since the generators $B_\alpha^{\mu_1 \cdots \mu_{D_\alpha}}$ commute with the momentum operators, according to the results obtained so far they act on one-particle states with matrices of the form

$$b_\alpha^{\mu_1 \cdots \mu_{D_\alpha}}(p) = b_\alpha^{\sharp\mu_1 \cdots \mu_{D_\alpha}} + a_\alpha^{\mu\mu_1 \cdots \mu_{D_\alpha}} p_\mu \, 1 \,, \tag{24.B.27}$$

where the $b_\alpha^{\sharp\mu_1 \cdots \mu_{D_\alpha}}$ are momentum-independent traceless Hermitian matrices generating an ordinary internal symmetry algebra, and the $a_\alpha^{\mu\mu_1 \cdots \mu_{D_\alpha}}$ are momentum-independent numerical constants, with both $b_\alpha^{\sharp\mu_1 \cdots \mu_{D_\alpha}}$ and $a_\alpha^{\mu\mu_1 \cdots \mu_{D_\alpha}}$ symmetric in the indices $\mu_1 \cdots \mu_{D_\alpha}$. Also, even though the A_α do not necessarily commute with P_μ, they cannot take one-particle states off the mass shell, so since Assumption 1 requires that the mass-squared operator $-P_\mu P^\mu$ has only discrete eigenvalues, the A_α must commute with $-P_\mu P^\mu$. It follows in particular that for $D \geq 1$

$$0 = [P^{\mu_1} P_{\mu_1}, [P^{\mu_2}, \ldots [P^{\mu_{D_\alpha}}, A_\alpha]\,] \ldots] = 2 P_{\mu_1} B_\alpha^{\mu_1 \cdots \mu_{D_\alpha}} \,,$$

so that

$$0 = p_{\mu_1} b_\alpha^{\mu_1 \cdots \mu_{D_\alpha}}(p) \,. \tag{24.B.28}$$

As long as the theory contains massive particles this must be satisfied for p in any timelike direction, so for $D_\alpha \geq 1$

$$b_\alpha^{\sharp\mu_1\cdots\mu_{D_\alpha}} = 0\,, \qquad (24.B.29)$$

and

$$a_\alpha^{\mu\mu_1\cdots\mu_{D_\alpha}} = -a_\alpha^{\mu_1\mu\cdots\mu_{D_\alpha}}\,. \qquad (24.B.30)$$

But for $D_\alpha \geq 2$ Eq. (24.B.30) together with the symmetry of $a_\alpha^{\mu\mu_1\cdots\mu_D}$ in the indices $\mu_1\cdots\mu_{D_\alpha}$ would require that $a_\alpha^{\mu\mu_1\cdots\mu_{D_\alpha}} = 0$. (For then $a_\alpha^{\mu\mu_1\mu_2\cdots\mu_{D_\alpha}} = a_\alpha^{\mu\mu_2\mu_1\cdots\mu_{D_\alpha}} = -a_\alpha^{\mu_2\mu\mu_1\cdots\mu_{D_\alpha}} = -a_\alpha^{\mu_2\mu_1\mu\cdots\mu_{D_\alpha}} = a_\alpha^{\mu_1\mu_2\mu\cdots\mu_{D_\alpha}} = a_\alpha^{\mu_1\mu\mu_2\cdots\mu_{D_\alpha}} = -a_\alpha^{\mu\mu_1\mu_2\cdots\mu_{D_\alpha}}$.) We are left with at most two kinds of non-vanishing symmetry generator: those with $D_\alpha = 0$, for which the generator A_α commutes with P_μ and therefore must be either an internal symmetry generator or some linear combination of the P_μ; and those with $D_\alpha = 1$, in which case

$$[P^\nu, A_\alpha] = a_\alpha^{\mu\nu} P_\mu\,, \qquad (24.B.31)$$

with $a_\alpha^{\mu\nu}$ some numerical constants antisymmetric in μ and ν. Eq. (24.B.31) requires that

$$A_\alpha = -\tfrac{1}{2} i\, a_\alpha^{\mu\nu} J_{\mu\nu} + B_\alpha\,, \qquad (24.B.32)$$

where $J_{\mu\nu}$ is the generator of proper Lorentz transformations, which according to Eq. (2.4.13) satisfies $[P^\nu, J^{\rho\sigma}] = -i\eta^{\nu\rho}P^\sigma + i\eta^{\nu\sigma}P^\rho$, and B_α commutes with P_μ. Since A_α and $J_{\mu\nu}$ are symmetry generators, so is B_α, which must therefore be a linear combination of internal symmetry generators and/or the components of P_μ. Eq. (24.B.32) therefore completes the proof of the Coleman–Mandula theorem.

* * *

In theories with only massless particles Eq. (24.B.30) does not necessarily follow from Eq. (24.B.28); since $p_\mu p^\mu = 0$, we can also have

$$a_\alpha^{\mu\mu_1\cdots\mu_{D_\alpha}} + a_\alpha^{\mu_1\mu\cdots\mu_{D_\alpha}} \propto \eta^{\mu\mu_1}\,. \qquad (24.B.33)$$

In this case the symmetry algebra consists of internal symmetries plus the algebra of the conformal group, which is spanned by generators K^μ and D together with the generators $J^{\mu\nu}$ and P^μ of the Poincaré group. The commutation relations are

$$\begin{aligned}
&[P^\mu, D] = iP^\mu\,, \qquad [K^\mu, D] = -iK^\mu\,,\\
&[P^\mu, K^\nu] = 2i\eta^{\mu\nu}D + 2iJ^{\mu\nu}\,, \qquad [K^\mu, K^\nu] = 0\,, \qquad (24.B.34)\\
&[J^{\rho\sigma}, K^\mu] = i\eta^{\mu\rho}K^\sigma - i\eta^{\mu\sigma}K^\rho\,, \qquad [J^{\rho\sigma}, D] = 0\,,
\end{aligned}$$

together with the commutation relations (2.4.12)–(2.4.14) of the Poincaré algebra

$$\begin{aligned} i\,[J^{\mu\nu}, J^{\rho\sigma}] &= \eta^{\nu\rho} J^{\mu\sigma} - \eta^{\mu\rho} J^{\nu\sigma} - \eta^{\sigma\mu} J^{\rho\nu} + \eta^{\sigma\nu} J^{\rho\mu}\,, \\ i\,[P^{\mu}\,,\, J^{\rho\sigma}] &= \eta^{\mu\rho} P^{\sigma} - \eta^{\mu\sigma} P^{\rho}\,, \\ [P^{\mu}\,,\, P^{\rho}] &= 0\,. \end{aligned} \tag{24.B.35}$$

The infinitesimal group element

$$U(1+\omega, \epsilon, \lambda, \rho) = 1 + (i/2)J_{\mu\nu}\omega^{\mu\nu} + iP_{\mu}\epsilon^{\mu} + i\lambda D + iK_{\mu}\rho^{\mu} \tag{24.B.36}$$

induces the infinitesimal spacetime transformation

$$x^{\mu} \to x^{\mu} + \omega^{\mu\nu} x_{\nu} + \epsilon^{\mu} + \lambda x^{\mu} + \rho^{\mu} x^{\nu} x_{\nu} - 2x^{\mu} \rho^{\nu} x_{\nu}\,. \tag{24.B.37}$$

These are the most general infinitesimal spacetime transformations that leave the light-cone invariant.

Problems

1. Show that the most general symmetry algebra allowed under the assumptions of the Coleman–Mandula theorem in the case where all particles are massless consists of internal symmetry generators plus either the Poincaré algebra or the conformal algebra (24.B.34), (24.B.35).

2. Show that the Gervais–Sakita action (24.2.5) is invariant under the worldsheet supersymmetry transformation (24.2.7).

3. Calculate the change in the Wess–Zumino Lagrangian density (24.2.9) under the spacetime supersymmetry transformation (24.2.8).

References

1. B. Sakita, *Phys. Rev.* **136**, B 1756 (1964); F. Gursey and L. A. Radicati, *Phys. Rev. Lett.* **13**, 173 (1964); A. Pais, *Phys. Rev. Lett.* **13**, 175 (1964); F. Gursey, A. Pais, and L. A. Radicati, *Phys. Rev. Lett.* **13**, 299 (1964). These articles are reprinted in *Symmetry Groups in Nuclear and Particle Physics*, F. J. Dyson, ed. (W. A. Benjamin, New York, 1966), along with a helpful set of Dyson's lecture notes on this subject.

2. E. P. Wigner, *Phys. Rev.* **51**. 106 (1937). This is reproduced in *Symmetry Groups in Nuclear and Particle Physics,* Ref. 1.

3. A. Salam, R. Delbourgo, and J. Strathdee, *Proc. Roy. Soc. (London)* **A 284**, 146 (1965); M. A. Beg and A. Pais, *Phys. Rev. Lett.* **14**, 267 (1965); B. Sakita and K. C. Wali, *Phys. Rev.* **139**, B 1355 (1965). These are reproduced in *Symmetry Groups in Nuclear and Particle Physics*, Ref. 1.
4. W. D. McGlinn, *Phys. Rev. Lett.* **12**, 467 (1964); O. W. Greenberg, *Phys. Rev.* **135**, B 1447 (1964); L. Michel, *Phys. Rev.* **137**, B 405 (1964); L. Michel and B. Sakita, *Ann. Inst. Henri-Poincaré* **2**, 167 (1965); M. A. B. Beg and A. Pais, *Phys. Rev. Lett.* **14**, 509, 577 (1965); S. Coleman, *Phys. Rev.* **138**, B 1262 (1965); S. Weinberg, *Phys. Rev.* **139**, B 597 (1965); L. O'Raifeartaigh, *Phys. Rev.* **139**, B 1052 (1965). These are reproduced in *Symmetry Groups in Nuclear and Particle Physics*, Ref. 1.
5. S. Coleman and J. Mandula, *Phys. Rev.* **159**, 1251 (1967).
6. For an introduction with references to the original literature, see M. B. Green, J. H. Schwarz, and E. Witten, *Superstring Theory* (Cambridge University Press, Cambridge, 1987); J. Polchinski, *String Theory* (Cambridge University Press, Cambridge, 1998).
7. P. Ramond, *Phys. Rev.* **D3**, 2415 (1971). This article is reprinted in *Superstrings — The First 15 Years of Superstring Theory*, J. H. Schwarz, ed. (World Scientific, Singapore, 1985).
8. A. Neveu and J. H. Schwarz, *Nucl. Phys.* **B31**, 86 (1971); *Phys. Rev.* **D4**, 1109 (1971). These articles are reprinted in *Superstrings — The First 15 Years of Superstring Theory*, Ref. 7. Also see Y. Aharonov, A. Casher, and L. Susskind, *Phys. Rev.* **D5**, 988 (1972).
9. J.-L. Gervais and B. Sakita, *Nucl. Phys.* **B34**, 632 (1971). This article is reprinted in *Superstrings — The First 15 Years of Superstring Theory*, Ref. 7.
10. J. Wess and B. Zumino, *Nucl. Phys.* **B70**, 39 (1974). This article is reprinted in *Supersymmetry*, S. Ferrara, ed. (North Holland/World Scientific, Amsterdam/Singapore, 1987).
11. J. Wess and B. Zumino, *Phys. Lett.* **49B**, 52 (1974). This article is reprinted in *Supersymmetry*, Ref. 10.

11*a*. F. Gliozzi, J. Scherk, and D. Olive, *Nucl. Phys.* **B122**, 253 (1977).

12. Yu. A. Gol'fand and E. P. Likhtman, *JETP Letters* **13**, 323 (1971). This article is reprinted in *Supersymmetry*, Ref. 10.
13. D. V. Volkov and V. P. Akulov, *Phys. Lett.* **46B**, 109 (1973). This article is reprinted in *Supersymmetry*, Ref. 10.
14. M. Gell-Mann, Cal. Tech. Synchotron Laboratory Report CTSL–20 (1961), unpublished. This is reproduced along with other articles on

$SU(3)$ symmetry in M. Gell-Mann and Y. Ne'eman, *The Eightfold Way* (Benjamin, New York, 1964).

15. R. Haag, J. T. Lopuszanski, and M. Sohnius, *Nucl. Phys.* **B88**, 257 (1975). This article is reprinted in *Supersymmetry*, Ref. 10.

25

Supersymmetry Algebras

This chapter will develop the form of the supersymmetry algebra from first principles, following the treatment of Haag, Lopuszanski, and Sohnius.[1] As we shall see, under conditions in which the Coleman–Mandula theorem applies, this structure is almost uniquely fixed by the requirements of Lorentz invariance. The supermultiplet structure of one-particle states will then be deduced directly from the supersymmetry algebra.

25.1 Graded Lie Algebras and Graded Parameters

We saw in Section 2.2 how to express any continuous symmetry transformation in terms of a Lie algebra of linearly independent symmetry generators t_a that satisfy commutation relations $[t_a, t_b] = i\sum_c C^c_{ab} t_c$. In much the same way, supersymmetry is expressed in terms of symmetry generators t_a that form a *graded* Lie algebra,[2] embodied in commutation *and* anticommutation relations of the form

$$t_a t_b - (-1)^{\eta_a \eta_b} t_b t_a = i \sum_c C^c_{ab} t_c \, . \tag{25.1.1}$$

(The summation convention is suspended in this section.) Here η_a for each a is either $+1$ or 0 and is known as the *grading* of the generator t_a, and the C^c_{ab} are a set of numerical structure constants. Generators t_a for which $\eta_a = 1$ are called *fermionic*; the others, for which $\eta_a = 0$, are called *bosonic*. Eq. (25.1.1) provides commutation relations for bosonic operators with each other and with fermionic operators, but anticommutation relations for fermionic operators with each other. We will come back soon to the motivation for Eq. (25.1.1); for the moment we will just take a look at its consequences for the structure constants.

According to Eq. (25.1.1), the structure constants must satisfy the conditions

$$C^c_{ba} = -(-1)^{\eta_a \eta_b} C^c_{ab} \, . \tag{25.1.2}$$

For any operator formed as a functional of field operators, the products of two bosonic or two fermionic operators are bosonic, and the products of a fermionic with a bosonic operator are fermionic, so that

$$C^c_{ab} = 0 \text{ unless } \eta^c = \eta^a + \eta^b \text{ (mod 2)}\,. \tag{25.1.3}$$

Also, for any operator formed in this way, the Hermitian adjoint of a bosonic or fermionic operator is, respectively, bosonic or fermionic. If the t_a are Hermitian operators, then the structure constants satisfy a reality condition

$$C^c_{ab}{}^* = -C^c_{ba}\,. \tag{25.1.4}$$

The structure constants also satisfy a non-linear constraint, which follows from a super-Jacobi identity

$$(-1)^{\eta_c\eta_a}[[t_a,t_b\},t_c\} + (-1)^{\eta_a\eta_b}[[t_b,t_c\},t_a\} + (-1)^{\eta_b\eta_c}[[t_c,t_a\},t_b\} = 0\,. \tag{25.1.5}$$

Here '[...}' denotes a commutator/anticommutator like that appearing on the left-hand side of Eq. (25.1.1), but here extended to any graded operators O, O', etc.

$$[O,O'\} \equiv OO' - (-1)^{\eta(O)\eta(O')}O'O = -(-1)^{\eta(O)\eta(O')}[O',O\}\,, \tag{25.1.6}$$

it now being understood that any product $O = t_a t_b t_c \cdots$ of generators is given a grading $\eta(O) \equiv \eta_a + \eta_b + \eta_c + \cdots$ (mod 2). (To prove Eq. (25.1.5), it is sufficient to prove that the coefficients of $t_a t_b t_c$ and $t_a t_c t_b$ vanish, for then the symmetry of the left-hand side of Eq. (25.1.5) under the cyclic permutations $abc \to bca \to cab$ will ensure that the coefficients of all other products of generators also vanish. The coefficient of $t_a t_b t_c$ in Eq. (25.1.5) is

$$(-1)^{\eta_c\eta_a} - (-1)^{\eta_a\eta_b}(-1)^{\eta_a(\eta_b+\eta_c)} = 0\,,$$

while the coefficient of $t_a t_c t_b$ is

$$(-1)^{\eta_a\eta_b}(-1)^{\eta_b\eta_c}(-1)^{\eta_a(\eta_b+\eta_c)} - (-1)^{\eta_b\eta_c}(-1)^{\eta_c\eta_a} = 0\,,$$

completing the proof.) By inserting Eq. (25.1.1) in Eq. (25.1.5), we find the constraint

$$\sum_d (-1)^{\eta_c\eta_a} C^d_{ab} C^e_{dc} + \sum_d (-1)^{\eta_a\eta_b} C^d_{bc} C^e_{da} + \sum_d (-1)^{\eta_b\eta_c} C^d_{ca} C^e_{db} = 0\,. \tag{25.1.7}$$

Of course, in the case where all generators are bosonic Eq. (25.1.5) is the usual Jacobi identity, and Eq. (25.1.7) is the usual non-linear condition (2.2.22) on the structure constants.

Eq. (25.1.1) can be taken as our starting point, but it can also be given a motivation like that given for ordinary Lie algebras in Section 2.2, as a necessary feature of finite continuous symmetry transformations. The difference is that now these transformations depend on continuous *graded* parameters. A set of graded c-number parameters can be thought of as 'numbers,' including Grassmann parameters (see Section 9.5) as well as ordinary numbers, which satisfy the associative and distributive rules of arithmetic, but which instead of simply commuting satisfy relations

$$\alpha^a\beta^b = (-1)^{\eta_a\eta_b}\beta^b\alpha^a\,, \tag{25.1.8}$$

where α^a, β^a, ... are used to distinguish different values of the ath parameter, in the same way that in vector algebra we might let v^a and u^a denote the a-components of different values of some real vector. Again the ath graded parameter is given a grading η_a equal to +1 or 0 when α^a is a fermionic or bosonic parameter, respectively. That is, these parameters commute if either is bosonic, and anticommute if both are fermionic. The product $\alpha^a\beta^b\gamma^c\cdots$ of a set of graded parameters is given the grading $\eta_a+\eta_b+\eta_c+\cdots$ (mod 2); that is, such a product is fermionic if it involves an odd number of fermionic parameters, and is otherwise bosonic. With this grading, it is easy to see that products of graded parameters satisfy a commutation or anticommutation rule just like Eq. (25.1.8).

Consider a continuous transformation $T(\alpha)$, given by a formal power series in the graded parameters α^a:

$$T(\alpha) = 1 + \sum_a \alpha^a t_a + \sum_{ab} \alpha^a\alpha^b t_{ab} + \cdots \quad , \tag{25.1.9}$$

where t_a, t_{ab}, etc. are a set of α-independent operator coefficients, not yet assumed to satisfy any algebraic relations like Eq. (25.1.1). Because the parameters α^a satisfy Eq. (25.1.8), the coefficients $t_{ab\cdots}$ must satisfy symmetry/antisymmetry conditions, such as

$$t_{ab} = (-1)^{\eta_a\eta_b}t_{ba}\,. \tag{25.1.10}$$

It is convenient also to assume that the transformation $T(\beta)$ commutes with any value α^a of any graded parameter, in which case the operator coefficients in (25.1.9) satisfy the conditions

$$\alpha^a t_b = (-1)^{\eta_a\eta_b}t_b\alpha^a\,, \tag{25.1.11}$$

$$\alpha^a t_{bc} = (-1)^{\eta_a(\eta_b+\eta_c)}t_{bc}\alpha^a\,. \tag{25.1.12}$$

That is, t_b and t_{bc} commute or anticommute with graded parameters as if they were graded parameters themselves, with grading η_b and $\eta_b+\eta_c$ (mod 2), respectively.

The other constraints on these operators follow from the requirement that the $T(\alpha)$ form a semi-group; that is, that the product of the operators for different values α and β of the graded parameters is itself a T-operator

$$T(\alpha)T(\beta) = T(f(\alpha,\beta))\,, \tag{25.1.13}$$

where $f^c(\alpha,\beta)$ is itself a formal power series in the graded parameters. Because $T(0)T(\beta) = T(\beta)$ and $T(\alpha)T(0) = T(\alpha)$, we must have

$$f^c(0,\beta) = \beta^c\,, \qquad f^c(\alpha,0) = \alpha^c\,, \tag{25.1.14}$$

and therefore the power series expansion for $f(\alpha,\beta)$ must take the form

$$f^c(\alpha,\beta) = \alpha^c + \beta^c + \sum_{ab} f^c_{ab}\,\alpha^a\,\beta^b + \cdots\,, \tag{25.1.15}$$

where f^c_{ab} is a set of ordinary (that is, bosonic) constants, and '$\cdots$' denotes terms of third or higher order in the graded parameters. In order for $f^c(\alpha,\beta)$ to be a graded parameter, it is necessary for each term in Eq. (25.1.15) to have the same grading, which implies that

$$f^c_{ab} = 0 \;\text{ unless }\; \eta^c = \eta^a + \eta^b \;(\text{mod } 2)\,. \tag{25.1.16}$$

Inserting the power series (25.1.9) and (25.1.15) into the product rule (25.1.13) gives

$$\begin{aligned}
&\Big[1+\sum_a \alpha^a t_a + \sum_{ab} \alpha^a\alpha^b t_{ab} + \cdots\Big]\Big[1+\sum_a \beta^a t_a + \sum_{ab} \beta^a\beta^b t_{ab} + \cdots\Big] \\
&\quad = 1 + \sum_c \Big(\alpha^c + \beta^c + \sum_{ab} f^c_{ab}\alpha^a\beta^b + \cdots\Big)t_c \\
&\quad\quad + \sum_{cd}\Big(\alpha^c + \beta^c + \cdots\Big)\Big(\alpha^d + \beta^d + \cdots\Big)t_{cd} + \cdots \qquad .
\end{aligned}$$

The coefficients of 1, α^a, β^a, $\alpha^a\alpha^b$, and $\beta^a\beta^b$ match on both sides of this equation, but the condition that the coefficients of $\alpha^a\beta^b$ are the same yields the non-trivial relation

$$(-1)^{\eta_a\eta_b} t_a t_b = \sum_c f^c_{ab} t_c + t_{ab} + (-1)^{\eta_a\eta_b} t_{ba} = \sum_c f^c_{ab} t_c + 2t_{ab}\,. \tag{25.1.17}$$

(The sign factor on the left-hand side arises from the interchange of t_a and β^b.) Together with higher-order relations of the same sort, this allows us to calculate the whole function (25.1.9) if we know the generators t_a and the group-composition function $f^a(\alpha,\beta)$. But for this to be possible, t_a must satisfy a constraint. Using Eq. (25.1.10), the difference or sum of Eq. (25.1.17) and the same equation with a and b interchanged yields the Lie superalgebra relations Eq. (25.1.1), with structure constants given by

$$i\,C^c_{ab} = (-1)^{\eta_a\eta_b} f^c_{ab} - f^c_{ba}\,. \tag{25.1.18}$$

Also, Eq. (25.1.3) follows immediately from Eqs. (25.1.16) and (25.1.18).

The complex conjugate α^* of an anticommuting c-number α is defined so that the Hermitian adjoint of the product of α and an arbitrary operator $\mathscr{O}$ is

$$(\alpha\mathscr{O})^* = \mathscr{O}^*\alpha^*\,. \tag{25.1.19}$$

It follows that products of c-numbers behave the same under complex conjugation as operators do under Hermitian conjugation:

$$(\alpha\beta)^* = \beta^*\alpha^* , \tag{25.1.20}$$

and that α^* has the same grading as α.

The graded Lie algebras of importance to physics are severely restricted by spacetime symmetries. We now turn to a consideration of these restrictions.

25.2 Supersymmetry Algebras

Consider a general graded Lie algebra of symmetry generators that commute with the S-matrix. If Q is any of the fermionic symmetry generators, then so will be $U^{-1}(\Lambda)\, Q\, U(\Lambda)$, where $U(\Lambda)$ is the quantum mechanical operator corresponding to an arbitrary homogeneous Lorentz transformation $\Lambda^\mu{}_\nu$. Therefore $U^{-1}(\Lambda)\, Q\, U(\Lambda)$ is a linear combination of the complete set of fermionic symmetry generators, and hence this set of generators must furnish a representation of the homogeneous Lorentz group. The individual generators may therefore be classified according to the irreducible representation of the homogeneous Lorentz group to which they belong.

As described in Section 5.6, the representations of the homogeneous Lorentz group furnished by any set of operators can be specified by giving their commutation relations with generators **A** and **B**, defined by

$$\mathbf{A} \equiv \tfrac{1}{2}\Big(\mathbf{J} + i\mathbf{K}\Big) , \qquad \mathbf{B} \equiv \tfrac{1}{2}\Big(\mathbf{J} - i\mathbf{K}\Big) , \tag{25.2.1}$$

where **J** and **K** are the Hermitian generators of rotations and boosts, respectively. These satisfy the commutation relations

$$[A_i, A_j] = \sum_k \epsilon_{ijk} A_k , \quad [B_i, B_j] = \sum_k \epsilon_{ijk} B_k , \quad [A_i, B_j] = 0 , \tag{25.2.2}$$

where i, j, and k run over the values 1, 2, 3, and ϵ_{ijk} is totally antisymmetric, with $\epsilon_{123} = +1$. Thus representations of the homogeneous Lorentz group are labelled, like states with two independent spins, by a pair of integers or half-integers A and B, with elements of the representation labelled by a pair of indices a and b, which run by unit steps from $-A$ to $+A$ and from $-B$ to $+B$, respectively. More specifically, a set of $(2A+1)(2B+1)$ operators Q^{AB}_{ab} that form an (A, B) representation of the homogeneous Lorentz group satisfies the commutation relations

$$[\mathbf{A}, Q^{AB}_{ab}] = -\sum_{a'} \mathbf{J}^{(A)}_{aa'}\, Q^{AB}_{a'b} , \qquad [\mathbf{B}, Q^{AB}_{ab}] = -\sum_{b'} \mathbf{J}^{(B)}_{bb'}\, Q^{AB}_{ab'} , \tag{25.2.3}$$

where $\mathbf{J}^{(j)}$ is the spin three-vector matrix for angular momentum j:

$$\left(J_1^{(j)} \pm iJ_2^{(j)}\right)_{\sigma'\sigma} = \delta_{\sigma',\sigma\pm1}\sqrt{(j\mp\sigma)(j\pm\sigma+1)}\,, \qquad \left(J_3^{(j)}\right)_{\sigma'\sigma} = \delta_{\sigma'\sigma}\sigma\,. \tag{25.2.4}$$

From Eq. (25.2.4), it follows that*

$$-\left(\mathbf{J}^{(j)}\right)^*_{\sigma',\sigma} = (-1)^{\sigma'-\sigma}\left(\mathbf{J}^{(j)}\right)_{-\sigma',-\sigma}\,. \tag{25.2.5}$$

Thus if Q^j_σ are a set of operators that transform according to the spin j representation of the rotation group, then so are $(-1)^{j-\sigma}Q^{j*}_{-\sigma}$. Also, Eq. (25.2.1) shows that $\mathbf{A}^* = \mathbf{B}$. By taking Hermitian adjoints in Eq. (25.2.3), we see that the Hermitian adjoint Q^{AB*}_{ab} of operators that transform according to the (A,B) representation of the homogeneous Lorentz group are related by a similarity transformation to operators $\bar{Q}^{BA}_{ba}$ that transform according to the (B,A) representation:

$$Q^{AB*}_{ab} = (-1)^{A-a}(-1)^{B-b}\,\bar{Q}^{BA}_{-b,-a}\,. \tag{25.2.6}$$

The Haag–Lopuszanski–Sohnius theorem[1] states in part that the fermion symmetry generators can only belong to the $(0,1/2)$ and $(1/2,0)$ representations. As we have seen, the Hermitian adjoint of a $(0,1/2)$ or $(1/2,0)$ operator is a linear combination respectively of $(1/2,0)$ or $(0,1/2)$ operators, so the complete set of fermionic symmetry operators may be divided into $(0,1/2)$ generators Q_{ar} (with the superscript $0\frac{1}{2}$ omitted) and their $(1/2,0)$ Hermitian adjoints Q^*_{ar}, where a is a spinor index running over the values $\pm1/2$, and r is used to distinguish different two-component generators with the same Lorentz transformation properties.** The theorem further states that the fermionic generators may be defined so as to satisfy the anticommutation relations

$$\{\mathrm{Q}_{ar}\,,\,\mathrm{Q}^*_{bs}\} = 2\delta_{rs}\,\sigma^\mu_{ab}\,P_\mu\,, \tag{25.2.7}$$
$$\{\mathrm{Q}_{ar}\,,\,\mathrm{Q}_{bs}\} = e_{ab}\,Z_{rs}\,, \tag{25.2.8}$$

where P_μ is the four-momentum operator, the $Z_{rs} = -Z_{sr}$ are bosonic symmetry generators, and σ_μ and e are the 2×2 matrices (with rows and

---

* We use an asterisk for the Hermitian adjoint of an operator or the complex conjugate of a number. The dagger † will be used for the transpose of the matrix formed from the Hermitian adjoints of operators or complex conjugates of numbers.

** We are using roman instead of italic letters for the two-component Weyl spinors Q_{ar}, to distinguish them from four-component Dirac spinors that will be introduced later in this section. There is a notation due to van der Waerden, according to which a $(0,1/2)$ operator like Q is written with dotted indices, as $\mathrm{Q}_{\dot{a}}$, while $(1/2,0)$ operators are written with undotted indices. We will not use this notation here, but will instead explicitly indicate which two-component spinors transform according to the $(0,1/2)$ or $(1/2,0)$ representations of the homogeneous Lorentz group.

columns labelled +1/2, −1/2):

$$\sigma_1 = \begin{pmatrix} 0 & 1 \\ 1 & 0 \end{pmatrix}, \quad \sigma_2 = \begin{pmatrix} 0 & -i \\ i & 0 \end{pmatrix}, \quad \sigma_3 = \begin{pmatrix} 1 & 0 \\ 0 & -1 \end{pmatrix}, \\ \sigma_0 = \begin{pmatrix} 1 & 0 \\ 0 & 1 \end{pmatrix}, \quad e = \begin{pmatrix} 0 & 1 \\ -1 & 0 \end{pmatrix}. \tag{25.2.9}$$

Finally, the fermionic generators commute with energy and momentum:

$$[P_\mu, Q_{ar}] = [P_\mu, Q^*_{ar}] = 0, \tag{25.2.10}$$

and the Z_{rs} and Z^*_{rs} are a set of central charges of this algebra, in the sense that

$$0 = [Z_{rs}, Q_{at}] = [Z_{rs}, Q^*_{at}] = [Z_{rs}, Z_{tu}] = [Z_{rs}, Z^*_{tu}] \\ = [Z^*_{rs}, Q_{at}] = [Z^*_{rs}, Q^*_{at}] = [Z^*_{rs}, Z^*_{tu}]. \tag{25.2.11}$$

To prove these results, let us start by considering non-vanishing fermionic symmetry generators that belong to some (A, B) irreducible representation of the homogeneous Lorentz group, and that therefore may be labelled Q^{AB}_{ab}, where a and b run by unit steps from $-A$ to $+A$ and from $-B$ to $+B$, respectively. As already mentioned, the Hermitian adjoint is related to operators belonging to the (B, A) representation by Eq. (25.2.6), so the anticommutator of these operators must take the form

$$\{Q^{AB}_{ab}, Q^{AB*}_{a'b'}\} = (-1)^{A-a'}(-1)^{B-b'} \sum_{C=|A-B|}^{A+B} \sum_{D=|A-B|}^{A+B} \sum_{c=-C}^{C} \sum_{d=-D}^{D} \\ \times C_{AB}(Cc; a, -b')\, C_{AB}(Dd; -a'b)\, X^{CD}_{cd}, \tag{25.2.12}$$

where $C_{AB}(j\sigma; ab)$ is the usual Clebsch–Gordan coefficient for coupling spins A and B to form spin j, and X^{CD}_{cd} is the (c, d)-component of an operator that transforms according to the (C, D) representation of the homogeneous Lorentz group. Using the well-known unitarity properties of the Clebsch–Gordan coefficients, we may express the operator X^{CD}_{cd} in terms of these anticommutators:

$$X^{CD}_{cd} = \sum_{a=-A}^{A} \sum_{b=-B}^{B} \sum_{a'=-A}^{A} \sum_{b'=-B}^{B} (-1)^{A-a'}(-1)^{B-b'} \\ \times C_{AB}(Cc; a - b')\, C_{AB}(Dd; -a'b)\, \{Q^{AB}_{ab}, Q^{AB*}_{a'b'}\}. \tag{25.2.13}$$

Not all of these operators are necessarily non-zero. But the only non-zero Clebsch–Gordan coefficients $C_{AB}(j\sigma, ab)$ for $j = \sigma = A + B$ and $j = -\sigma = A + B$ are the ones for $a = A$, $b = B$ and $a = -A$, $b = -B$, respectively, which both have the value unity, so by taking $C = D = c = -d = A + B$ in Eq. (25.2.13), we find that

$$X^{A+B, A+B}_{A+B, -A-B} = (-1)^{2B}\, \{Q^{AB}_{A,-B}, Q^{AB*}_{A,-B}\}. \tag{25.2.14}$$

This cannot vanish unless $Q^{AB}_{A,-B} = 0$, which would imply (taking commutators with the 'lowering' operators $A_1 - iA_2$ and 'raising' operators $B_1 + iB_2$) that all the Q^{AB}_{ab} vanish. Hence if there are any non-vanishing (A, B) fermionic generators, then their anticommutators with their adjoints must at least involve non-zero bosonic symmetry generators belonging to the $(A + B, A + B)$ representation.

Now, the Coleman–Mandula theorem tells us that the bosonic symmetry generators consist of the $(1/2, 1/2)$ generators P_μ of translations, the $(1, 0) + (0, 1)$ generators $J_{\mu\nu}$ of proper Lorentz transformations, and perhaps the $(0, 0)$ generators T_A of various internal symmetries. (Recall that symmetric traceless tensors of rank N transform according to the representation $(N/2, N/2)$, antisymmetric tensors of rank 2 transform according to the representation $(1, 0) + (0, 1)$, while Dirac fields transform according to the representation $(1/2, 0) + (0, 1/2)$.) Hence the fermionic symmetry generators can only belong to representations (A, B) with $A + B \leq 1/2$. These operators turn bosons into fermions and vice-versa, so they cannot be scalars, leaving only the $(1/2, 0)$ and $(0, 1/2)$ representations, as was to be shown. Labelling the linearly independent $(0, 1/2)$ fermionic generators as Q_{ar}, the anticommutator $\{\mathrm{Q}_{ar}, \mathrm{Q}^*_{bs}\}$ belongs to the representation $(0, 1/2) \times (1/2, 0) = (1/2, 1/2)$, and therefore must be proportional to the only $(1/2, 1/2)$ bosonic symmetry generator, the momentum four-vector P_μ. Lorentz invariance dictates that the form of this relation must be

$$\{\mathrm{Q}_{ar}, \mathrm{Q}^*_{bs}\} = 2N_{rs}\, \sigma^\mu_{ab}\, P_\mu \,, \tag{25.2.15}$$

where N_{rs} is a numerical matrix.

To see this, we use the isomorphism of the Lorentz group (or more properly its covering group) with the group $SL(2, C)$ of two-dimensional unimodular complex matrices λ, discussed in Section 2.7. The effect of a Lorentz transformation $\Lambda^\mu{}_\nu$ on the $(0, 1/2)$ fermionic generators is

$$U^{-1}(\Lambda)\, \mathrm{Q}_{ar}\, U(\Lambda) = \sum_b \lambda_{ab}\, \mathrm{Q}_{br} \,, \tag{25.2.16}$$

where Λ is the Lorentz transformation defined by

$$\lambda\, \sigma_\mu\, \lambda^\dagger = \Lambda^\nu{}_\mu \sigma_\nu \,. \tag{25.2.17}$$

We can check that Eq. (25.2.16) applies for $(0, 1/2)$ operators by noting that for an infinitesimal Lorentz transformation $\Lambda^\mu{}_\nu = \delta^\mu{}_\nu + \omega^\mu{}_\nu$ with $\omega_{\mu\nu} = -\omega_{\nu\mu}$, Eq. (25.2.17) is satisfied for

$$\lambda = 1 + \tfrac{1}{2}\Big[\tfrac{1}{2}\, i\, \epsilon_{ijk}\omega_{ij} + \omega_{k0}\Big]\sigma_k \,,$$

while[†]

$$U(\Lambda) = 1 + \tfrac{1}{2}\, i\,\omega_{\mu\nu}\, J^{\mu\nu} = 1 + \tfrac{1}{2}\, i\,\epsilon_{ijk}\omega_{ij}J_k - i\,\omega_{i0}\, K_i\,.$$

(Repeated latin indices i, j, k are summed over values $1, 2, 3$.) In this case, by equating the coefficients of ω_{ij} and ω_{i0} in Eq. (25.2.16), we find that

$$[\mathbf{J}, \mathrm{Q}_a] = -\tfrac{1}{2}\sum_b \boldsymbol{\sigma}_{ab}\, \mathrm{Q}_b\,, \qquad [\mathbf{K}, \mathrm{Q}_a] = -\tfrac{1}{2}\, i\sum_b \boldsymbol{\sigma}_{ab}\, \mathrm{Q}_b\,,$$

or, equivalently,

$$[\mathbf{B}, \mathrm{Q}_a] = -\tfrac{1}{2}\sum_b \boldsymbol{\sigma}_{ab}\mathrm{Q}_b\,, \qquad [\mathbf{A}, \mathrm{Q}_a] = 0\,,$$

which shows that an operator satisfying Eq. (25.2.16) belongs to the $(0, 1/2)$ representation. Now, the σ_μ form a complete set of 2×2 matrices, so we can put the anticommutator $\{\mathrm{Q}_{ar}, \mathrm{Q}^*_{bs}\}$ in the form $N^\mu_{rs}\,(\sigma_\mu)_{ab}$, where N^μ is some matrix of operators. Eqs. (25.2.16) and (25.2.17) show that these operators are four-vectors, in the sense that $U^{-1}(\Lambda)N^\mu U(\Lambda) = \Lambda^\mu{}_\nu\, N^\nu$, and so by the Coleman–Mandula theorem they must be proportional to P^μ, the only four-vector of bosonic symmetry operators. Setting $N^\mu_{rs} = 2P^\mu N_{rs}$ then gives Eq. (25.2.15).

Now we will apply a linear transformation to the Q_{ar}, to put their anticommutators in the form (25.2.7). For this purpose, we need to establish that the matrix N_{rs} is Hermitian and positive-definite. That it is Hermitian follows immediately by taking the Hermitian adjoint of Eq. (25.2.15). To see that it is positive-definite, recall that the Q_{ar} are taken linearly independent, so for any non-zero linear combination $\mathrm{Q} \equiv \sum_r d_a\, c_r\, \mathrm{Q}_{ar}$ there must be some state $|\Psi\rangle$ that Q does not annihilate. Taking the expectation value of Eq. (25.2.15) in this state implies that

$$2\langle\Psi|\sum_{ab}\sigma^\mu_{ab}P_\mu d_a d^*_b\,|\Psi\rangle \sum_{rs} c_r c^*_s N_{rs} = \langle\Psi|\{\mathrm{Q},\mathrm{Q}^*\}|\Psi\rangle > 0\,.$$

This shows immediately that $\sum_{rs} c_r c^*_s N_{rs}$ cannot vanish for any c_r that are not all zero, so N_{rs} is either positive-definite or negative-definite. The operator $\sum_{ab}(\sigma_\mu)_{ab}P^\mu d_a d^*_b$ is positive on the space of physical states with $-P^\mu P_\mu \geq 0$ and $P^0 > 0$, so the matrix N_{rs} must be positive-definite.[††]

[†] Here K_i is defined as J_{i0}. There was a mistake in the first two printings of Volume I: K_i was defined as J^{i0} in Sections 2.4, 3.3, and 3.5, but as J_{i0} in Sections 5.6 and 5.9, with **A** and **B** throughout given by Eq. (25.2.1).

[††] This argument may be turned around. Assuming a supersymmetry with N_{rs} positive-definite, as in Eq. (25.2.7), we can deduce that $P^0 > 0$ for all states.[2] However, this conclusion is not valid when gravitation is taken into account, and unless gravitation is taken into account a shift in the energy of all states by the same amount would have no physical effects.

We may now define new fermionic generators

$$Q'_{ar} \equiv \sum_s N_{rs}^{-1/2} Q_{as} ,$$

for which the anticommutators take the form

$$\{Q'_{ar}, Q'^*_{bs}\} = 2\delta_{rs}\, \sigma^{\mu}_{ab}\, P_{\mu} .$$

From now on we shall assume that the fermionic generators are defined in this way, and drop the primes, so that Eq. (25.2.7) is satisfied.

Next we must show that the Q_{ar} commute with the momentum four-vector P_μ. The commutator of a $(1/2, 1/2)$ operator like P_μ with a $(0, 1/2)$ operator like Q can only be a $(1/2, 0)$ or $(1/2, 1)$ operator, but we have seen that there are no $(1/2, 1)$ symmetry generators, so the commutator of P_μ with Q can only be proportional to the $(1/2, 0)$ symmetry generator Q^*. Lorentz invariance requires this relation to take the form

$$[\mathscr{M}_{ab}, Q_{cr}] = \sum_s e_{ac}\, K_{rs}\, Q^*_{bs} , \tag{25.2.18}$$

where K is a numerical matrix, and $\mathscr{M}$ is the matrix of operators

$$\mathscr{M} \equiv \sigma_\mu P^\mu . \tag{25.2.19}$$

(The matrix e_{ac} is the Clebsch–Gordan coefficient for coupling two spins $1/2$ to give zero spin.) It is straightforward then to calculate that

$$[\mathscr{M}_{-\frac{1}{2}-\frac{1}{2}}, [\mathscr{M}_{-\frac{1}{2}-\frac{1}{2}}, \{Q_{\frac{1}{2}r}, Q^*_{\frac{1}{2}s}\}]] = -4(\mathscr{M})_{-\frac{1}{2}-\frac{1}{2}}(KK^\dagger)_{rs} . \tag{25.2.20}$$

Using Eq. (25.2.7), the left-hand side is a linear combination of multiple commutators $[P_\mu, [P_\nu, P_\lambda]]$, all of which vanish, while $(\mathscr{M})_{-1/2-1/2}$ is non-zero for generic momenta, so $KK^\dagger = 0$, and therefore $K = 0$, which with Eq. (25.2.18) shows that $[P_\mu, Q_{ar}] = 0$. The complex conjugate gives $[P_\mu, Q^*_{ar}] = 0$.

Now we can take up the anticommutator of two Qs. The anticommutator of two $(0, 1/2)$ symmetry generators must be a linear combination of $(0, 1)$ and $(0, 0)$ symmetry generators. The Coleman–Mandula theorem tells us that the only $(0, 1)$ symmetry generators are linear combinations of the generators $J_{\nu\lambda}$ of proper homogeneous Lorentz transformations, but since the Qs commute with P_μ so must their anticommutators, and Eq. (2.4.13) shows that no linear combination of the $J_{\nu\lambda}$ commutes with P_μ. This leaves just $(0, 0)$ operators, which commute with both P_μ and $J_{\mu\nu}$. Lorentz invariance then requires the anticommutators of the Qs with each other to take the form of Eq. (25.2.8). The internal symmetry generators Z_{rs} are antisymmetric in r and s because the whole expression must be symmetric under interchange of r and a with s and b, and the matrix e_{ab} is antisymmetric in a and b.

It now only remains to show that the Zs are central charges. It follows immediately from Eqs. (25.2.8) and (25.2.10) that

$$[P_\mu, Z_{rs}] = 0 . \tag{25.2.21}$$

Next consider the generalized Jacobi identity (25.1.5) involving two Qs and a Q*:

$$0 = [\{Q_{ar}, Q_{bs}\}, Q^*_{ct}] + [\{Q_{bs}, Q^*_{ct}\}, Q_{ar}] + [\{Q^*_{ct}, Q_{ar}\}, Q_{bs}] .$$

Eqs. (25.2.7) and (25.2.10) show that the second and third terms vanish, so

$$[Z_{rs}, Q^*_{ct}] = 0 . \tag{25.2.22}$$

Finally, consider the generalized Jacobi identity for a Z, a Q, and a Q*:

$$0 = -[Z_{rs}, \{Q_{at}, Q^*_{bu}\}] + \{Q^*_{bu}, [Z_{rs}, Q_{at}]\} - \{Q_{at}, [Q^*_{bu}, Z_{rs}]\} .$$

The first and third terms vanish because of Eqs. (25.2.21) and (25.2.22), respectively, so we are left with the second term

$$\{Q^*_{bu}, [Z_{rs}, Q_{at}]\} = 0 . \tag{25.2.23}$$

Now, $[Z_{rs}, Q_{at}]$ is a (0, 1/2) symmetry generator, so it must be a linear combination of the Qs:

$$[Z_{rs}, Q_{at}] = \sum_u M_{rstu}\, Q_{au} . \tag{25.2.24}$$

Eq. (25.2.23) then reads

$$\sigma^\mu_{ab} P_\mu M_{rstu} = 0 ,$$

for all a, b, r, s, t, and u. Since the operator $\sigma^\mu_{ab} P_\mu$ is not zero, we conclude that $M_{rstu} = 0$, so that

$$[Z_{rs}, Q_{at}] = 0 . \tag{25.2.25}$$

Using the anticommutation relation (25.2.8) and its adjoint together with the commutation relations (25.2.22) and (25.2.25) and their adjoints then gives

$$[Z_{rs}, Z_{tu}] = [Z_{rs}, Z^*_{tu}] = [Z^*_{rs}, Z^*_{tu}] = 0 , \tag{25.2.26}$$

which finishes the proof of Eq. (25.2.11), and with it the proof of the Haag–Lopuszanski–Sohnius theorem.

Of course, the fact that the Z_{rs} are central charges of the supersymmetry algebra does not rule out the possibility that there may be *other* Abelian or non-Abelian internal symmetries. Let T_A span the complete Lie algebra of bosonic internal symmetries. Then $[T_A, Q_{ar}]$ is a (0, 1/2) symmetry generator, so it must be a linear combination of the Qs:

$$[T_A, Q_{ar}] = -\sum_s (t_A)_{rs} Q_{as} . \tag{25.2.27}$$

From the Jacobi identity for two Ts and a Q, we learn that the t_A matrices furnish a representation of the internal symmetry algebra

$$[t_A\,,\,t_B] = i\sum_C C^C_{AB}\,t_C\,, \tag{25.2.28}$$

where the coefficients C^C_{AB} are the structure constants of the internal symmetry algebra

$$[T_A\,,\,T_B] = i\sum_C C^C_{AB}\,T_C\,. \tag{25.2.29}$$

The Z_{rs} will then be central charges not only of the superalgebra consisting of the Qs, Q*s, P_μ, Zs, and Z^*s, but also of the larger symmetry superalgebra that in addition contains all the T_A. To see this, note from Eqs. (25.2.27) and (25.2.8) that

$$[T_A,\,Z_{rs}] = -\sum_{r'}(t_A)_{rr'}Z_{r's} - \sum_{s'}(t_A)_{ss'}Z_{rs'}\,,$$

so the Z_{rs} form an *invariant* Abelian subalgebra of the whole bosonic symmetry algebra. But recall that in proving the Coleman–Mandula theorem we found that the complete Lie algebra of internal bosonic symmetries, which in our case is spanned by the T_A, is isomorphic to a direct sum of a compact semi-simple Lie algebra and several $U(1)$ algebras. The only invariant Abelian subalgebras of such a Lie algebra are spanned by $U(1)$ generators, so the Z_{rs} must be $U(1)$ generators, and hence commute with all the T_A.

Even though the Zs commute with all symmetry generators, they are not just numbers; they are quantum operators, whose value may vary from state to state. In fact, the Zs must obviously take the value zero for a supersymmetric vacuum state, which is annihilated by all supersymmetry generators, but they need not vanish in general. In Section 27.9 we will see how the Zs may be calculated in gauge theories with extended supersymmetry.

In the absence of central charges, the supersymmetry algebra (25.2.7), (25.2.8) is invariant under a group $U(N)$ of internal symmetries

$$Q_{ar} \to \sum_s V_{rs} Q_{as}\,, \tag{25.2.30}$$

with V_{rs} an $N \times N$ unitary (not necessarily unimodular) matrix. This is known as *R-symmetry.* This symmetry may or may not be a good symmetry of the action, and if it is then it may be violated by anomalies, or it may be spontaneously broken, or it may be a good symmetry of nature.

A supersymmetry algebra where r, s, etc. run over $N > 1$ values is known as an *N-extended supersymmetry.* Where there is just one Q, the condition $Z_{rs} = -Z_{sr}$ tells us that the Zs vanish, yielding a simpler form of the anticommutation relations

$$\{\mathrm{Q}_a, \mathrm{Q}_b^*\} = 2\sigma^\mu_{ab} P_\mu \,, \tag{25.2.31}$$

$$\{\mathrm{Q}_a, \mathrm{Q}_b\} = 0 \,. \tag{25.2.32}$$

This is known as the case of *simple supersymmetry*, or $N = 1$ supersymmetry. In this case the R-symmetry transformations are $U(1)$ phase transformations

$$\mathrm{Q}_a \to \exp(i\varphi)\, \mathrm{Q}_a \,, \tag{25.2.33}$$

with φ a real phase.

For several purposes it is convenient to combine the $(0, 1/2)$ operators Q_{ar} together with $(1/2, 0)$ operators, which according to Eq. (25.2.6) may be taken as $e_{ab}\mathrm{Q}^*_{br}$, into four-component Majorana spinor generators $Q_{\alpha r}$, defined as

$$Q_r \equiv \begin{pmatrix} e\mathrm{Q}_r^* \\ \mathrm{Q}_r \end{pmatrix} \,, \tag{25.2.34}$$

or more explicitly

$$Q_{1r} = \mathrm{Q}^*_{-\frac{1}{2}r} \,, \quad Q_{2r} = -\mathrm{Q}^*_{\frac{1}{2}r} \,, \quad Q_{3r} = \mathrm{Q}_{\frac{1}{2}r} \,, \quad Q_{4r} = \mathrm{Q}_{-\frac{1}{2}r} \,.$$

This is a Majorana spinor in the sense that

$$Q_r = -\beta\epsilon\gamma_5 Q_r^* \,,$$

where β, ϵ, and γ_5 are 4×4 matrices that may be written as the 2×2 block matrices:

$$\beta = \begin{pmatrix} 0 & 1 \\ 1 & 0 \end{pmatrix} \qquad \epsilon = \begin{pmatrix} e & 0 \\ 0 & e \end{pmatrix} \qquad \gamma_5 = \begin{pmatrix} 1 & 0 \\ 0 & -1 \end{pmatrix} \,.$$

(The properties of Majorana spinors are reviewed in the appendix to Chapter 26.) The form (25.2.34) is chosen in accordance with the usual notation for the four-component Dirac representation of the homogeneous Lorentz group, in which according to Eq. (5.4.4) the rotation and boost generators are represented according to Eqs. (5.4.19) and (5.4.20) by

$$\mathscr{J}_i = \frac{1}{2}\begin{bmatrix} \sigma_i & 0 \\ 0 & \sigma_i \end{bmatrix} \,, \qquad \mathscr{K}_i = -\frac{i}{2}\begin{bmatrix} \sigma_i & 0 \\ 0 & -\sigma_i \end{bmatrix} \,. \tag{25.2.35}$$

With Eq. (25.2.1), this shows that the operators $\mathbf{A}$ and $\mathbf{B}$ act only on the top two and bottom two components of the Dirac spinor, respectively, which is why we use the $(0, 1/2)$ operators Q_{ar} as the bottom rather than the top components in Eq. (25.2.34).

In this four-component notation, the fundamental anticommutation relations (25.2.31) and (25.2.32) for simple supersymmetry read

$$\{Q\,,\overline{Q}\} = 2\begin{pmatrix} 0 & -e\,(\sigma_\mu P^\mu)^{\rm T}\,e \\ \sigma_\mu P^\mu & 0 \end{pmatrix} = -2i\,P_\mu\,\gamma^\mu\,. \qquad (25.2.36)$$

Our notation for Dirac matrices is reviewed in the Preface to this volume; here we need only recall that

$$\gamma^0 = -i\beta = -i\begin{pmatrix} 0 & \sigma_0 \\ \sigma_0 & 0 \end{pmatrix}\,, \qquad \boldsymbol{\gamma} = -i\begin{pmatrix} 0 & \boldsymbol{\sigma} \\ -\boldsymbol{\sigma} & 0 \end{pmatrix}\,, \qquad (25.2.37)$$

while $e\boldsymbol{\sigma}^{\rm T}e = \boldsymbol{\sigma}$, $e\sigma_0 e = -\sigma_0$, and as usual $\overline{Q} \equiv Q^\dagger\beta$. The presence of central charges will change this formula in the case of extended supersymmetry; instead of Eq. (25.2.36), we have

$$\{Q_r\,,\overline{Q}_s\} = -2i\,P_\mu\,\gamma^\mu\delta_{rs} + \left(\frac{1+\gamma_5}{2}\right)Z^*_{sr} + \left(\frac{1-\gamma_5}{2}\right)Z_{rs}\,. \qquad (25.2.38)$$

The analysis presented here for the case of four spacetime dimensions will be repeated, in a somewhat less explicit form, for general spacetime dimensions in Chapter 32. As we will see there, the supersymmetry generators always belong to the fundamental spinor representation of the higher-dimensional Lorentz group, even in theories with extended objects that allow the construction of bosonic symmetry generators other than those allowed by the Coleman–Mandula theorem.

* * *

In theories of massless particles that are invariant under the conformal symmetry algebra (24.B.34)–(24.B.35) there are two additional bosonic symmetry generators D and K_μ that can appear on the right-hand side of the supersymmetry anticommutation relations. These new generators have the Lorentz transformation properties of a scalar and a vector, respectively, just like Z_{rs} and P_μ, so once again the fermionic generators must belong to the fundamental $(1/2, 0)$ spinor representation of the Lorentz algebra, and its Hermitian adjoint, the $(0, 1/2)$ representation. It is convenient to classify all generators also according to their commutators with the dilation generator D; an operator X is said to have dimensionality a if

$$[X\,,D] = iaX\,. \qquad (25.2.39)$$

Inspection of Eq. (24.B.34) shows that the bosonic symmetry generators $J^{\mu\nu}$, P^μ, K^μ, and D have dimensionalities 0, +1, −1, and 0, respectively. Also, the generators of any Lie group of internal symmetries have dimensionality 0. The anticommutator of a fermionic generator of dimensionality a with its adjoint is a positive-definite bosonic operator of dimensionality $2a$, so since the only positive-definite bosonic symmetry generators are linear combinations of the components of P_μ and K_μ, the only fermionic symmetry generators have dimensionalities +1/2 and −1/2. The (0, 1/2) fermionic symmetry generators of dimensionality 1/2 and their adjoints may again be assembled into Majorana spinors $Q_{r\alpha}$, with

$$\{Q_{r\alpha}\,,\,\bar{Q}_{s\beta}\} = -2iP_\mu(\gamma^\mu)_{\alpha\beta}\delta_{rs}\,, \tag{25.2.40}$$

$$[P_\mu\,,\,Q_{r\alpha}] = 0\,, \tag{25.2.41}$$

$$[D\,,\,Q_{r\alpha}] = -\tfrac{1}{2}i\,Q_{r\alpha}\,. \tag{25.2.42}$$

(Note that central charges are not allowed here, because their dimensionality would be 0, not +1.) The commutators of the K_μ with the $Q_{r\alpha}$ are linear combinations of Majorana fermionic symmetry generators $Q^\#_{r\alpha}$, which Lorentz invariance allows us to write in the form

$$[K^\mu\,,\,Q_{r\alpha}] = i\,(\gamma^\mu)_{\alpha\beta}Q^\#_{r\beta}\,. \tag{25.2.43}$$

(An arbitrary factor on the right-hand side has been absorbed into the normalization of $Q^\#_{r\beta}$. The phase of the right-hand side is chosen so that $Q^\#_{r\beta}$ will satisfy the standard reality condition (26.A.2) for Majorana spinors.) The $Q^\#_{r\beta}$ have dimensionality $+1/2 - 1 = -1/2$, so

$$[D\,,\,Q^\#_{r\alpha}] = +\tfrac{1}{2}i\,Q^\#_{r\alpha}\,. \tag{25.2.44}$$

Taking the commutator of Eq. (25.2.43) with P^ν and using the commutator of K^μ with P^ν given by Eq. (24.B.34) gives

$$[P^\nu\,,\,Q^\#_{r\alpha}] = -i(\gamma^\nu)_{\alpha\beta}Q_{r\beta}\,. \tag{25.2.45}$$

We see that the Qs and $Q^\#$s are paired. By taking the commutator of the anticommutation relation (25.2.40) with K_μ we find the anticommutators of the $Q^\#$s and Qs:

$$\{Q^\#_{r\alpha}\,,\,\bar{Q}_{s\beta}\} = 2iD\delta_{rs}\delta_{\alpha\beta} + 2J_{\mu\nu}\delta_{rs}\mathscr{J}^{\mu\nu}_{\alpha\beta} + O_{rs}\delta_{\alpha\beta} + O'_{rs}(\gamma_5)_{\alpha\beta}\,, \tag{25.2.46}$$

where $\mathscr{J}^{\mu\nu} \equiv -i[\gamma^\mu\,,\,\gamma^\nu]/4$, and O_{rs} and O'_{rs} are Lorentz invariant operators of zero dimensionality with

$$O_{rs} = -O_{sr}\,, \qquad O'_{rs} = +O'_{sr}\,. \tag{25.2.47}$$

Taking the commutator of Eq. (25.2.43) with K_ν and using the fact that $[K_\nu, K_\mu] = 0$, we find that $(\gamma^\mu)_{\alpha\beta}[K^\nu, Q_{r\beta}]$ is symmetric in μ and ν, which with a little algebra tells us that

$$[K^\nu, Q_{r\beta}] = 0\,. \tag{25.2.48}$$

Also, taking the commutator of Eq. (25.2.46) with K_ν gives

$$\{Q^\#_{r\alpha}\,,\,\overline{Q^\#}_{s\beta}\} = +2iK_\mu(\gamma^\mu)_{\alpha\beta}\delta_{rs}\,. \tag{25.2.49}$$

Finally, taking the commutator of Eq. (25.2.46) with $Q_{t\gamma}$ shows that O_{rs} and O'_{rs} act as the generators of the R-symmetry group $U(N)$, with the left- and right-handed parts of $Q_{r\alpha}$ transforming according to the representations $\mathbf{N}$ and $\bar{\mathbf{N}}$, respectively, while the P_μ, K_μ, and D are $U(N)$-invariant. The $U(N)$ commutation relations of these generators with each other and with other generators together with Eqs. (24.B.34), (24.B.35), (25.2.40)–(25.2.49) and the commutators of $J_{\mu\nu}$ and D with the various generators constitute the *superconformal algebra.* One of the outstanding differences between this algebra and that of ordinary simple or N-extended supersymmetry is that the $U(N)$ symmetry is not merely an outer automorphism of the algebra that may or may not be a symmetry of the action — it is a part of the superconformal algebra, which therefore must be a symmetry of any action that is supersymmetric and conformally invariant.

25.3 Space Inversion Properties of Supersymmetry Generators

In theories that respect the conservation of parity, the result $\mathsf{P}^{-1}\mathrm{Q}_{ar}\mathsf{P}$ of acting on the fermionic symmetry generator Q_{ar} with the parity operator P must also be a fermionic symmetry generator. Since J_i and K_i are respectively even and odd under space inversion, Eq. (25.2.1) shows that the effect of acting on A_i with the parity operator is

$$\mathsf{P}^{-1}A_i\mathsf{P} = B_i \,. \tag{25.3.1}$$

According to Eq. (25.2.3), the definition of Q_{ar} as a $(0, 1/2)$ operator means that

$$[B_i\,, \mathrm{Q}_{ar}] = -\tfrac{1}{2}\sum_b \left(\sigma_i\right)_{ab}\mathrm{Q}_{br}\,, \qquad [A_i\,, \mathrm{Q}_{ar}] = 0\,. \tag{25.3.2}$$

Applying the parity operator yields

$$[A_i\,, \mathsf{P}^{-1}\mathrm{Q}_{ar}\mathsf{P}] = -\tfrac{1}{2}\sum_b \left(\sigma_i\right)_{ab}\mathsf{P}^{-1}\mathrm{Q}_{br}\mathsf{P}\,, \qquad [B_i\,, \mathsf{P}^{-1}\mathrm{Q}_{ar}\mathsf{P}] = 0\,, \tag{25.3.3}$$

so $\mathsf{P}^{-1}\mathrm{Q}_{ar}\mathsf{P}$ is a $(1/2, 0)$ symmetry generator, and therefore must be a linear combination of the Q^*_{ar}. According to Eq. (25.2.6), Lorentz invariance dictates the form of this relation as

$$\mathsf{P}^{-1}\mathrm{Q}_{ar}\mathsf{P} = \sum_{bs} \mathscr{P}_{rs}\, e_{ab}\mathrm{Q}^*_{bs}\,, \tag{25.3.4}$$

where $\mathscr{P}$ is a numerical matrix, and the matrix e is given by Eq. (25.2.9).

We can learn something about the properties of the matrix $\mathscr{P}$ by requiring that Eq. (25.3.4) be consistent with the fundamental anticommutation relation (25.2.7). Eq. (25.3.4) and its adjoint yield

$$\mathsf{P}^{-1}\{\mathrm{Q}_{ar}\,, \mathrm{Q}^*_{bs}\}\mathsf{P} = \sum_{cdtu} \mathscr{P}_{rt}\, e_{ac}\, \mathscr{P}^*_{su}\, e_{bd}\, \{\mathrm{Q}^*_{ct}\,, \mathrm{Q}_{du}\}\,.$$

Inserting Eq. (25.2.7), this becomes

$$\delta_{rs}\sigma^{\mu}_{ab}\,\mathsf{P}^{-1}P_{\mu}\mathsf{P} = \sum_{cdtu} \mathscr{P}_{rt}\, e_{ac}\, \mathscr{P}^{*}_{su}\, e_{bd}\, \delta_{tu}\, \sigma^{\mu}_{dc}\, P_{\mu}\,.$$

But $e\sigma_i^{\mathrm{T}}e^{-1} = -\sigma_i$ and $e\sigma_0^{\mathrm{T}}e^{-1} = +\sigma_0$, while $\mathsf{P}^{-1}P_i\mathsf{P} = -P_i$ and $\mathsf{P}^{-1}P_0\mathsf{P} = P_0$, so this reduces to the statement that $\mathscr{P}$ is unitary

$$\mathscr{P}\mathscr{P}^{\dagger} = 1\,. \tag{25.3.5}$$

The matrix $\mathscr{P}$ is to some extent arbitrary, because for any set of fermionic generators Q_{ar} that satisfies Eqs. (25.3.2) and (25.2.7), we construct another set Q'_{ar} that also satisfies Eqs. (25.3.2) and (25.2.7) by a unitary transformation

$$\mathsf{Q}'_{ar} = \sum_{s} \mathscr{U}_{rs}\, \mathsf{Q}_{as}\,, \qquad \mathscr{U}^{\dagger} = \mathscr{U}^{-1}\,, \tag{25.3.6}$$

so that the parity transformation rule (25.3.4) becomes

$$\mathsf{P}^{-1}\mathsf{Q}'_{ar}\mathsf{P} = \sum_{bs} \mathscr{P}'_{rs}\, e_{ab}\mathsf{Q}'^{*}_{bs}\,, \tag{25.3.7}$$

where

$$\mathscr{P}' = \mathscr{U}\,\mathscr{P}\,\mathscr{U}^{-1*} = \mathscr{U}\,\mathscr{P}\,\mathscr{U}^{\mathrm{T}}\,. \tag{25.3.8}$$

For simple supersymmetry $\mathscr{P}$ is just a 1×1 phase factor, and Eq. (25.3.4) reads

$$\mathsf{P}^{-1}\mathsf{Q}_a\mathsf{P} = \mathscr{P}\sum_{b} e_{ab}\mathsf{Q}^{*}_{b}\,. \tag{25.3.9}$$

Combining this with its adjoint gives

$$\mathsf{P}^{-2}\mathsf{Q}_a\,\mathsf{P}^{2} = -\,\mathsf{Q}_a\,, \tag{25.3.10}$$

independently of the value chosen for the phase factor $\mathscr{P}$. This has the striking consequence that if a boson in a supermultiplet of particles has a real intrinsic parity, then the fermions obtained by acting on this boson state with Q_a have *imaginary* intrinsic parities.

Because for simple supersymmetry $\mathscr{U}$ and $\mathscr{P}$ are just phase factors, it is obvious from Eq. (25.3.8) that by a suitable choice of $\mathscr{U}$, the phase factor $\mathscr{P}'$ can be made anything we like. It will be convenient to choose $\mathscr{P}' = +i$, so that Eq. (25.3.7) takes the simple form (now dropping primes)

$$\mathsf{P}^{-1}\mathsf{Q}_a\mathsf{P} = i\sum_{b} e_{ab}\,\mathsf{Q}^{*}_{b}\,. \tag{25.3.11}$$

Just as for spinor field operators, the representation of space inversion is simpler if we combine the $(0,1/2)$ operators Q_a and the $(1/2,0)$ operators $\sum_b e_{ab}\mathsf{Q}^{*}_{b}$ into the four-component Dirac spinor generator Q_α defined by

Eq. (25.2.34). In these terms, Eq. (25.3.11) and its adjoint read

$$\mathsf{P}^{-1}Q\,\mathsf{P} = i\beta\,Q\,. \tag{25.3.12}$$

(We are using the notation for Dirac matrices given in Section 5.4 and in the Preface to this volume, which gives

$$\beta = \begin{pmatrix} 0 & 1 \\ 1 & 0 \end{pmatrix},$$

where 1 and 0 are understood as 2×2 submatrices.)

For extended supersymmetry it is not always possible to choose $\mathscr{U}$ so that $\mathscr{P}'$ is diagonal. However, a theorem in matrix algebra (proved in Appendix C of Chapter 2) shows that it is possible to choose $\mathscr{U}$ so that $\mathscr{P}'$ is block diagonal, in general with some of the matrices on the diagonal equal to 1×1 submatrices that can be chosen equal to i (or any other phase factors we like), and other submatrices on the diagonal given by 2×2 matrices that can be chosen to have the form

$$\begin{pmatrix} 0 & \exp(i\phi) \\ \exp(-i\phi) & 0 \end{pmatrix},$$

with various phases ϕ. Correspondingly, with this choice of $\mathscr{U}$ (and now dropping primes) the two-component Qs are of two types. The Qs of the first type also satisfy Eq. (25.3.11):

$$\mathsf{P}^{-1}\mathrm{Q}_{ar}\mathsf{P} = i\sum_b e_{ab}\,\mathrm{Q}^*_{br}\,. \tag{25.3.13}$$

The two-component Qs of the second type come in pairs, which we will call $\mathrm{Q}_{a\,s1}$ and $\mathrm{Q}_{a\,s2}$, with the sth pair having the parity transformation rule

$$\mathsf{P}^{-1}\,\mathrm{Q}_{a\,s1}\,\mathsf{P} = e^{i\phi_s}\sum_b e_{ab}\,\mathrm{Q}^*_{b\,s2}\,,\quad \mathsf{P}^{-1}\,\mathrm{Q}_{a\,s2}\,\mathsf{P} = e^{-i\phi_s}\sum_b e_{ab}\,\mathrm{Q}^*_{b\,s1}\,. \tag{25.3.14}$$

In particular, we now have

$$\mathsf{P}^{-2}\mathrm{Q}_{a\,s1}\mathsf{P}^2 = -\,e^{2i\phi_s}\mathrm{Q}_{a\,s1}\,,\qquad \mathsf{P}^{-2}\mathrm{Q}_{a\,s2}\mathsf{P}^2 = -\,e^{-2i\phi_s}\mathrm{Q}_{a\,s2}\,. \tag{25.3.15}$$

This shows that, unless $\phi_s = 0 \pmod{\pi}$, it is not possible to form supersymmetry generators of the first type out of linear combinations of extended supersymmetry generators of the second type.

In terms of the four-component spinors (25.2.34), the effect of the parity operator on extended supersymmetry generators of the first type is

$$\mathsf{P}^{-1}Q_r\,\mathsf{P} = i\beta\,Q_r\,, \tag{25.3.16}$$

while for generators of the second type

$$\mathsf{P}^{-1}\,Q_{s1}\,\mathsf{P} = \beta\,\gamma_5\,\exp(i\gamma_5\phi_s)\,Q_{s2}\,,\qquad \mathsf{P}^{-1}\,Q_{s2}\,\mathsf{P} = \beta\,\gamma_5\,\exp(-i\gamma_5\phi_s)\,Q_{s1}\,. \tag{25.3.17}$$

25.4 Massless Particle Supermultiplets

Supersymmetry requires that the known particles be accompanied in irreducible representations of the supersymmetry algebra by 'sparticles': bosonic 'squarks' and 'sleptons' accompanying quarks and leptons, and fermionic 'gauginos' accompanying gauge bosons. None of these sparticles have been observed, so supersymmetry is certainly broken, with the masses of the sparticles almost certainly much larger than the quark, lepton, and gauge boson masses produced by the spontaneous breakdown of the electroweak $SU(2) \times U(1)$ gauge group, and hence of the same order of magnitude as the splittings within supermultiplets. Thus it is very likely that at energy scales that are large enough so that we can neglect supersymmetry breaking and these mass splittings, we can also treat the known quarks, leptons, and gauge bosons and their superpartners as massless. Therefore we will be specially interested in supermultiplets of massless particles.

Consider a state containing a single massless particle belonging to some supermultiplet. We obtain the other states in the same supermultiplet by applying operators Q_{ar} and/or Q^*_{ar} to this state. Since Q_{ar} and Q^*_{ar} commute with P_μ, all these states have the same value of the four-momentum. We will work in a Lorentz frame in which the four-momentum of these states is $p^1 = p^2 = 0$ and $p^3 = p^0 = E$. With this choice of four-momentum, we have

$$\sigma_\mu p^\mu = E(\sigma_0 + \sigma_3) = 2E \begin{pmatrix} 1 & 0 \\ 0 & 0 \end{pmatrix}, \tag{25.4.1}$$

which aside from the factor $2E$ is the projection matrix onto the subspace with helicity $+1/2$. The anticommutation relation (25.2.7) therefore shows that $\{Q_{(-1/2)r}, Q^*_{(-1/2)r}\}$ gives zero when acting on any state in a supermultiplet with this momentum, and thus so do $Q_{(-1/2)r}$ and $Q^*_{(-1/2)r}$. We must therefore construct the states of the supermultiplet by acting only with $Q_{(1/2)r}$ and $Q^*_{(1/2)r}$. Furthermore, we are labelling the Qs with their J_3 values, in the sense that

$$[J_3, Q_{ar}] = -a\, Q_{ar}\,, \tag{25.4.2}$$

so $Q_{(1/2)r}$ and $Q^*_{(1/2)r}$ respectively lower and raise the helicity by 1/2.

We will first consider the case of simple supersymmetry. Consider a supermultiplet with maximum helicity λ_{max}, and let $|\lambda_{\text{max}}\rangle$ be any one-particle state with this helicity and four-momentum p^μ. Then

$$Q^*_{\frac{1}{2}} |\lambda_{\text{max}}\rangle = 0\,, \tag{25.4.3}$$

while acting on this state with $Q_{1/2}$ gives a state $|\lambda_{\text{max}} - 1/2\rangle$ with helicity

$\lambda_{\rm max} - 1/2$. We will define this state as

$$|\lambda_{\rm max} - 1/2\rangle \equiv (4E)^{-1/2}\, Q_{\frac{1}{2}}\, |\lambda_{\rm max}\rangle \,. \tag{25.4.4}$$

The fundamental anticommutation relation (25.2.7) together with Eqs. (25.4.1) and (25.4.3) shows that this state is normalized in the same way as $|\lambda_{\rm max}\rangle$

$$\langle \lambda_{\rm max} - 1/2 | \lambda_{\rm max} - 1/2\rangle = \langle \lambda_{\rm max} | \lambda_{\rm max}\rangle \,, \tag{25.4.5}$$

and in particular this state cannot vanish. Eq. (25.2.32) shows that $Q_{1/2}^2 = 0$, so acting with $Q_{1/2}$ on $|\lambda_{\rm max} - 1/2\rangle$ gives zero:

$$Q_{\frac{1}{2}} |\lambda_{\rm max} - 1/2\rangle = (4E)^{-1/2} Q_{\frac{1}{2}}^2 |\lambda_{\rm max}\rangle = 0 \,. \tag{25.4.6}$$

On the other hand, acting with $Q_{1/2}^*$ on this state gives the state with which we started. That is,

$$Q_{\frac{1}{2}}^* |\lambda_{\rm max} - 1/2\rangle = (4E)^{-1/2} Q_{\frac{1}{2}}^*\, Q_{\frac{1}{2}} |\lambda_{\rm max}\rangle = (4E)^{-1/2} \{Q_{\frac{1}{2}}^*\,,\, Q_{\frac{1}{2}}\} |\lambda_{\rm max}\rangle \,,$$

so that Eq. (25.4.1) and the anticommutation relation (25.2.31) yield

$$Q_{\frac{1}{2}}^* |\lambda_{\rm max} - 1/2\rangle = (4E)^{1/2} |\lambda_{\rm max}\rangle \,. \tag{25.4.7}$$

Thus the supermultiplet consists of just two states, with helicities $\lambda_{\rm max}$ and $\lambda_{\rm max} - 1/2$. In the basis provided by these two states, the operators $Q_{1/2}$ and $Q_{1/2}^*$ are represented by the matrices

$$q_{\frac{1}{2}} = \sqrt{4E} \begin{pmatrix} 0 & 0 \\ 1 & 0 \end{pmatrix} \,, \qquad q_{\frac{1}{2}}^\dagger = \sqrt{4E} \begin{pmatrix} 0 & 1 \\ 0 & 0 \end{pmatrix} \,, \tag{25.4.8}$$

while the operators $Q_{-1/2}$ and $Q_{-1/2}^*$ are represented by zero.

It is worth emphasizing that this is the *only* kind of massless supermultiplet in theories with simple supersymmetry. There are no massless particles that are not accompanied with a superpartner, and none that have more than one superpartner. Of course, CPT invariance implies that for every supermultiplet of massless particles of helicities λ and $\lambda - 1/2$ there must be an antimultiplet with helicities $-\lambda + 1/2$ and $-\lambda$. In particular, a massless particle and antiparticle with helicities $+1/2$ and $-1/2$ must be accompanied by a massless particle and antiparticle either with helicities $+1$ and -1 or with helicities both zero.

How might the known quarks, leptons, and gauge bosons fit into this picture? We will assume that the supersymmetry generator commutes with the generators of the $SU(3) \times SU(2) \times U(1)$ gauge group.* The quarks

* In simple supersymmetry the generator Q_α must in any case commute with the $SU(3) \times SU(2)$ generators, because semi-simple algebras like $SU(3) \times SU(2)$ have no non-trivial one-dimensional representations.

and leptons belong to different representations of the gauge group from the gauge bosons, so they cannot be in the same supermultiplets. We have to conclude then that in the limit of high energy where $SU(2) \times U(1)$ symmetry breaking may be neglected, the massless quarks and leptons of each color and flavor are in supermultiplets with *pairs* of massless squarks and sleptons of zero helicity and the same color and flavor, while the massless gauge bosons are accompanied by massless gauginos of helicity $\pm 1/2$ comprising an adjoint representation of $SU(3) \times SU(2) \times U(1)$.

Because gravity exists we know that in addition to the particles of the standard model there must also exist a massless particle of helicity ± 2, the *graviton*. Massless particles with helicity λ having $|\lambda| > 1/2$ must couple at low momentum to conserved quantities.** Soft massless particles of helicity ± 1 can couple to various internal symmetry generators, soft massless particles of helicity $\pm 3/2$ can couple to the supersymmetry generators Q_a, and a soft particle of helicity ± 2 can couple to a single conserved quantity, the momentum four-vector P_μ, but there are no conserved quantities to which a soft massless particle with $|\lambda| > 2$ could couple. We conclude that the graviton cannot be in a supermultiplet with particles of helicity $\pm 5/2$, so it must be in a supermultiplet with a massless particle of helicity $\pm 3/2$, known as a *gravitino*, coupled to the supersymmetry generators themselves. The field theory of this supermultiplet is known as *supergravity*, and will be discussed in Chapter 31.

Now let us consider the case of extended supersymmetry, with N supersymmetry generators. We first note that because the $Q_{(-1/2)r}$ all give zero when acting on the states of a supermultiplet (including a state obtained by letting $Q_{(1/2)s}$ act on any other state of the multiplet), the central charges Z_{rs} must also annihilate any state of the multiplet. With central charges out of the picture, the supersymmetry generators $Q_{(1/2)r}$ all anticommute when acting on a massless particle supermultiplet, so applying n of them to a one-particle state of maximum helicity $\lambda_{\max}$ and four-momentum p^μ gives $N!/n!(N-n)!$ one-particle states of the same four-momentum and helicity $\lambda_{\max} - n/2$, forming a rank n antisymmetric tensor representation of the $SU(N)$ R-symmetry† (25.2.30). The maximum value of n that gives a non-zero state is $n = N$, so the minimum helicity in a supermultiplet is given by

$$\lambda_{\min} = \lambda_{\max} - N/2 \,. \tag{25.4.9}$$

If we wish to exclude massless particle helicities λ with $|\lambda| > 2$, then

** This is discussed for the case of integer helicity in Section 13.1. The argument for half-integer helicity was given by Grisaru and Pendleton.[3]

† The $U(1)$ part of $U(N)$ R-symmetry is often violated by quantum mechanical anomalies.

$\lambda_{\max} - \lambda_{\min} \leq 4$, so extended supersymmetries are allowed only with $N \leq 8$.

For $N = 8$, with helicities with $|\lambda| > 2$ excluded, there is only one possible supermultiplet, consisting of: 1 graviton with each helicity ± 2; 8 gravitinos with each helicity $\pm 3/2$; 28 gauge bosons with each helicity ± 1; 56 fermions with each helicity $\pm 1/2$; and 70 bosons with helicity zero.

Compare this with the case $N = 7$, again excluding helicities with $|\lambda| > 2$. Here there are two supermultiplets. One supermultiplet contains: 1 graviton with helicity $+2$; 7 gravitinos with helicity $+3/2$; 21 gauge bosons with helicity $+1$; 35 fermions with helicity $+1/2$; 35 bosons with helicity zero; 21 fermions with helicity $-1/2$; 7 gauge bosons with helicity -1; and 1 gravitino with helicity $-3/2$. The other is the CPT-conjugate supermultiplet, with all helicities reversed. Adding the numbers of particles in these two supermultiplets, we have 1 graviton with each helicity ± 2; $7 + 1 = 8$ gravitinos with each helicity $\pm 3/2$; $21 + 7 = 28$ gauge bosons with each helicity ± 1; $35 + 21 = 56$ fermions with each helicity $\pm 1/2$; and $35 + 35 = 70$ bosons with helicity zero. Extended supergravity theories with $N = 8$ and $N = 7$ thus have precisely the same particle content and are in fact identical.

On the other hand, extended supergravity theories with $N \leq 6$ have just N gravitinos of each helicity $\pm 3/2$ and are therefore all distinct.

For $N \leq 4$ there is also the possibility of *global* supersymmetry theories, theories with supermultiplets that do not include gravitons or gravitinos. For global $N = 4$ supersymmetry there is just one supermultiplet, containing: 1 gauge boson of each helicity ± 1; 4 fermions of each helicity $\pm 1/2$; and 6 bosons of helicity zero. This is equivalent to the global supersymmetry theory with $N = 3$, which has two supermultiplets: 1 supermultiplet with one gauge boson of helicity $+1$, 3 fermions with helicity $+1/2$; 3 bosons with helicity zero; and 1 fermion with helicity $-1/2$; and the other the CPT conjugate supermultiplet with opposite helicities. Adding the numbers of particles of each helicity in these two $N = 3$ supermultiplets gives the same particle content as for $N = 4$ global supersymmetry. The gauge field theory with $N = 4$ supersymmetry has remarkable properties, which will be discussed in Section 27.9.

For $N = 2$ extended global supersymmetry there are supermultiplets of two different types, apart from those related by CPT. There are *gauge* supermultiplets, each containing one gauge boson of helicity $+1$, two fermions of helicity $+1/2$ forming a doublet under the $SU(2)$ R-symmetry (25.2.30), and one boson of helicity zero, together with their CPT-conjugate supermultiplets, with helicities reversed. Together each gauge supermultiplet and its antimultiplet contain one gauge boson with each helicity ± 1, an $SU(2)$ doublet of fermions of each helicity $\pm 1/2$, and two $SU(2)$ singlet bosons of helicity zero. Then there are *hypermultiplets*, which contain one fermion of each helicity $\pm 1/2$ and an $SU(2)$ doublet of bosons of helicity zero, together with its CPT-conjugate. (In quantum field theory a hypermultiplet cannot be its own antimultiplet, because then the helicity zero particles would be described by just two *real* scalar fields, which cannot form an $SU(2)$ doublet.) Of course, in the real world there would also have to be a graviton supermultiplet, containing a graviton of helicity $+2$, an $SU(2)$ doublet of gravitinos with helicity $+3/2$, and one gauge boson with helicity $+1$, together with their CPT-conjugates, with opposite helicity. Gauge theories with $N = 2$ supersymmetry are constructed in Section 27.9, and explored non-perturbatively in Section 29.5.

The particle content of these supermultiplets reveals a difficulty in incorporating extended supersymmetry in realistic theories of particles at accessible energies. In all cases but one, the helicity $+1/2$ fermions belong to supermultiplets along with helicity $+1$ gauge bosons. Gauge bosons belong to the adjoint representation of the gauge group, so if the supersymmetry generators are invariant under the gauge group then the helicity $+1/2$ fermions must also belong to the adjoint representation, which is real. This is in conflict with the fact that the known quarks and leptons belong to a representation of $SU(3) \times SU(2) \times U(1)$ which is *chiral* — that is, for which the helicity $+1/2$ fermions belong to a complex representation, which is then necessarily different from the representation furnished by their CPT-conjugates, the helicity $-1/2$ fermions. The one exception, where helicity $+1/2$ fermions are not in a supermultiplet with gauge bosons, is the $N = 2$ hypermultiplet discussed above. But in this case particles of both helicities $+1/2$ and $-1/2$ are in the same supermultiplet, and therefore must transform the same under any gauge transformations that leave the supersymmetry generators invariant. They may belong to a complex representation of this gauge group, but then the CPT-conjugate of this hypermultiplet belongs to the complex-conjugate representation, and the fermions of each helicity then belong to the sum of the two representations, which is real, again in conflict with the chiral nature of the known quarks and leptons.

In contrast, for simple supersymmetry there are supermultiplets containing just helicity $+1/2$ and helicity zero, which may be in a complex representation of the gauge group, distinct from the representation furnished by the CPT-conjugate supermultiplet. Here there is no conflict with chirality. For this reason, most discussions of supersymmetry as a symmetry that remains unbroken at accessible energies have focused on simple rather than extended supersymmetry.

25.5 Massive Particle Supermultiplets

Although the known quarks, leptons, and gauge bosons and their superpartners may probably be treated as massless at energies where supersymmetry breaking is negligible, this is not necessarily true for other particles, including the extra gauge bosons of large mass required by theories that unify the strong and electroweak interactions. Also, ever since the Wess–Zumino model, massive particle theories have been useful test cases for studying supersymmetry theories. It will therefore be worthwhile for us briefly to consider the implications of unbroken supersymmetry for massive particles.

As in the previous section, we obtain the various one-particle states in a supermultiplet by acting on any one of them with the operators Q_{ar} and Q^*_{ar}, and all of these states have the same four-momentum. Unlike the case of zero mass, for mass $M > 0$ we may now take this to be the four-momentum of a particle at rest, with $p^i = 0$ for $i = 1, 2, 3$ and $p^0 = M$. In this frame of reference, we have

$$\sigma_\mu p^\mu = M\sigma_0 = M \begin{pmatrix} 1 & 0 \\ 0 & 1 \end{pmatrix} . \tag{25.5.1}$$

Thus, acting on any state $|\,\rangle$ in a supermultiplet with this four-momentum, the anticommutation relation (25.2.7) yields

$$\{Q_{ar}\,, Q^*_{bs}\}|\,\rangle = 2M\,\delta_{ab}\,\delta_{rs}\,|\,\rangle\,. \tag{25.5.2}$$

In contrast with the case of zero mass, here no component of Q_{ar} or Q^*_{ar} can vanish on the whole multiplet, so we have two sets of raising and lowering operators: both $Q_{(1/2)r}$ and $Q^*_{(-1/2)r}$ lower the spin 3-component by 1/2, while both $Q_{(-1/2)r}$ and $Q^*_{(1/2)r}$ raise the spin 3-component by 1/2. However, as we shall see, for extended supersymmetry it is possible for certain linear combinations of the Qs and Q^*s to vanish.

We will first consider the case of simple supersymmetry. By using the supersymmetry algebra (25.2.31), (25.2.32), we shall show that the general massive supermultiplet consists of a particle of spin $j + 1/2$, a *pair* of particles of spin j, and a particle of spin $j - 1/2$. Where parity is conserved, the particles of spin $j \pm 1/2$ have equal intrinsic parity, given by some phase η, while the two particles of spin j have parities $+i\eta$ and $-i\eta$. Here j is any integer or half-integer greater than zero. There is also a collapsed supermultiplet, consisting of two particles of spin zero and a particle of spin 1/2. When parity is conserved the particles of spin zero have parities $i\eta$ and $-i\eta$, where η is the parity of the particle of spin 1/2.

Here is the proof. We first show that any supermultiplet will contain at least one spin multiplet of states $|j, \sigma\rangle$ with spin 3-component σ running by unit steps from $-j$ to $+j$, having the special property that, for all such

σ and for $a = \pm 1/2$,

$$\mathrm{Q}_a \, |j, \sigma\rangle = 0. \tag{25.5.3}$$

Starting with any non-zero state $|\psi\rangle$ in the supermultiplet, we can define non-zero states

$$|\psi'\rangle \equiv \begin{cases} (2M)^{-1/2}\, \mathrm{Q}_{1/2}\, |\psi\rangle & \mathrm{Q}_{1/2}\, |\psi\rangle \neq 0 \\ |\psi\rangle & \mathrm{Q}_{1/2}|\psi\rangle = 0 \end{cases} ,$$

and

$$|\psi''\rangle \equiv \begin{cases} (2M)^{-1/2}\, \mathrm{Q}_{-1/2}\, |\psi'\rangle & \mathrm{Q}_{-1/2}\, |\psi'\rangle \neq 0 \\ |\psi'\rangle & \mathrm{Q}_{-1/2}\, |\psi'\rangle = 0 \end{cases} .$$

Because the Q_a anticommute, $\mathrm{Q}_{1/2}|\psi'\rangle = 0$, and so $\mathrm{Q}_a|\psi''\rangle = 0$ for $a = \pm 1/2$. If any state $|\psi''\rangle$ satisfies the condition that $\mathrm{Q}_a|\psi''\rangle = 0$, then so does $U(R)|\psi''\rangle$, where $U(R)$ is the unitary operator representing an arbitrary spatial rotation. It follows that the states that satisfy this condition may be decomposed into complete spin multiplets $|j, \sigma\rangle$, satisfying condition (25.5.3).

Now focus on any one of these spin multiplets satisfying Eq. (25.5.3), normalized so that

$$\langle j, \sigma' | j, \sigma\rangle = \delta_{\sigma'\sigma} \, . \tag{25.5.4}$$

For $j > 0$, by applying the spin 1/2 operators* Q_a^* to these states we can construct states of spin $j \pm 1/2$:

$$|j \pm 1/2, \sigma\rangle = \frac{1}{\sqrt{2M}} \sum_a C_{\frac{1}{2}\, j}\Big(j \pm 1/2\, ,\, \sigma\, ;\, a, \sigma - a\Big)\, \mathrm{Q}_a^*\, |j, \sigma - a\rangle \, , \tag{25.5.5}$$

where $C_{jj'}(j'', \sigma''; \sigma, \sigma')$ is the conventional Clebsch–Gordan coefficient for coupling spins j and j' with 3-components σ and σ' to make spin j'' with 3-component σ''. Using Eqs. (25.5.2)–(25.5.5) and the orthonormality properties of the Clebsch–Gordan coefficients, we can show that these states are properly normalized:

$$\langle j \pm 1/2, \sigma | j \pm 1/2, \sigma'\rangle = \delta_{\sigma\sigma'} \, , \qquad \langle j \pm 1/2, \sigma | j \mp 1/2, \sigma'\rangle = 0 \, , \tag{25.5.6}$$

so none of the states $|j \pm 1/2, \sigma\rangle$ can vanish. The only exception is for $j = 0$, in which case of course there is no state $|j - 1/2, \sigma\rangle$. We can also construct other states by applying *two* Q^*s to $|j, \sigma\rangle$. Since each Q_a^*

* Since Q_a transforms under rotations like a field that destroys a particle of spin 1/2 and spin 3-component a, it is Q_a^* that transforms like a field that creates such a particle, and hence transforms like the particle itself. To put this formally, $[J_i, \mathrm{Q}_a] = -\sum_b \frac{1}{2}(\sigma_i)_{ab}\, \mathrm{Q}_b$, so $[J_i, \mathrm{Q}_a^*] = \sum_b \frac{1}{2}(\sigma_i)_{ba}\, \mathrm{Q}_b^*$, which may be compared with the transformation property of a spin 1/2 particle, $J_i\, |a\rangle = \sum_b \frac{1}{2}(\sigma_i)_{ba}\, |b\rangle$.

anticommutes with itself, the only such non-zero states are formed by applying the operator $Q^*_{1/2}Q^*_{-1/2} = -Q^*_{-1/2}Q^*_{1/2}$. This operator may be written as $\frac{1}{2}e_{ab}Q^*_aQ^*_b$, which shows that it is rotationally invariant, so this gives a second spin multiplet with spin j:

$$|j,\sigma\rangle^\flat = \frac{1}{2M}\,Q^*_{1/2}\,Q^*_{-1/2}|j,\sigma\rangle\,, \tag{25.5.7}$$

which is distinguished from $|j,\sigma\rangle$ by the fact that instead of Eq. (25.5.3), we have

$$Q^*_a|j,\sigma\rangle^\flat = 0\,. \tag{25.5.8}$$

Again using Eqs. (25.5.2)–(25.5.4), we find that these are also normalized states:

$${}^\flat\langle j,\sigma'|j,\sigma\rangle^\flat = \delta_{\sigma'\sigma}\,, \qquad \langle j,\sigma'|j,\sigma\rangle^\flat = 0\,. \tag{25.5.9}$$

It is then easy to show that the states constructed so far form a complete representation of the supersymmetry algebra. The orthonormality property of the Clebsch–Gordan coefficients allows us to rewrite Eq. (25.5.5) as

$$Q^*_a\,|j,\sigma\rangle = \sqrt{2M}\sum_\pm C_{1/2j}\Big(j\pm 1/2,\sigma+a;a,\sigma\Big)\,|j\pm 1/2,\sigma+a\rangle\,. \tag{25.5.10}$$

Also, Eq. (25.5.2) shows that, for any state $|\,\rangle$ in the supermultiplet,

$$\Big[Q_a\,,\,Q^*_{\frac{1}{2}}Q^*_{-\frac{1}{2}}\Big]\,|\,\rangle = 2M\sum_b e_{ab}\,Q^*_b\,|\,\rangle\,, \tag{25.5.11}$$

so Eqs. (25.5.7) and (25.5.3) give

$$\begin{aligned}Q_a\,|j,\sigma\rangle^\flat &= \sum_b e_{ab}\,Q^*_b\,|j,\sigma\rangle\\ &= \sqrt{2M}\sum_b e_{ab}\sum_\pm C_{\frac{1}{2}j}\Big(j\pm 1/2,\sigma+b;b,\sigma\Big)\,|j\pm 1/2,\sigma+b\rangle\,.\end{aligned} \tag{25.5.12}$$

From Eqs. (25.5.2), (25.5.3), and (25.5.5) we have

$$Q_a\,|j\pm 1/2,\sigma\rangle = \sqrt{2M}C_{\frac{1}{2}\,j}\Big(j\pm 1/2\,,\,\sigma\,;\,a\,,\,\sigma-a\Big)\,|j,\sigma-a\rangle\,, \tag{25.5.13}$$

while Eqs. (25.5.5), (25.2.31), and (25.5.7) yield

$$Q^*_a\,|j\pm 1/2,\sigma\rangle = \sqrt{2M}\sum_b e_{ab}\,C_{\frac{1}{2}\,j}\Big(j\pm 1/2,\sigma;b,\sigma-b\Big)\,|j,\sigma-b\rangle^\flat\,. \tag{25.5.14}$$

Eqs. (25.5.3), (25.5.8), (25.5.10), and (25.5.12)–(25.5.14) give the action of the Qs and Q*s on all the states of the supermultiplet.

For $j = 0$ we have the collapsed supermultiplet: Eqs. (25.5.3), (25.5.8), (25.5.10), and (25.5.12)–(25.5.14) become

$$\begin{aligned} &Q_a\,|0,0\rangle = 0\,, && Q^*_a\,|0,0\rangle^\flat = 0\,, \\ &Q^*_a\,|0,0\rangle = \sqrt{2M}\,|1/2,a\rangle\,, && Q_a\,|0,0\rangle^\flat = \sqrt{2M}\textstyle\sum_b e_{ab}\,|1/2,b\rangle\,, \\ &Q_a\,|1/2,b\rangle = \sqrt{2M}\delta_{ab}|0,0\rangle\,, && Q^*_a|1/2,b\rangle = \sqrt{2M}e_{ab}\,|0,0\rangle^\flat\,. \end{aligned} \tag{25.5.15}$$

Now suppose that parity is conserved. Recall that the phase of the supersymmetry generator may be chosen so that the action of the parity operator on these generators is given by Eq. (25.3.13). Then Q^*_a acting on $\mathsf{P}|j,\sigma\rangle$ is a linear combination of the states $\mathsf{P}Q_a|j,\sigma\rangle$, which vanish, and since $\mathsf{P}|j,\sigma\rangle$ has the same rotation properties as $|j,\sigma\rangle^\flat$, it must simply be proportional to it

$$\mathsf{P}|j,\sigma\rangle = -\eta|j,\sigma\rangle^\flat\,. \tag{25.5.16}$$

Since P is unitary, η is a phase factor, with $|\eta| = 1$. A corresponding argument shows that $\mathsf{P}|j,\sigma\rangle^\flat$ is proportional to $|j,\sigma\rangle$. To find the proportionality coefficient, we note that

$$\begin{aligned} \mathsf{P}|j,\sigma\rangle^\flat &= (2M)^{-1}\mathsf{P}\,Q^*_{\frac{1}{2}}\,Q^*_{-\frac{1}{2}}\,|j,\sigma\rangle = -\eta(2M)^{-1}\,Q_{-\frac{1}{2}}\,Q_{\frac{1}{2}}\,|j,\sigma\rangle^\flat \\ &= -\eta(2M)^{-2}\,Q_{-\frac{1}{2}}\,Q_{\frac{1}{2}}\,Q^*_{\frac{1}{2}}\,Q^*_{-\frac{1}{2}}\,|j,\sigma\rangle = -\eta|j,\sigma\rangle\,. \end{aligned}$$

We can then define states of spin j

$$|j,\sigma\rangle^\pm \equiv \frac{1}{\sqrt{2}}\Big(|j,\sigma\rangle \pm i\,|j,\sigma\rangle^\flat\Big)\,, \tag{25.5.17}$$

with definite parity

$$\mathsf{P}|j,\sigma\rangle^\pm = \pm i\eta\,|j,\sigma\rangle^\pm\,. \tag{25.5.18}$$

Finally, applying the parity operator to Eq. (25.5.5) and using Eqs. (25.3.13) and (25.5.16) gives

$$\mathsf{P}\,|j\pm1/2,\sigma\rangle = -\frac{\eta}{\sqrt{2M}}\sum_a C_{\frac{1}{2}\,j}\Big(j\pm1/2,\,\sigma\,;\,a,\,\sigma-a\Big)\sum_b e_{ab}Q_b\,|j,\sigma-a\rangle^\flat\,.$$

Eq. (25.5.12) and the orthonormality property of the Clebsch–Gordan coefficients then yield

$$\mathsf{P}\,|j\pm1/2,\sigma\rangle = \eta\,|j\pm1/2,\sigma\rangle\,, \tag{25.5.19}$$

as was to be shown.

We now turn briefly to the case of extended supersymmetry, with N supersymmetry generators. As mentioned in the previous section, there can be no massless particle with a non-vanishing eigenvalue for any central charge. We can go further and show that the eigenvalues of the central charge operators set a lower bound on the mass of any supermultiplet.

Because the central charges Z_{rs} and Z^*_{rs} commute with each other and with P_μ, one-particle states can be chosen to be eigenstates of all the central charges as well as of P_μ, and because the central charges commute with the Q_{ar} and Q^*_{ar}, all states in a supermultiplet have the same eigenvalues.

To derive an inequality relating the mass M of a supermultiplet and the eigenvalues of the central charges on this multiplet, we use the anti-commutation relations (25.2.7) and (25.2.8) to write

$$\sum_{ar}\left\{\left(Q_{ar}-\sum_{bs} e_{ab}U_{rs}Q^*_{bs}\right),\left(Q^*_{ar}-\sum_{ct} e_{ac}U^*_{rt}Q_{ct}\right)\right\}$$
$$=8NP^0-2\mathrm{Tr}\left(ZU^\dagger+UZ^\dagger\right), \tag{25.5.20}$$

where U_{rs} is an arbitrary $N\times N$ unitary matrix. The left-hand side is a positive-definite operator, so by letting this act on the states of the supermultiplet at rest, we find

$$M\geq\frac{1}{4N}\mathrm{Tr}\left(ZU^\dagger+UZ^\dagger\right), \tag{25.5.21}$$

where now Z_{rs} denotes the values of the central charges for the supermultiplet of mass M. The polar decomposition theorem tells us that any square matrix Z may be written as $H\,V$, where H is a positive Hermitian matrix and V is unitary. We can obtain a useful inequality (which is in fact optimal) by setting $U=V$, in which case Eq. (25.5.21) becomes

$$M\geq\frac{1}{2N}\mathrm{Tr}\,H=\frac{1}{2N}\mathrm{Tr}\,\sqrt{Z^\dagger Z}\,. \tag{25.5.22}$$

States for which M equals the minimum value allowed by this inequality are known as *BPS* states, by analogy with the Bogomol'nyi–Prasad–Sommerfeld magnetic monopole configurations discussed in Section 23.3, whose mass also equals a lower bound on general monopole masses. In fact, this is more than an analogy; we will see in Section 27.9 that the lower bound on monopole masses in gauge theories with extended supersymmetry is a special case of the lower bound (25.5.22).

As we can see from this derivation of Eq. (25.5.22), for BPS supermultiplets the operator $Q_{ar}-\sum_{bs}e_{ab}U_{rs}Q^*_{bs}$ gives zero when acting on any state of a supermultiplet, so there are only N independent helicity-lowering operators $Q_{(1/2)r}$ and only N independent helicity-raising operators $Q_{(-1/2)r}$, just as in the case of massless supermultiplets. This leads to smaller supermultiplets than would be found in the general case.

For instance, for $N=2$ supersymmetry the central charge is given by a single complex number**

$$Z = \begin{pmatrix} 0 & Z_{12} \\ -Z_{12} & 0 \end{pmatrix} . \qquad (25.5.23)$$

The inequality (25.5.22) here reads

$$M \geq |Z_{12}|/2 . \qquad (25.5.24)$$

Where $M = |Z_{12}|/2$, the helicity contents of the massive particle supermultiplets are the same as for those of zero mass: there are gauge supermultiplets, consisting of one particle of spin 1, an $SU(2)$ R-symmetry doublet of spin 1/2, and one particle of spin 0 (the other helicity zero state belonging to the spin one particle), and hypermultiplets, consisting of one particle of spin 1/2 and an $SU(2)$ R-symmetry doublet of spin 0. These are sometimes called 'short' supermultiplets, to distinguish them from the larger supermultiplets encountered when $M > |Z_{12}|/2$.

Problems

1. Find a set of 2×2 matrices that form a graded Lie algebra containing fermionic as well as bosonic generators.

2. Following the approach of Haag, Lopuszanski, and Sohnius, derive the form of the most general symmetry superalgebra in $2+1$ spacetime dimensions. (Hint: With the generators of the Lorentz group in $2+1$ spacetime dimensions labelled $A_1 = -iJ_{10}$, $A_2 = -iJ_{20}$, $A_3 = J_{12}$, the commutation relations of the Poincaré algebra are $[A_i, , A_j] = i\sum_k \epsilon_{ijk} A_k$, so the representations of the homogeneous Lorentz group in $2+1$ spacetime dimensions are labelled with a *single* positive integer or half-integer A.) Assume that the conditions for the Coleman–Mandula theorem are satisfied here.

3. Suppose that there were no massless particles with helicity greater than $+3/2$ or less than $-3/2$. Find the most general massless particle multiplets for $N = 6$ extended supersymmetry and (using CPT symmetry) $N = 5$ extended supersymmetry. What does the comparison between the particle content you find suggest about these two extended supersymmetries?

4. What are the possible parities of the particles in the short supermultiplets of extended $N = 2$ supersymmetry?

** In some articles on $N = 2$ supersymmetry the central charge Z is what we would call $Z/2\sqrt{2}$.

References

1. R. Haag, J. T. Lopuszanski, and M. Sohnius, *Nucl. Phys.* **B88**, 257 (1975). This article is reprinted in *Supersymmetry*, S. Ferrara, ed. (North Holland/World Scientific, Amsterdam/Singapore, 1987).
2. B. Zumino, *Nucl. Phys.* **B89**, 535 (1975). This article is reprinted in *Supersymmetry*, Ref. 1.
3. M. T. Grisaru and H. N. Pendleton, *Phys. Lett.* **67B**, 323 (1977).

26
Supersymmetric Field Theories

Now we know the structure of the most general supersymmetry algebras, and we have seen how to work out the implications of this symmetry for the particle spectrum. In order to learn what supersymmetry has to say about particle interactions, we need to see how to construct supersymmetric field theories.

Originally the construction of field supermultiplets was done directly, by a repeated use of the Jacobi identities, much as in the construction of supermultiplets of one-particle states in Sections 25.4 and 25.5. Section 26.1 presents one example of this technique, used here to construct supermultiplets containing only scalar and Dirac fields. Fortunately there is an easier technique, invented by Salam and Strathdee,[1] in which supermultiplets of fields are gathered into 'superfields,' which depend on fermionic coordinates as well as on the usual four coordinates of space-time. Superfields are introduced in Section 26.2, and used to construct supersymmetric field theories and to study some of their consequences in Sections 26.3–26.8. This chapter will be concerned only with $N = 1$ supersymmetry, where the superfield formalism has been chiefly useful. At the end of the next chapter we will construct theories with N-extended supersymmetry by imposing the $U(N)$ R-symmetry on theories of $N = 1$ superfields.

26.1 Direct Construction of Field Supermultiplets

To illustrate the direct construction of a field supermultiplet, we will consider fields that can destroy the particles belonging to the simplest supermultiplet of arbitrary mass discussed in Section 25.5: two spinless particles and one particle of spin 1/2. We saw in Eq. (25.5.15) that the one-particle zero spin state $|0,0\rangle$ is annihilated by the supersymmetry generator Q_a but not by Q_a^*, so we would expect the scalar field $\phi(x)$ that creates this particle from the vacuum (which is assumed to be annihilated by all supersymmetry generators) to commute with Q_a but not with Q_a^*.

That is,

$$[Q_a\,,\,\phi(x)]=0\,, \tag{26.1.1}$$

$$-i\sum_b e_{ab}[Q^*_b\,,\,\phi(x)]\equiv\zeta_a(x)\neq 0\,. \tag{26.1.2}$$

The antisymmetric 2×2 matrix e_{ab} (with $e_{1/2,-1/2}\equiv +1$) is introduced here because it is $\sum_b e_{ab}Q^*_b$ that transforms under the homogeneous Lorentz group according to the $(1/2,0)$ representation. It follows that $\zeta_a(x)$ is a two-component spinor field that also belongs to the $(1/2,0)$ representation of the homogeneous Lorentz group.*

From Eqs. (26.1.1)–(26.1.2) and the anticommutation relation (25.2.31), we find

$$\{Q_b\,,\,\zeta_a\}=-i\sum_c e_{ac}[\{Q_b\,,\,Q^*_c\}\,,\,\phi(x)]=2i(\sigma^\mu e)_{ba}\,[P_\mu\,,\,\phi]\,,$$

and so

$$\{Q_b\,,\,\zeta_a(x)\}=-2(\sigma^\mu e)_{ba}\partial_\mu\phi(x)\,. \tag{26.1.3}$$

On the other hand, Eq. (26.1.2) and the anticommutation relation (25.2.32) gives

$$-i\sum_c e_{ac}\{Q^*_b\,,\,\zeta_c\}=\{Q^*_b\,,\,[Q^*_a\,,\,\phi]\}=-\{Q^*_a\,,\,[Q^*_b\,,\,\phi]\}=i\sum_c e_{bc}\{Q^*_a\,,\,\zeta_c\}\,,$$

so $\sum_c e_{ac}\{Q^*_b\,,\,\zeta_c\}$ is antisymmetric, and therefore proportional to the antisymmetric 2×2 matrix e_{ab}:

$$i\{Q^*_b\,,\,\zeta_a(x)\}=2\delta_{ab}\,\mathscr{F}(x)\,. \tag{26.1.4}$$

Lorentz invariance requires that the coefficient $\mathscr{F}(x)$ is a scalar field.

We must now go one step further, and calculate the commutators of the supersymmetry generators with $\mathscr{F}(x)$. Using Eqs. (26.1.4), (26.1.2), and (25.2.32), we have

$$\delta_{ab}\,[Q^*_c\,,\,\mathscr{F}]=\tfrac{1}{2}i[Q^*_c\,,\,\{Q^*_b\,,\,\zeta_a\}]=\tfrac{1}{2}i[\{Q^*_c\,,\,\zeta_a\}\,,\,Q^*_b]=-\,\delta_{ac}\,[Q^*_b\,,\,\mathscr{F}]\,.$$

By taking $a=b\neq c$, we find that this commutator vanishes:

$$[Q^*_c\,,\,\mathscr{F}(x)]=0\,. \tag{26.1.5}$$

* At this point we are not assuming anything about the masses or interactions of the particles described by these fields, but it may be noted that, as explained in Section 5.9, a $(1/2,0)$ free field can create massless particles only of helicity $+1/2$, in agreement with the result (25.5.15) that the massless spinless one-particle state $|0,0\rangle$ that is annihilated by Q_a is in a supermultiplet with a state of helicity $+1/2$.

Finally, using Eqs. (26.1.4), (25.2.31), and (26.1.3), we have

$$\begin{aligned}\delta_{ab}\,[Q_c\,,\mathscr{F}] &= \tfrac{1}{2}i[Q_c\,,\{Q^*_b\,,\zeta_a\}] = \tfrac{1}{2}i[\{Q_c\,,Q^*_b\}\,,\zeta_a] - \tfrac{1}{2}i[Q^*_b\,,\{Q_c\,,\zeta_a\}]\\ &= -\,\sigma^\mu_{cb}\,\partial_\mu\zeta_a + i\,(\sigma^\mu e)_{ca}\,[Q^*_b\,,\partial_\mu\phi]\\ &= -\,\sigma^\mu_{cb}\,\partial_\mu\zeta_a + \sum_d e_{bd}\,(\sigma^\mu e)_{ca}\,\partial_\mu\zeta_d\,.\end{aligned}$$

Contracting with δ_{ab}, this becomes

$$[Q_c\,,\mathscr{F}(x)] = -\sum_a \sigma^\mu_{ca}\,\partial_\mu\zeta_a(x)\,. \tag{26.1.6}$$

Eqs. (26.1.1)–(26.1.6) show that the fields $\phi(x)$, $\zeta_a(x)$, and $\mathscr{F}(x)$ furnish a complete representation of the supersymmetry algebra. These fields are not Hermitian, so their complex conjugates furnish another supermultiplet:

$$[Q^*_a\,,\phi^*(x)] = 0\,, \tag{26.1.7}$$

$$-i\sum_b e_{ab}[Q_b\,,\phi^*(x)] = \zeta^*_a(x)\,, \tag{26.1.8}$$

$$\{Q^*_b\,,\zeta^*_a(x)\} = 2(e\sigma^\mu)_{ab}\partial_\mu\phi^*(x)\,, \tag{26.1.9}$$

$$-i\{Q_b\,,\zeta^*_a(x)\} = 2\delta_{ab}\,\mathscr{F}^*(x)\,, \tag{26.1.10}$$

$$[Q_c\,,\mathscr{F}^*(x)] = 0\,, \tag{26.1.11}$$

$$[Q^*_c\,,\mathscr{F}^*(x)] = \sum_a \sigma^\mu_{ac}\,\partial_\mu\zeta^*_a(x)\,. \tag{26.1.12}$$

We can express these commutation and anticommutation relations as transformation rules under a supersymmetry transformation, which shifts any bosonic or fermionic field operator $\mathscr{O}(x)$ by the infinitesimal amount

$$\delta\mathscr{O}(x) \equiv \left[\sum_a (\epsilon^*_a Q_a + \epsilon_a Q^*_a)\;,\;\mathscr{O}(x)\right]\,, \tag{26.1.13}$$

where ϵ_a is an infinitesimal fermionic c-number spinor. (Because ϵ_a and ϵ^*_a anticommute with Q_a and Q^*_a, the quantity $\epsilon^*_a Q_a + \epsilon_a Q^*_a$ is *anti-Hermitian*, so Eq. (26.1.13) gives $(\delta\mathscr{O})^* = \delta\mathscr{O}^*$.) The commutation and anticommutation rules (26.1.1)–(26.1.6) are equivalent to the transformation rules

$$\delta\phi(x) = -i\sum_{ab}\epsilon_a\,e_{ab}\,\zeta_b(x)\,, \tag{26.1.14}$$

$$\delta\zeta_a(x) = -2\sum_b \epsilon^*_b\,(\sigma^\mu e)_{ba}\,\partial_\mu\phi(x) - 2i\epsilon_a\,\mathscr{F}(x)\,, \tag{26.1.15}$$

$$\delta\mathscr{F}(x) = -\sum_{ab}\epsilon^*_b\,\sigma^\mu_{ba}\,\partial_\mu\zeta_a(x)\,. \tag{26.1.16}$$

This may be put into a four-component Dirac notation by introducing an infinitesimal, Majorana,** four-component, spinor transformation parameter

$$\alpha \equiv -i \begin{pmatrix} \epsilon_a \\ \sum_b e_{ab}\epsilon_b^* \end{pmatrix} , \tag{26.1.17}$$

so that Eq. (26.1.13) becomes

$$i\,\delta\mathscr{O}(x) \equiv [\bar{\alpha}Q\,,\, \mathscr{O}(x)] \ . \tag{26.1.18}$$

The transformation rules (26.1.14)–(26.1.16) and their complex conjugates may be put into a convenient covariant form by introducing a set of real bosonic fields A, B, F, and G, defined by

$$\frac{A+iB}{\sqrt{2}} \equiv \phi\,, \qquad \frac{F-iG}{\sqrt{2}} \equiv \mathscr{F}\,, \tag{26.1.19}$$

and a four-component Majorana spinor ψ, defined by

$$\psi \equiv \frac{1}{\sqrt{2}} \begin{pmatrix} \zeta_a \\ -\sum_b e_{ab}\zeta_b^* \end{pmatrix} . \tag{26.1.20}$$

Let's also recall the relation between the 4×4 Dirac matrices and the 2×2 matrices σ_μ:

$$\gamma_\mu = \begin{pmatrix} 0 & -i\,e\sigma_\mu^{\mathrm{T}} e \\ i\,\sigma_\mu & 0 \end{pmatrix} .$$

The transformation rules now take the form

$$\begin{aligned} &\delta A = \bar{\alpha}\,\psi\,, \qquad \delta B = -i\,\bar{\alpha}\,\gamma_5\,\psi\,, \\ &\delta\psi = \partial_\mu(A + i\gamma_5 B)\gamma^\mu\alpha + (F - i\gamma_5 G)\alpha\,, \\ &\delta F = \bar{\alpha}\,\gamma^\mu\,\partial_\mu\psi\,, \qquad \delta G = -i\,\bar{\alpha}\,\gamma_5\gamma^\mu\,\partial_\mu\psi\,. \end{aligned} \tag{26.1.21}$$

A direct but tedious calculation shows that this transformation leaves

** With the phase convention we will be using here, a Majorana four-component spinor is formed from a $(1/2,0)$ two-component spinor u_a as

$$\begin{pmatrix} u \\ -e u^* \end{pmatrix} .$$

Eq. (26.1.17) fits this definition, with $u = -i\epsilon$. Equivalently, a Majorana spinor can be formed from a two-component $(0,1/2)$ spinor v_a as

$$\begin{pmatrix} e v^* \\ v \end{pmatrix} .$$

An example is provided by Eq. (25.2.34). Properties of Majorana spinors are considered in detail in the appendix to this chapter.

invariant the action

$$I = \int d^4x \left\{ -\tfrac{1}{2}\partial_\mu A\, \partial^\mu A - \tfrac{1}{2}\partial_\mu B\, \partial^\mu B - \tfrac{1}{2}\bar{\psi}\gamma^\mu\partial_\mu\psi \right.$$
$$+ \tfrac{1}{2}(F^2 + G^2) + m\, \left[FA + GB - \tfrac{1}{2}\bar{\psi}\psi\right]$$
$$\left. + g\Big[F(A^2 + B^2) + 2GAB - \bar{\psi}(A + i\gamma_5 B)\psi\Big]\right\}. \qquad (26.1.22)$$

Eqs. (26.1.21) and (26.1.22) agree with the transformation rules (24.2.8) and Lagrangian density (24.2.9) found in the original work of Wess and Zumino. In the next three sections we will explore a convenient technique for checking the supersymmetry of Eq. (26.1.22) and for deriving more general supersymmetric actions.

Where the fermion field $\psi(x)$ satisfies the free-field Dirac equation $(\gamma^\mu\partial_\mu + m)\psi = 0$, these transformation rules show that $F + mA$ and $G + mB$ are invariant, and therefore commute with Q_a and Q^*_a, and hence also with P_μ. This does not prove that $F = -mA$ and $G = -mB$, but without changing any of the commutation and anticommutation rules (26.1.1)–(26.1.6) or transformation rules (26.1.21), we can redefine the fields F and G by subtracting the constants $F + mA$ and $B + mG$, respectively, so that the new fields F and G are given by $F = -mA$ and $G = -mB$, and therefore $\mathscr{F} = -m\phi^*$. This is not true in the presence of interactions, but even in the interacting case $\mathscr{F}(x)$, $F(x)$, and $G(x)$ are typically auxiliary fields that can be expressed in terms of the other fields of the supermultiplet, as is the case for the action (26.1.22).

26.2 General Superfields

It is straightforward to construct supermultiplets of fields by the direct technique illustrated in the previous section, but in order to construct supersymmetric actions we also need to know how to multiply field supermultiplets to make other supermultiplets. A great deal of work can be saved by using a formalism invented by Salam and Strathdee,[1] in which the fields in any supermultiplet are assembled into a single superfield.

Just as the four-momentum operators P_μ are defined as the generators of translations of the ordinary spacetime coordinates x^μ, the four supersymmetry generators Q_a and Q^*_a may be regarded as the generators of translations of four fermionic c-number superspace coordinates, which anticommute with each other and with fermionic fields but commute with the x^μ and all bosonic fields. We aim at constructing Lorentz-invariant Lagrangian densities, so it will be convenient to adopt the four-component Dirac formalism described in Section 25.2. The supersymmetry generators are gathered into a four-component Majorana spinor Q_α, and correspondingly, the superspace coordinates are gathered into another four-component Majorana spinor θ_α. (Various properties of Majorana spinors are outlined in the appendix to this chapter.) The supersymmetry generators have non-vanishing anticommutators, so we cannot take them as simply proportional to the supercoordinate translation operators $\partial/\partial\theta_\alpha$. Instead, Salam and Strathdee found that the supersymmetry algebra would be satisfied if we suppose that the commutator or anticommutator of the supersymmetry generator Q with any bosonic or fermionic superfield $S(x,\theta)$ is

$$[Q\,,\,S\} = i\mathscr{Q}\,S\,, \tag{26.2.1}$$

where $\mathscr{Q}$ is the superspace differential operator

$$\mathscr{Q} \equiv -\frac{\partial}{\partial\bar{\theta}} + \gamma^\mu\,\theta\,\frac{\partial}{\partial x^\mu}\,. \tag{26.2.2}$$

(As usual $\bar{\theta} \equiv \theta^\dagger\beta$. All derivatives with respect to fermionic c-number variables should be understood as *left*-derivatives, calculated by moving the variable to the left of any expression before differentiating with respect to it.) For Majorana spinors $\bar{\theta} = \theta^{\rm T}\gamma_5\epsilon$, with the 4×4 matrix ϵ given by Eq. (26.A.3), so Eq. (26.2.1) may be expressed more explicitly as

$$\mathscr{Q}_\alpha = \sum_\gamma (\gamma_5\epsilon)_{\alpha\gamma}\,\frac{\partial}{\partial\theta_\gamma} + \sum_\gamma \gamma^\mu_{\alpha\gamma}\theta_\gamma\,\frac{\partial}{\partial x^\mu}\,. \tag{26.2.3}$$

Likewise,

$$\overline{\mathscr{Q}}_\beta = \sum_\gamma \mathscr{Q}_\gamma\,(\gamma_5\,\epsilon)_{\gamma\beta} = \frac{\partial}{\partial\theta_\beta} - \sum_\gamma (\gamma_5\,\epsilon\,\gamma^\mu)_{\beta\gamma}\,\theta_\gamma\,\frac{\partial}{\partial x^\mu}\,. \tag{26.2.4}$$

It is straightforward to calculate that

$$\left\{\mathscr{Q}_\alpha\,,\,\overline{\mathscr{Q}}_\beta\right\} = (\gamma_5\,\epsilon\,\gamma^\mu\,\gamma_5\,\epsilon)_{\beta\alpha}\,\frac{\partial}{\partial x^\mu} + \gamma^\mu_{\alpha\beta}\,\frac{\partial}{\partial x^\mu}\,. \tag{26.2.5}$$

But Eq. (5.4.35) shows that $\gamma_\mu^{\rm T} = -\mathscr{C}\gamma_\mu\mathscr{C}^{-1}$, where $\mathscr{C}$ is the matrix $\mathscr{C} = -\gamma_5\epsilon$, so both terms on the right-hand side of Eq. (26.2.5) are equal, and therefore

$$\left\{\mathscr{Q}_\alpha\,,\,\overline{\mathscr{Q}}_\beta\right\} = 2\gamma^\mu_{\alpha\beta}\,\frac{\partial}{\partial x^\mu}\,. \tag{26.2.6}$$

Eqs. (26.2.6) and (26.2.1) together with the generalized Jacobi identities (25.1.5) show that

$$[\{Q_\alpha\,,\,\overline{Q}_\beta\}\,,\,S] = \{\mathscr{Q}_\alpha\,,\,\overline{\mathscr{Q}}_\beta\}\,S = 2\gamma^\mu_{\alpha\beta}\partial_\mu S = -2i\gamma^\mu_{\alpha\beta}[P_\mu\,,\,S]\,, \tag{26.2.7}$$

in agreement with the anticommutation relation (25.2.36).

It is often more convenient to express the commutation and anticommutation relations (26.2.1) as transformation rules under infinitesimal supersymmetry transformations. Combining Eqs. (26.1.18), (26.2.1), and (26.2.2) shows that a supersymmetry transformation with infinitesimal Majorana spinor parameter α changes a superfield $S(x,\theta)$ by an amount

$$\delta S = (\bar{\alpha}\,\mathscr{Q})\,S = -\left(\bar{\alpha}\,\frac{\partial S}{\partial \bar{\theta}}\right) + (\bar{\alpha}\,\gamma^\mu\,\theta)\,\frac{\partial S}{\partial x^\mu}\,. \tag{26.2.8}$$

Recall that $\partial/\partial\bar{\theta}$ here acts on the left of any expression. In particular, where M is any linear combination of the matrices 1, $\gamma_5\gamma_\mu$, and γ_5 for which $\bar{\theta}M\theta$ does not vanish, we have $\bar{\theta}'M\,\theta'' = \bar{\theta}''M\,\theta'$, so

$$\frac{\partial}{\partial\bar{\theta}}\,(\bar{\theta}M\theta) = 2M\theta\,. \tag{26.2.9}$$

The components of θ anticommute, so any product of their components vanishes if two of them are the same component. But θ has only four components, so any function of θ has a power series that terminates with its quartic term. Furthermore, as shown in the appendix to this chapter, a product of two θs is proportional to a linear combination of $(\bar{\theta}\theta)$, $(\bar{\theta}\gamma_\mu\gamma_5\theta)$, and $(\bar{\theta}\gamma_5\theta)$; a product of three θs is proportional to $(\bar{\theta}\gamma_5\theta)\theta$; and a product of four θs is proportional to $(\bar{\theta}\gamma_5\theta)^2$. The most general function of x^μ and θ may therefore be expressed as

$$\begin{aligned} S(x,\theta) = {} & C(x) - i\big(\bar{\theta}\,\gamma_5\,\omega(x)\big) - \frac{i}{2}\big(\bar{\theta}\,\gamma_5\,\theta\big)M(x) - \frac{1}{2}\big(\bar{\theta}\,\theta\big)N(x) \\ & + \frac{i}{2}\big(\bar{\theta}\,\gamma_5\,\gamma_\mu\,\theta\big)V^\mu(x) - i\big(\bar{\theta}\,\gamma_5\,\theta\big)\left(\bar{\theta}\left[\lambda(x) + \frac{1}{2}\,\partial\!\!\!/\omega(x)\right]\right) \\ & - \frac{1}{4}\big(\bar{\theta}\,\gamma_5\,\theta\big)^2\left(D(x) + \frac{1}{2}\Box C(x)\right). \end{aligned} \tag{26.2.10}$$

(The terms $\frac{1}{2}\partial\!\!\!/\omega$ and $\frac{1}{2}\Box C(x)$ are separated from $\lambda(x)$ and $D(x)$, respectively, for later convenience.) If $S(x,\theta)$ is a scalar then $C(x)$, $M(x)$, $N(x)$, and $D(x)$ are scalar (or pseudoscalar) fields; $\omega(x)$ and $\lambda(x)$ are four-component spinor fields; and $V^\mu(x)$ is a vector field. Also, using the reality properties of bilinear products of Majorana fields given in the appendix to this chapter, we see that if $S(x,\theta)$ is real, then $C(x)$, $M(x)$, $N(x)$, $V^\mu(x)$, and $D(x)$ are all real, while $\omega(x)$ and $\lambda(x)$ are Majorana spinors satisfying the phase convention $s^* = -\beta\epsilon\gamma_5 s$.

Now we must work out the supersymmetry transformation properties of the component fields in Eq. (26.2.10). Applying Eqs. (26.2.8) and (26.2.9) to the expansion (26.2.10) gives

$$\begin{aligned}
\delta S = & \ (\bar{\alpha}\gamma^\mu\theta)\frac{\partial C}{\partial x^\mu} \\
& + i\,(\bar{\alpha}\gamma_5\omega) - i\,(\bar{\alpha}\gamma^\mu\theta)\left(\bar{\theta}\gamma_5\frac{\partial\omega}{\partial x^\mu}\right) \\
& + i\,(\bar{\alpha}\gamma_5\theta)\,M - \frac{i}{2}(\bar{\alpha}\gamma^\mu\theta)\left(\bar{\theta}\gamma_5\theta\right)\frac{\partial M}{\partial x^\mu} \\
& + (\bar{\alpha}\theta)\,N - \frac{1}{2}(\bar{\alpha}\gamma^\mu\theta)\left(\bar{\theta}\theta\right)\frac{\partial N}{\partial x^\mu} \\
& - i\,(\bar{\alpha}\gamma_5\gamma_\nu\theta)\,V^\nu + \frac{i}{2}(\bar{\alpha}\gamma^\mu\theta)\left(\bar{\theta}\gamma_5\gamma_\nu\theta\right)\frac{\partial V^\nu}{\partial x^\mu} \\
& + 2i\,(\bar{\alpha}\gamma_5\theta)\left(\bar{\theta}[\lambda + \tfrac{1}{2}\,\partial\!\!\!/\omega]\right) + i\left(\bar{\theta}\gamma_5\theta\right)(\bar{\alpha}[\lambda + \tfrac{1}{2}\,\partial\!\!\!/\omega]) \\
& - i\,(\bar{\alpha}\gamma^\mu\theta)\left(\bar{\theta}\gamma_5\theta\right)\left(\bar{\theta}\partial_\mu[\lambda + \tfrac{1}{2}\,\partial\!\!\!/\omega]\right) + \left(\bar{\theta}\gamma_5\theta\right)(\bar{\alpha}\gamma_5\theta)\,[D + \tfrac{1}{2}\Box C]\,.
\end{aligned}$$

We need to put each term in the standard form of Eq. (26.2.10). For this purpose, we note that: Eq. (26.A.9) gives

$$(\bar{\alpha}\gamma^\mu\theta)(\bar{\theta}\gamma_5\partial_\mu\omega) = -\tfrac{1}{4}(\bar{\theta}\theta)(\bar{\alpha}\,\partial\!\!\!/\gamma_5\omega) - \tfrac{1}{4}(\bar{\theta}\gamma_5\gamma^\nu\theta)(\bar{\alpha}\,\partial\!\!\!/\gamma_\nu\omega) - \tfrac{1}{4}(\bar{\theta}\gamma_5\theta)(\bar{\alpha}\,\partial\!\!\!/\omega)\;;$$

Eq. (26.A.16) gives

$$(\bar{\alpha}\gamma^\mu\theta)(\bar{\theta}\theta) = -(\bar{\alpha}\gamma^\mu\gamma_5\theta)(\bar{\theta}\gamma_5\theta)\;;$$

Eq. (26.A.17) gives

$$(\bar{\alpha}\gamma^\mu\theta)(\bar{\theta}\gamma_5\gamma_\nu\theta) = -(\bar{\alpha}\gamma^\mu\gamma_\nu\theta)(\bar{\theta}\gamma_5\theta)\;;$$

Eq. (26.A.9) gives

$$\begin{aligned}
(\bar{\alpha}\gamma_5\theta)(\bar{\theta}[\lambda + \tfrac{1}{2}\,\partial\!\!\!/\omega]) = & -\tfrac{1}{4}(\bar{\theta}\theta)(\bar{\alpha}\gamma_5[\lambda + \tfrac{1}{2}\,\partial\!\!\!/\omega]) + \tfrac{1}{4}(\bar{\theta}\gamma_5\gamma^\mu\theta)(\bar{\alpha}\gamma_\mu[\lambda + \tfrac{1}{2}\,\partial\!\!\!/\omega]) \\
& - \tfrac{1}{4}(\bar{\theta}\gamma_5\theta)(\bar{\alpha}[\lambda + \tfrac{1}{2}\,\partial\!\!\!/\omega])\;;
\end{aligned}$$

and Eq. (26.A.19) gives

$$(\bar{\alpha}\gamma^\mu\theta)(\bar{\theta}\gamma_5\theta)(\bar{\theta}\partial_\mu[\lambda + \tfrac{1}{2}\,\partial\!\!\!/\omega]) = -\tfrac{1}{4}(\bar{\alpha}\,\partial\!\!\!/\gamma_5[\lambda + \tfrac{1}{2}\,\partial\!\!\!/\omega])(\bar{\theta}\gamma_5\theta)^2\,.$$

Using these relations and rearranging terms in order of increasing numbers of θ factors, we have

$$\begin{aligned}
\delta S = & \ i\,(\bar{\alpha}\gamma_5\omega) + (\bar{\alpha}[\partial\!\!\!/ C + i\gamma_5 M + N - i\gamma_5\,V\!\!\!/\,]\theta) \\
& - \tfrac{1}{2}i\left(\bar{\theta}\theta\right)(\bar{\alpha}\gamma_5[\lambda + \partial\!\!\!/\omega]) + \tfrac{1}{2}i\left(\bar{\theta}\gamma_5\theta\right)(\bar{\alpha}[\lambda + \partial\!\!\!/\omega]) \\
& + \tfrac{1}{2}i\left(\bar{\theta}\gamma_5\gamma^\mu\theta\right)(\bar{\alpha}\gamma_\mu\lambda) + \tfrac{1}{2}i\left(\bar{\theta}\gamma_5\gamma^\nu\theta\right)(\bar{\alpha}\partial_\nu\omega) \\
& + \tfrac{1}{2}\left(\bar{\theta}\gamma_5\theta\right)(\bar{\alpha}[-i\,\partial\!\!\!/ M - \gamma_5\,\partial\!\!\!/ N - i\,\partial\!\!\!/\,V\!\!\!/ + \gamma_5(D + \tfrac{1}{2}\Box C)]\theta) \\
& - \tfrac{1}{4}i\left(\bar{\theta}\gamma_5\theta\right)^2(\bar{\alpha}\gamma_5[\partial\!\!\!/\lambda + \tfrac{1}{2}\Box\omega])\,,
\end{aligned}$$

or, using the symmetry property (26.A.7),

$$\begin{aligned}\delta S = & \; i(\bar{\alpha}\gamma_5\omega) + \Big(\bar{\theta}[-\not\partial C + i\gamma_5 M + N - i\gamma_5 \not V]\alpha\Big) \\ & -\tfrac{1}{2}i\Big(\bar{\theta}\theta\Big)(\bar{\alpha}\gamma_5[\lambda + \not\partial\omega]) + \tfrac{1}{2}i\Big(\bar{\theta}\gamma_5\theta\Big)(\bar{\alpha}[\lambda + \not\partial\omega]) \\ & + \tfrac{1}{2}i\Big(\bar{\theta}\gamma_5\gamma^\mu\theta\Big)(\bar{\alpha}\gamma_\mu\lambda) + \tfrac{1}{2}i\Big(\bar{\theta}\gamma_5\gamma^\nu\theta\Big)(\bar{\alpha}\partial_\nu\omega) \\ & + \tfrac{1}{2}\Big(\bar{\theta}\gamma_5\theta\Big)\Big(\bar{\theta}[i\not\partial M - \gamma_5\not\partial N - i\partial_\mu \not V\gamma^\mu + \gamma_5(D + \tfrac{1}{2}\Box C)]\alpha\Big) \\ & -\tfrac{1}{4}i\Big(\bar{\theta}\gamma_5\theta\Big)^2(\bar{\alpha}\gamma_5[\not\partial\lambda + \tfrac{1}{2}\Box\omega]) \, .\end{aligned}$$

If we now compare this with the terms up to second order in θ in the expansion (26.2.10), we find the transformation rules:

$$\delta C = i\Big(\bar{\alpha}\gamma_5\omega\Big) \, , \tag{26.2.11}$$

$$\delta\omega = \Big(-i\gamma_5\not\partial C - M + i\gamma_5 N + \not V\Big)\alpha \, , \tag{26.2.12}$$

$$\delta M = -\Big(\bar{\alpha}\,[\lambda + \not\partial\omega]\Big) \, , \tag{26.2.13}$$

$$\delta N = i\Big(\bar{\alpha}\gamma_5\,[\lambda + \not\partial\omega]\Big) \, , \tag{26.2.14}$$

$$\delta V_\mu = \Big(\bar{\alpha}\gamma_\mu\lambda\Big) + \Big(\bar{\alpha}\partial_\mu\omega\Big) \, . \tag{26.2.15}$$

The terms of third and fourth order in θ give

$$\delta[\lambda + \tfrac{1}{2}\not\partial\omega] = \tfrac{1}{2}\Big[-\not\partial M - i\gamma_5\not\partial N + \partial_\mu \not V\gamma^\mu + i\gamma_5\Big(D + \tfrac{1}{2}\Box C\Big)\Big]\alpha \, ,$$

$$\delta[D + \tfrac{1}{2}\Box C] = i\Big(\bar{\alpha}\gamma_5[\not\partial\lambda + \tfrac{1}{2}\Box\omega]\Big) \, .$$

Combining the latter two transformation rules with the transformation rules (26.2.11) and (26.2.12) for C and ω yields much simpler transformation rules for λ and D:

$$\delta\lambda = \left(\tfrac{1}{2}\Big[\partial_\mu \not V \, , \gamma^\mu\Big] + i\gamma_5 D\right)\alpha \, , \tag{26.2.16}$$

$$\delta D = i\Big(\bar{\alpha}\gamma_5\not\partial\lambda\Big) . \tag{26.2.17}$$

It is to achieve this simplification that the terms $\frac{1}{2}\not\partial\omega$ and $\frac{1}{2}\Box C$ were separated from λ and D in the expansion (26.2.10).

The whole point of the superfield formalism is to simplify the task of making supermultiplets out of other supermultiplets. Given two superfields S_1 and S_2 that both satisfy the transformation rule (26.2.8), their product $S \equiv S_1 S_2$ satisfies

$$\begin{aligned}\delta S &\equiv [(\bar{\alpha}Q) \, , S_1 S_2] = (\delta S_1)S_2 + S_1(\delta S_2) \\ &= \Big((\bar{\alpha}\,\mathscr{Q})S_1\Big)S_2 + S_1\Big((\bar{\alpha}\,\mathscr{Q})\Big)S_2 = (\bar{\alpha}\,\mathscr{Q})\,S \, ,\end{aligned} \tag{26.2.18}$$

and is therefore also a superfield. A straightforward calculation using Eqs. (26.A.7), (26.A.16), (26.A.18), and (26.A.19) gives its components as

$$C = C_1 C_2 \,, \tag{26.2.19}$$

$$\omega = C_1 \omega_2 + C_2 \omega_1 \,, \tag{26.2.20}$$

$$M = C_1 M_2 + C_2 M_1 + \tfrac{1}{2} i \left(\overline{\omega_1} \gamma_5 \, \omega_2 \right) , \tag{26.2.21}$$

$$N = C_1 N_2 + C_2 N_1 - \tfrac{1}{2} \left(\overline{\omega_1} \, \omega_2 \right) , \tag{26.2.22}$$

$$V^\mu = C_1 V_2^\mu + C_2 V_1^\mu - \tfrac{1}{2} i \left(\overline{\omega_1} \gamma_5 \gamma^\mu \omega_2 \right) , \tag{26.2.23}$$

$$\begin{aligned} \lambda = {} & C_1 \lambda_2 + C_2 \lambda_1 - \tfrac{1}{2} \gamma^\mu \omega_1 \partial_\mu C_2 - \tfrac{1}{2} \gamma^\mu \omega_2 \partial_\mu C_1 + \tfrac{1}{2} i \, \not{V}_1 \gamma_5 \, \omega_2 + \tfrac{1}{2} i \, \not{V}_2 \gamma_5 \, \omega_1 \\ & + \tfrac{1}{2} (N_1 - i \gamma_5 M_1) \, \omega_2 + \tfrac{1}{2} (N_2 - i \gamma_5 M_2) \, \omega_1 \,, \end{aligned} \tag{26.2.24}$$

$$\begin{aligned} D = {} & -\partial_\mu C_1 \, \partial^\mu C_2 + C_1 D_2 + C_2 D_1 + M_1 M_2 + N_1 N_2 \\ & - \left(\overline{\omega_1} \, [\lambda_2 + \tfrac{1}{2} \, \not{\partial} \omega_2] \right) - \left(\overline{\omega_2} \, [\lambda_1 + \tfrac{1}{2} \, \not{\partial} \omega_1] \right) - V_{1\mu} V_2^\mu \,. \end{aligned} \tag{26.2.25}$$

It is trivial that linear combinations of superfields are superfields, in the same sense, and that spacetime derivatives and complex conjugates of superfields are superfields. But multiplying a superfield by some function of θ or differentiating it with respect to θ does not in general yield a superfield. (For instance, θ itself is evidently not a superfield, because θ is a fermionic c-number and therefore commutes with $\bar{\alpha} Q$, while $\mathscr{Q}\theta \neq 0$.) There is, however, a way of combining a derivative of a superfield with respect to θ and multiplying it by a factor θ which does yield another superfield.

Consider the superspace differential operator $\mathscr{D}_\alpha$ defined by

$$\mathscr{D} \equiv -\frac{\partial}{\partial \bar{\theta}} - \gamma^\mu \, \theta \, \frac{\partial}{\partial x^\mu} \,, \tag{26.2.26}$$

or more explicitly

$$\mathscr{D}_\alpha = \sum_\gamma (\gamma_5 \epsilon)_{\alpha\gamma} \, \frac{\partial}{\partial \theta_\gamma} - \sum_\gamma \gamma^\mu_{\alpha\gamma} \theta_\gamma \, \frac{\partial}{\partial x^\mu} \,. \tag{26.2.27}$$

The only difference between the definitions of $\mathscr{D}$ and $\mathscr{Q}$ is a change of sign of the terms involving the spacetime derivative. In consequence of this change of sign, in the anticommutator of $\mathscr{D}_\beta$ with $\mathscr{Q}_\alpha$, instead of getting two equal terms like those in Eq. (26.2.5) with the same sign, we get them with opposite signs, so that they cancel:

$$\{ \mathscr{D}_\beta \,, \, \mathscr{Q}_\alpha \} = 0 \,. \tag{26.2.28}$$

Since α is fermionic, it follows that $(\bar{\alpha} \mathscr{Q})$ commutes with $\mathscr{D}_\beta$, so if $S(x, \theta)$ is a superfield, then

$$\delta \mathscr{D}_\beta S \equiv -i [(\bar{\alpha} Q) \,, \, \mathscr{D}_\beta S] = -i \mathscr{D}_\beta [(\bar{\alpha} Q) \,, \, S] = \mathscr{D}_\beta (\bar{\alpha} \mathscr{Q}) S = (\bar{\alpha} \mathscr{Q}) \mathscr{D}_\beta S \,, \tag{26.2.29}$$

so that $\mathscr{D}_\beta S$ is also a superfield. *Thus an arbitrary polynomial function of superfields S and their superderivatives* $\mathscr{D}_\beta S$, $\mathscr{D}_\beta \mathscr{D}_\gamma S$, *etc. is also a superfield.*

It is not necessary to add here that in constructing a superfield out of other superfields we can also include their spacetime derivatives, because these can be obtained from second superderivatives. Since the only difference between $\mathscr{D}_\beta$ and $\mathscr{Q}_\beta$ is in the sign of the term involving ∂_μ, the anticommutators of the $\mathscr{D}$s are the same as those of the $\mathscr{Q}$s, except for a change of sign:

$$\left\{\mathscr{D}_\alpha, \overline{\mathscr{D}}_\beta\right\} = -2\gamma^\mu_{\alpha\beta} \frac{\partial}{\partial x^\mu} . \tag{26.2.30}$$

Now let us consider how to construct a supersymmetric action out of superfields. There is no such thing as a supersymmetric Lagrangian density, because the anticommutation relation (26.2.6) shows that if $\delta\mathscr{L} = 0$ then $\mathscr{L}$ must be a constant. Even if the Lagrangian density is not supersymmetric, the action will still be supersymmetric if $\delta\mathscr{L}(x)$ is a derivative, which would not contribute to $\delta \int \mathscr{L}\, d^4x$. In general, the Lagrangian density $\mathscr{L}$ can be written as a sum of terms, each of which is some component of a superfield that is constructed out of elementary superfields and their superderivatives. Inspection of the transformation rules (26.2.11)–(26.2.17) for the individual components shows that in the absence of any special conditions on a general superfield, the only component of such a superfield whose variation is a derivative is the D-component. Also, for the D-component of any superfield to be a scalar, the superfield itself must be a scalar. Therefore, unless there are special conditions on the individual superfields from which the Lagrangian density is constructed, a supersymmetric action can only be the integral of the D-term of a scalar superfield Λ:

$$I = \int d^4x\, [\Lambda]_D . \tag{26.2.31}$$

But in fact no action of this sort would be physically satisfactory without special conditions on the superfields from which it is constructed. For a general superfield $S(x,\theta)$, the only sort of supersymmetric kinematic action I_0 that is bilinear in S and S^* and involves no more than two derivatives of the component fields is of the form

$$I_0 \propto \int d^4x \left[S^* S\right]_D . \tag{26.2.32}$$

From Eq. (26.2.25), we see that $S^* S$ has the D-component

$$\begin{aligned}\left[S^* S\right]_D = &-\partial_\mu C^* \partial^\mu C - \tfrac{1}{2}\left(\bar\omega \gamma^\mu \partial_\mu \omega\right) + \tfrac{1}{2}\left((\partial_\mu \bar\omega)\gamma^\mu \omega\right) \\ &+ C^* D + D^* C - \left(\bar\omega \lambda\right) - \left(\bar\lambda \omega\right) \\ &+ M^* M + N^* N - V^*_\mu V^\mu .\end{aligned} \tag{26.2.33}$$

The terms quadratic in C or ω look promising as the kinematic Lagrangians for massless fields of spins zero and 1/2; the final three terms are harmless; but the terms involving D or λ have the disastrous effect in path integrals of constraining C and ω to vanish. Fortunately, as we shall see in the following section, there are constrained superfields from which we *can* construct physically sensible actions. The introduction of these constrained superfields will also open up ways of constructing supersymmetric terms in the action that are not the D-components of functions of superfields.

If parity is conserved, then the space inversion properties of the component fields of a superfield will be related by supersymmetry. To work out this relation, we apply the parity operator P to the commutation/anticommutation relation (26.2.1) and use the transformation property (25.3.16) of supersymmetry generators, which gives

$$i\beta\left[Q, \mathsf{P}^{-1}S(x,\theta)\,\mathsf{P}\right\} = \mathscr{Q}\,\mathsf{P}^{-1}S(x,\theta)\,\mathsf{P}\,. \tag{26.2.34}$$

The solution of Eq. (26.2.34) for a scalar superfield is of the form

$$\mathsf{P}^{-1}S(x,\theta)\,\mathsf{P} = \eta\, S(\Lambda_P x, -i\beta\theta)\,, \tag{26.2.35}$$

where η is some phase (the intrinsic parity of the superfield) and $\Lambda_P x \equiv (-\mathbf{x}, +x^0)$. (To check that Eq. (26.2.35) satisfies Eq. (26.2.34), note that Eq. (26.2.35) gives the left-hand side of Eq. (26.2.34) as

$$i\eta\beta\left(-\frac{\partial}{\partial(-i\beta\theta)} + \gamma^\mu(-i\beta\theta)\frac{\partial}{\partial(\Lambda_P x)^\mu}\right) S(\Lambda_P x, \theta) = \eta\,\mathscr{Q}\, S(\Lambda_P x, -i\beta\theta)\,,$$

in agreement with what Eq. (26.2.35) gives for the right-hand side of Eq. (26.2.34).) Using the expansion (26.2.10) in Eq. (26.2.35) then gives the space inversion properties of the component fields:

$$\begin{aligned}
&\mathsf{P}^{-1}C(x)\,\mathsf{P} = \eta\, C(\Lambda_P x)\,,\\
&\mathsf{P}^{-1}\omega(x)\,\mathsf{P} = -i\eta\,\beta\,\omega(\Lambda_P x)\,,\\
&\mathsf{P}^{-1}M(x)\,\mathsf{P} = -\eta\, M(\Lambda_P x)\,,\\
&\mathsf{P}^{-1}N(x)\,\mathsf{P} = \eta\, N(\Lambda_P x)\,,\\
&\mathsf{P}^{-1}V^\mu(x)\,\mathsf{P} = -\eta\,(\Lambda_P)^\mu{}_\nu\, V^\nu(\Lambda_P x)\,,\\
&\mathsf{P}^{-1}\lambda(x)\,\mathsf{P} = i\eta\,\beta\,\lambda(\Lambda_P x)\,,\\
&\mathsf{P}^{-1}D(x)\,\mathsf{P} = \eta\, D(\Lambda_P x)\,.
\end{aligned} \tag{26.2.36}$$

* * *

The general real superfield S involves four real spinless fields C, M, N, and D, plus one real four-vector field V_μ, for a total of eight independent bosonic field components. For comparison, there are two four-component Majorana spinor fields ω and λ, also for a total of eight independent field components. The equality of the numbers of independent bosonic and fermionic field components holds in general not only for the unconstrained general superfield studied in this section, but also for all superfields obtained from the general superfield by imposing supersymmetric constraints, such as the chiral and other constrained superfields discussed in the next section.

To see this in general, suppose that we have a representation of the supersymmetry algebra provided by N_B linearly independent real bosonic field operators $b_n(x)$ and N_F linearly independent fermionic field operators $f_k(x)$. We will assume that these fields satisfy only non-trivial field equations, so that no linear combination of the b_n or f_k with non-vanishing coefficients satisfies a homogeneous linear field equation. Consider a real supersymmetry generator $Q(u)$, defined as

$$Q(u) \equiv \left(\bar{u}\, Q\right) = \left(\bar{Q}\, u\right), \tag{26.2.37}$$

where u is some ordinary numerical Majorana spinor (*not* an anticommuting c-number). (For extended supersymmetry, in place of Q_α we could use any one of the $Q_{r\alpha}$, say $Q_{1\alpha}$.) In order for the b_n and f_k to furnish a representation of the supersymmetry algebra, we must have

$$[\, Q(u)\,,\, b_n\,] = i \sum_k q_{nk}(\partial)\, f_k\,, \tag{26.2.38}$$

$$\{\, Q(u)\,,\, f_k\,\} = \sum_n p_{kn}(\partial)\, b_n\,, \tag{26.2.39}$$

for some matrix differential operators $q(\partial)$ and $p(\partial)$. Taking the anticommutator of Eq. (26.2.38) and the commutator of Eq. (26.2.39) with $Q(u)$ gives

$$[\, Q^2(u)\,,\, b_n\,] = i \sum_m \Big(q(\partial)\, p(\partial)\Big)_{nm} b_m\,, \tag{26.2.40}$$

$$[\, Q^2(u)\,,\, f_k\,] = i \sum_\ell \Big(p(\partial)\, q(\partial)\Big)_{k\ell} f_\ell\,. \tag{26.2.41}$$

The anticommutation relation (25.2.36) or (25.2.38) gives the square of $Q(u)$ as $Q^2(u) = -iP_\mu \left(\bar{u}\gamma^\mu u\right)$. Hence the square matrices $p(\partial)q(\partial)$ and $q(\partial)p(\partial)$ must both be non-singular, because if there were any non-vanishing coefficients $c_n(\partial)$ or $d_k(\partial)$ for which $\sum_n c_n(\partial)(q(\partial)p(\partial))_{nm} = 0$ or $\sum_k d_k(\partial)(p(\partial)q(\partial))_{k\ell} = 0$ then b_n or f_k would satisfy the homogeneous linear field equations

$$\left(\bar{u}\gamma^\mu u\right)\partial_\mu \sum_n c_n(\partial) b_n = 0 \quad \text{or} \quad \left(\bar{u}\gamma^\mu u\right)\partial_\mu \sum_k d_k(\partial) f_k = 0\,,$$

in contradiction with our assumption that the fields do not satisfy such field equations. In order for qp to be non-singular we must have $N_F \geq N_B$, and in order for pq to be non-singular we must have $N_B \geq N_F$, so we can conclude that $N_B = N_F$. Also the square matrices q and p must both be non-singular, so the complex conjugate of Eq. (26.2.38) tells us that $f^* = q^{*-1}qf$, so that the number of independent fermion fields is N_F rather than $2N_F$, and is therefore equal to the number N_B of independent boson fields, as was to be shown.

26.3 Chiral and Linear Superfields

In the previous section we found that the presence of D and λ components in a general superfield stood in the way of using such superfields in a physically satisfactory Lagrangian density. Suppose then that we consider a superfield with

$$\lambda = D = 0\,. \tag{26.3.1}$$

Are these conditions preserved by supersymmetry transformations? According to Eqs. (26.2.17) and (26.2.16), the condition $D = 0$ is invariant if $\lambda = 0$, but the condition $\lambda = 0$ is only invariant if we also impose the condition that $\partial_\mu V_\nu - \partial_\nu V_\mu = 0$, which requires that V_μ be a pure gauge:

$$V_\mu(x) = \partial_\mu Z(x)\,. \tag{26.3.2}$$

Eq. (26.2.15) shows that, with $\lambda = 0$, this condition is preserved by a supersymmetry transformation. We have thus arrived at a reduced superfield, subject to the constraints (26.3.1) and (26.3.2), with component fields having the transformation properties

$$\delta C = i\left(\bar{\alpha}\gamma_5\,\omega\right), \tag{26.3.3}$$

$$\delta\omega = \left(-i\gamma_5\,\not{\partial}C - M + i\gamma_5 N + \not{\partial}Z\right)\alpha\,, \tag{26.3.4}$$

$$\delta M = -\left(\bar{\alpha}\,\not{\partial}\omega\right), \tag{26.3.5}$$

$$\delta N = i\left(\bar{\alpha}\gamma_5\,\not{\partial}\omega\right), \tag{26.3.6}$$

$$\delta Z = \left(\bar{\alpha}\,\omega\right). \tag{26.3.7}$$

Comparing with Eq. (26.1.21), we see that this is the same as the supermultiplet constructed by direct methods in Section 26.1, with the identifications

$$C = A\,, \qquad \omega = -i\gamma_5\psi\,, \qquad M = G\,, \qquad N = -F\,, \qquad Z = B\,. \tag{26.3.8}$$

A superfield satisfying the conditions (26.3.1) and (26.3.2) is said to be *chiral.**

To distinguish a chiral superfield $X(x,\theta)$ from the general superfield $S(x,\theta)$ of the previous section, we will use A, B, F, G, and ψ for its components, instead of C, M, N, Z, and ω. By using Eqs. (26.3.1), (26.3.2), and (26.3.8) in Eq. (26.2.10), we find the form of a general chiral superfield to be

$$\begin{aligned}X(x,\theta) = {} & A(x) - \left(\bar{\theta}\,\psi(x)\right) + \frac{1}{2}\left(\bar{\theta}\,\theta\right)F(x) - \frac{i}{2}\left(\bar{\theta}\,\gamma_5\,\theta\right)G(x) \\ & + \frac{i}{2}\left(\bar{\theta}\,\gamma_5\,\gamma_\mu\,\theta\right)\partial^\mu B(x) + \frac{1}{2}\left(\bar{\theta}\,\gamma_5\,\theta\right)\left(\bar{\theta}\gamma_5\,\partial\!\!\!/\psi(x)\right) \\ & - \frac{1}{8}\left(\bar{\theta}\,\gamma_5\,\theta\right)^2\Box A(x)\,. \end{aligned} \tag{26.3.9}$$

(We could just as well have taken $C = -B$, $\omega = \psi$, $M = -F$, $N = -G$, and $Z = A$. We make the identifications (26.3.8) because, as we see here, for a scalar superfield they are consistent with the usual convention that A and F are scalars while B and G are pseudoscalars.)

The chiral superfield (26.3.9) may be further decomposed, as

$$X(x,\theta) = \frac{1}{\sqrt{2}}\left[\Phi(x,\theta) + \tilde{\Phi}(x,\theta)\right], \tag{26.3.10}$$

where

$$\begin{aligned}\Phi(x,\theta) = {} & \phi(x) - \sqrt{2}\left(\bar{\theta}\psi_L(x)\right) + \mathscr{F}(x)\left(\bar{\theta}\left(\frac{1+\gamma_5}{2}\right)\theta\right) + \frac{1}{2}\left(\bar{\theta}\gamma_5\gamma_\mu\theta\right)\partial^\mu\phi(x) \\ & - \frac{1}{\sqrt{2}}\left(\bar{\theta}\gamma_5\theta\right)\left(\bar{\theta}\,\partial\!\!\!/\psi_L(x)\right) - \frac{1}{8}\left(\bar{\theta}\gamma_5\theta\right)^2\Box\phi(x)\,, \end{aligned} \tag{26.3.11}$$

$$\begin{aligned}\tilde{\Phi}(x,\theta) = {} & \tilde{\phi}(x) - \sqrt{2}\left(\bar{\theta}\psi_R(x)\right) + \tilde{\mathscr{F}}(x)\left(\bar{\theta}\left(\frac{1-\gamma_5}{2}\right)\theta\right) - \frac{1}{2}\left(\bar{\theta}\gamma_5\gamma_\mu\theta\right)\partial^\mu\tilde{\phi}(x) \\ & + \frac{1}{\sqrt{2}}\left(\bar{\theta}\gamma_5\theta\right)\left(\bar{\theta}\,\partial\!\!\!/\psi_R(x)\right) - \frac{1}{8}\left(\bar{\theta}\gamma_5\theta\right)^2\Box\tilde{\phi}(x)\,, \end{aligned} \tag{26.3.12}$$

with component fields defined by

$$\phi \equiv \frac{A+iB}{\sqrt{2}}\,, \qquad \psi_L \equiv \left(\frac{1+\gamma_5}{2}\right)\psi\,, \qquad \mathscr{F} \equiv \frac{F-iG}{\sqrt{2}}\,, \tag{26.3.13}$$

* Some authors use the term 'chiral' to describe a special case of such superfields, introduced below, which are here called left-chiral or right-chiral. Our use here of the term chiral may at first seem strange, because it has no counterpart for Dirac spinors. Any Dirac spinor is the sum of Dirac spinors that are left-chiral and right-chiral, in the sense that they are respectively proportional to $1+\gamma_5$ and $1-\gamma_5$, so no special term is needed for such sums of Dirac spinors. In contrast, it is only superfields satisfying Eqs. (26.3.1) and (26.3.2) that can be expressed as the sum of a left-chiral and a right-chiral superfield.

$$\tilde{\phi} \equiv \frac{A - iB}{\sqrt{2}}\,, \qquad \psi_R \equiv \left(\frac{1-\gamma_5}{2}\right)\psi\,, \qquad \tilde{\mathscr{F}} \equiv \frac{F + iG}{\sqrt{2}}\,. \tag{26.3.14}$$

The component fields of either Φ or $\tilde{\Phi}$ furnish complete representations of the supersymmetry algebra:

$$\delta\psi_L = \sqrt{2}\partial_\mu\phi\,\gamma^\mu\,\alpha_R + \sqrt{2}\mathscr{F}\,\alpha_L\,, \tag{26.3.15}$$

$$\delta\mathscr{F} = \sqrt{2}\left(\overline{\alpha_L}\,\partial\!\!\!/\psi_L\right), \tag{26.3.16}$$

$$\delta\phi = \sqrt{2}\left(\overline{\alpha_R}\psi_L\right), \tag{26.3.17}$$

$$\delta\psi_R = \sqrt{2}\partial_\mu\tilde{\phi}\,\gamma^\mu\,\alpha_L + \sqrt{2}\tilde{\mathscr{F}}\,\alpha_R\,, \tag{26.3.18}$$

$$\delta\tilde{\mathscr{F}} = \sqrt{2}\left(\overline{\alpha_R}\,\partial\!\!\!/\psi_R\right), \tag{26.3.19}$$

$$\delta\tilde{\phi} = \sqrt{2}\left(\overline{\alpha_L}\psi_R\right), \tag{26.3.20}$$

where as usual

$$\alpha_L = \left(\frac{1+\gamma_5}{2}\right)\alpha\,, \qquad \alpha_R = \left(\frac{1-\gamma_5}{2}\right)\alpha\,,$$

and likewise for θ. A superfield of the form (26.3.11) or (26.3.12) is known as *left-chiral* or *right-chiral*, respectively. In the special case where a chiral superfield $X(x,\theta)$ is *real*, its left-chiral and right-chiral parts Φ and $\tilde{\Phi}$ are complex conjugates, so that $\tilde{\phi} = \phi^*$, $\tilde{\mathscr{F}} = \mathscr{F}^*$, and ψ is a Majorana field. However, if we do not require $X(x,\theta)$ to be real then in general there is no relation between Φ and $\tilde{\Phi}$; it is even possible that one of the two vanishes.

The component fields of the superfield Φ include two complex bosonic components ϕ and $\mathscr{F}$, or four independent real bosonic components, and one Majorana fermion field ψ, which has four independent fermionic components. This is another example of the general result, derived at the end of the previous section, that any set of fields that furnish a representation of the supersymmetry algebra must have an equal number of independent bosonic and fermionic components.

We can use Eqs. (26.A.5), (26.A.17), and (26.A.18) to rewrite Eqs. (26.3.11) and (26.3.12) in a form that clarifies the way that these superfields depend on θ_L and θ_R:

$$\Phi(x,\theta) = \phi(x_+) - \sqrt{2}\left(\theta_L^{\mathrm{T}}\epsilon\,\psi_L(x_+)\right) + \mathscr{F}(x_+)\left(\theta_L^{\mathrm{T}}\epsilon\,\theta_L\right), \tag{26.3.21}$$

$$\tilde{\Phi}(x,\theta) = \tilde{\phi}(x_-) + \sqrt{2}\left(\theta_R^{\mathrm{T}}\epsilon\,\psi_R(x_-)\right) - \tilde{\mathscr{F}}(x_-)\left(\theta_R^{\mathrm{T}}\epsilon\,\theta_R\right), \tag{26.3.22}$$

where

$$x_\pm^\mu \equiv x^\mu \pm \tfrac{1}{2}\left(\bar{\theta}\gamma_5\gamma^\mu\theta\right) = x^\mu \pm \left(\theta_R^{\mathrm{T}}\epsilon\gamma^\mu\theta_L\right). \tag{26.3.23}$$

The expansions of $\phi(x_+)$ and $\tilde{\phi}(x_-)$ in powers of $x^\mu - x^\mu_\pm$ terminate with quadratic terms, the expansions of $\psi_{L,R}(x_\pm)$ terminate with linear terms, while the expansions of $\mathscr{F}(x_+)$ and $\tilde{\mathscr{F}}(x_-)$ terminate at zeroth order, because all higher terms make contributions in Eqs. (26.3.21) and (26.3.22) that contain three or more factors of θ_L or θ_R and therefore vanish. For the same reason, it is easy to see that any superfield that depends only on θ_L and x^μ_+ but not otherwise on θ_R must take the form (26.3.21), and any superfield that depends only on θ_R and x^μ_- but not otherwise on θ_L must take the form (26.3.22).

We have seen that for a superfield to be left-chiral or right-chiral is entirely a matter of what the superfield is allowed to depend on. It follows immediately that *any function of left-chiral superfields (or of right-chiral superfields), but not their complex conjugates or spacetime derivatives, is a left-(or right-)chiral superfield.* This can also be shown in a more formal way. Because $\Phi(x,\theta)$ depends on θ_R only through its dependence on x_+, and $\tilde{\Phi}(x,\theta)$ depends on θ_L only through its dependence on x_-, they satisfy the conditions

$$\mathscr{D}_{R\alpha}\Phi = \mathscr{D}_{L\alpha}\tilde{\Phi} = 0\,, \tag{26.3.24}$$

where $\mathscr{D}_R$ and $\mathscr{D}_L$ are the right- and left-handed parts of the superderivative (26.2.26):

$$\mathscr{D}_{R\alpha} \equiv \left[\left(\frac{1-\gamma_5}{2}\right)\mathscr{D}\right]_\alpha = -\sum_\beta \epsilon_{\alpha\beta}\frac{\partial}{\partial\theta_{R\beta}} - (\gamma^\mu\theta_L)_\alpha\frac{\partial}{\partial x^\mu}\,, \tag{26.3.25}$$

$$\mathscr{D}_{L\alpha} \equiv \left[\left(\frac{1+\gamma_5}{2}\right)\mathscr{D}\right]_\alpha = +\sum_\beta \epsilon_{\alpha\beta}\frac{\partial}{\partial\theta_{L\beta}} - (\gamma^\mu\theta_R)_\alpha\frac{\partial}{\partial x^\mu}\,, \tag{26.3.26}$$

for which

$$\mathscr{D}_{R\alpha}x^\mu_+ = \mathscr{D}_{L\alpha}x^\mu_- = 0\,.$$

Conversely, if a superfield Φ satisfies $\mathscr{D}_R\Phi = 0$ it is left-chiral, and if it satisfies $\mathscr{D}_L\Phi = 0$ it is right-chiral. Any function $f(\Phi)$ of superfields Φ_n that all satisfy $\mathscr{D}_R\Phi_n = 0$ or all satisfy $\mathscr{D}_L\Phi_n = 0$ will satisfy $\mathscr{D}_R f(\Phi) = 0$ or $\mathscr{D}_L f(\Phi) = 0$, and hence be left-chiral or right-chiral, respectively. But a function of left-chiral *and* right-chiral superfields is in general not chiral at all.

Using the representation (26.3.21) for left-chiral superfields makes it easy to work out their multiplication properties. For instance, if Φ_1 and Φ_2 are two left-chiral superfields, then their product $\Phi = \Phi_1\Phi_2$ is a left-chiral superfield, with components

$$\phi = \phi_1\phi_2\,, \tag{26.3.27}$$

$$\psi_L = \phi_1\psi_{2L} + \phi_2\psi_{1L}\,, \tag{26.3.28}$$

$$\mathscr{F} = \phi_1 \mathscr{F}_2 + \phi_2 \mathscr{F}_1 - \left(\psi_{1L}^{\mathrm{T}}\, \epsilon\, \psi_{2L}\right) . \tag{26.3.29}$$

The presence of chiral superfields in a theory opens up an additional possibility for constructing supersymmetric actions. Inspection of the transformation rule (26.3.16) shows that a supersymmetry transformation changes the $\mathscr{F}$-term of a left-chiral superfield Φ by a derivative, so that the integral of the $\mathscr{F}$-term of any left-chiral superfield is supersymmetric. Thus we can form a supersymmetric action as

$$I = \int d^4x \left[f\right]_{\mathscr{F}} + \int d^4x \left[f\right]_{\mathscr{F}}^{*} + \frac{1}{2}\int d^4x \left[K\right]_D , \tag{26.3.30}$$

where f and K are any left-chiral superfield and general real superfield, respectively, formed from the elementary superfields.

On what can f and K depend? The function f will be left-chiral if it depends only on left-chiral elementary superfields Φ_n, but not their right-chiral complex conjugates. On the other hand, the superderivative of a chiral superfield is not chiral, so we cannot freely include superderivatives of the Φ_n in f. It is true that, acting on a superfield S that is not left-chiral (such as one involving complex conjugates of left-chiral superfields), a pair of right superderivatives gives a left-chiral superfield, because there are only two independent right superderivatives and they anticommute:

$$\mathscr{D}_{R\alpha}(\mathscr{D}_{R\beta}\mathscr{D}_{R\gamma}S) = 0 .$$

However, the $\mathscr{F}$-term of any function f that is constructed in this way makes a contribution to the action that is the same as that of a D term of some other composite superfield. Since the $\mathscr{D}$s anticommute, the most general left-chiral superfield formed by acting on a general superfield S with two $\mathscr{D}_R$s may be expressed in terms of $(\mathscr{D}_R^{\mathrm{T}}\epsilon\mathscr{D}_R)S$. If one of the left-chiral superfields in the superpotential is of this form, then since each $\mathscr{D}_R$ annihilates all the other superfields in the superpotential, we can write the whole superpotential as $f = (\mathscr{D}_R^{\mathrm{T}}\epsilon\mathscr{D}_R)h$ for some other superfield h. Now,

$$\left(\mathscr{D}_R^{\mathrm{T}}\epsilon\mathscr{D}_R\right)\left(\theta_R^{\mathrm{T}}\epsilon\theta_R\right) = -4 ,$$

so, apart from spacetime derivatives that do not contribute to the action, $(\mathscr{D}_R^{\mathrm{T}}\epsilon\mathscr{D}_R)h$ is the coefficient of $-(\theta_R^{\mathrm{T}}\epsilon\theta_R)/4$ in h. But, again apart from spacetime derivatives, $[f]_{\mathscr{F}}$ is the coefficient of $(\theta_L^{\mathrm{T}}\epsilon\theta_L)$ in f, so $[(\mathscr{D}_R^{\mathrm{T}}\epsilon\mathscr{D}_R)h]_{\mathscr{F}}$ equals the coefficient of $-(\theta_L^{\mathrm{T}}\epsilon\theta_L)(\theta_R^{\mathrm{T}}\epsilon\theta_R)/4 = -(\bar{\theta}\gamma_5\theta)^2/4$ in h, and therefore

$$\int d^4x\, [(\mathscr{D}_R^{\mathrm{T}}\epsilon\mathscr{D}_R)h]_{\mathscr{F}} = 2\int d^4x\, [h]_D . \tag{26.3.31}$$

Thus we do not need to include terms in f that depend on left-chiral superfields of the form $\mathscr{D}_{R\beta}\mathscr{D}_{R\gamma}S$ — any such terms will be included in the list of all possible D-terms. When f is expressed as a function only of elementary left-chiral superfields and not their superderivatives or spacetime derivatives, it is known as the *superpotential.*

In contrast, the function K is in general a real scalar function of both left-chiral superfields Φ_n and their right-chiral complex conjugates Φ_n^*, as well as their superderivatives and spacetime derivatives, known as the *Kahler potential.* (Any right-chiral superfield is the complex conjugate of a left-chiral superfield, so there is no loss of generality in supposing K to depend only on left-chiral superfields and their complex conjugates.) However, not all Ks obtained in this way will yield distinct actions. For one thing, chiral superfields have no D-terms, so two Ks that differ by a chiral superfield make the same contribution to the action.

It is also possible to change the form of K without changing the action by a partial integration in superspace. The D-term of the superderivative $\mathscr{D}_\alpha S$ of an arbitrary superfield makes no contribution to the action because

$$\int d^4x\, [\mathscr{D}_\alpha S]_D = 0\,. \tag{26.3.32}$$

To see this, recall that

$$\mathscr{D}_\alpha S = \sum_\beta \Big(\gamma_5\epsilon\Big)_{\alpha\beta}\frac{\partial S}{\partial\theta_\beta} - (\gamma^\mu\theta)_\alpha\frac{\partial S}{\partial x^\mu}\,.$$

Since S is a polynomial at most of fourth order in θ, the first term in $\mathscr{D}_\alpha S$ is a polynomial in θ at most of third order and therefore cannot have a non-zero D-term that is not a derivative, while the second term is also a spacetime derivative, so that its D-term is also a spacetime derivative, and therefore neither the first nor the second terms in $\mathscr{D}_\alpha S$ can contribute to the integral in Eq. (26.3.32). Also, the superderivative acts distributively, so it follows from Eq. (26.3.32) that we can integrate by parts in superspace: for any two bosonic superfields S_1 and S_2,

$$\int d^4x\, [S_1\mathscr{D}_\alpha S_2]_D = -\int d^4x\, [S_2\mathscr{D}_\alpha S_1]_D\,. \tag{26.3.33}$$

In Sections 26.4 and 26.8 we will consider in detail the case where f and K depend only on elementary superfields, but not their superderivatives or ordinary derivatives.

We saw in the previous section that in theories in which parity is conserved, the effect of the space inversion operator on a general scalar superfield is to subject its arguments to the transformations $x^\mu \to (\Lambda_P)^\mu{}_\nu x^\nu$ and $\theta \to -i\beta\theta$, and perhaps to multiply the superfield by a phase η. Under these transformations, the arguments $x_\pm^\mu$ in Eqs. (26.3.21) and (26.3.22) are changed by

$$x_\pm^\mu \to (\Lambda_P x)^\mu \pm \tfrac{1}{2}\Big(\bar\theta\beta\gamma_5\gamma^\mu\beta\theta\Big) = (\Lambda_P x_\mp)^\mu\,, \tag{26.3.34}$$

while $\theta_L \to -i\beta\theta_R$ and $\theta_R \to -i\beta\theta_L$. Thus space inversion takes left-chiral superfields into right-chiral superfields, and vice-versa. The only right-handed scalar superfield whose component fields involve creation and annihilation operators for the same particles that are created and destroyed by a left-handed scalar superfield Φ is $\tilde{\Phi} \propto \Phi^*$, so $\mathsf{P}^{-1}\Phi\,\mathsf{P}$ must be proportional to Φ^*. By a suitable choice of phase of Φ, we can arrange that this transformation rule reads

$$\mathsf{P}^{-1}\Phi(x,\theta)\,\mathsf{P} = \Phi^*(\Lambda_P x, -i\beta\theta)\,. \tag{26.3.35}$$

In terms of component fields, this transformation is

$$\begin{aligned}\mathsf{P}^{-1}\phi(x)\,\mathsf{P} &= \phi^*(\Lambda_P x)\,,\\ \mathsf{P}^{-1}\psi_L(x)\,\mathsf{P} &= -i\epsilon\gamma_5\beta\psi_L^*(\Lambda_P x)\,,\\ \mathsf{P}^{-1}\mathscr{F}(x)\,\mathsf{P} &= \mathscr{F}^*(\Lambda_P x)\,.\end{aligned} \tag{26.3.36}$$

There is another type of possible symmetry, known as *R-symmetry*, that is important in some models of spontaneous supersymmetry breaking discussed in Section 26.5, and that will also be used in proving no-renormalization theorems in Section 27.6. As mentioned in Section 25.2, in theories of simple N=1 supersymmetry an R-symmetry is invariance under a $U(1)$ transformation under which the left-handed components of the supersymmetry generator (called Q_a in Section 25.2) carry a non-vanishing quantum number, say -1, in which case their adjoints, the right-handed components of the supersymmetry generator carry the opposite quantum number $+1$. Inspection of Eq. (26.2.2) shows that the θ superspace coordinate has a non-trivial transformation property under R-symmetry transformations: θ_L carries R-quantum number $+1$ and θ_R, which is proportional to θ_L^*, carries R quantum number -1. In addition, the whole superfield may be given an R quantum number. If we give a left-chiral superfield Φ the R quantum number R_Φ, then its scalar component ϕ has the same R quantum number, while the left-handed spinor component ψ_L has $R_\psi = R_\Phi - 1$, and the auxiliary field $\mathscr{F}$ has $R_\mathscr{F} = R_\Phi - 2$. In particular, in order for the superpotential term $\int d^4x\,[f]_\mathscr{F}$ to conserve R, the superpotential itself must have $R_f = +2$, so if f depends on a single left-chiral superfield Φ, then it must be proportional to Φ^{2/R_Φ}. To put this another way, if $f(\Phi)$ is a pure mass term proportional to Φ^2 then we must choose $R_\Phi = +1$, while if $f(\Phi)$ is a pure interaction term proportional to Φ^3, then we must choose $R_\Phi = 2/3$. On the other hand, inspection of Eq. (26.2.10) shows that the D-term of a superfield has the same R value as the superfield, so in order for the term $\int d^4x[g]_D$ in the action to conserve R it is only necessary that K have $R = 0$, which will be the case if each term in K contains equal numbers of Φ and Φ^* factors, whatever R value we give to Φ. Of course, there is no general reason why R-symmetry *should* be respected by the action, or why it should not be spontaneously broken.

* * *

There are other ways of constraining superfields, to yield other types of supermultiplets of fields. Among the more common are the *linear* superfields. To learn the conditions defining this sort of superfield, we note that if S is a general superfield, then we can form a chiral superfield

$$S' \equiv \frac{1}{4}\left(\bar{\mathscr{D}}\mathscr{D}\right) S\,. \tag{26.3.37}$$

This is a chiral superfield because it can be written as a sum of $\frac{1}{4}(\bar{\mathscr{D}}_L\mathscr{D}_L)\,S$, which is right-chiral, and $\frac{1}{4}(\bar{\mathscr{D}}_R\mathscr{D}_R)\,S$, which is left-chiral. Its components are given in terms of the components of S by

$$C' = N\,, \tag{26.3.38}$$

$$\omega' = \lambda + \not{\partial}\omega\,, \tag{26.3.39}$$

$$M' = -\partial_\mu V^\mu\,, \tag{26.3.40}$$

$$N' = D + \Box C\,, \tag{26.3.41}$$

$$V'_\mu = -\partial_\mu M\,, \tag{26.3.42}$$

$$\lambda' = D' = 0\,. \tag{26.3.43}$$

A multiplet S is said to be linear if the superfield S' defined in this way vanishes

$$\left(\bar{\mathscr{D}}\mathscr{D}\right) S = 0\,, \tag{26.3.44}$$

or in terms of its components,

$$N = M = \partial_\mu V^\mu = 0\,, \quad \lambda = -\not{\partial}\omega\,, \quad D = -\Box C\,. \tag{26.3.45}$$

This leaves four independent bosonic fields, C and the three components of V_μ subject to the condition $\partial_\mu V^\mu = 0$, and four independent fermionic fields, the components of the Majorana four-spinor ω. We will see in Section 26.6 that the current superfields, whose V_μ-terms are the conserved currents associated with symmetry transformations, are linear superfields.

26.4 Renormalizable Theories of Chiral Superfields

We will now work out the details of a general renormalizable theory of scalar chiral superfields. This will provide some insight into the implications of supersymmetry, and the theory we obtain will be part of the supersymmetric standard model to be discussed in Chapter 28.

As discussed in Section 12.2, the Lagrangian density in a renormalizable theory can contain only operators with dimensionality (counting powers of energy or momentum, with $\hbar = c = 1$) four or less. Eq. (26.2.6) shows that $\mathscr{Q}_\alpha$ and hence $\partial/\partial\theta_\alpha$ has dimensionality 1/2, so $\mathscr{D}_\alpha$ has dimensionality +1/2, and θ_α has dimensionality −1/2. The $\mathscr{F}$- and D-terms of a superfield S are the coefficients of two or four factors of θ, respectively, so if the superfield has dimensionality $d(S)$ then its $\mathscr{F}$- and D-terms have dimensionalities $d(\mathscr{F}^S) = d(S) + 1$ and $d(D^S) = d(S) + 2$. Thus in a renormalizable theory the functions f and K in Eq. (26.3.30) consist of operators with dimensionality at most three and two, respectively.

The dimensionality of an elementary scalar superfield Φ_n is that of an elementary scalar field, or +1, so in order for each term in the function f to have dimensionality three or less, it can contain at most three factors of Φ_n and/or derivatives $\partial/\partial x^\mu$ and/or pairs of spinor superderivatives $\mathscr{D}_\alpha$. As discussed in the previous section, any left-chiral term in f that involves superderivatives could be replaced with a term in K, so superderivatives can be omitted in f. Eq. (26.2.30) shows that spacetime derivatives can be expressed in terms of superderivatives, so these too can be omitted. (In any case, Lorentz invariance would rule out terms with one spacetime derivative, and terms with two derivatives in a renormalizable theory could involve only one Φ_n factor, on which these derivatives would have to act, so such terms would not contribute to the action.) We conclude that $f(\Phi)$ is at most a cubic polynomial in the Φ_n, without spacetime derivatives or superderivatives.

The same dimensional analysis shows that in a renormalizable theory K is at most a quadratic function of Φ_n and Φ_n^*, without derivatives. But any term in $K(\Phi, \Phi^*)$ that involves only Φ_n or only Φ_n^* would be a chiral superfield, and chiral superfields by definition have no D-terms, so $[K(\Phi, \Phi^*)]_D$ receives contributions only from terms in $K(\Phi, \Phi^*)$ that involve *both* Φ_n and Φ_n^*. Thus $K(\Phi, \Phi^*)$ must be of the form

$$K(\Phi, \Phi^*) = \sum_{mn} g_{nm}\, \Phi_n^* \Phi_m \,, \tag{26.4.1}$$

with constant coefficients g_{nm} forming a Hermitian matrix.

We must now calculate the $\mathscr{F}$- and D-components of $f(\Phi)$ and $K(\Phi, \Phi^*)$, respectively. To calculate the D-component of $K(\Phi, \Phi^*)$, we note that the term in $\Phi_n^* \Phi_m$ of fourth order in θ is

$$\begin{aligned}\left[\Phi_n^* \Phi_m\right]_{\theta^4} = &-\frac{1}{8}\left(\bar\theta\gamma_5\theta\right)^2 \left[\phi_n^* \Box \phi_m + \left(\Box\phi_m^*\right)\phi_n\right] \\ &+\left(\bar\theta\gamma_5\theta\right)\left[\left(\overline{\psi_n}\,\theta\right)\left(\bar\theta\gamma^\mu\partial_\mu\psi_m\right) + \left(\left(\partial_\mu\overline{\psi_n}\right)\gamma^\mu\theta\right)\left(\bar\theta\,\psi_m\right)\right] \\ &+\frac{1}{4}\mathscr{F}_n^*\mathscr{F}_m\left(\bar\theta(1-\gamma_5)\theta\right)\left(\bar\theta(1+\gamma_5)\theta\right) \\ &-\frac{1}{4}\partial^\mu\phi_n^*\,\partial^\nu\phi_m\left(\bar\theta\gamma_5\gamma_\mu\theta\right)\left(\bar\theta\gamma_5\gamma_\nu\theta\right) .\end{aligned}$$

Using (26.A.18) and (26.A.19) allows us to convert the θ dependence of this expression to an over-all factor $(\bar{\theta}\gamma_5\theta)^2$:

$$\left[\Phi_n^*\Phi_m\right]_{\theta^4} = -\frac{1}{4}\left(\bar{\theta}\gamma_5\theta\right)^2\left[\frac{1}{2}\phi_n^*\Box\phi_m + \frac{1}{2}\left(\Box\phi_m^*\right)\phi_n - \left(\overline{\psi_n}\gamma^\mu\partial_\mu\psi_m\right)\right.$$
$$\left. + \left((\partial_\mu\overline{\psi_n})\gamma^\mu\psi_m\right) + 2\mathscr{F}_n^*\mathscr{F}_m - \partial^\mu\phi_n^*\partial_\mu\phi_m\right].$$

The D-term of a superfield is the coefficient of $-\frac{1}{4}\left(\bar{\theta}\gamma_5\theta\right)^2$ minus $\frac{1}{2}\Box$ acting on the θ-independent term, which for $\Phi_n^*\Phi_m$ is $\phi_n^*\phi_m$, so

$$\frac{1}{2}\left[K(\Phi,\Phi^*)\right]_D = \sum_{nm} g_{nm}\left[-\partial_\mu\phi_n^*\partial^\mu\phi_m + \mathscr{F}_n^*\mathscr{F}_m\right.$$
$$\left. -\frac{1}{2}\left(\overline{\psi_{nL}}\gamma^\mu\partial_\mu\psi_{mL}\right) + \frac{1}{2}\left(\partial_\mu(\overline{\psi_{nL}})\gamma^\mu\psi_{mL}\right)\right]. \quad (26.4.2)$$

If we write Φ_n as a linear combination $\sum_m N_{nm}\Phi'_m$ of new superfields Φ'_m, then $K(\Phi,\Phi^*)$ is given in terms of the new superfields by a formula which is the same as Eq. (26.4.1), except that g_{nm} is replaced with $g'_{nm} = (N^\dagger g N)_{nm}$. In order for the kinematic terms for the scalar and spinor fields to have a sign consistent with the quantum commutation and anticommutation relations, it is necessary that the Hermitian matrix g_{nm} be positive-definite and, as shown in Section 12.5, this means that we can choose N so that $g'_{nm} = \delta_{nm}$. Dropping primes, the term (26.4.2) is now

$$\frac{1}{2}\left[K(\Phi,\Phi^*)\right]_D = \sum_{n}\left[-\partial_\mu\phi_n^*\partial^\mu\phi_n + \mathscr{F}_n^*\mathscr{F}_n\right.$$
$$\left. -\frac{1}{2}\left(\overline{\psi_{nL}}\gamma^\mu\partial_\mu\psi_{nL}\right) + \frac{1}{2}\left(\partial_\mu(\overline{\psi_{nL}})\gamma^\mu\psi_{nL}\right)\right]. \quad (26.4.3)$$

We can still use a unitary transformation to redefine the superfields without changing the form of Eq. (26.4.3), a freedom we will need to exercise shortly.

The terms in Eq. (26.4.3) involving ϕ_n and ψ_{nL} are the correct kinematic Lagrangians for conventionally normalized complex scalar and Majorana spinor fields. We will rewrite the fermion terms in a more familiar form after we have had a chance to consider mass terms.

To calculate the $\mathscr{F}$-term of $f(\Phi)$, it is most convenient to use the representation (26.3.21) of the superfields, and pick out the term of second order in θ_L:

$$\Big[f\Big(\Phi(x,\theta)\Big)\Big]_{\theta_L^2} = \sum_{nm}\Big(\theta_L^{\mathrm{T}}\epsilon\,\psi_{nL}(x)\Big)\Big(\theta_L^{\mathrm{T}}\epsilon\,\psi_{mL}(x)\Big)\frac{\partial^2 f\Big(\phi(x)\Big)}{\partial\phi_n(x)\,\partial\phi_m(x)}$$
$$+\sum_n \mathscr{F}_n(x)\,\frac{\partial f\Big(\phi(x)\Big)}{\partial\phi_n(x)}\Big(\theta_L^{\mathrm{T}}\epsilon\theta_L\Big)\,.$$

(We have replaced x_+ here with x because the term $(\theta_R^{\mathrm{T}}\epsilon\gamma^\mu\theta_L)$ in Eq. (26.3.21) vanishes when multiplied by an expression with two θ_L factors.) The θ dependence of the first term on the right-hand side can be put into a standard form by using Eq. (26.A.11) to write*

$$\Big(\theta_L^{\mathrm{T}}\epsilon\,\psi_{nL}\Big)(\theta_L^{\mathrm{T}}\epsilon\,\psi_{mL}\Big) = \left(\psi_{nL}^{\mathrm{T}}\epsilon\left(\frac{1+\gamma_5}{2}\right)\theta\right)\left(\theta^{\mathrm{T}}\epsilon\left(\frac{1+\gamma_5}{2}\right)\psi_{mL}\right)$$
$$= -\frac{1}{2}\Big(\overline{\psi}_{nL}\,\psi_{mL}\Big)\Big(\theta_L^{\mathrm{T}}\epsilon\,\theta_L\Big)\,.$$

The $\mathscr{F}$-term of any left-chiral superfield is the coefficient of $(\theta_L^{\mathrm{T}}\epsilon\theta_L)$, so here

$$\Big[f(\Phi)\Big]_{\mathscr{F}} = -\frac{1}{2}\sum_{nm}\frac{\partial^2 f(\phi_n)}{\partial\phi_n\,\partial\phi_m}\Big(\overline{\psi}_{nL}\,\psi_{mL}\Big) + \sum_n \mathscr{F}_n\,\frac{\partial f(\phi)}{\partial\phi_n}\,. \tag{26.4.4}$$

The complete Lagrangian density is the sum of the terms (26.4.3), (26.4.4), and the complex conjugate of (26.4.4):

$$\mathscr{L} = \sum_n\Bigg[-\partial_\mu\phi_n^*\,\partial^\mu\phi_n + \mathscr{F}_n^*\mathscr{F}_n$$
$$-\frac{1}{2}\Big(\overline{\psi_{nL}}\,\gamma^\mu\partial_\mu\,\psi_{nL}\Big) + \frac{1}{2}\Big((\partial_\mu\overline{\psi_{nL}})\,\gamma^\mu\,\psi_{nL}\Big)\Bigg]$$
$$-\frac{1}{2}\sum_{nm}\frac{\partial^2 f(\phi)}{\partial\phi_n\,\partial\phi_m}\Big(\overline{\psi}_{nL}\,\psi_{mL}\Big) - \frac{1}{2}\sum_{nm}\left(\frac{\partial^2 f(\phi)}{\partial\phi_n\,\partial\phi_m}\right)^*\Big(\overline{\psi}_{nL}\,\psi_{mL}\Big)^*$$
$$+\sum_n \mathscr{F}_n\,\frac{\partial f(\phi)}{\partial\phi_n} + \sum_n \mathscr{F}_n^*\left(\frac{\partial f(\phi)}{\partial\phi_n}\right)^*\,. \tag{26.4.5}$$

The auxiliary fields $\mathscr{F}_n$ enter quadratically in the action, with constant coefficients for the second-order terms, so they can be eliminated by setting $\mathscr{F}_n$ equal to the value at which the Lagrangian density (26.4.5) is stationary with respect to $\mathscr{F}_n$ and $\mathscr{F}_n^*$:

$$\mathscr{F}_n = -\left(\frac{\partial f(\phi)}{\partial\phi_n}\right)^*\,. \tag{26.4.6}$$

* Note that $\overline{\psi}_{nL}$ is the left-handed component of $\overline{\psi}_n$, not $\overline{\psi_{nL}}$.

Inserting this in Eq. (26.4.5) gives

$$\mathscr{L} = \sum_n \Big[-\partial_\mu \phi_n^* \partial^\mu \phi_n - \frac{1}{2}\Big(\overline{\psi_{nL}}\gamma^\mu \partial_\mu \psi_{nL}\Big) + \frac{1}{2}\Big((\partial_\mu \overline{\psi_{nL}})\gamma^\mu \psi_{nL}\Big)\Big]$$
$$-\frac{1}{2}\sum_{nm} \frac{\partial^2 f(\phi)}{\partial\phi_n \partial\phi_m}\Big(\overline{\psi}_{nL}\,\psi_{mL}\Big) - \frac{1}{2}\sum_{nm}\left(\frac{\partial^2 f(\phi)}{\partial\phi_n \partial\phi_m}\right)^* \Big(\overline{\psi}_{nL}\,\psi_{mL}\Big)^*$$
$$-\sum_n \left(\frac{\partial f(\phi)}{\partial\phi_n}\right)^* \frac{\partial f(\phi)}{\partial \phi_n}\,. \tag{26.4.7}$$

Thus the scalar field potential is $V(\phi) = \sum_n |\partial f(\phi)/\partial\phi_n|^2$.

With the auxiliary fields eliminated in this way, the action is no longer invariant under the supersymmetry transformations (26.3.15), (26.3.17) of the remaining fields ψ_{nL} and ϕ_n:

$$\delta\psi_{nL} = \sqrt{2}\partial_\mu\phi_n\gamma^\mu\alpha_R - \sqrt{2}\left(\frac{\partial f(\phi)}{\partial\phi_n}\right)^* \alpha_L\,, \qquad \delta\phi_n = \sqrt{2}\Big(\overline{\alpha_R}\psi_{nL}\Big)\,.$$

This is because the expression (26.4.6) does not obey the transformation rule $\delta\mathscr{F}_n = \sqrt{2}(\overline{\alpha_L}\,\partial\!\!\!/\psi_{nL})$ for $\mathscr{F}_n$ given by Eq. (26.3.16), but instead

$$\delta\left(-\frac{\partial f(\phi)}{\partial\phi_n}\right)^* = -\sum_m \left(\frac{\partial^2 f(\phi)}{\partial\phi_n\partial\phi_m}\right)^* \delta\phi_m^* = -\sqrt{2}\sum_m \left(\frac{\partial^2 f(\phi)}{\partial\phi_n \partial\phi_m}\right)^* \Big(\overline{\alpha_L}\psi_{mR}\Big)\,.$$

For the same reason, after eliminating the auxiliary fields the commutators of the supersymmetry transformations of ϕ_n and ψ_{nL} are no longer given by the supersymmetry anticommutation relations, and in fact do not form a closed Lie superalgebra. But this is not inconsistent with the existence of quantum mechanical operators Q_α that satisfy the anticommutation relations of supersymmetry. These operators generate supersymmetry transformations, in the sense that the commutator of $-i(\bar\alpha Q)$ with any Heisenberg-picture quantum field ϕ_n or ψ_{nL} equals the change in that field under a supersymmetry transformation with infinitesimal parameter α. With $\mathscr{F}_n$ given by Eq. (26.4.6) the commutator of $-i(\bar\alpha Q)$ with $\mathscr{F}_n$ *is* given by $\delta\mathscr{F}_n = \sqrt{2}(\overline{\alpha_L}\,\partial\!\!\!/\psi_{nL})$, because in the Heisenberg picture the quantum field ψ_{nL} satisfies the field equation derived from the Lagrangian (26.4.7):

$$\partial\!\!\!/\psi_{nL} = -\sum_m \left(\frac{\partial^2 f(\phi)}{\partial\phi_n\partial\phi_m}\right)^* \psi_{mR}\,.$$

Likewise, the supersymmetry transformations of the quantum fields ϕ_n and ψ_{nL} do form a closed Lie superalgebra when the field equations are taken into account. Such algebras are often called *on-shell*.

The zeroth-order expectation values ϕ_{n0} of the scalar fields ϕ_n must be at the maximum of the last term in Eq. (26.4.7). Since this term is always negative or zero, the maximum will be at spacetime-independent field values ϕ_{n0} at which this term vanishes, so that

$$\left.\frac{\partial f(\phi)}{\partial \phi_n}\right|_{\phi=\phi_0} = 0\,, \tag{26.4.8}$$

provided of course that a solution of this equation exists. Eq. (26.4.8) not only maximizes the last term in Eq. (26.4.7) — it is also the condition for supersymmetry to be unbroken. Invariance of the vacuum under supersymmetry transformations requires that the vacuum expectation value of the change in any field under a supersymmetry transformation should vanish. The change in a bosonic field is a fermionic field, which of course has zero expectation value anyway, but Eq. (26.3.15) shows that the vacuum expectation value of $\delta\psi_{nL}$ is proportional to the vacuum expectation value of the auxiliary field $\mathscr{F}_n$, which therefore must vanish if supersymmetry is unbroken. According to Eq. (26.4.6), in zeroth-order perturbation theory this condition requires that Eq. (26.4.8) must be satisfied. We will see in Section 27.6 that if Eq. (26.4.8) is satisfied then supersymmetry is unbroken to all orders of perturbation theory.

For a single left-chiral scalar superfield Φ, the fundamental theorem of algebra tells us that the polynomial $\partial f(\phi)/\partial\phi$ always has at least one zero somewhere in the complex plane. This is not necessarily true when there is more than one superfield. If we *assume* that there is a solution ϕ_{n0} of Eq. (26.4.8), we can evaluate the physical degrees of freedom of the theory by setting

$$\phi_n = \phi_{n0} + \varphi_n\,, \tag{26.4.9}$$

and expanding in powers of φ_n. The masses of the particles of this theory can be calculated by inspecting the terms of second order in φ and ψ:

$$\begin{aligned}\mathscr{L}_0 = \sum_n \Big[&-\partial_\mu\varphi_n^*\partial^\mu\varphi_n - \frac{1}{2}\Big(\overline{\psi_{nL}}\,\gamma^\mu\partial_\mu\,\psi_{nL}\Big) + \frac{1}{2}\Big(\partial_\mu(\overline{\psi_{nL}})\,\gamma^\mu\,\psi_{nL}\Big)\Big]\\ &-\frac{1}{2}\sum_{nm}\mathscr{M}_{nm}\Big(\overline{\psi}_{nL}\,\psi_{mL}\Big) - \frac{1}{2}\sum_{nm}\mathscr{M}^*_{nm}\Big(\overline{\psi}_{nL}\,\psi_{mL}\Big)^*\\ &-\sum_{nm}\Big(\mathscr{M}^\dagger\mathscr{M}\Big)_{mn}\varphi_m^*\varphi_n\,,\end{aligned} \tag{26.4.10}$$

where $\mathscr{M}$ is the symmetric complex matrix

$$\mathscr{M}_{mn} \equiv \left(\frac{\partial^2 f(\phi)}{\partial\phi_n\partial\phi_m}\right)_{\phi=\phi_0}\,. \tag{26.4.11}$$

Now if we redefine the fields by a unitary transformation

$$\varphi_n = \sum_m \mathscr{U}_{nm}\varphi'_m\,, \qquad \psi_{nL} = \sum_m \mathscr{U}_{nm}\psi'_{mL}\,, \tag{26.4.12}$$

then the free-field Lagrangian (26.4.10) will take the same form, but with $\mathscr{M}$ replaced with $\mathscr{M}'$, where

$$\mathscr{M}' = \mathscr{U}^{\mathrm{T}} \mathscr{M} \mathscr{U} . \tag{26.4.13}$$

According to a theorem of matrix algebra, for any complex symmetric matrix $\mathscr{M}$ it is always possible to find a unitary matrix $\mathscr{U}$ such that the matrix $\mathscr{M}'$ defined by Eq. (26.4.13) is diagonal, with real positive elements m_n on the diagonal. (For future use, we note that $\mathscr{M}'^{\dagger}\mathscr{M}' = \mathscr{U}^{\dagger}\mathscr{M}^{\dagger}\mathscr{M}\mathscr{U}$, so the quantities m_n^2 are just the eigenvalues of the positive Hermitian matrix $\mathscr{M}^{\dagger}\mathscr{M}$.) Redefining the fields in this way and dropping the primes, the quadratic part of the Lagrangian is now

$$\begin{aligned}\mathscr{L}_0 = \sum_n & \left[-\partial_\mu \varphi_n^* \partial^\mu \varphi_n - \frac{1}{2}\left(\overline{\psi_{nL}}\, \gamma^\mu \partial_\mu \psi_{nL}\right) + \frac{1}{2}\left(\partial_\mu (\overline{\psi_{nL}})\, \gamma^\mu \psi_{nL}\right)\right] \\ & -\frac{1}{2}\sum_n m_n \left(\bar{\psi}_{nL}\, \psi_{nL}\right) - \frac{1}{2}\sum_n m_n \left(\bar{\psi}_{nL}\, \psi_{nL}\right)^* \\ & -\sum_n m_n^2\, \varphi_n^* \varphi_n .\end{aligned} \tag{26.4.14}$$

To put the fermion mass terms in a more familiar form, we introduce fields $\psi_n(x)$ defined as the *Majorana* fields whose left-handed components are $\psi_{nL}(x)$. Then, using the symmetry properties (26.A.7) of Majorana bilinears:

$$\begin{aligned} & -\frac{1}{2}\left(\overline{\psi_{nL}}\, \gamma^\mu \partial_\mu \psi_{nL}\right) + \frac{1}{2}\left(\partial_\mu (\overline{\psi_{nL}})\, \gamma^\mu \psi_{nL}\right) \\ & = -\frac{1}{2}\left(\overline{\psi_n}\, \gamma^\mu \left(\frac{1+\gamma_5}{2}\right) \partial_\mu \psi_n\right) + \frac{1}{2}\left(\partial_\mu (\overline{\psi_n})\, \gamma^\mu \left(\frac{1+\gamma_5}{2}\right) \psi_n\right) \\ & = -\frac{1}{2}\left(\overline{\psi_n}\, \gamma^\mu \left(\frac{1+\gamma_5}{2}\right) \partial_\mu \psi_n\right) - \frac{1}{2}\left(\overline{\psi_n}\, \gamma^\mu \left(\frac{1-\gamma_5}{2}\right) \partial_\mu \psi_n\right) \\ & = -\frac{1}{2}\left(\overline{\psi_n}\, \gamma^\mu \partial_\mu \psi_n\right) ,\end{aligned}$$

while the reality properties (26.A.21) give

$$\left(\bar{\psi}_{nL}\, \psi_{nL}\right) + \left(\bar{\psi}_{nL}\, \psi_{nL}\right)^* = 2\,\mathrm{Re}\left(\overline{\psi_n} \left(\frac{1+\gamma_5}{2}\right) \psi_n\right) = \left(\overline{\psi_n}\, \psi_n\right) .$$

The complete quadratic Lagrangian is then

$$\begin{aligned}\mathscr{L}_0 = \sum_n & \left[-\partial_\mu \varphi_n^* \partial^\mu \varphi_n - \sum_n m_n^2\, \varphi_n^* \varphi_n \right. \\ & \left. -\frac{1}{2}\left(\overline{\psi_n}\, \gamma^\mu \partial_\mu \psi_n\right) - \frac{m_n}{2}\left(\overline{\psi_n}\, \psi_n\right)\right] .\end{aligned} \tag{26.4.15}$$

The factor 1/2 in the fermion terms is correct because these are Majorana fermion fields, while there is no factor 1/2 in the scalar terms because these are complex scalars. We see that the spinless and spin 1/2 particles have equal masses m_n, as required by the unbroken supersymmetry of the theory.

The interaction part $\mathscr{L}'$ of the Lagrangian density is given by the terms in Eq. (26.4.7) of higher than second order in φ_n and ψ_n. Since the superpotential $f(\phi_0+\varphi)$ is being assumed to be a cubic polynomial and stationary at $\varphi_n = 0$, with φ defined so that the second-order terms are $\frac{1}{2}\sum_n m_n\varphi_n^2$, we may write the superpotential (apart from an inconsequential constant term) as

$$f(\phi_0+\varphi) = \frac{1}{2}\sum_n m_n\,\varphi_n^2 + \frac{1}{6}\sum_{nm\ell} f_{nm\ell}\,\varphi_n\,\varphi_m\,\varphi_\ell\,. \tag{26.4.16}$$

Using this in Eq. (26.4.7) gives the interaction as

$$\begin{aligned}\mathscr{L}' = &-\frac{1}{2}\sum_{nm\ell} f_{nm\ell}\,\varphi_n\left(\overline{\psi_m}\left(\frac{1+\gamma_5}{2}\right)\psi_\ell\right)\\ &-\frac{1}{2}\sum_{nm\ell} f^*_{nm\ell}\,\varphi^*_n\left(\overline{\psi_m}\left(\frac{1-\gamma_5}{2}\right)\psi_\ell\right)\\ &-\frac{1}{2}\sum_{nm\ell} m_n\,f_{nm\ell}\varphi^*_n\varphi_m\varphi_\ell - \frac{1}{2}\sum_{nm\ell} m_n\,f^*_{nm\ell}\varphi_n\varphi^*_m\varphi^*_\ell\\ &-\frac{1}{4}\sum_{nm\ell m'\ell'} f_{nm\ell}f^*_{nm'\ell'}\varphi_m\varphi_\ell\varphi^*_{m'}\varphi^*_{\ell'}\,.\end{aligned} \tag{26.4.17}$$

We see that a knowledge of the masses m_n and the 'Yukawa' couplings $f_{nm\ell}$ of the scalars and fermions is enough to determine all the cubic and quartic self-couplings of the spinless fields.

As an illustration, consider the case of a single left-chiral superfield. For comparison with earlier results, let us write the single coefficient f in Eq. (26.4.16) as

$$f \equiv 2\sqrt{2}\,e^{i\alpha}\,\lambda\,, \tag{26.4.18}$$

where λ is real and α is some real phase. We will also introduce a pair of real spinless fields $A(x)$ and $B(x)$ by writing the single complex scalar here as

$$\varphi \equiv e^{-i\alpha}\left(\frac{A+iB}{\sqrt{2}}\right)\,. \tag{26.4.19}$$

The total Lagrangian density is then given by the sum of Eqs. (26.4.15) and (26.4.17) as

$$\begin{aligned}\mathscr{L} = &-\tfrac{1}{2}\partial_\mu A\,\partial^\mu A - \tfrac{1}{2}\partial_\mu B\,\partial^\mu B - \tfrac{1}{2}m^2\,(A^2+B^2)\\ &-\tfrac{1}{2}\left(\overline{\psi}\gamma^\mu\partial_\mu\psi\right) - \tfrac{1}{2}m\left(\overline{\psi}\psi\right)\end{aligned}$$

$$-\lambda A\left(\bar{\psi}\psi\right) - i\lambda B\left(\bar{\psi}\gamma_5\psi\right)$$
$$-m\lambda A(A^2+B^2) - \tfrac{1}{2}\lambda^2(A^2+B^2)^2 \,. \tag{26.4.20}$$

which is the same as the Lagrangian density (24.2.9) originally found by Wess and Zumino.[2] It is noteworthy that in this simple case the Lagrangian turns out to be invariant under a space inversion transformation

$$A(x) \to A(\Lambda_P x)\,, \qquad B(x) \to -B(\Lambda_P x)\,, \qquad \psi(x) \to i\beta\psi(\Lambda_P x)\,, \tag{26.4.21}$$

even though we did not assume parity conservation in deriving it. The appearance of parity conservation as an 'accidental' symmetry is a familiar feature of various renormalizable gauge theories (see Sections 12.5 and 18.7) but not of theories involving spinless fields, so this is a special consequence of supersymmetry in the renormalizable theory of a single scalar superfield.

26.5 Spontaneous Supersymmetry Breaking in the Tree Approximation

We saw in the previous section that supersymmetry is unbroken (at least in the tree approximation) in renormalizable theories of chiral superfields if Eq. (26.4.8) has a solution, that is, if there is a value ϕ_0 of the fields at which the superpotential $f(\phi)$ is stationary:

$$\left.\frac{\partial f(\phi)}{\partial \phi_n}\right|_{\phi=\phi_0} = 0\,. \tag{26.5.1}$$

There are as many independent variables here as there are equations to satisfy, so we generally expect there to be solutions of Eq. (26.5.1). In order for supersymmetry to be spontaneously broken in these theories, it is necessary to impose restrictions on the form of the superpotential.

To see how a choice of superpotential may allow supersymmetry to be spontaneously broken, we will consider a generalization of a class of models due to O'Raifeartaigh.[3] Suppose that the superpotential is a linear combination of a set Y_i of left-chiral superfields, with coefficients given by functions $f_i(X)$ of a second set of left-chiral superfields X_n:

$$f(X,Y) = \sum_i Y_i f_i(X)\,. \tag{26.5.2}$$

The conditions for supersymmetry to be unbroken by the values x_n and y_i of the scalar components of these superfields are that

$$0 = \frac{\partial f(x,y)}{\partial y_i} = f_i(x)\,, \tag{26.5.3}$$

$$0 = \frac{\partial f(x,y)}{\partial x_n} = \sum_i y_i \frac{\partial f_i(x)}{\partial x_n} \,. \tag{26.5.4}$$

Eq. (26.5.4) can always be solved by taking $y_i = 0$, with no effect on the problem of solving Eq. (26.5.3). On the other hand, if the number of X_n superfields is smaller than the number of Y_i superfields, then Eq. (26.5.3) imposes more conditions on the x_n than there are variables, so without fine-tuning a solution is impossible, and supersymmetry is broken.

It may appear that the initial assumption (26.5.2) itself represents a radical form of fine-tuning, but in fact this form can be imposed on the superpotential by assuming a suitable R-symmetry. As discussed in Section 26.3, in theories with $N = 1$ supersymmetry an R-symmetry is a $U(1)$ symmetry for which the θ superspace coordinate has a non-trivial transformation property. If we assume an R-symmetry for which θ_L carries quantum number $+1$, then the $\mathscr{F}$-term of any superpotential has a quantum number equal to that of the superpotential itself minus 2, so R invariance requires the superpotential itself to have $R = 2$. We can therefore impose the structure (26.5.2) by requiring R invariance, with the Y_i and X_n superfields given R quantum numbers $+2$ and 0, respectively.

The scalar fields in this sort of model have a potential

$$V(x,y) = \sum_i |f_i(x)|^2 + \sum_n \left| \sum_i y_i \frac{\partial f_i(x)}{\partial x_n} \right|^2 \,. \tag{26.5.5}$$

The potential is always minimized by choosing the x_n to minimize the first term; whatever values this gives the x_n, the second term can always be minimized by taking $y_i = 0$. Whether or not supersymmetry can be spontaneously broken, these models have the peculiar feature that there are always directions in the space of the fields in which the minimum of the potential is flat. Whatever values x_{n0} of the x_n minimize the first term in Eq. (26.5.5), the second term vanishes not only for $y_i = 0$, but for any vector y_i in a direction orthogonal to all the vectors $(v^n)_i = (\partial f_i(x)/\partial x_n)_{x=x_0}$. If there are N_X superfields X_n and N_Y superfields Y_i with $N_Y > N_X$, then the v^n cannot span the space of the ys, and there will be at least $N_Y - N_X$ of these flat directions. For any non-vanishing values $y_i = y_{0i}$ along any one of these flat directions, the R-symmetry of the Lagrangian density is spontaneously broken, and the Goldstone boson field ϕ associated with this global symmetry breakdown corresponds to a term $\propto y_{0i}$ in the y_i.

The simplest example of this class of models is provided by the case where there is just one X superfield and two Y superfields. Renormalizability requires the coefficient functions $f_i(X)$ to be quadratic functions of X, and by taking suitable linear combinations of the Y_i and shifting and rescaling X we can choose these functions so that

$$f_1(X) = X - a\,, \qquad\qquad f_2(X) = X^2\,, \tag{26.5.6}$$

with an arbitrary constant a. There clearly is no simultaneous solution of the two equations (26.5.1) unless the superpotential is fine-tuned so that $a = 0$. The potential (26.5.5) here is

$$V(x, y) = |x|^4 + |x - a|^2 + |y_1 + 2xy_2|^2\,. \tag{26.5.7}$$

The sum of the first two terms has a unique global minimum x_0. The flat direction here is the one for which $y_1 + 2x_0y_2 = 0$. For $a = 0$ we have $x_0 = 0$, and the minima of the potential are along the line with $y_1 = 0$ and y_2 arbitrary.

Whatever the reason for a spontaneous breakdown of supersymmetry, this phenomenon always entails the existence of a massless spin 1/2 particle, the *goldstino*, analogous to the Goldstone bosons associated with the spontaneous breakdown of ordinary global symmetries. (The one exception, discussed in Section 31.3, is that in supergravity theories, where supersymmetry is a local symmetry, the goldstino appears as the helicity $\pm 1/2$ states of a massive particle of spin 3/2, the gravitino.) In renormalizable theories of chiral superfields, the tree-approximation vacuum expectation values ϕ_{n0} of the scalar fields must be at a minimum of the potential $\sum_n |\partial f(\phi)/\partial\phi_n|^2$ in Eq. (26.4.7), so

$$\sum_m \mathscr{M}_{nm}\left(\frac{\partial f(\phi)}{\partial\phi_m}\bigg|_{\phi=\phi_0}\right)^* = 0\,, \tag{26.5.8}$$

where

$$\mathscr{M}_{nm} \equiv \frac{\partial^2 f(\phi)}{\partial\phi_n\partial\phi_m}\bigg|_{\phi=\phi_0}\,. \tag{26.5.9}$$

If Eq. (26.5.1) is not satisfied, then Eq. (26.5.8) tells us that the matrix $\mathscr{M}_{nm}$ has at least one eigenvector with eigenvalue zero, so according to Eq. (26.4.10) there must be at least one linear combination of the spin 1/2 particles described by ψ_n with zero mass. For instance, for the model defined by Eqs. (26.5.2) and (26.5.6), the matrix $\mathscr{M}$ has the non-vanishing components

$$\mathscr{M}_{xy_1} = \mathscr{M}_{y_1x} = 1\,, \qquad \mathscr{M}_{xy_2} = \mathscr{M}_{y_2x} = 2x_0\,, \tag{26.5.10}$$

so this matrix has eigenvalues $\pm 2x_0$ and 0, with the last eigenvalue corresponding to the goldstino mode. In Chapter 29 we will show without the use of perturbation theory that the spontaneous breakdown of supersymmetry requires the existence of goldstinos, and explore their general properties.

26.6 Superspace Integrals, Field Equations, and the Current Superfield

The '$\mathscr{F}$-terms' and 'D-terms' from which we construct the Lagrangian density may be expressed as integrals over the superspace coordinates θ_α. The rules for integrals over fermionic parameters originally given by Berezin[4] are derived in Section 9.5. Briefly, because the square of any fermionic parameter vanishes, any function of a set of N fermionic parameters ξ_n may be expressed as

$$f(\xi) = \left(\prod_{n=1}^{N} \xi_n\right) c \;+\; \text{terms with fewer } \xi \text{ factors}\,, \tag{26.6.1}$$

and its integral over the ξs is defined simply by

$$\int d^N\xi\, f(\xi) \equiv c\,. \tag{26.6.2}$$

The coefficient c may itself depend on other unintegrated c-number variables that anticommute with the ξs over which we integrate, in which case it is important to standardize the definition of c by moving all ξs to the left of c before integrating over them, as we have done in Eq. (26.6.1). With this definition, integration over fermionic variables is a linear operation. It resembles the integral over a real variable in the sense that, since a shift $\xi_n \to \xi_n + a_n$ of the variable ξ_n by a constant a_n changes the product $\prod_n \xi_n$ only by terms involving fewer ξ factors, it does not affect the value of the integral

$$\int d^N\xi\, f(\xi + a) = \int d^N\xi\, f(\xi)\,. \tag{26.6.3}$$

Also, as a special case of Eq. (26.6.2), the integral over N fermionic parameters of a polynomial of order $< N$ vanishes. Integrals over fermionic and bosonic parameters are strikingly different in the way they respond to change of variables: for bosonic parameters x_n we have $d^N x' = \text{Det}\,(\partial x'/\partial x)\, d^N x$, while for fermionic parameters

$$d^N\xi' = [\text{Det}\,(\partial\xi'/\partial\xi)]^{-1} d^N\xi\,. \tag{26.6.4}$$

In particular, the dimensionality of $d\xi$ is *opposite* to the dimensionality of ξ.

According to Eq. (26.2.10), the D-term of a general superfield $S(x,\theta)$ (which may be elementary or composite) is equal, up to a derivative term, to the coefficient of $-(\bar\theta\gamma_5\theta)^2/4 = -(\theta^{\rm T}\epsilon\theta)^2/4$. Any one of the four θs can be θ_1, and each possibility gives an equal contribution, so we may assume that θ_1 is the left-most, and pick up a factor of 4. Then θ_2 must be the next-to-left-most. Any of the two remaining θs may be θ_3, and each possibility gives an equal contribution, so we may assume that θ_3 is third from the left and pick up a factor of 2, and θ_4 must then be the right-most. That is,

$$-\tfrac{1}{4}(\bar{\theta}\gamma_5\theta)^2 = -\tfrac{1}{4} \times 4 \times 2 \times \theta_1\theta_2\theta_3\theta_4\,,$$

so the coefficient of this function of θ is $-1/2$ the integral over $d^4\theta$. Since this is the D-term up to a derivative, we have then

$$\int d^4x\, [S]_D = -\frac{1}{2}\int d^4x \int d^4\theta\, S(x,\theta)\,. \tag{26.6.5}$$

In the same way, using Eq. (26.3.11), we find that the spacetime integral of the $\mathscr{F}$-term of a general left-chiral superfield Φ (again, either elementary or composite) may be expressed as

$$\int d^4x\, [\Phi]_{\mathscr{F}} = \frac{1}{2}\int d^4x \int d^2\theta_L\, \Phi(x,\theta)\,. \tag{26.6.6}$$

Since we are now integrating over θs, it is convenient to introduce a delta function, defined as usual by the condition that for an arbitrary function $f(\theta)$

$$\int d^4\theta'\, \delta^4(\theta'-\theta)\, f(\theta') = f(\theta)\,. \tag{26.6.7}$$

According to Eq. (9.5.40), this condition is satisfied by

$$\begin{aligned}\delta^4(\theta'-\theta) &= (\theta'_1-\theta_1)(\theta'_2-\theta_2)(\theta'_3-\theta_3)(\theta'_4-\theta_4)\\ &= \frac{1}{4}\left[\left(\theta_L-\theta'_L\right)^{\rm T}\epsilon\left(\theta_L-\theta'_L\right)\right]\left[\left(\theta_R-\theta'_R\right)^{\rm T}\epsilon\left(\theta_R-\theta'_R\right)\right].\end{aligned} \tag{26.6.8}$$

The representation of the action as an integral over superspace allows an easy derivation of the field equations in superfield form. Consider, for instance, the action for a set of left-chiral scalar superfields Φ_n (which includes as a special case the general renormalizable theory of left-chiral superfields Φ_n):

$$I = \frac{1}{2}\int d^4x\, \left[K(\Phi,\Phi^*)\right]_D + 2\,{\rm Re}\int d^4x\, [f(\Phi)]_{\mathscr{F}}\,, \tag{26.6.9}$$

with K an arbitrary function of Φ_n and Φ^*_n without derivatives, and f an arbitrary function of Φ_n, also without derivatives. (The motivation for this form of the action and the expression of the action in terms of component fields is described in Section 26.8.) We cannot derive correct field equations by simply demanding that this is stationary with respect to arbitrary variations in Φ, because the Φ_n are constrained by the requirement that they satisfy the requirement $\mathscr{D}_R\Phi_n = 0$ for left-chiral superfields. To make sure that this condition is preserved by any variation, we can use a trick that will also turn out to be useful in deriving superspace Feynman rules in Chapter 30. We write the Φ_n in terms of *potential superfields* $S_n(x, \theta)$ as

$$\Phi_n = \mathscr{D}_R^2 S_n \,, \tag{26.6.10}$$

from which it follows (using Eq. (26.A.21)) that

$$\Phi_n^* = -\mathscr{D}_L^2 S_n^* \,, \tag{26.6.11}$$

where $\mathscr{D}_R^2$ and $\mathscr{D}_L^2$ are abbreviations for $(\mathscr{D}_R^{\rm T}\epsilon\mathscr{D}_R) = -(\bar{\mathscr{D}}_R\mathscr{D}_R)$ and $(\mathscr{D}_L^{\rm T}\epsilon\mathscr{D}_L) = (\bar{\mathscr{D}}_L\mathscr{D}_L)$, respectively. To see that it is always possible to find an S_n (not necessarily local) that satisfies Eq. (26.6.10), note that for any left-chiral superfields Φ_n,

$$\mathscr{D}_R^2\mathscr{D}_L^2\Phi_n = -16\Box\Phi_n \,, \tag{26.6.12}$$

so that Eq. (26.6.10) is satisfied by the solution of

$$-16\Box S_n = \mathscr{D}_L^2\Phi_n \,. \tag{26.6.13}$$

The expression $\mathscr{D}_R^2 S$ is left-chiral for any S, so the action must be stationary with respect to arbitrary variations in the S_n. Using Eq. (26.6.5), the action may be expressed in terms of the S_n and S_n^* as

$$I = -\frac{1}{4}\int d^4x \int d^4\theta \, K(-\mathscr{D}_L^2 S^* \,, \mathscr{D}_R^2 S) + 2\,\mathrm{Re}\int d^4x \left[f(\mathscr{D}_R^2 S)\right]_{\mathscr{F}} \,. \tag{26.6.14}$$

The variation of the first term under infinitesimal changes δS_n in the S_n (but not the S_n^*) is easily calculated by an integration by parts in superspace:

$$-\delta\frac{1}{4}\int d^4x \int d^4\theta \, K(-\mathscr{D}_L^2 S^* \,, \mathscr{D}_R^2 S)$$
$$= -\sum_n \int d^4\theta \, \delta S_n \, \mathscr{D}_R^2 \, \frac{\delta K(-\mathscr{D}_L^2 S^* \,, \mathscr{D}_R^2 S)}{\delta \mathscr{D}_R^2 S_n} \,.$$

Eqs. (26.3.31) and (26.6.5) allow us to express the variation of the integral in the superpotential term under infinitesimal changes δS_n in the S_n as

$$\delta\int d^4x \left[f(\mathscr{D}_R^2 S)\right]_{\mathscr{F}} = \sum_n \int d^4x \left[\frac{\partial f(\Phi)}{\partial \Phi_n}\bigg|_{\Phi=\mathscr{D}_R^2 S} \mathscr{D}_R^2 \delta S_n\right]_{\mathscr{F}}$$
$$= \sum_n \int d^4x \left[\mathscr{D}_R^2\left(\frac{\partial f(\Phi)}{\partial \Phi_n}\bigg|_{\Phi=\mathscr{D}_R^2 S} \delta S_n\right)\right]_{\mathscr{F}}$$
$$= 2\sum_n \int d^4x \left[\frac{\partial f(\Phi)}{\partial \Phi_n}\bigg|_{\Phi=\mathscr{D}_R^2 S} \delta S_n\right]_D$$
$$= -\sum_n \int d^4x \int d^4\theta \, \frac{\partial f(\Phi)}{\partial \Phi_n}\bigg|_{\Phi=\mathscr{D}_R^2 S} \delta S_n \,.$$

The condition for Eq. (26.6.14) to be stationary with respect to arbitrary variations in S_n is then

$$\mathscr{D}_R^2 \frac{\delta K(-\mathscr{D}_L^2 S^*, \mathscr{D}_R^2 S)}{\delta \mathscr{D}_R^2 S_n} = -4 \left. \frac{\partial f(\Phi)}{\partial \Phi_n} \right|_{\Phi = \mathscr{D}_R^2 S} ,$$

or in terms of the chiral superfields

$$\mathscr{D}_R^2 \frac{\delta K(\Phi, \Phi^*)}{\delta \Phi_n} = -4 \frac{\partial f(\Phi)}{\partial \Phi_n} . \tag{26.6.15}$$

The complex conjugate yields

$$\mathscr{D}_L^2 \frac{\delta K(\Phi, \Phi^*)}{\delta \Phi_n^*} = 4 \left(\frac{\partial f(\Phi)}{\partial \Phi_n} \right)^* . \tag{26.6.16}$$

It can readily be checked that the components of these equations yield the field equations for the components of Φ_n^* and Φ_n. For instance, recalling that $\mathscr{D}_R^2(\theta_R^{\mathrm{T}} \epsilon \theta_R) = -4$, the θ-independent part of $\mathscr{D}_R^2 \Phi_n^*$ is $4\mathscr{F}_n^*$, while the θ-independent part of $\partial f(\Phi)/\partial \Phi_n$ is $\partial f(\phi)/\partial \phi_n$, so the θ-independent part of Eq. (26.6.15) for $K = \sum_n \Phi_n^* \Phi_n$ yields the relation $\mathscr{F}_n^* = -\partial f(\phi)/\partial \phi_n$, in agreement with Eq. (26.4.6).

As an example of the use of this formalism, let us consider the superfields to which conserved currents belong. Suppose that the superpotential and Kahler potential in the action (26.6.9) are invariant under an infinitesimal global transformation

$$\delta \Phi_n = i\epsilon \sum_m \mathscr{T}_{nm} \Phi_m , \qquad \delta \Phi_n^* = -i\epsilon \sum_m \mathscr{T}_{mn} \Phi_m^* , \tag{26.6.17}$$

with ϵ a real infinitesimal constant and $\mathscr{T}_{nm}$ a Hermitian matrix, perhaps part of a Lie algebra of similar transformation matrices. Since the superpotential depends only on the Φ_n, it is automatically also invariant under the extended transformations

$$\delta \Phi_n = i\epsilon \Lambda \sum_m \mathscr{T}_{nm} \Phi_m , \qquad \delta \Phi_n^* = -i\epsilon \Lambda^* \sum_m \mathscr{T}_{mn} \Phi_m^* , \tag{26.6.18}$$

where $\Lambda(x, \theta)$ is a superfield that must be taken left-chiral in order that the $\delta \Phi_n$ be left-chiral, but is otherwise unrestricted. On the other hand, other terms such as the Kahler potential are not in general invariant under these transformations, because $\Lambda \neq \Lambda^*$. For general fields the change in the action must therefore be of the form

$$\delta I = i\epsilon \int d^4x \int d^4\theta \, [\Lambda - \Lambda^*] \mathscr{J} , \tag{26.6.19}$$

where $\mathscr{J}(x, \theta)$ is some real superfield, known as *the current superfield*. But if the field equations are satisfied then the action is stationary under *any* variation in the superfields, so the integral (26.6.19) must vanish for any left-chiral superfield $\Lambda(x,\theta)$. Any such Λ may be put in the form $\Lambda = \mathscr{D}_R^2 S$, so this means that the current superfield must satisfy

$$\mathscr{D}_R^2 \mathscr{J} = \mathscr{D}_L^2 \mathscr{J} = 0\,. \tag{26.6.20}$$

That is, $\mathscr{J}$ is a *linear* superfield. As we saw in Section 26.3, this means that its components satisfy

$$N^{\mathscr{J}} = M^{\mathscr{J}} = \partial^\mu V_\mu^{\mathscr{J}} = 0\,, \quad \lambda^{\mathscr{J}} = -\not\partial\omega^{\mathscr{J}}\,, \quad D^{\mathscr{J}} = -\Box C^{\mathscr{J}}\,. \tag{26.6.21}$$

This allows us to identify the V-component $V_\mu^{\mathscr{J}}$ as the conserved current associated with this symmetry.

For the particular action (26.6.9), the current superfield takes the form

$$\mathscr{J} = \sum_{nm} \frac{\partial K(\Phi,\Phi^*)}{\partial\Phi_n}\mathscr{T}_{nm}\Phi_m = \sum_{nm} \frac{\partial K(\Phi,\Phi^*)}{\partial\Phi_n^*}\mathscr{T}_{mn}\Phi_m^*\,, \tag{26.6.22}$$

with the equality of these two expressions a consequence of symmetry under the transformation (26.6.17). Then, using the field equation (26.6.15),

$$\mathscr{D}_R^2\mathscr{J} = \sum_{nm}\left[\mathscr{D}_R^2\frac{\partial K(\Phi,\Phi^*)}{\partial\Phi_n}\right]\mathscr{T}_{nm}\Phi_m = -4\sum_{nm}\frac{\partial f(\Phi)}{\partial\Phi_n}\mathscr{T}_{nm}\Phi_m\,, \tag{26.6.23}$$

which vanishes because of the assumed invariance of the superpotential under the transformation (26.6.17). In the same way, using the second of the two expressions for $\mathscr{J}$ and the field equation (26.6.16), we find that $\mathscr{D}_L^2\mathscr{J} = 0$, thus verifying the conservation condition (26.6.20).

26.7 The Supercurrent

Like any other continuous global symmetry, supersymmetry leads to the existence of a conserved current.[5] The conservation and commutation properties of the supersymmetry current are operator equations that will remain valid even when supersymmetry is spontaneously broken, and that will therefore be useful to us in Chapter 29, when we consider theories of spontaneous supersymmetry breaking in a non-perturbative context. Also, the supersymmetry current is related to components of a superfield known as the *supercurrent*,[6] that will be of fundamental importance in our treatment of supergravity in Chapter 31.

As we saw in Section 7.3, the existence of an ordinary global symmetry of the Lagrangian density under an infinitesimal transformation $\chi^\ell \to \chi^\ell + \epsilon\mathscr{F}^\ell$ (with χ^ℓ a generic canonical or auxiliary boson or fermion field and $\mathscr{F}^\ell$ a function of the canonical and auxiliary fields) leads to the existence of a current

$$J^\mu(x) \propto \sum_\ell \frac{\partial\mathscr{L}(x)}{\partial(\partial\chi^\ell(x)/\partial x^\mu)}\mathscr{F}^\ell(x)\,,$$

which is conserved for fields satisfying the field equations, and generates the symmetry in the sense that the canonical commutation relations give

$$\left[\int d^3x\, J^0(x)\,,\, \chi^\ell(y)\right] \propto \mathscr{F}^\ell(y)\,.$$

The supersymmetry current requires a somewhat more complicated treatment for two reasons. One is that supersymmetry is only a symmetry of the action, not of the Lagrangian density or the Lagrangian. Instead, the variation of the Lagrangian density under an infinitesimal supersymmetry transformation is a spacetime derivative, which we may write in the form

$$\delta\mathscr{L} = \sum_\ell \left(\bar{\alpha}\,\partial_\mu K^\mu\right), \tag{26.7.1}$$

with K^μ a four-vector of Majorana spinors. In consequence, the supersymmetry current is not the usual Noether current. The Noether current is a four-vector N^μ of Majorana spinors defined by

$$\sum_\ell \frac{\partial_R \mathscr{L}}{\partial(\partial_\mu \chi^\ell)}\delta\chi^\ell \equiv -\left(\bar{\alpha}\, N^\mu\right), \tag{26.7.2}$$

whose divergence is given by the Euler–Lagrange equations as

$$\begin{aligned}\left(\bar{\alpha}\,\partial_\mu N^\mu\right) &= -\sum_\ell \frac{\partial_R \mathscr{L}}{\partial \chi^\ell}\delta\chi^\ell - \sum_\ell \frac{\partial_R \mathscr{L}}{\partial(\partial_\mu \chi^\ell)}\partial_\mu \delta\chi^\ell \\ &= -\,\delta\mathscr{L}\,.\end{aligned} \tag{26.7.3}$$

(Here ∂_R denotes the right partial derivative, defined by moving differentiated fermionic variables to the right before differentiating.) Instead, we must define the supersymmetry current as

$$S^\mu \equiv N^\mu + K^\mu\,, \tag{26.7.4}$$

which Eqs. (26.7.1) and (26.7.3) tell us *is* conserved:

$$\partial_\mu S^\mu = 0\,. \tag{26.7.5}$$

The second complication is that the change $\delta\chi^\ell$ in a canonical field χ^ℓ under a supersymmetry transformation is not just a function of canonical fields, but also involves their canonical conjugates. For instance, Eq. (26.3.15) shows that the change in the ψ-component of a chiral scalar superfield involves the time derivative of the ϕ-component. As a result the commutator of the Noether charge $\int d^3x\, N^0$ with a general canonical field does not give the supersymmetry transformation of that field. Fortunately, this complication is cancelled by the first complication: when the commutators of $\int d^3x\, K^0$ as well as $\int d^3x\, N^0$ with the fields are taken into account, the operator $\int d^3x\, S^0$ does generate supersymmetry transformations* in the sense that

$$\left[\int d^3x\left(\bar{\alpha}\,S^0\right),\,\chi^\ell\right] = i\,\delta\chi^\ell\,, \tag{26.7.6}$$

consistently with Eqs. (26.2.1) and (26.2.8).

For example, we may derive an explicit formula for the supersymmetry current in the general renormalizable theory of left-chiral superfields Φ_n, which can be used to check that it does generate supersymmetry transformations, in the sense of Eq. (26.7.6). The Lagrangian density (26.4.7) for

---

* This is a general result for currents constructed in this way. For instance, consider a Lagrangian L (not Lagrangian density) that depends on a set of canonical variables q^n and their time derivatives $\dot{q}^n$, with no constraints of any class. In quantum field theory the label n includes spatial coordinates as well as discrete spin and species labels, and $L = \int d^3x\,\mathscr{L}$. Our assumption here that the Lagrangian density is invariant up to spacetime derivatives under some infinitesimal transformation δ means that δL is the time derivative of some functional F. That is,

$$\sum_n \frac{\partial L}{\partial q^n}\delta q^n + \sum_n \frac{\partial L}{\partial \dot{q}^n}\delta\dot{q}^n = \frac{d}{dt}F\,.$$

Using the canonical equations of motion, this can be written as a conservation law $\dot{Q} = 0$, where the conserved charge is

$$Q = -\sum_n \frac{\partial L}{\partial \dot{q}^n}\delta q^n + F\,.$$

In our case here, $Q = \int d^3x\,[N^0 + K^0]$. We assume the usual unconstrained commutation relations

$$\left[\frac{\partial L}{\partial \dot{q}^n},\,q^m\right] = -i\,\delta^m_n\,, \qquad \left[q^n,\,q^m\right] = 0\,,$$

and find the commutator

$$\left[Q,\,q^m\right] = i\,\delta q^m - \sum_{nl} \frac{\partial L}{\partial \dot{q}^l}\frac{\partial \delta q^l}{\partial \dot{q}^n}\left[\dot{q}^n,\,q^m\right] + \sum_n \frac{\partial F}{\partial \dot{q}^n}\left[\dot{q}^n,\,q^m\right].$$

To evaluate the second and third terms, we note that the second time derivatives $\ddot{q}^n$ appear linearly in the invariance condition, so their coefficients must match: even without using the equations of motion, we have

$$\sum_l \frac{\partial L}{\partial \dot{q}^l}\frac{\partial \delta q^l}{\partial \dot{q}^n} = \frac{\partial F}{\partial \dot{q}^n}\,.$$

The second and third terms in the commutator therefore cancel, leaving us with the desired result

$$\left[Q,\,q^m\right] = i\,\delta q^m\,.$$

Taking the time derivative gives also

$$\left[Q,\,\dot{q}^m\right] = i\,\delta\dot{q}^m\,.$$

This result has been extended to theories with constraints.[7]

this theory may be put in the form

$$\mathscr{L} = \sum_n \left[-\partial_\mu \phi_n^* \partial^\mu \phi_n - \frac{1}{2}\left(\overline{\psi_{Ln}}\gamma^\mu \partial_\mu \psi_{Ln}\right) - \frac{1}{2}\left(\overline{\psi_{Rn}}\gamma^\mu \partial_\mu \psi_{Rn}\right)\right]$$
$$+ \text{non-derivative terms}\,. \tag{26.7.7}$$

Using the transformation rules (26.3.15), (26.3.17), (26.3.18), and (26.3.20) (with $\tilde{\phi} = \phi^*$), the Noether current defined by Eq. (26.7.2) is

$$N^\mu = \frac{1}{\sqrt{2}}\sum_n \Big[2\,(\partial^\mu \phi_n^*)\,\psi_{nL} + 2\,(\partial^\mu \phi_n)\,\psi_{nR} + (\not\partial \phi_n)\,\gamma^\mu \psi_{nR} + (\not\partial \phi_n^*)\,\gamma^\mu \psi_{nL}$$
$$-\mathscr{F}_n\,\gamma^\mu \psi_{nR} - \mathscr{F}_n^*\,\gamma^\mu \psi_{nL}\Big]\,. \tag{26.7.8}$$

We can calculate the change in the Lagrangian density either directly, or, more easily, by noting that the changes of D-terms and $\mathscr{F}$-terms under supersymmetry transformations are given respectively by Eqs. (26.2.17) and (26.3.16). Either way, we find that the current K^μ in Eq. (26.7.1) is

$$K^\mu = \frac{1}{\sqrt{2}}\sum_n \gamma^\mu \Bigg[-(\not\partial \phi_n)\,\psi_{nR} - (\not\partial \phi_n^*)\,\psi_{nL} + \mathscr{F}_n^*\,\psi_{nL} + \mathscr{F}_n\,\psi_{nR}$$
$$+2\left(\frac{\partial f(\phi)}{\partial \phi_n}\right)\psi_{nL} + 2\left(\frac{\partial f(\phi)}{\partial \phi_n}\right)^*\psi_{nR}\Bigg]\,. \tag{26.7.9}$$

Adding (26.7.8) and (26.7.9) gives the supersymmetry current for this class of theories

$$S^\mu = \sqrt{2}\sum_n \left[(\not\partial \phi_n)\gamma^\mu \psi_{nR} + (\not\partial \phi_n^*)\gamma^\mu \psi_{nL} + \left(\frac{\partial f}{\partial \phi_n}\right)\gamma^\mu \psi_{nL} + \left(\frac{\partial f}{\partial \phi_n}\right)^*\gamma^\mu \psi_{nR}\right]\,. \tag{26.7.10}$$

It is straightforward then to use the canonical commutation and anti-commutation relations to verify that $\int d^3x\, S^0$ satisfies the commutation relations (26.7.6).

There is another definition of symmetry currents, in terms of the response of the matter action to a local symmetry transformation, which is particularly useful when the associated symmetries are 'gauged,' as supersymmetry will be when we turn to supergravity theories in Chapter 31. In the absence of supergravity fields, the action is not invariant under local supersymmetry transformations. If we make such a transformation with a spacetime-dependent parameter $\alpha(x)$, the action will change by an amount that, in order to vanish when $\alpha(x)$ is constant, must (even when the field equations are not satisfied) be of the form

$$\delta I = -\int d^4x\,\Big((\partial_\mu \bar{\alpha}(x))\,S^\mu(x)\Big)\,, \tag{26.7.11}$$

with $S^\mu(x)$ a four-vector of Majorana spinor operator coefficients. This does not define $S^\mu(x)$ uniquely, because when we generalize global supersymmetry transformations to local transformations, we might in general give the change $\delta\chi$ of a field χ under a local supersymmetry transformation an arbitrary dependence on the derivatives of $\alpha(x)$. There is, however, one way of defining local supersymmetry transformations that guarantees that the coefficient $S^\mu(x)$ in Eq. (26.7.11) is the same as the current defined by Eq. (26.7.4), which as we have seen generates the symmetry transformations in the sense of Eq. (26.7.6). It is to specify that *derivatives of* $\alpha(x)$ *do not appear in the supersymmetry transformation of the canonical or auxiliary fields* χ^ℓ. For instance, the local versions of the transformation rules (26.3.15)–(26.3.17) for the components of a left-chiral superfield are

$$\delta\psi_L(x) = \sqrt{2}\partial_\mu\phi(x)\gamma^\mu\alpha_R(x)\,\phi(x) + \sqrt{2}\mathscr{F}(x)\,\alpha_L(x)\,, \tag{26.7.12}$$

$$\delta\mathscr{F}(x) = \sqrt{2}\Big(\overline{\alpha_L}(x)\,\not{\partial}\psi_L(x)\Big)\,, \tag{26.7.13}$$

$$\delta\phi(x) = \sqrt{2}\Big(\overline{\alpha_R}(x)\psi_L(x)\Big)\,. \tag{26.7.14}$$

Eq. (26.3.21) shows that the superfield may be expressed in terms of its component fields at x_+^μ without derivatives, so the transformation rule for the superfield may be expressed as

$$\delta\Phi(x,\theta) = \Big(\bar{\alpha}(x_+)\,\mathscr{Q}\Big)\Phi(x,\theta)\,, \tag{26.7.15}$$

where $\mathscr{Q}$ is the operator (26.2.2).

With local supersymmetry transformations defined in this way, the change of the action that they induce consists of two terms. First, although the variation in canonical fields under a supersymmetry transformation does not involve derivatives of $\alpha(x)$, the variation in derivatives of canonical fields does. This produces a change in the Lagrangian density which is the same as Eq. (26.7.2), except that $\bar{\alpha}$ is replaced with $\partial_\mu\bar{\alpha}$:

$$\delta_1 I = -\int d^4x\,\Big([\partial_\mu\bar{\alpha}(x)]\,N^\mu(x)\Big).$$

The second term in the change of the action arises from the fact that the Lagrangian density is not invariant even under the part of the supersymmetry transformation that does not involve derivatives of $\alpha(x)$. According to Eq. (26.7.1), this produces a change in the action

$$\delta_2 I = \int d^4x\,\Big(\bar{\alpha}(x)\,\partial_\mu K^\mu(x)\Big) = -\int d^4x\,\Big((\partial_\mu\bar{\alpha}(x))\,K^\mu(x)\Big)\,.$$

Adding $\delta_1 I$ and $\delta_2 I$ gives a total change in the action of the form (26.7.11), with $S^\mu(x)$ given by Eq. (26.7.4), as was to be shown.

Even with this specification of the transformation property of the component fields, the supersymmetry current $S^\mu(x)$ is not uniquely specified by Eq. (26.7.11), because we could always introduce a modified current

$$S^\mu_{\rm new} = S^\mu + \partial_\nu A^{\mu\nu} \,, \tag{26.7.16}$$

where $A^{\mu\nu} = -A^{\nu\mu}$ is an arbitrary antisymmetric tensor of Majorana spinors. The term $\partial_\nu A^{\mu\nu}$ is conserved whether or not the field equations are satisfied, and its time component is a space derivative, so $\int d^3x\, S^0_{\rm new} = \int d^3x\, S^0$, leaving Eq. (26.7.6) unchanged.

There is in fact a particular choice of $A^{\mu\nu}$ with the convenient feature that $\gamma_\mu S^\mu_{\rm new}$ turns out to be a measure of the violation of scale invariance by the theory. By using the Dirac equations derived from the Lagrangian density (26.4.7):

$$\partial\!\!\!/ \psi_{mL} = -\sum_n \left(\frac{\partial^2 f(\phi)}{\partial\phi_m \partial\phi_n}\right)^* \psi_{nR} \,, \qquad \partial\!\!\!/ \psi_{mR} = -\sum_n \left(\frac{\partial^2 f(\phi)}{\partial\phi_m \partial\phi_n}\right) \psi_{nL} \,, \tag{26.7.17}$$

it is straightforward to calculate that

$$\begin{aligned}\gamma_\mu S^\mu = -2\sqrt{2}\sum_n \Bigg\{ & \partial\!\!\!/ \Big(\phi_n\psi_{nR} + \phi_n^*\psi_{nL}\Big) \\ & + \left(\sum_m \phi_m \frac{\partial^2 f(\phi)}{\partial\phi_n\partial\phi_m} - 2\frac{\partial f(\phi)}{\partial\phi_n}\right)\psi_{nL} \\ & + \left(\sum_m \phi_m \frac{\partial^2 f(\phi)}{\partial\phi_n\partial\phi_m} - 2\frac{\partial f(\phi)}{\partial\phi_n}\right)^* \psi_{nR}\Bigg\} \,.\end{aligned}$$

We can eliminate the first term by introducing a modified supersymmetry current of the general type of Eq. (26.7.16):

$$S^\mu_{\rm new} = S^\mu + \frac{\sqrt{2}}{3}[\gamma^\mu, \gamma^\nu]\sum_n \partial_\nu\Big(\phi_n\psi_{nR} + \phi_n^*\psi_{nL}\Big) \,, \tag{26.7.18}$$

for which

$$\begin{aligned}\gamma_\mu S^\mu_{\rm new} = -2\sqrt{2}\sum_n \Bigg\{ & \left(\sum_m \phi_m \frac{\partial^2 f(\phi)}{\partial\phi_n\partial\phi_m} - 2\frac{\partial f(\phi)}{\partial\phi_n}\right)\psi_{nL} \\ & + \left(\sum_m \phi_m \frac{\partial^2 f(\phi)}{\partial\phi_n\partial\phi_m} - 2\frac{\partial f(\phi)}{\partial\phi_n}\right)^* \psi_{nR}\Bigg\} \,.\end{aligned} \tag{26.7.19}$$

The right-hand side vanishes for a scale-invariant Lagrangian density, with $f(\Phi)$ a homogeneous polynomial of third order in the Φ_n.

We now turn to the supersymmetry transformation property of the supersymmetry current. It is straightforward to check that the current given by Eqs. (26.7.18) and (26.7.10) is related to the ω-component ω^Θ_μ of a real non-chiral superfield Θ_μ by**

$$S^\mu_{\rm new} = -2\omega^{\Theta\,\mu} + 2\gamma^\mu\gamma^\nu\,\omega^\Theta_\nu\,, \tag{26.7.20}$$

where

$$\Theta_\mu = \frac{i}{12}\sum_n\left[4\Phi_n^*\partial_\mu\Phi_n - 4\Phi_n\partial_\mu\Phi_n^* + \Big((\bar{\mathscr{D}}\Phi_n^*)\gamma_\mu(\mathscr{D}\Phi_n)\Big)\right]. \tag{26.7.21}$$

The superfield Θ^μ is known as the *supercurrent.*

The supercurrent obeys a conservation law that incorporates the conservation of the supersymmetry current (26.7.20) and much else besides. To derive it, we may use the anticommutation relation (26.2.30) to write†

$$[\mathscr{D}_R\,,(\bar{\mathscr{D}}_L\mathscr{D}_L)] = -4\,\not{\partial}\mathscr{D}_L\,.$$

Together with the chirality conditions $\mathscr{D}_R\Phi_n = \mathscr{D}_L\Phi_n^* = 0$, this gives

$$\gamma^\mu\mathscr{D}_L\sum_n\left[\Phi_n^*\,\partial_\mu\Phi_n - \Phi_n\,\partial_\mu\Phi_n^*\right] = -\tfrac{1}{4}\sum_n\Phi_n^*\mathscr{D}_R\Big(\bar{\mathscr{D}}_L\mathscr{D}_L\Big)\Phi_n - \sum_n(\not{\partial}\Phi_n^*)\,\mathscr{D}_L\Phi_n$$

and

$$\gamma^\mu\mathscr{D}_L\sum_n\Big((\bar{\mathscr{D}}\Phi_n^*)\,\gamma_\mu(\mathscr{D}\Phi_n)\Big) = 4\sum_n(\not{\partial}\Phi_n^*)\,\mathscr{D}\Phi_n + 2\sum_n\mathscr{D}\Phi_n^*\Big(\bar{\mathscr{D}}_L\mathscr{D}_L\Big)\Phi_n\,,$$

so that the superfield (26.7.21) satisfies

$$\gamma_\mu\,\mathscr{D}_L\Theta^\mu = \tfrac{1}{6}i\sum_n(\mathscr{D}_R\Phi_n^*)\,\Big(\bar{\mathscr{D}}_L\mathscr{D}_L\Big)\Phi_n - \tfrac{1}{12}i\sum_n\Phi_n^*\mathscr{D}_R\Big(\bar{\mathscr{D}}_L\mathscr{D}_L\Big)\Phi_n\,. \tag{26.7.22}$$

We saw in Section 26.6 that the field equations for the Lagrangian density (26.4.7) may be expressed in the form

$$\Big(\bar{\mathscr{D}}_L\mathscr{D}_L\Big)\Phi_n = -4\left(\frac{\partial f(\Phi)}{\partial\Phi_n}\right)^*\,. \tag{26.7.23}$$

** Here we introduce a notation that will be used extensively in Chapter 31; following Eq. (26.2.10), the components C^S, ω^S, M^S, N^S, V^S_ν, λ^S, and D^S of an arbitrary superfield $S(x,\theta)$ are defined by the expansion

$$\begin{aligned}S(x,\theta) = {}& C^S(x) - i\Big(\bar\theta\,\gamma_5\,\omega^S(x)\Big) - \frac{i}{2}\Big(\bar\theta\,\gamma_5\,\theta\Big)M^S(x) - \frac{1}{2}\Big(\bar\theta\,\theta\Big)N^S(x)\\ &+\frac{i}{2}\Big(\bar\theta\,\gamma_5\,\gamma^\nu\,\theta\Big)V^S_\nu(x) - i\Big(\bar\theta\,\gamma_5\,\theta\Big)\left(\bar\theta\left[\lambda^S(x) + \frac{1}{2}\not{\partial}\omega^S(x)\right]\right)\\ &-\frac{1}{4}\Big(\bar\theta\,\gamma_5\,\theta\Big)^2\left[D^S(x) + \frac{1}{2}\Box C^S(x)\right].\end{aligned}$$

† It should be noted that $\bar{\mathscr{D}}_L$ and $\bar{\mathscr{D}}_R$ are the left- and right-handed components of the covariant adjoint $\bar{\mathscr{D}}$, rather than the covariant adjoints $\overline{\mathscr{D}_L}$ and $\overline{\mathscr{D}_R}$ of $\mathscr{D}_L$ and $\mathscr{D}_R$.

Using this in Eq. (26.7.22) gives finally

$$\gamma^\mu \mathscr{D}_L \Theta_\mu = -\frac{2}{3} i \sum_n (\mathscr{D}_R \Phi_n^*) \left(\frac{\partial f}{\partial \Phi_n}\right)^* + \frac{1}{3} i \sum_n \Phi_n^* \mathscr{D}_R \left(\frac{\partial f}{\partial \Phi_n}\right)^*$$
$$= \frac{1}{3} i \mathscr{D}_R \left[\sum_n \Phi_n \frac{\partial f(\Phi)}{\partial \Phi_n} - 3 f(\Phi)\right]^* . \tag{26.7.24}$$

The Hermitian adjoint of Eq. (26.7.24) is

$$\gamma^\mu \mathscr{D}_R \Theta_\mu = -\frac{1}{3} i \mathscr{D}_L \left[\sum_n \Phi_n \frac{\partial f(\Phi)}{\partial \Phi_n} - 3 f(\Phi)\right] . \tag{26.7.25}$$

The sum of this and Eq. (26.7.24) then gives the conservation law

$$\gamma^\mu \mathscr{D} \Theta_\mu = \mathscr{D} X , \tag{26.7.26}$$

where X is a real chiral superfield given (up to an additive constant) in this class of theories by

$$X = \frac{2}{3} \text{Im} \left[\sum_n \Phi_n \frac{\partial f(\Phi)}{\partial \Phi_n} - 3 f(\Phi)\right] . \tag{26.7.27}$$

Although it has been derived here only for renormalizable theories of chiral superfields, we may expect that the conservation law (26.7.26) holds more generally, though of course with X not necessarily given by Eq. (26.7.27), because of the other conservation laws that it incorporates. (A generalized formula for X will be given in Section 31.4.) To derive these relations, we must use Eq. (26.2.10) to express Θ_μ in terms of components C_μ^Θ, ω_μ^Θ, etc. and use Eq. (26.3.9) to express the chiral superfield X in terms of components A^X, ψ^X, etc. With the aid of Eqs. (26.A.9), (26.A.16), (26.A.17), and the Dirac matrix identities

$$[\gamma^\rho , \gamma^\sigma] = -\tfrac{1}{2} i \epsilon^{\rho\sigma\mu\nu} \gamma_5 [\gamma_\mu , \gamma_\nu] , \tag{26.7.28}$$

$$\gamma^\mu \gamma^\rho \gamma^\nu = \eta^{\mu\rho} \gamma^\nu - \eta^{\mu\nu} \gamma^\rho + \eta^{\nu\rho} \gamma^\mu + i \gamma_5 \epsilon^{\mu\nu\rho\sigma} \gamma_\sigma , \tag{26.7.29}$$

we can then expand both sides of Eq. (26.7.26) in the terms

$$1 , \ \theta , \ \gamma_5\theta , \ \gamma^\nu\theta , \ \gamma_5\gamma^\nu\theta , \ \gamma_5[\gamma^\mu,\gamma^\nu]\theta ,$$
$$\left(\bar\theta\theta\right) , \ \left(\bar\theta\gamma_5\theta\right) , \ \left(\bar\theta\gamma_5\gamma^\nu\theta\right) ,$$
$$\theta\left(\bar\theta\gamma_5\theta\right) , \ \gamma_5\theta\left(\bar\theta\gamma_5\theta\right) , \ \gamma^\nu\theta\left(\bar\theta\gamma_5\theta\right) ,$$
$$\gamma^\nu\gamma_5\theta\left(\bar\theta\gamma_5\theta\right) , \ [\gamma^\rho,\gamma^\sigma]\theta\left(\bar\theta\gamma_5\theta\right) , \ \left(\bar\theta\gamma_5\theta\right)^2 .$$

Matching the coefficients of 1, θ, $\gamma_5\theta$, $\gamma^\nu\theta$, $\gamma_5\gamma^\nu\theta$, $\gamma_5[\gamma^\mu, \gamma^\nu]\theta$, respectively, yields the results[††]

$$\psi^X = -i\gamma_5\gamma^\mu\omega^\Theta_\mu \,, \tag{26.7.30}$$

$$F^X = \partial^\mu C^\Theta_\mu \,, \tag{26.7.31}$$

$$G^X = (V^\Theta)^\mu{}_\mu \,, \tag{26.7.32}$$

$$\partial_\mu A^X = -N^\Theta_\mu \,, \tag{26.7.33}$$

$$\partial_\mu B^X = M^\Theta_\mu \,, \tag{26.7.34}$$

$$0 = V^\Theta_{\mu\nu} - V^\Theta_{\nu\mu} + \epsilon_{\mu\nu\rho\sigma}\partial^\sigma C^{\Theta\rho} \,. \tag{26.7.35}$$

Matching the coefficients of either $(\bar\theta\theta)$ or $(\bar\theta\gamma_5\theta)$ yields the same result:

$$0 = \gamma^\mu \lambda^\Theta_\mu \,, \tag{26.7.36}$$

and matching the coefficients of $(\bar\theta\gamma_5\gamma^\nu\theta)$ yields the result

$$-i\gamma_5 [\gamma^\nu , \not\partial]\, \psi^X = 2\gamma^\mu\gamma^\nu\lambda^\Theta_\mu + \gamma^\mu [\gamma^\nu , \not\partial]\, \omega^\Theta_\mu \,. \tag{26.7.37}$$

From Eqs. (26.7.30), (26.7.36), and (26.7.37) we obtain the conservation of the supersymmetry current (26.7.20):

$$0 = \partial_\mu S^\mu_{\text{new}} = -2\partial^\mu\omega^\Theta_\mu + 2\,\not\partial\gamma^\mu\omega^\Theta_\mu \,, \tag{26.7.38}$$

and a relation among λ^Θ_μ and ω^Θ_μ:

$$\lambda^\Theta_\nu = -\,\not\partial\omega^\Theta_\nu + \partial_\nu\gamma^\mu\omega^\Theta_\mu \,. \tag{26.7.39}$$

Matching the coefficients of $\theta(\bar\theta\gamma_5\theta)$ and $\gamma_5\theta(\bar\theta\gamma_5\theta)$ gives relations that can be obtained by taking the divergence of Eqs. (26.7.34) and (26.7.33), respectively. Matching the coefficients of $\gamma^\rho\theta(\bar\theta\gamma_5\theta)$ gives

$$\partial_\rho G^X = \partial^\mu V^\Theta_{\mu\rho} + \partial^\mu V^\Theta_{\rho\mu} - \partial_\rho V^{\Theta\lambda}{}_\lambda \,, \tag{26.7.40}$$

which, combined with Eq. (26.7.32), yields the conservation law

$$\partial_\mu T^{\mu\nu} = 0 \,, \tag{26.7.41}$$

where $T^{\mu\nu}$ is the symmetric tensor

$$T_{\mu\nu} \equiv -\tfrac{1}{2}V^\Theta_{\mu\nu} - \tfrac{1}{2}V^\Theta_{\nu\mu} + \eta_{\mu\nu}V^{\Theta\lambda}{}_\lambda \,. \tag{26.7.42}$$

Matching the coefficients of $\gamma^\rho\gamma_5\theta(\bar\theta\gamma_5\theta)$ gives

$$\partial_\mu F^X = 2D^\Theta_\mu + \Box C^\Theta_\mu + \epsilon_{\rho\nu\sigma\mu}\partial^\nu V^{\Theta\rho\sigma} \,, \tag{26.7.43}$$

which, with Eqs. (26.7.31) and (27.6.35), yields a relation between D^Θ_μ and C^Θ_μ:

$$D^\Theta_\mu = -\Box C^\Theta_\mu + \partial_\mu\partial^\nu C^\Theta_\nu \,. \tag{26.7.44}$$

[††] Note that $V^\Theta_{\mu\nu}$ is the V_ν-component of Θ_μ, not the V_μ-component of Θ_ν.

Matching the coefficients of $[\gamma^\rho, \gamma^\sigma]\theta(\bar{\theta}\gamma_5\theta)$ and $(\bar{\theta}\gamma_5\theta)^2$ gives results that already follow from Eq. (26.7.34) and from Eqs. (26.7.38) and (26.7.39), respectively.

The conserved symmetric tensor $T^{\mu\nu}$ may be identified as the energy-momentum tensor of the system. To check this, we use Eqs. (26.1.18) and (26.2.12) to write the change in $\omega_\mu^\Theta(x)$ under a supersymmetry transformation with infinitesimal parameter α as

$$\delta\omega_\mu^\Theta = -i\Big[(\bar{Q}\alpha),\, \omega_\mu^\Theta\Big] = +i\Big[\omega_\mu^\Theta\,,(\bar{Q}\alpha)\Big]$$
$$= \Big(-i\gamma_5\,\partial\!\!\!/ C_\mu^\Theta - M_\mu^\Theta + i\gamma_5 N_\mu^\Theta + \gamma^\nu V_{\mu\nu}^\Theta\Big)\alpha\,.$$

Eqs. (26.7.33)–(26.7.35) allow us to put this in the form

$$i\Big\{\omega_\mu^\Theta\,,\bar{Q}\Big\} = \tfrac{1}{2}\gamma^\nu(V_{\mu\nu}^\Theta + V_{\nu\mu}^\Theta) - \partial_\mu(B^X + \gamma_5 A^X) - i\gamma_5\,\partial\!\!\!/ C_\mu^\Theta + \tfrac{1}{2}\epsilon_{\mu\nu\kappa\sigma}\gamma^\nu\partial^\kappa C^{\Theta\sigma}\,.$$

In terms of the currents (26.7.20) and (26.7.42), this reads

$$i\,\{S^\mu_{\rm new}, \bar{Q}\} = 2\gamma_\nu T^{\mu\nu} + 2(\partial^\mu - \gamma^\mu\,\partial\!\!\!/)(B^X + \gamma_5 A^X) - \epsilon^{\mu\nu\kappa\sigma}\gamma_\nu\partial_\kappa C_\sigma^\Theta$$
$$+2i\gamma_5\Big(\,\partial\!\!\!/ C^{\Theta\mu} - \gamma^\mu\gamma^\lambda\,\partial\!\!\!/ C_\lambda^\Theta - \tfrac{1}{2}\gamma^\mu[\partial\!\!\!/, \gamma^\sigma]C_\sigma^\Theta\Big)\,. \tag{26.7.45}$$

For $\mu = 0$ all the terms on the right except the first are space derivatives, and therefore vanish when we integrate over space, leaving us with

$$i\Big\{\int d^3x\, S^0_{\rm new}\,,\bar{Q}\Big\} = 2\gamma_\nu \int d^3x\, T^{0\nu}\,. \tag{26.7.46}$$

We have defined the supersymmetry current $S^\mu_{\rm new}$ to give $\int d^3x\, S^0_{\rm new} = Q$, so the fundamental anticommutation relation (25.2.36) tells us that

$$\int d^3x\, T^{0\nu} = P^\nu\,, \tag{26.7.47}$$

which, with the conservation condition (26.7.41), allows us to identify $T^{\mu\nu}$ as the energy-momentum tensor.

It is important to note *which* energy-momentum tensor we have constructed in this way. Either directly from Eq. (26.7.21), or by considering the supersymmetry transformation of the current (26.7.18), we can calculate that the energy-momentum tensor $T^{\mu\nu}$ for renormalizable theories of chiral superfields is

$$T^{\mu\nu} = \sum_n \Big[\partial^\mu\phi_n^*\partial^\nu\phi_n + \partial^\nu\phi_n^*\partial^\mu\phi_n\Big] - \eta^{\mu\nu}\sum_n\left[\partial^\lambda\phi_n^*\partial_\lambda\phi_n + \left|\frac{\partial f(\phi)}{\partial\phi_n}\right|^2\right]$$
$$+\frac{1}{3}\Big(\eta^{\mu\nu}\Box - \partial^\mu\partial^\nu\Big)\sum_n|\phi_n|^2 + \cdots\,, \tag{26.7.48}$$

where the dots denote terms involving fermions, which do not concern us here. We see that the last term, which is related by supersymmetry to the correction term in Eq. (26.7.18), has the effect that the energy-momentum tensor is traceless for the massless free-field theory with no superpotential, in which case $\Box\phi_n = 0$. A simple calculation shows that $T^{\mu\nu}$ is also traceless more generally for scale-invariant theories, with $f(\phi)$ a homogeneous polynomial of third order in the ϕ_n.

Supersymmetry also imposes an interesting relation among violations of scale invariance and R conservation. Eqs. (26.7.30)–(26.7.32) show that $\gamma_\mu S^\mu_{\rm new} = 6\gamma_\mu \omega^{\Theta\,\mu}$, $\partial^\mu C^\Theta_\mu$, and $T^\lambda{}_\lambda = 2V^{\Theta\,\mu}{}_\mu$ (which measures the violation of scale invariance) are proportional to components of a chiral superfield X, so if any one of these vanishes as an operator equation (that is, not just for some particular field configuration), then they all do. In this case, we can show that $C^{\Theta\,\rho}$ is proportional to the current of an R quantum number. To see this, note that Eq. (26.2.11) gives

$$\delta C^\Theta_\sigma = i\left[C^\Theta_\sigma\,,\,(\bar\alpha\, Q)\right] = i\left(\bar\alpha\,\gamma_5\,\omega^\Theta_\sigma\right)\,,$$

so that in general

$$\left[C^\Theta_\sigma\,,\,Q\right] = \gamma_5\omega^\Theta_\sigma\,. \tag{26.7.49}$$

We have seen that if C^Θ_σ is conserved then $\gamma_\mu S^\mu = 0$, so that Eq. (26.7.20) gives $S_\sigma = -2\omega^\Theta_\sigma$. Setting $\sigma = 0$ and integrating over $\mathbf{x}$ in Eq. (26.7.49) then gives

$$\left[\int d^3x\; C^{0\Theta}\,,\,Q\right] = -\tfrac{1}{2}\gamma_5\, Q\,. \tag{26.7.50}$$

We can therefore introduce a current

$$\mathscr{R}^\mu \equiv 2\,C^{\mu\Theta}\,, \tag{26.7.51}$$

which *if* conserved is the current of a quantum number $\mathscr{R} \equiv \int d^3x\,\mathscr{R}^0$ for which Q_L and Q_R destroy the values $+1$ and -1, respectively. Since the commutator of Q_L with a scalar superfield Φ involves a term $\partial\Phi/\partial\theta_L$, this means that θ_L carries the $\mathscr{R}$ value $+1$, in accord with the usual definition. A theory in which the superfield X vanishes, or, equivalently, in which $T^\mu{}_\mu$, $\gamma_\mu S^\mu$, and $\partial_\mu\mathscr{R}^\mu$ all vanish, is invariant under an enlarged set of supersymmetry transformations, generated by the superconformal algebra described at the end of Section 25.2.

In scale-invariant theories the value of the $\mathscr{R}$ quantum number carried by various superfields is fixed by the structure of the Lagrangian. For instance, in a scale-invariant theory of chiral scalar superfields, the superpotential must be a homogeneous polynomial of third order in the superfields. The $\mathscr{F}$-term of the superpotential is proportional to the coefficient of θ_L^2, which has $\mathscr{R}$ quantum number $+2$, so the $\mathscr{R}$ quantum number of the $\mathscr{F}$-term of the superpotential is the $\mathscr{R}$ quantum number of the superpotential itself minus two. $\mathscr{R}$ invariance then requires that we give the scalar superfields an $\mathscr{R}$ quantum number +2/3, so that the superpotential will have $\mathscr{R}$ quantum number +2, and its $\mathscr{F}$-term will have $\mathscr{R}$ quantum number zero. That is, the scalar components ϕ_n have $\mathscr{R} = 2/3$ and the spinor components ψ_{nL} (proportional to the coefficient of θ_L in the superfield) have $\mathscr{R} = -1/3$. This can be verified by calculating the current $\mathscr{R}^\mu$ from the C-term of the supercurrent (26.7.21) for this class of theories:

$$\mathscr{R}_\mu = \tfrac{2}{3} i [\phi^* \partial_\mu \phi - \phi \partial_\mu \phi^*] - \tfrac{1}{6} i \left(\bar{\psi} \gamma_\mu \gamma_5 \psi\right) . \tag{26.7.52}$$

(The second term contains an extra factor of 1/2 because ψ is a Majorana spinor.)

Quantum corrections can introduce violations of $\mathscr{R}$ invariance (through Adler–Bell–Jackiw anomalies) and of scale invariance (through the renormalization group running of the coupling constants) but supersymmetry continues to impose a relation between these symmetry violations despite these corrections.[7a] We will see an example of this in Section 29.3.

* * *

The conservation condition (26.7.26) does not uniquely determine either the supercurrent Θ^μ or the associated chiral superfield X. In particular, we may add to Θ^μ a change

$$\Delta\Theta^\mu = \partial^\mu Y , \tag{26.7.53}$$

with Y an arbitrary chiral superfield. The left-hand side of Eq. (26.7.26) is then changed by

$$\gamma_\mu \mathscr{D} \Delta\Theta^\mu = \partial\!\!\!/ \mathscr{D} Y .$$

For a left-chiral superfield Y_L the chirality condition $\mathscr{D}_R Y_L = 0$ and the anticommutation relations (26.2.30) give

$$\begin{aligned} \partial\!\!\!/ \mathscr{D}_\alpha Y &= -\tfrac{1}{2} \left[\left\{\mathscr{D}_L, \bar{\mathscr{D}}_R\right\} \mathscr{D}_R\right]_\alpha Y_L \\ &= -\tfrac{1}{2} \left[\mathscr{D}_{L\alpha} \left(\bar{\mathscr{D}}_R \mathscr{D}_R\right) Y_L + \sum_\beta \bar{\mathscr{D}}_{L\beta} \mathscr{D}_{R\alpha} \mathscr{D}_{L\beta} Y_L\right] . \end{aligned}$$

The matrix $\epsilon\gamma_5$ in $\bar{\mathscr{D}}_{L\beta}$ in the second term on the right of the above expression may be moved to the final operator $\mathscr{D}_{L\beta}$, so that the chirality condition and anticommutation relations give

$$\sum_\beta \bar{\mathscr{D}}_{L\beta} \mathscr{D}_{R\alpha} \mathscr{D}_{L\beta} Y_L = -\sum_\beta \mathscr{D}_{L\beta} \mathscr{D}_{R\alpha} \bar{\mathscr{D}}_{L\beta} Y_L = 2(\partial\!\!\!/ \mathscr{D})_\alpha Y_L ,$$

and hence

$$\partial\!\!\!/ \mathscr{D} Y_L = -\tfrac{1}{2} \left[\mathscr{D}\left(\bar{\mathscr{D}}\mathscr{D}\right) Y_L + 2\, \partial\!\!\!/ \mathscr{D} Y_L\right] = -\tfrac{1}{4} \mathscr{D}\left(\bar{\mathscr{D}}\mathscr{D}\right) Y_L .$$

The same result can be derived in the same way for any right-chiral superfield, and hence it holds also for an arbitrary sum Y of a left-chiral and a right-chiral superfield

$$\gamma_\mu \mathscr{D} \Delta\Theta^\mu = \not{\partial}\mathscr{D} Y = -\tfrac{1}{4}\mathscr{D}\left(\bar{\mathscr{D}}\mathscr{D}\right) Y \,. \tag{26.7.54}$$

This is of the same form as the conservation condition (26.7.26), with the associated chiral superfield X changed by the chiral superfield

$$\Delta X = -\tfrac{1}{4}(\bar{\mathscr{D}}\mathscr{D}) Y \,. \tag{26.7.55}$$

It is easy to check that the addition of $\Delta\Theta^\mu$ to Θ^μ changes $T^{\mu 0}$ and $S^0_{\rm new}$ only by space derivatives, and therefore does not change the energy-momentum four-vector P^μ or the supercharge Q.

We saw in Section 26.6 that any chiral superfield X may be expressed in the form $X = (\bar{\mathscr{D}}\mathscr{D})S$, and hence can be removed by adding a term of the form (26.7.55) to Θ^μ with $Y = 4S$. But in general S and the new Θ^μ constructed in this way will not be local. This situation is already familiar from our experience with triangle anomalies, discussed in Chapter 22 — we saw there that, although it is always possible to construct terms that if added to the Lagrangian density would cancel these anomalies, in general these terms would not be local, and hence must be excluded from the Lagrangian density. There are chiral superfields that can be expressed as $(\bar{\mathscr{D}}\mathscr{D})S$ with S local, and that therefore, if present in the associated chiral superfield X, could be eliminated by adding local terms of the form (26.7.53) to Θ^μ. They include, for instance, a term of the form $\mathrm{Re}\,(k\partial f(\Phi)/\partial\Phi)$, with k an arbitrary complex constant, because the field equations (26.6.15) and (26.6.16) show that $(\bar{\mathscr{D}}\mathscr{D})\mathrm{Re}\,(k^*\Phi) = 4\mathrm{Re}\,(k\partial f(\Phi)/\partial\Phi)$. But in general the changes in X that can be made in this way are quite limited.

26.8 General Kahler Potentials*

There are several circumstances in which it is necessary to consider non-renormalizable Lagrangian densities of the general form (26.3.30)

$$\mathscr{L} = 2\,\mathrm{Re}\left[f(\Phi)\right]_{\mathscr{F}} + \tfrac{1}{2}\left[K(\Phi,\Phi^*)\right]_D \,, \tag{26.8.1}$$

where the superpotential f is an arbitrary function of left-chiral scalar superfields Φ_n but not their derivatives, and the Kahler potential K is an arbitrary function of the Φ_n and Φ^*_n but not their derivatives.

* This section lies somewhat out of the book's main line of development, and may be omitted in a first reading.

This situation arises in effective field theories whose symmetries rule out any renormalizable interactions, or in which the renormalizable interactions all happen to be small. It is often then possible to calculate scattering amplitudes at low energy from tree graphs, using a Lagrangian with the smallest value for some combination of the numbers of derivatives, fermion fields, and any small renormalizable couplings. In Section 19.5 we examined such an effective field theory with no renormalizable couplings, involving nucleons and soft pions. The dynamically broken gauge theories discussed in Section 21.4 provide examples of effective field theories of this sort with small renormalizable couplings. This circumstance also arises in supersymmetric theories whose symmetries do not allow a superpotential, or where the superpotential is for some reason small. We will encounter an example of this sort when we consider the extended $N = 2$ supersymmetric theory of Abelian gauge superfields and gauge-neutral chiral scalar superfields in Section 29.5. We will show there that low-energy scattering amplitudes in this theory are generated by tree graphs using a Lagrangian density of the form (26.8.1), with $f = 0$ and K a function only of the Φ_n and Φ_n^* but no derivatives, plus $\mathscr{F}$-terms quadratic in gauge superfields. The inclusion of Kahler potentials having arbitrary dependence on Φ_n and Φ_n^* but not on their derivatives is particularly important in effective field theories in which some scalar fields are of the same order of magnitude as the fundamental energy scale of the underlying theory, although all other field values and all energies are much smaller. This will be of interest, for instance, in connection with theories of gravity-mediated supersymmetry breaking, discussed in Section 31.6.

Let us consider how to express the Lagrangian density (26.8.1) in terms of component fields. We did not use the assumption that $f(\Phi)$ is a *cubic* polynomial in deriving Eq. (26.4.4), so this continues to give the $\mathscr{F}$-terms in the Lagrangian contributed by an arbitrary superpotential. To derive the D-terms, we note that the term in the Kahler potential of fourth order in θ is

$$
\begin{aligned}
K(\Phi, \Phi^*)_{\theta^4} = &-\frac{1}{8}\left(\bar{\theta}\gamma_5\theta\right)^2 \sum_n \left[\frac{\partial K(\phi,\phi^*)}{\partial \phi_n}\,\Box\phi_n + \frac{\partial K(\phi,\phi^*)}{\partial \phi_n^*}\,\Box\phi_n^*\right] \\
&+\sum_{nm} \frac{\partial^2 K(\phi,\phi^*)}{\partial\phi_n \partial\phi_m^*}\left(\bar{\theta}\gamma_5\theta\right)\Big[\left(\bar{\theta}\psi_{mR}\right)\left(\bar{\theta}\ \partial\!\!\!/\psi_{nL}\right) \\
&\qquad -\left(\bar{\theta}\psi_{nL}\right)\left(\bar{\theta}\ \partial\!\!\!/\psi_{mR}\right)\Big] \\
&+2\,\mathrm{Re}\sum_{nml} \frac{\partial^3 K(\phi,\phi^*)}{\partial\phi_n\,\partial\phi_m\,\partial\phi_l^*}\left(\theta_L^{\mathrm{T}}\epsilon\psi_{nL}\right)\left(\theta_L^{\mathrm{T}}\epsilon\psi_{mL}\right)\left(\theta_L^{\mathrm{T}}\epsilon\theta_L\right)^*\mathscr{F}_l^*
\end{aligned}
$$

$$+2\,\mathrm{Re}\sum_{nml}\frac{\partial^3 K(\phi,\phi^*)}{\partial\phi_n\,\partial\phi_m\,\partial\phi_l^*}\Big(\bar\theta\psi_{mL}\Big)\Big(\bar\theta\psi_{lR}\Big)\Big(\bar\theta\gamma_5\gamma_\mu\theta\Big)\partial^\mu\phi_n$$
$$+\sum_{nmlk}\frac{\partial^4 K(\phi,\phi^*)}{\partial\phi_n\,\partial\phi_m\,\partial\phi_k^*\,\partial\phi_\ell^*}\Big(\bar\theta\psi_{nL}\Big)\Big(\bar\theta\psi_{mL}\Big)\Big(\bar\theta\psi_{lR}\Big)\Big(\bar\theta\psi_{kR}\Big)$$
$$-\frac{1}{4}\sum_{nm}\frac{\partial^2 K(\phi,\phi^*)}{\partial\phi_n\,\partial\phi_m^*}\mathscr{F}_n\mathscr{F}^*_m\Big(\bar\theta(1+\gamma_5)\theta\Big)\Big(\bar\theta(1-\gamma_5)\theta\Big)$$
$$+\frac{1}{4}\Big(\bar\theta\gamma_5\gamma^\mu\theta\Big)\Big(\bar\theta\gamma_5\gamma^\nu\theta\Big)\sum_{mn}\Bigg[-\frac{\partial^2 K(\phi,\phi^*)}{\partial\phi_n\partial\phi_m^*}\partial_\mu\phi_n\,\partial_\nu\phi_m^*$$
$$+\frac{1}{2}\frac{\partial^2 K(\phi,\phi^*)}{\partial\phi_n\phi_m}\partial_\mu\phi_n\,\partial_\nu\phi_m+\frac{1}{2}\frac{\partial^2 K(\phi,\phi^*)}{\partial\phi_n^*\partial\phi_m^*}\partial_\mu\phi_n^*\,\partial_\nu\phi_m^*\Bigg]\,. \tag{26.8.2}$$

We can again use Eqs. (26.A.18) and (26.A.19) as well as Eq. (26.A.9) to put the θ dependence of this expression in the form of an over-all factor $(\bar\theta\gamma_5\theta)^2$, and find

$$K(\Phi,\Phi^*)_{\theta^4}=\frac{1}{4}\Big(\bar\theta\gamma_5\theta\Big)^2\Bigg\{-\frac{1}{2}\sum_n\frac{\partial K(\phi,\phi^*)}{\partial\phi_n}\Box\phi_n-\frac{1}{2}\sum_n\frac{\partial K(\phi,\phi^*)}{\partial\phi_n^*}\Box\phi_n^*$$
$$+\sum_{nm}\frac{\partial^2 K(\phi,\phi^*)}{\partial\phi_n\,\partial\phi_m^*}\Big[\Big(\overline{\psi_m}\,\partial\!\!\!/\psi_{nL}\Big)+\Big(\overline{\psi_n}\,\partial\!\!\!/\psi_{mR}\Big)-2\mathscr{F}_n\mathscr{F}^*_m\Big]$$
$$+2\,\mathrm{Re}\sum_{nml}\frac{\partial^3 K(\phi,\phi^*)}{\partial\phi_n\,\partial\phi_m\,\partial\phi_l^*}\Big(\overline{\psi_n}\psi_{mL}\Big)\mathscr{F}^*_\ell$$
$$-2\,\mathrm{Re}\sum_{nml}\frac{\partial^3 K(\phi,\phi^*)}{\partial\phi_n\,\partial\phi_m\,\partial\phi_l^*}\Big(\overline{\psi_m}\gamma^\mu\psi_{\ell R}\Big)\partial_\mu\phi_n$$
$$-\frac{1}{2}\sum_{nmlk}\frac{\partial^4 K(\phi,\phi^*)}{\partial\phi_n\,\partial\phi_m\,\partial\phi_l^*\,\partial\phi_k^*}\Big(\overline{\psi_n}\psi_{mL}\Big)\Big(\overline{\psi_k}\psi_{lR}\Big)$$
$$+\sum_{nm}\frac{\partial^2 K(\phi,\phi^*)}{\partial\phi_n\,\partial\phi_m^*}\partial_\mu\phi_n\,\partial^\mu\phi_m^*-\frac{1}{2}\sum_{nm}\frac{\partial^2 K(\phi,\phi^*)}{\partial\phi_n\,\partial\phi_m}\partial_\mu\phi_n\,\partial^\mu\phi_m$$
$$-\frac{1}{2}\sum_{nm}\frac{\partial^2 K(\phi,\phi^*)}{\partial\phi_n^*\,\partial\phi_m^*}\partial_\mu\phi_n^*\,\partial^\mu\phi_m^*\Bigg\}\,. \tag{26.8.3}$$

To make the reality properties of the fermion kinematic terms transparent, we can use Eq. (26.A.21) to write

$$\Big(\overline{\psi_n}\,\partial\!\!\!/\psi_{mR}\Big)=\Big(\overline{\psi_n}\,\partial\!\!\!/\psi_{mL}\Big)^*\,.$$

The D-term of $K(\Phi,\Phi^*)$ is the coefficient of $-(\bar\theta\gamma_5\theta)^2/4$ minus half the d'Alembertian of the θ-independent term in $K(\Phi,\Phi^*)$, which is just

$K(\phi, \phi^*)$, so

$$\frac{1}{2}\Big[K(\Phi, \Phi^*)\Big]_D = \mathrm{Re}\sum_{nm}\mathscr{G}_{nm}\Bigg[-\frac{1}{2}\Big(\overline{\psi_m}\ \not\partial(1+\gamma_5)\psi_n\Big) + \mathscr{F}_n\mathscr{F}^*_m - \partial_\mu\phi_n\,\partial^\mu\phi^*_m\Bigg]$$
$$-\mathrm{Re}\sum_{nml}\frac{\partial^3 K(\phi,\phi^*)}{\partial\phi_n\,\partial\phi_m\,\partial\phi^*_l}\Big(\overline{\psi_n}\psi_{mL}\Big)\mathscr{F}^*_\ell$$
$$+\mathrm{Re}\sum_{nml}\frac{\partial^3 K(\phi,\phi^*)}{\partial\phi_n\,\partial\phi_m\,\partial\phi^*_l}\Big(\overline{\psi_m}\gamma^\mu\psi_{\ell R}\Big)\partial_\mu\phi_n$$
$$+\frac{1}{4}\sum_{nmlk}\frac{\partial^4 K(\phi,\phi^*)}{\partial\phi_n\,\partial\phi_m\,\partial\phi^*_l\,\partial\phi^*_k}\Big(\overline{\psi_n}\psi_{mL}\Big)\Big(\overline{\psi_k}\psi_{lR}\Big)\,, \tag{26.8.4}$$

where $\mathscr{G}(\phi, \phi^*)$ is the *Kahler metric*

$$\mathscr{G}_{nm}(\phi, \phi^*) \equiv \frac{\partial^2 K(\phi,\phi^*)}{\partial\phi_n\,\partial\phi^*_m}\,. \tag{26.8.5}$$

Note that the constant matrix g_{nm} in Eq. (26.4.2) is replaced here with the Kahler metric $\mathscr{G}_{nm}(\phi, \phi^*)$. Because the Kahler metric is field-dependent, we cannot in general make it equal to the unit matrix by a field redefinition, so the total Lagrangian must be left in the form

$$\mathscr{L} = \mathrm{Re}\sum_{nm}\mathscr{G}_{nm}\Bigg[-\frac{1}{2}\Big(\overline{\psi_m}\ \not\partial(1+\gamma_5)\psi_n\Big) + \mathscr{F}_n\mathscr{F}^*_m - \partial_\mu\phi_n\,\partial^\mu\phi^*_m\Bigg]$$
$$-\mathrm{Re}\sum_{nml}\frac{\partial^3 K(\phi,\phi^*)}{\partial\phi_n\,\partial\phi_m\,\partial\phi^*_l}\Big(\overline{\psi_n}\psi_{mL}\Big)\mathscr{F}^*_\ell$$
$$+\mathrm{Re}\sum_{nml}\frac{\partial^3 K(\phi,\phi^*)}{\partial\phi_n\,\partial\phi_m\,\partial\phi^*_l}\Big(\overline{\psi_m}\gamma^\mu\psi_{\ell R}\Big)\partial_\mu\phi_n$$
$$+\frac{1}{4}\sum_{nmlk}\frac{\partial^4 K(\phi,\phi^*)}{\partial\phi_n\,\partial\phi_m\,\partial\phi^*_l\,\partial\phi^*_k}\Big(\overline{\psi_n}\psi_{mL}\Big)\Big(\overline{\psi_k}\psi_{lR}\Big)$$
$$-\mathrm{Re}\sum_{nm}\frac{\partial^2 f(\phi)}{\partial\phi_n\,\partial\phi_m}\Big(\overline{\psi_n}\,\psi_{mL}\Big) + 2\,\mathrm{Re}\sum_n\mathscr{F}_n\frac{\partial f(\phi)}{\partial\phi_n}\,. \tag{26.8.6}$$

The bilinear $(\overline{\psi_m}\ \not\partial\gamma_5\psi_n)$ is a total derivative, and could be discarded if $\mathscr{G}_{nm}$ were a constant, but must be kept for general Kahler potentials. This result will be extended to include gauge superfields at the end of Section 27.4.

* * *

As discussed in Section 19.6, the spontaneous breakdown of a global symmetry group G to a subgroup H entails the existence of a set of real massless Goldstone bosons with scalar fields π_k, for which the term in the Lagrangian with the minimum number of derivatives takes the form

$$\mathscr{L}_{G/H} = -\sum_{k\ell} G_{k\ell}(\pi)\partial_\mu \pi_k \partial^\mu \pi_\ell \,, \tag{26.8.7}$$

where $G_{k\ell}(\pi)$ is the metric of the coset space G/H. (Theories with Lagrangian densities of this general form are known as *non-linear σ-models.*) By writing the complex fields ϕ_n in terms of their real and imaginary parts, the term $-\sum_{nm} \mathscr{G}_{nm}(\phi, \phi^*)\partial_\mu \phi_n \partial^\mu \phi^*_m$ in the Lagrangian density (26.8.6) can be put in the form (26.8.7), but the converse is not generally true: the condition that a set of real coordinates like the Goldstone boson fields π_k can be interpreted as the real and imaginary parts of a set of complex coordinates like the fields ϕ_n with the metric in these coordinates locally given by Eq. (26.8.5) defines what is called a *Kahler manifold.*** But it should not be thought that in the common cases where G/H is *not* a Kahler manifold, it is impossible for G to be spontaneously broken to H while keeping supersymmetry unbroken. What happens in these cases is the appearance of extra massless bosons, which, together with the Goldstone bosons, do form a Kahler manifold.

This is because the superpotential $f(\phi)$ depends on ϕ but not ϕ^*, so if the whole Lagrangian is invariant under a global symmetry group G then the superpotential is automatically invariant under a group $G_{\mathbb{C}}$, the complexification of G: if G consists of transformations $\exp(i\sum_A \theta_A t_A)$ with generators t_A and arbitrary real parameters θ_A, then $G_{\mathbb{C}}$ consists of transformations $\exp(i\sum_A z_A t_A)$ with the same generators and arbitrary complex parameters z_A. (For instance, if G is $U(n)$ then $G_{\mathbb{C}}$ is $GL(n,\mathbb{C})$, the group of all complex non-singular matrices, while if G is $SU(n)$ then $G_{\mathbb{C}}$ is $SL(n,\mathbb{C})$, the group of all complex non-singular matrices with unit determinant.) Likewise, if some stationary point $\phi^{(0)}$ of $f(\phi)$ is left invariant by some subgroup H of G, then it will also be left invariant by a

** The significance of Kahler manifolds in this context was pointed out in an early paper by Zumino.[8] Note that it is not necessary for the metric to be expressible in the form (26.8.5) over the whole manifold with a *single* Kahler potential $K(\phi, \phi^*)$; it is only necessary that the manifold can be covered in finite overlapping patches in which this holds, with different Kahler potentials in each patch. The simplest example of a Kahler manifold is the flat complex plane, with Kahler potential $|z|^2$. As an example of a coset space G/H that is a Kahler manifold, Zumino gave the case where $G = GL(p,\mathbb{C}) \times GL(p+q,\mathbb{C})$ and $H = GL(p,\mathbb{C})$, with p and q arbitrary positive-definite integers, and $GL(N,\mathbb{C})$ the group of complex non-singular $N \times N$ matrices. The coset space G/H here has complex coordinates ϕ_n that may be taken as the components of a complex $p \times (p+q)$ matrix A, which under G and H undergoes the transformations $A \to BAC$ and $A \to BA$, respectively, where B and C are square non-singular complex matrices of dimensionality p and $p+q$, respectively. The Kahler potential in this case is simply $K \propto \ln \operatorname{Det} A\, A^\dagger$.

subgroup $H_{\mathbb{C}}$ of $G_{\mathbb{C}}$, the complexification of H. Whether or not G/H is a Kahler manifold, the complexified coset space $G_{\mathbb{C}}/H_{\mathbb{C}}$ is always a Kahler manifold. This follows because $G_{\mathbb{C}}/H_{\mathbb{C}}$ is a complex submanifold of the flat complex space of the ϕ_n, which is a Kahler manifold, and it is a theorem that any complex submanifold of a Kahler manifold is a Kahler manifold.[9] If parameterized by the values of $\phi_n(z) = [\exp(i\sum_A z_A t_A)\phi^{(0)}]_n$, the Kahler manifold $G_{\mathbb{C}}/H_{\mathbb{C}}$ has the metric obtained by embedding it in the flat complex space of the ϕ_n, usually taken with line element $\sum_n d\phi_n\, d\phi_n^*$.

It is true that $G_{\mathbb{C}}$ is not a symmetry of the whole Lagrangian, but the Goldstone bosons associated with the breakdown of $G_{\mathbb{C}}$ to $H_{\mathbb{C}}$ are nevertheless exactly massless. This is guaranteed by the non-renormalization theorem of Section 27.6, or, more simply, by the result of Section 25.4, that massless spin zero particles must come in *pairs* that are related by supersymmetry transformations and that therefore have the same transformation under any global symmetry group G that commutes with supersymmetry.

Appendix Majorana Spinors

This appendix summarizes some algebraic properties of Majorana spinors that are needed in dealing with superfields.

Consider a four-component fermionic Majorana spinor s that like Q or θ may be expressed in the form

$$s = \begin{pmatrix} e\varsigma^* \\ \varsigma \end{pmatrix}, \tag{26.A.1}$$

where ς is some two-component spinor and e is the 2×2 matrix

$$e \equiv \begin{pmatrix} 0 & 1 \\ -1 & 0 \end{pmatrix} = i\sigma_2 .$$

Such a spinor is related to its complex conjugate by

$$s^* = \begin{pmatrix} 0 & e \\ -e & 0 \end{pmatrix} s = -\beta\,\gamma_5\,\epsilon\, s\,, \tag{26.A.2}$$

where ϵ is the 4×4 matrix

$$\epsilon \equiv \begin{pmatrix} e & 0 \\ 0 & e \end{pmatrix} \tag{26.A.3}$$

and as usual γ_5 and β are 4×4 matrices

$$\gamma_5 = \begin{pmatrix} 1 & 0 \\ 0 & -1 \end{pmatrix}, \qquad \beta = \begin{pmatrix} 0 & 1 \\ 1 & 0 \end{pmatrix},$$

with 1 and 0 understood here as 2×2 submatrices. Taking the transpose of Eq. (26.A.2) and multiplying on the right by β gives the equivalent formula

$$\bar{s} \equiv s^{\dagger}\beta = s^{\mathrm{T}}\epsilon\gamma_5 \,. \tag{26.A.4}$$

The anticommutation of the spinor components limits the variety of covariants that can be formed from a Majorana spinor. To see this, it will be convenient first to consider the symmetry properties of bilinear covariants, which will be of some interest in themselves. For a pair of Majorana spinors s_1 and s_2 and any 4×4 numerical matrix M, Eq. (26.A.4) gives

$$\begin{aligned}\overline{s_1}\, M\, s_2 &= \sum_{\alpha\beta} s_{1\alpha}\, s_{2\beta}\, (\epsilon\,\gamma_5\, M)_{\alpha\beta} = -\sum_{\alpha\beta} s_{2\alpha}\, s_{1\beta}\, (\epsilon\,\gamma_5\, M)_{\beta\alpha} \\ &= +\sum_{\alpha\beta} s_{2\alpha}\, s_{1\beta}\, (M^{\mathrm{T}}\epsilon\,\gamma_5)_{\alpha\beta} = \overline{s_2}\,(\epsilon\gamma_5)^{-1}M^{\mathrm{T}}\epsilon\gamma_5\, s_1 \,,\end{aligned}$$

with the minus sign following the second equal sign arising from the fermionic nature of these spinors. In Section 5.4 we found that the 16 covariant matrices formed from Dirac matrices satisfy

$$M^{\mathrm{T}} = \begin{cases} +\mathscr{C}M\mathscr{C}^{-1} & M = 1,\ \gamma_5\gamma_\mu,\ \gamma_5 \\ -\mathscr{C}M\mathscr{C}^{-1} & M = \gamma_\mu,\ [\gamma_\mu, \gamma_\nu] \end{cases} \,, \tag{26.A.5}$$

where $\mathscr{C}$ is the matrix

$$\mathscr{C} = \gamma_2\beta = -\epsilon\gamma_5 = \begin{pmatrix} -e & 0 \\ 0 & e \end{pmatrix} . \tag{26.A.6}$$

It follows that

$$(\overline{s_1}\, M\, s_2) = \begin{cases} +(\overline{s_2}\, M\, s_1) & M = 1,\ \gamma_5\gamma_\mu,\ \gamma_5 \\ -(\overline{s_2}\, M\, s_1) & M = \gamma_\mu,\ [\gamma_\mu, \gamma_\nu] \end{cases} . \tag{26.A.7}$$

In particular, setting $s_1 = s_2 = s$, we find that

$$\bar{s}\,\gamma_\mu\, s = \bar{s}\,[\gamma_\mu\,,\,\gamma_\nu]\, s = 0 \,, \tag{26.A.8}$$

so the only bilinear covariants formed from a single Majorana spinor s are $\bar{s}\,s$, $\bar{s}\,\gamma_5\gamma_\mu\, s$, and $\bar{s}\,\gamma_5\, s$.

In considering the form of the most general superfields, we need to have expressions for the products of two or more Majorana spinors. For two spinors, we recall that any 4×4 matrix may be expanded as a sum of the 16 covariant matrices 1, γ_μ, $[\gamma_\mu\,,\,\gamma_\nu]$, $\gamma_5\gamma_\mu$, γ_5. Lorentz invariance tells us that for the matrix $s_\alpha\,\bar{s}_\beta$ this expansion must take the form

$$\begin{aligned} s\,\bar{s} &= k_S\,(\bar{s}\,s) + k_V\,\gamma_\mu\,(\bar{s}\,\gamma^\mu\, s) + k_T\,[\gamma_\mu\,,\,\gamma_\nu]\,(\bar{s}\,[\gamma^\mu\,,\,\gamma^\nu]\, s) \\ &\quad + k_A\,\gamma_5\gamma_\mu\,(\bar{s}\,\gamma_5\gamma^\mu\, s) + k_P\,\gamma_5\,(\bar{s}\,\gamma_5\, s) \,, \end{aligned}$$

where the ks are constants to be determined. Eq. (26.A.8) shows that we may take $k_V = k_T = 0$. The remaining coefficients may be calculated by multiplying on the right with 1, $\gamma_5\gamma^\mu$, and γ_5 and taking the trace, which gives $k_S = -1/4$, $k_A = +1/4$, and $k_P = -1/4$. In this way we find that

$$s\,\bar{s} = -\tfrac{1}{4}\,(\bar{s}\,s) + \tfrac{1}{4}\,\gamma_5\gamma_\mu\,(\bar{s}\,\gamma_5\gamma^\mu\,s) - \tfrac{1}{4}\,\gamma_5\,(\bar{s}\,\gamma_5\,s)\,. \tag{26.A.9}$$

By multiplying on the right with $-\epsilon\gamma_5$ and using Eq. (26.A.4), we may put this in the form

$$s_\alpha\,s_\beta = \tfrac{1}{4}(\epsilon\gamma_5)_{\alpha\beta}\,(\bar{s}\,s) + \tfrac{1}{4}\,(\gamma_\mu\epsilon)_{\alpha\beta}\,(\bar{s}\,\gamma_5\gamma^\mu\,s) + \tfrac{1}{4}\,\epsilon_{\alpha\beta}\,(\bar{s}\,\gamma_5\,s)\,, \tag{26.A.10}$$

or, equivalently,

$$s_\alpha\,s_\beta = \tfrac{1}{4}(\epsilon\gamma_5)_{\alpha\beta}\,(s^{\mathrm{T}}\epsilon\,\gamma_5\,s) + \tfrac{1}{4}\,(\gamma_\mu\epsilon)_{\alpha\beta}\,(s^{\mathrm{T}}\epsilon\,\gamma^\mu\,s) + \tfrac{1}{4}\,\epsilon_{\alpha\beta}\,(s^{\mathrm{T}}\epsilon\,s)\,. \tag{26.A.11}$$

Now consider the product $s_\alpha s_\beta s_\gamma$ of three components of a Majorana spinor s. We can divide s into left- and right-handed parts

$$s = s_L + s_R\,, \qquad s_L = \tfrac{1}{2}(1+\gamma_5)s\,, \qquad s_R = \tfrac{1}{2}(1-\gamma_5)s\,. \tag{26.A.12}$$

Each of s_L and s_R has only two independent components, so since the square of any fermionic c-number vanishes, we have $s_{L\alpha}s_{L\beta}s_{L\gamma} = 0$ and $s_{R\alpha}s_{R\beta}s_{R\gamma} = 0$ for all α, β, and γ, and therefore

$$s_\alpha s_\beta s_\gamma = s_{L\alpha}s_{L\beta}s_{R\gamma} + s_{L\alpha}s_{R\beta}s_{L\gamma} + s_{R\alpha}s_{L\beta}s_{L\gamma} + L \leftrightarrow R\,,$$

with '$L \leftrightarrow R$' denoting the sum of previous terms with labels L and R interchanged. To evaluate this expression, we multiply Eq. (26.A.11) by suitable factors of $(1+\gamma_5)/2$, and find

$$s_{L\alpha}\,s_{L\beta} = \tfrac{1}{4}[\epsilon(1+\gamma_5)]_{\alpha\beta}\,(s_L^{\mathrm{T}}\epsilon s_L)\,.$$

If we now multiply this with $s_{R\gamma}$, since $(s_R^{\mathrm{T}}\epsilon s_R)\,s_{R\gamma} = 0$ we can drop the label L on the spinors in the bilinear $(s_L^{\mathrm{T}}\epsilon s_L)$:

$$s_{L\alpha}\,s_{L\beta}\,s_{R\gamma} = \tfrac{1}{4}[\epsilon(1+\gamma_5)]_{\alpha\beta}\,(s^{\mathrm{T}}\epsilon\,s)\,s_{R\gamma}\,.$$

The same arguments also yield

$$s_{R\alpha}\,s_{R\beta}\,s_{L\gamma} = \tfrac{1}{4}[\epsilon(1-\gamma_5)]_{\alpha\beta}\,(s^{\mathrm{T}}\epsilon\,s)\,s_{L\gamma}\,.$$

Adding the sum of these two expressions to the same quantity with γ replaced with α or β yields finally

$$\begin{aligned} s_\alpha s_\beta s_\gamma = \; & \tfrac{1}{4}\left(s^{\mathrm{T}}\epsilon s\right)\Big[\epsilon_{\alpha\beta}\,s_\gamma - (\epsilon\gamma_5)_{\alpha\beta}\,(\gamma_5 s)_\gamma - \epsilon_{\alpha\gamma}\,s_\beta \\ & +(\epsilon\gamma_5)_{\alpha\gamma}\,(\gamma_5 s)_\beta + \epsilon_{\beta\gamma}\,s_\alpha - (\epsilon\gamma_5)_{\beta\gamma}\,(\gamma_5 s)_\alpha\Big]\,. \end{aligned} \tag{26.A.13}$$

To calculate the products of four Majorana spinor components, we note that $(s^{\mathrm{T}}\epsilon s)$ contains only terms with two s_Ls or two s_Rs, so

$$(s^{\mathrm{T}}\epsilon s)s_\gamma s_\delta = (s^{\mathrm{T}}\epsilon s)[s_{R\gamma}s_{R\delta} + s_{L\gamma}s_{L\delta}]\,.$$

Using Eq. (26.A.11) to evaluate the sum in square brackets, and noting that

$$(s^{\mathrm{T}}\epsilon s)\,(s^{\mathrm{T}}\epsilon\gamma_5 s) = (s_L^{\mathrm{T}}\epsilon s_L)(s_R^{\mathrm{T}}\epsilon s_R) - (s_R^{\mathrm{T}}\epsilon s_R)(s_L^{\mathrm{T}}\epsilon s_L) = 0\,,$$

we find that

$$(s^{\mathrm{T}}\epsilon s)\, s_\gamma s_\delta = \tfrac{1}{4}\,\epsilon_{\gamma\delta}(s^{\mathrm{T}}\epsilon s)^2\,. \tag{26.A.14}$$

Multiplying Eq. (26.A.13) with s_δ therefore yields the result

$$\begin{aligned} s_\alpha s_\beta s_\gamma s_\delta = \; & \tfrac{1}{16}\left(s^{\mathrm{T}}\epsilon s\right)^2 \Big[\epsilon_{\alpha\beta}\,\epsilon_{\gamma\delta} - (\epsilon\gamma_5)_{\alpha\beta}\,(\epsilon\gamma_5)_{\gamma\delta} - \epsilon_{\alpha\gamma}\,\epsilon_{\beta\delta} \\ & + (\epsilon\gamma_5)_{\alpha\gamma}\,(\epsilon\gamma_5)_{\beta\delta} + \epsilon_{\beta\gamma}\,\epsilon_{\alpha\delta} - (\epsilon\gamma_5)_{\beta\gamma}\,(\epsilon\gamma_5)_{\alpha\delta}\Big]\,. \end{aligned} \tag{26.A.15}$$

Any product of five components of s vanishes, so this completes the list of formulas for products of components of a Majorana spinor.

We can use these formulas to derive some additional relations that will be useful in dealing with superfields. By contracting Eq. (26.A.13) with $(\epsilon\gamma_5)_{\beta\gamma}$ and $(\epsilon\gamma_\mu)_{\beta\gamma}$, we find

$$s_\alpha\left(\bar{s}\,s\right) = -(\gamma_5\,s)_\alpha\left(\bar{s}\,\gamma_5\,s\right) \tag{26.A.16}$$

and

$$s_\alpha\left(\bar{s}\,\gamma_5\gamma_\mu\,s\right) = -(\gamma_\mu\,s)_\alpha\left(\bar{s}\,\gamma_5\,s\right)\,. \tag{26.A.17}$$

From Eqs. (26.A.16) and (26.A.17) we may derive 'Fierz' identities

$$\left(\bar{s}\,s\right)^2 = -\left(\bar{s}\,\gamma_5\,s\right)^2, \qquad \left(\bar{s}\,\gamma_5\gamma_\mu\,s\right)\left(\bar{s}\,\gamma_5\gamma_\nu\,s\right) = -\eta_{\mu\nu}\left(\bar{s}\,\gamma_5\,s\right)^2. \tag{26.A.18}$$

Also, Eq. (26.A.14) may be put in a covariant form

$$(\bar{s}\,\gamma_5\,s)\,s\,\bar{s} = -\tfrac{1}{4}\gamma_5\,(\bar{s}\,\gamma_5\,s)^2\,. \tag{26.A.19}$$

It will also be useful to record the reality properties of bilinear products of Majorana spinors. For any pair of Majorana spinors s_1, s_2 satisfying the phase convention (26.A.1), Eqs. (26.A.2) and (26.A.4) give

$$(\overline{s_1}\,M\,s_2)^* = -(s_1^\dagger\epsilon\gamma_5\,M^*\,s_2^*) = (\overline{s_1}\,\beta\,\epsilon\,\gamma_5\,M^*\,\beta\,\epsilon\,\gamma_5\,s_2)\,.$$

(The minus sign in the middle expression arises from reversing the interchange of s_1 and s_2 that occurs when we take the complex conjugate.) But Eqs. (5.4.40) and (26.A.6) give $\beta\epsilon\gamma_5\gamma_\mu^*\beta\epsilon\gamma_5 = \gamma_\mu$, so

$$\beta\,\epsilon\,\gamma_5\,M^*\,\beta\,\epsilon\,\gamma_5 = \begin{cases} +M & M = 1\,,\,\gamma_\mu\,,\,[\gamma_\mu\,,\,\gamma_\mu] \\ -M & M = \gamma_\mu\gamma_5\,,\,\gamma_5 \end{cases}\,, \tag{26.A.20}$$

and therefore

$$(\overline{s_1}\,M\,s_2)^* = \begin{cases} +(\overline{s_1}\,M\,s_2) & M = 1\,,\,\gamma_\mu\,,\,[\gamma_\mu\,,\,\gamma_\mu] \\ -(\overline{s_1}\,M\,s_2) & M = \gamma_\mu\gamma_5\,,\,\gamma_5 \end{cases}\,. \tag{26.A.21}$$

Finally, we mention that any spinor u may be written in terms of a pair of Majorana spinors $s_\pm$ as

$$u = s_+ + i s_- \,, \tag{26.A.22}$$

where

$$s_+ \equiv \frac{1}{2}\Big(u - \beta\epsilon\gamma_5 u^*\Big)\,, \qquad s_- \equiv \frac{1}{2i}\Big(u + \beta\epsilon\gamma_5 u^*\Big)\,. \tag{26.A.23}$$

To check that $s_\pm$ are Majorana spinors satisfying Eq. (26.A.2), it is only necessary to recall that $\beta\epsilon\gamma_5$ is real, and that $(\beta\epsilon\gamma_5)^2 = 1$.

Problems

1. Using the direct technique of Section 26.1 in the case of $N = 2$ supersymmetry, find the supersymmetry transformation rules of the massive field supermultiplet with just one Majorana spinor field and two complex scalars.

2. Calculate the component fields of a time-reversed superfield

$$\mathsf{T}^{-1} S(x,\theta) \mathsf{T}$$

in terms of the components of the superfield $S(x,\theta)$. What sort of superfield do we get by the time reversal of a left-chiral superfield? Of a linear superfield?

3. Consider the $N = 1$ supersymmetric theory of a single left-chiral superfield Φ. In superfield notation, list all the terms of dimensionality 5 involving Φ and/or Φ^* that might be added to the Lagrangian density.

4. Consider the theory of three left-chiral scalar superfields Φ_1, Φ_2, and Φ_3, with a conventional kinematic term, and with a superpotential

$$f(\Phi_1,\Phi_2,\Phi_3) = \Phi_1\Phi_3^2 + \Phi_2\Big(\Phi_3^2 + a\Big)\,,$$

where a is a non-zero real constant. Show that this is a theory with spontaneously broken supersymmetry. Find the minimum value of the potential. Express the field of the goldstino in terms of the fermionic components of Φ_1, Φ_2, and Φ_3.

5. Find all the components of the current superfield for the action (26.6.9), in terms of the components of the left-chiral superfields Φ_n and derivatives of the superpotential f and the Kahler potential K.

6. Check that the supersymmetry current given by Eqs. (26.7.18) and (26.7.10) is related to the ω-component of the superfield (26.7.21) by Eq. (26.7.20).

References

1. A. Salam and J. Strathdee, *Nucl. Phys.* **B76**, 477 (1974). This article is reprinted in *Supersymmetry*, S. Ferrara, ed. (North Holland/World Scientific, Amsterdam/Singapore, 1987).
2. J. Wess and B. Zumino, *Nucl. Phys.* **B70**, 13 (1974). This article is reprinted in *Supersymmetry*, Ref. 1.
3. L. O'Raifeartaigh, *Nucl. Phys.* **B96**, 331 (1975). This article is reprinted in *Supersymmetry*, Ref. 1.
4. F. A. Berezin, *The Method of Second Quantization* (Academic Press, New York, 1966).
5. J. Iliopoulos and B. Zumino, *Nucl. Phys.* **B76**, 310 (1974); S. Ferrara and B. Zumino, *Nucl. Phys.* **B87**, 207 (1975). These articles are reprinted in *Supersymmetry*, Ref. 1.
6. This section follows the approach of S. Ferrara and B. Zumino, Ref. 5.
7. X. Gràcia and J. Pons, *J. Phys.* **A25**, 6357 (1992). I am grateful to J. Gomis for suggesting the use of the equation matching the coefficients of $\ddot{q}^n$.

7*a*. M. Grisaru, in *Recent Developments in Gravitation – Cargèse 1978*, M. Lévy and S. Deser, eds. (Plenum Press, New York, 1979): 577.

8. B. Zumino, *Phys. Lett.* **87B**, 203 (1979). This article is reprinted in *Supersymmetry*, Ref. 1.
9. P. Griffiths and J. Harris, *Principles of Algebraic Geometry* (Wiley, New York, 1978): 109. I thank D. Freed for telling me of this application of the general theorem.

27

Supersymmetric Gauge Theories

The successful theories of strong, weak, and electromagnetic interactions described in the first two volumes are all gauge theories. In order to see how simple supersymmetry may make contact with reality, we must therefore consider how to construct actions that satisfy both supersymmetry and gauge invariance.[1]

27.1 Gauge-Invariant Actions for Chiral Superfields

Consider a set of Abelian or non-Abelian gauge transformations that leave the supersymmetry generator Q invariant. (For simple supersymmetry there is just one Majorana spinor supersymmetry generator, which can only furnish a trivial representation of any semi-simple gauge group.) Each component field in a supermultiplet must transform in the same way under such gauge transformations. In particular, for a left-chiral superfield we have

$$\begin{aligned}
\phi_n(x) &\rightarrow \sum_m \left[\exp\left(i\sum_A t_A\Lambda^A(x)\right)\right]_{nm} \phi_m(x)\,, \\
\psi_{nL}(x) &\rightarrow \sum_m \left[\exp\left(i\sum_A t_A\Lambda^A(x)\right)\right]_{nm} \psi_{mL}(x)\,, \\
\mathscr{F}_n(x) &\rightarrow \sum_m \left[\exp\left(i\sum_A t_A\Lambda^A(x)\right)\right]_{nm} \mathscr{F}_m(x)\,,
\end{aligned} \tag{27.1.1}$$

where t_A are Hermitian matrices representing the generators of the gauge algebra, and $\Lambda^A(x)$ are real functions of x^μ that parameterize a finite gauge transformation. (We are using the same notation for gauge transformations as in Section 15.1, except that in order to avoid confusion with Dirac indices we label the gauge generators and gauge transformation parameters with the letters A, B, etc. instead of α, β, etc.)

The left-chiral superfield (26.3.11) involves derivatives of some of the component fields, so its transformation is more complicated than

Eq. (27.1.1). However, Eq. (26.3.21) shows that the superfield does not involve derivatives if expressed in terms of θ_L and the variable x_+ defined by Eq. (26.3.23). It therefore has the transformation property

$$\Phi_n(x,\theta) \to \sum_m \left[\exp\left(i\sum_A t_A \Lambda^A(x_+)\right)\right]_{nm} \Phi_m(x,\theta)\,. \qquad (27.1.2)$$

If a term in the action depends only on left-chiral superfields, and not their derivatives or complex conjugates, like the term $\int d^4x\, [f(\Phi)]_{\mathscr{F}}$ in Eq. (26.3.30), then it (and its complex conjugate) will be invariant under the local transformation (27.1.2) if it is invariant under global transformations with $\Lambda^A(x)$ independent of x^μ. The need for introducing gauge fields in renormalizable theories of chiral superfields arises only in the D-terms, which involve both Φ_n and Φ_n^*. Because the matrices t_A are Hermitian, the Hermitian adjoint of Eq. (27.1.2) is

$$\Phi_n^\dagger(x,\theta) \to \sum_m \Phi_m^\dagger(x,\theta) \left[\exp\left(-i\sum_A t_A \Lambda^A(x_+)^*\right)\right]_{mn}\,. \qquad (27.1.3)$$

If it were not for the difference between $\Lambda^A(x_+)^* = \Lambda^A(x_-)$ and $\Lambda^A(x_+)$, this would just say that $\Phi^\dagger$ transforms according to the representation of the gauge group that is contragredient to the representation furnished by Φ, and any function of Φ and $\Phi^\dagger$ that is invariant under global gauge transformations would also be invariant under local gauge transformations. Because x_+ and x_- are different, we must introduce a gauge connection matrix $\Gamma_{nm}(x,\theta)$, with the transformation property

$$\Gamma(x,\theta) \to \exp\left(+i\sum_A t_A \Lambda^A(x_+)^*\right) \Gamma(x,\theta) \exp\left(-i\sum_A t_A \Lambda^A(x_+)\right)\,. \quad (27.1.4)$$

Then by multiplying $\Phi^\dagger$ on the right with Γ we obtain a superfield that transforms as

$$\left[\Phi^\dagger(x,\theta)\,\Gamma(x,\theta)\right]_n \to \sum_m \left[\Phi^\dagger(x,\theta)\,\Gamma(x,\theta)\right]_m \left[\exp\left(-i\sum_A t_A \Lambda^A(x_+)\right)\right]_{mn}\,, \qquad (27.1.5)$$

so that any globally gauge-invariant function constructed from Φ and $\Phi^\dagger\Gamma$ (and not their derivatives or complex conjugates) will also be locally gauge-invariant. One obvious example is the gauge-invariant version $(\Phi^\dagger\Gamma\Phi)_D$ of the D-term in the Lagrangian constructed in Section 26.4.

Any $\Gamma(x,\theta)$ that transforms as in Eq. (27.1.4) will allow us to construct gauge-invariant Lagrangians of chiral superfields. The choice is not unique; if Γ transforms as in Eq. (27.1.4), and we multiply on the right with any left-chiral superfield Υ_L with the transformation rule

$$\Upsilon_L(x,\theta) \to \exp\left(i\sum_A t_A \Lambda^A(x_+)\right) \Upsilon_L(x,\theta) \exp\left(-i\sum_A t_A \Lambda^A(x_+)\right)\,,$$

then we obtain a new gauge connection that also satisfies Eq. (27.1.4). One simplification is to take $\Gamma(x,\theta)$ Hermitian:

$$\Gamma^\dagger(x,\theta) = \Gamma(x,\theta)\,. \tag{27.1.6}$$

This is always possible if there is any $\Gamma(x,\theta)$ that satisfies Eq. (27.1.4), for then by taking the Hermitian adjoint of Eq. (27.1.4) we easily see that $\Gamma^\dagger(x,\theta)$ transforms in the same way as $\Gamma(x,\theta)$, so if $\Gamma(x,\theta)$ is not Hermitian then we can replace it with its Hermitian part $(\Gamma+\Gamma^\dagger)/2$ (or, if that vanishes, by its anti-Hermitian part $(\Gamma-\Gamma^\dagger)/2i$.) Another simplification of great physical importance is to express $\Gamma(x,\theta)$ in terms of fields whose gauge transformation properties do not depend on the specific representation t_A of the gauge algebra under which the chiral superfield $\Phi(x,\theta)$ transforms, so that these fields can be used to form a suitable matrix $\Gamma(x,\theta)$ for chiral superfields that transform according to any representation of the gauge group. For this purpose, it is useful to recall the Baker–Hausdorff formula, which states that, for arbitrary matrices a and b,

$$e^a\,e^b = \exp\Big(a+b+\tfrac{1}{2}[a,b]+\tfrac{1}{12}[a,[a,b]]+\tfrac{1}{12}[b,[b,a]]+\cdots\Big)\,, \tag{27.1.7}$$

where '$\cdots$' denotes higher-order terms that can be written as multiple commutators of as and bs, like the second- and third-order terms that are shown explicitly. It follows from this that for any representation of a Lie algebra, we have

$$\exp\Big(\sum_A a^A t_A\Big)\exp\Big(\sum_A b^A t_A\Big) = \exp\Big(\sum_A f^A(a,b)\,t_A\Big)\,, \tag{27.1.8}$$

where

$$\begin{aligned} f^A(a,b) = a^A + b^A + \tfrac{1}{2}i\sum_{BC} C^A{}_{BC}a^Bb^C - \tfrac{1}{12}\sum_{BCDE} C^A{}_{BC}C^C{}_{DE}a^Ba^Db^E \\ -\tfrac{1}{12}\sum_{BCDE} C^A{}_{BC}C^C{}_{DE}b^Bb^Da^E + \cdots\,, \end{aligned} \tag{27.1.9}$$

which depends on the Lie algebra through its structure constants $C^A{}_{BC}$, defined as usual by

$$[t_B,t_C] = i\sum_A C^A{}_{BC}\,t_A\,,$$

but does not depend on the particular representation furnished by the t_A. We will therefore take $\Gamma(x,\theta)$ in the form

$$\Gamma(x,\theta) = \exp\left(-2\sum_A t_A\,V^A(x,\theta)\right)\,, \tag{27.1.10}$$

where $V^A(x,\theta)$ are a set of real superfields (so that Γ is Hermitian), not depending on the representation of the gauge algebra furnished by the t_A.

We can achieve an important further simplification by noting an additional symmetry of supersymmetric gauge theories. If some function of Φ and $\Phi^\dagger\Gamma$ is invariant under global gauge transformations, then it will automatically be invariant not only under the local gauge transformations (27.1.2)–(27.1.4), but also under the larger group of extended gauge transformations

$$\Phi_{nL}(x,\theta) \to \sum_m \left[\exp\left(i\sum_A t_A \Omega^A(x,\theta)\right)\right]_{nm} \Phi_{mL}(x,\theta) \tag{27.1.11}$$

and

$$\Gamma(x,\theta) \to \exp\left(-i\sum_A t_A\Omega^A(x,\theta)\right)\Gamma(x,\theta)\exp\left(+i\sum_A t_A\Omega^A(x,\theta)^*\right), \tag{27.1.12}$$

where $\Omega^A(x,\theta)$ is an arbitrary left-chiral superfield — that is, an arbitrary function of θ_L as well as x_+. Under this transformation,

$$V^A(x,\theta) \to V^A(x,\theta) + \frac{i}{2}\left[\Omega^A(x,\theta) - \Omega^A(x,\theta)^*\right] + \cdots\,, \tag{27.1.13}$$

where '$\cdots$' denotes terms arising from the commutators in Eq. (27.1.7), which are of first or higher order in gauge coupling constants. As a general left-chiral superfield, Ω may be written in the form (26.3.11)

$$\begin{aligned}\Omega^A(x,\theta) = {}& W^A(x) - \sqrt{2}\left(\bar\theta\left(\frac{1+\gamma_5}{2}\right)w^A(x)\right) + \mathcal{W}^A(x)\left(\bar\theta\left(\frac{1+\gamma_5}{2}\right)\theta\right)\\ &+\frac{1}{2}\left(\bar\theta\gamma_5\gamma_\mu\theta\right)\partial^\mu W^A(x) - \frac{1}{\sqrt{2}}\left(\bar\theta\gamma_5\theta\right)\left(\bar\theta\,\not\partial\left(\frac{1+\gamma_5}{2}\right)w^A(x)\right)\\ &-\frac{1}{8}\left(\bar\theta\gamma_5\theta\right)^2\Box W^A(x)\,,\end{aligned} \tag{27.1.14}$$

in which $W^A(x)$ and $\mathcal{W}^A(x)$ are arbitrary complex functions of x^μ, and we introduced Majorana spinors $w^A(x)$ defined so that the left-handed spinor components of the superfields are $\frac{1}{2}(1+\gamma_5)w^A(x)$. Using the complex conjugation properties (26.A.21) of Majorana bilinears, the complex conjugate of Eq. (27.1.14) gives

$$\begin{aligned}\Omega^A(x,\theta)^* = {}& W^{A*}(x) - \sqrt{2}\left(\bar\theta\left(\frac{1-\gamma_5}{2}\right)w^A(x)\right) + \mathcal{W}^{A*}(x)\left(\bar\theta\left(\frac{1-\gamma_5}{2}\right)\theta\right)\\ &-\frac{1}{2}\left(\bar\theta\gamma_5\gamma_\mu\theta\right)\partial^\mu W^{A*}(x)\frac{1}{\sqrt{2}}\left(\bar\theta\gamma_5\theta\right)\left(\bar\theta\,\not\partial\left(\frac{1-\gamma_5}{2}\right)w^A(x)\right)\\ &-\frac{1}{8}\left(\bar\theta\gamma_5\theta\right)^2\Box W^{A*}(x)\,.\end{aligned} \tag{27.1.15}$$

We write the real superfields $V^A(x,\theta)$ in terms of component fields as in

Eq. (26.2.10):

$$\begin{aligned}V^A(x,\theta) = {} & C^A(x) - i\Big(\bar\theta\gamma_5\,\omega^A(x)\Big) - \frac{i}{2}\Big(\bar\theta\gamma_5\,\theta\Big)M^A(x) - \frac{1}{2}\Big(\bar\theta\,\theta\Big)N^A(x) \\ & + \frac{i}{2}\Big(\bar\theta\gamma_5\gamma^\mu\,\theta\Big)V^A_\mu(x) - i\Big(\bar\theta\gamma_5\,\theta\Big)\left(\bar\theta\Big[\lambda^A(x) + \frac{1}{2}\not\partial\omega^A(x)\Big]\right) \\ & - \frac{1}{4}\Big(\bar\theta\gamma_5\,\theta\Big)^2\left(D^A(x) + \frac{1}{2}\Box C^A(x)\right), \end{aligned} \tag{27.1.16}$$

where $C^A(x)$, $M^A(x)$, $N^A(x)$, and $V^A_\mu(x)$ are all real, and $\omega^A(x)$ and $\lambda^A(x)$ are Majorana spinors. Using Eqs. (27.1.14)–(27.1.16) in Eq. (27.1.13), we find that the component fields of the gauge superfield undergo the extended gauge transformation

$$\begin{aligned} & C^A(x) \to C^A(x) - \mathrm{Im}\,W^A(x) + \cdots, \\ & \omega^A(x) \to \omega^A(x) + \frac{1}{\sqrt{2}}w^A(x) + \cdots, \\ & V^A_\mu(x) \to V^A_\mu(x) + \partial_\mu \mathrm{Re}\,W^A(x) + \cdots, \\ & M^A(x) \to M^A(x) - \mathrm{Re}\,\mathscr{W}^A(x) + \cdots, \\ & N^A(x) \to N^A(x) + \mathrm{Im}\,\mathscr{W}^A(x) + \cdots, \\ & \lambda^A(x) \to \lambda^A(x) + \cdots, \\ & D^A(x) \to D^A(x) + \cdots, \end{aligned} \tag{27.1.17}$$

where again '$\cdots$' denotes terms that arise from the structure constants in Eq. (27.1.9) and that are therefore proportional to one or more factors of gauge coupling constants. We can use such an extended gauge transformation to put the gauge superfields into a convenient form, known as *Wess–Zumino gauge*,[1] in which

$$C^A(x) = \omega^A(x) = M^A(x) = N^A(x) = 0\,, \tag{27.1.18}$$

so that

$$\begin{aligned}V^A(x,\theta) = {} & \frac{i}{2}\Big(\bar\theta\gamma_5\gamma^\mu\,\theta\Big)V^A_\mu(x) - i\Big(\bar\theta\gamma_5\,\theta\Big)\Big(\bar\theta\lambda^A(x)\Big) \\ & - \frac{1}{4}\Big(\bar\theta\gamma_5\,\theta\Big)^2 D^A(x)\,. \end{aligned} \tag{27.1.19}$$

To accomplish this to zeroth order in the coupling constants, it is only necessary to set $\mathrm{Im}\,W^A(x) = C^A(x)$, $w^A(x) = -\sqrt{2}\omega^A(x)$, and $\mathscr{W}^A(x) = M^A(x) - iN^A(x)$. For Abelian gauge theories, in which the structure constants vanish, this ends our task. For non-Abelian gauge theories it is necessary to add terms to $\mathrm{Im}\,W^A(x)$, $w^A(x)$, and $\mathscr{W}^A(x)$ of first order in gauge coupling constants to cancel the terms arising from commutators of zeroth-order terms, then add terms to $\mathrm{Im}\,W^A(x)$, $w^A(x)$, and $\mathscr{W}^A(x)$

of second order in gauge coupling constants to cancel the terms arising from commutators of first-order terms with zeroth-order terms, and so on. It is not easy to calculate the series of terms in Im $W^A(x)$, $w^A(x)$, and $\mathscr{W}^A(x)$ needed to satisfy the gauge conditions (27.1.18) to all orders in gauge couplings, but there is no need — the important thing is that it is possible.

Inspection of the transformation rules (26.2.11)–(26.2.14) shows that the Wess–Zumino gauge condition (27.1.18) is not invariant under supersymmetry transformations unless $V^A_\mu = \lambda^A = 0$, and the condition $\lambda^A = 0$ is not supersymmetric unless also $D^A = 0$, in which case the whole superfield vanishes. Once we adopt Wess–Zumino gauge, the action is no longer invariant under either general extended gauge transformations or under supersymmetry, but it is invariant under supersymmetry transformations, which take us out of Wess–Zumino gauge, followed by suitable extended gauge transformations that take us back to Wess–Zumino gauge. (We will go into this explicitly in Section 27.8.) As we shall now see, it is also invariant under the ordinary gauge transformations (27.1.2)–(27.1.4), which preserve Wess–Zumino gauge.

With the gauge superfield satisfying the Wess–Zumino gauge condition (27.1.18), it becomes relatively easy to calculate its behavior under ordinary infinitesimal gauge transformations. In this case, $\Omega^A(x_+)$ are left-chiral superfields of the form (26.3.11), but with no ψ_L- or $\mathscr{F}$-components, and with ϕ-components given by *real* infinitesimal functions $\Lambda^A(x)$:

$$\Omega^A(x_+) = \Lambda^A(x) + \frac{1}{2}\left(\bar{\theta}\gamma_5\gamma_\mu\theta\right)\partial^\mu\Lambda^A(x) - \frac{1}{8}\left(\bar{\theta}\gamma_5\theta\right)^2\Box\Lambda^A(x)\,. \qquad (27.1.20)$$

To calculate the product of exponentials in the transformation rule (27.1.4), we use a version of the Baker–Hausdorff formula:

$$\exp(a)\exp(X)\exp(b) = \exp\left[X + L_X\cdot(b-a) + (L_X\coth L_X)\cdot(b+a) + \cdots\right], \qquad (27.1.21)$$

where a, b, and X are arbitrary matrices, L_X is the operator

$$L_X \cdot f = \tfrac{1}{2}[X, f]\,, \qquad (27.1.22)$$

and '$\cdots$' here denotes terms of second and higher order in a and/or b. In our case we have

$$b + a = 2\sum_A t_A \,\mathrm{Im}\,\Lambda^A(x_+) = -i\left(\bar{\theta}\gamma_5\gamma_\mu\theta\right)\sum_A t_A\,\partial^\mu\Lambda^A(x)\,,$$

$$b - a = -2i\sum_A t_A \,\mathrm{Re}\,\Lambda^A(x_+) = -2i\sum_A t_A\left[\Lambda^A(x) - \frac{1}{8}\left(\bar{\theta}\gamma_5\theta\right)^2\Box\Lambda^A(x)\right],$$

$$X = -2\sum_A t_A V^A(x,\theta) = -2\sum_A t_A \left[\frac{i}{2}\left(\bar\theta\gamma_5\gamma^\mu\theta\right)V^A_\mu(x) -i\left(\bar\theta\gamma_5\theta\right)\left(\bar\theta\lambda^A(x)\right) - \frac{1}{4}\left(\bar\theta\gamma_5\theta\right)^2 D^A(x)\right].$$

Now, every term in X involves at least one factor of θ_L and at least one factor of θ_R, while $a+b$ has just one factor of θ_L and one factor of θ_R, so we can drop any terms in $L_X \coth L_X$ of second or higher order in L_X. Since $L_X \coth L_X$ is an *even* function of L_X, this means that we can replace it with its term of zeroth order in L_X, which is just unity. Also, we may drop the term in $b-a$ proportional to $(\bar\theta\gamma_5\theta)^2$, since when acted on by L_X it would yield at least three factors of either θ_L or θ_R. Thus the argument of the exponential on the right-hand side of Eq. (27.1.21) may be replaced with

$$X + \tfrac{1}{2}[X, b-a] + b + a = -2\sum_A t_A \left[V^A(x,\theta) + \sum_{BC} C^A{}_{BC}\, V^B(x,\theta)\,\Lambda^C(x) + \tfrac{1}{2}i\left(\bar\theta\gamma_5\gamma_\mu\theta\right)\partial^\mu\Lambda^A(x)\right].$$

Thus for infinitesimal gauge transformations, the transformation rule (27.1.4) yields

$$V^A(x,\theta) \to V^A(x,\theta) + \sum_{BC} C^A{}_{BC}\, V^B(x,\theta)\,\Lambda^C(x) + \frac{1}{2}i\left(\bar\theta\gamma_5\gamma_\mu\theta\right)\partial^\mu\Lambda^A(x)\,. \tag{27.1.23}$$

It is important to note that under ordinary gauge transformations, a gauge superfield in Wess–Zumino gauge remains in Wess–Zumino gauge. In terms of the component fields in Eq. (27.1.19), Eq. (27.1.23) reads

$$V^A_\mu(x) \to \sum_{BC} C^A{}_{BC}\, V^B_\mu(x)\,\Lambda^C(x) + \partial_\mu\Lambda^A(x)\,, \tag{27.1.24}$$

$$\lambda^A(x) \to \sum_{BC} C^A{}_{BC}\, \lambda^B(x)\,\Lambda^C(x)\,, \tag{27.1.25}$$

$$D^A(x) \to \sum_{BC} C^A{}_{BC}\, D^B(x)\,\Lambda^C(x)\,. \tag{27.1.26}$$

We recognize Eq. (27.1.24) as the usual Yang–Mills gauge transformation rule (15.1.9) of a gauge field, while Eqs. (27.1.25) and (27.1.26) tell us that the fields $\lambda^A(x)$ and $D^A(x)$ transform as 'matter' fields that belong to the adjoint representation of the gauge group. The Majorana spinors λ^A are known as *gaugino* fields, while the real scalars D^A will turn out to be another set of auxiliary fields.

Next we must evaluate the matrix Γ that is needed in constructing gauge-invariant functions of chiral superfields. Because all terms with more than four factors of θ vanish, the expansion of the exponential is very simple in Wess–Zumino gauge:

$$\begin{aligned}\Gamma(x,\theta) &= \exp\Big(-2\sum_A t_A\, V^A(x,\theta)\Big)\\ &= 1 - i\Big(\bar\theta\,\gamma_5\,\gamma^\mu\,\theta\Big)\sum_A t_A\, V^A_\mu(x)\\ &\quad -\frac{1}{2}\Big(\bar\theta\,\gamma_5\,\gamma^\mu\,\theta\Big)\Big(\bar\theta\,\gamma_5\,\gamma^\nu\,\theta\Big)\sum_{AB} t_A\, t_B\, V^A_\mu(x)\, V^B_\nu(x)\\ &\quad +2i\Big(\bar\theta\,\gamma_5\,\theta\Big)\sum_A t_A\Big(\bar\theta\,\lambda^A(x)\Big) + \frac{1}{2}\Big(\bar\theta\,\gamma_5\,\theta\Big)^2\sum_A t_A\, D^A(x)\,.\end{aligned}$$

We can construct a gauge-invariant density by multiplying this on the right with a column vector of left-chiral superfields of the form (26.3.11):

$$\begin{aligned}\Phi_n(x,\theta) =\; &\phi_n(x) - \sqrt{2}\Big(\bar\theta\psi_{nL}(x)\Big) + \mathscr{F}_n(x)\left(\bar\theta\left(\frac{1+\gamma_5}{2}\right)\theta\right)\\ &+\frac{1}{2}\Big(\bar\theta\gamma_5\gamma_\mu\theta\Big)\partial^\mu\phi_n(x) - \frac{1}{\sqrt{2}}\Big(\bar\theta\gamma_5\theta\Big)\Big(\bar\theta\,\partial\!\!\!/\psi_{nL}(x)\Big)\\ &-\frac{1}{8}\Big(\bar\theta\gamma_5\theta\Big)^2\Box\phi_n(x)\,,\end{aligned}$$

and multiplying on the left with the column

$$\begin{aligned}\Phi_n(x,\theta)^* =\; &\phi^*_n(x) - \sqrt{2}\Big(\overline{\psi_{nL}(x)}\theta\Big) + \mathscr{F}^*_n(x)\left(\bar\theta\left(\frac{1-\gamma_5}{2}\right)\theta\right)\\ &-\frac{1}{2}\Big(\bar\theta\gamma_5\gamma_\mu\theta\Big)\partial^\mu\phi^*_n(x) - \frac{1}{\sqrt{2}}\Big(\bar\theta\gamma_5\theta\Big)\partial_\mu\Big(\overline{\psi_{nL}(x)}\gamma^\mu\theta\Big)\\ &-\frac{1}{8}\Big(\bar\theta\gamma_5\theta\Big)^2\Box\phi^*_n(x)\,.\end{aligned}$$

The term in this product of fourth order in θ is

$$\begin{aligned}\Big[\Phi^\dagger\,\Gamma\,\Phi\Big]_{\theta^4} = &-\frac{1}{8}\Big(\bar\theta\gamma_5\theta\Big)^2\Big\{\Big[\phi^\dagger\Box\phi\Big] + \Big[\Big(\Box\phi^\dagger\Big)\phi\Big]\Big\}\\ &+\Big(\bar\theta\gamma_5\,\theta\Big)\Big\{\Big[\Big(\overline{\psi_L}\,\theta\Big)\Big(\bar\theta\gamma^\mu\partial_\mu\psi_L\Big)\Big] + \Big[\Big((\partial_\mu\overline{\psi_L})\,\gamma^\mu\theta\Big)\Big(\bar\theta\,\psi_L\Big)\Big]\Big\}\\ &+\frac{1}{4}\Big(\bar\theta(1-\gamma_5)\theta\Big)\Big(\bar\theta(1+\gamma_5)\theta\Big)\Big[\mathscr{F}^\dagger\,\mathscr{F}\Big]\\ &-\frac{1}{4}\Big(\bar\theta\gamma_5\gamma^\mu\theta\Big)\Big(\bar\theta\gamma_5\gamma^\nu\theta\Big)\Big[\partial_\mu\phi^\dagger\,\partial_\nu\phi\Big]\end{aligned}$$

$$
\begin{aligned}
&-\frac{i}{2}\left(\bar{\theta}\gamma_5\gamma^\mu\theta\right)\left(\bar{\theta}\gamma_5\gamma^\nu\theta\right)\sum_A V^A_\mu\left\{\left[\phi^\dagger\, t_A\,\partial_\nu\phi\right]-\left[(\partial_\nu\phi^\dagger)\,t_A\,\phi\right]\right\}\\
&-\frac{1}{2}\left(\bar{\theta}\gamma_5\gamma^\mu\theta\right)\left(\bar{\theta}\gamma_5\gamma^\nu\theta\right)\sum_{AB} V^A_\mu V^B_\nu\left[\phi^\dagger\, t_A\, t_B\,\phi\right]\\
&-2i\left(\bar{\theta}\gamma_5\gamma_\mu\theta\right)\sum_A V^A_\mu\left[\left(\overline{\psi_L}\,\theta\right)t_A\left(\bar{\theta}\,\psi_L\right)\right]\\
&-2i\sqrt{2}\left(\bar{\theta}\,\gamma_5\,\theta\right)\sum_A\left[\left(\overline{\psi_L}\,\theta\right)t_A\left(\bar{\theta}\lambda^A\right)\phi\right]\\
&-2i\sqrt{2}\left(\bar{\theta}\,\gamma_5\,\theta\right)\sum_A\left[\phi^\dagger\left(\overline{\lambda^A}\,\theta\right)t_A\left(\bar{\theta}\,\psi_L\right)\right]\\
&+\frac{1}{2}\left(\bar{\theta}\gamma_5\,\theta\right)^2\sum_A D_A\left[\phi^\dagger t_A\,\phi\right],
\end{aligned}
$$

in which we use square brackets to indicate scalar products in the flavor indices n, m, and continue to use round brackets to indicate scalar products in Dirac indices. Just as in Section 26.4, we may use the identities (26.A.17)–(26.A.19) to put all θ dependence in the form of an over-all factor $(\bar{\theta}\gamma_5\theta)^2$:

$$
\begin{aligned}
\left[\Phi^\dagger\,\Gamma\,\Phi\right]_{\theta^4}=\left(\bar{\theta}\gamma_5\theta\right)^2\Bigg\{&-\frac{1}{8}\left[\phi^\dagger\Box\phi\right]-\frac{1}{8}\left[\left(\Box\phi^\dagger\right)\phi\right]\\
&+\frac{1}{4}\left[\left(\overline{\psi_L}\,\gamma^\mu\partial_\mu\psi_L\right)\right]-\frac{1}{4}\left[\left((\partial_\mu\overline{\psi_L})\,\gamma^\mu\,\psi_L\right)\right]\\
&-\frac{1}{2}\left[\mathscr{F}^\dagger\,\mathscr{F}\right]+\frac{1}{4}\left[\partial_\mu\phi^\dagger\,\partial^\mu\phi\right]\\
&+\frac{i}{2}\sum_A V^A_\mu\left[\phi^\dagger\, t_A\,\partial^\mu\phi\right]-\frac{i}{2}\sum_A V^A_\mu\left[(\partial^\mu\phi^\dagger)\,t_A\,\phi\right]\\
&+\frac{1}{2}\sum_{AB} V^A_\mu V^{B\mu}\left[\phi^\dagger\, t_A\, t_B\,\phi\right]-\frac{i}{2}\sum_A V^A_\mu\left[\left(\overline{\psi_A}\gamma^\mu t_A\,\psi_A\right)\right]\\
&-\frac{i}{\sqrt{2}}\sum_A\left[\left(\overline{\psi_L}\,t_A\,\lambda^A\right)\phi\right]+\frac{i}{\sqrt{2}}\sum_A\left[\phi^\dagger\left(\overline{\lambda^A}\,t_A\,\psi_L\right)\right]\\
&+\frac{1}{2}\sum_A D_A\left[\phi^\dagger\, t_A\,\phi\right]\Bigg\}.
\end{aligned}
$$

The D-term is the coefficient of $-\frac{1}{4}(\bar{\theta}\gamma_5\theta)^2$ minus $\frac{1}{2}\Box$ acting on the θ-independent term, which for $[\Phi^\dagger\Gamma\Phi]$ is $[\phi^\dagger\phi]$, so

$$
\begin{aligned}
\left[\Phi^\dagger\,\Gamma\,\Phi\right]_D&=-2[\partial_\mu\phi^\dagger\,\partial^\mu\phi]\\
&-\left[\left(\overline{\psi_L}\,\gamma^\mu\partial_\mu\psi_L\right)\right]+\left[\left((\partial_\mu\overline{\psi_L})\,\gamma^\mu\,\psi_L\right)\right]+2\left[\mathscr{F}^\dagger\,\mathscr{F}\right]
\end{aligned}
$$

$$
\begin{aligned}
&-2i\sum_A V^A_\mu\Big[\phi^\dagger\, t_A\, \partial^\mu \phi\Big] + 2i\sum_A V^A_\mu\Big[(\partial^\mu\phi^\dagger)\, t_A\, \phi\Big] \\
&-2\sum_{AB} V^A_\mu V^{B\mu}\Big[\phi^\dagger\, t_A\, t_B\, \phi\Big] + 2i\sum_A V^A_\mu\Big[\Big(\overline{\psi_A}\gamma^\mu t_A\, \psi_A\Big)\Big] \\
&+2i\sqrt{2}\sum_A\Big[\Big(\overline{\psi_L}\, t_A\, \lambda^A\Big)\phi\Big] - 2i\sqrt{2}\sum_A\Big[\phi^\dagger\Big(\overline{\lambda^A}\, t_A\, \psi_L\Big)\Big] \\
&-2\sum_A D_A\Big[\phi^\dagger\, t_A\, \phi\Big]\,.
\end{aligned}
$$

To see that this *is* gauge-invariant, we note that it may be written as

$$
\begin{aligned}
\frac{1}{2}\Big[\Phi^\dagger\,\Gamma\,\Phi\Big]_D = &-\Big[(D_\mu\phi)^\dagger\, D^\mu\phi\Big] \\
&-\frac{1}{2}\Big[\Big(\overline{\psi_L}\,\gamma^\mu D_\mu\psi_L\Big)\Big] + \frac{1}{2}\Big[\Big(\overline{(D_\mu\psi_L)}\,\gamma^\mu\,\psi_L\Big)\Big] + \Big[\mathscr{F}^\dagger\,\mathscr{F}\Big] \\
&+i\sqrt{2}\sum_A\Big[\Big(\overline{\psi_L}\, t_A\, \lambda^A\Big)\phi\Big] - i\sqrt{2}\sum_A\Big[\phi^\dagger\Big(\overline{\lambda^A}\, t_A\, \psi_L\Big)\Big] \\
&-\sum_A D_A\Big[\phi^\dagger\, t_A\, \phi\Big]\,,
\end{aligned}
\tag{27.1.27}
$$

where D_μ is the gauge-invariant derivative (15.1.10):

$$
D_\mu\psi_L \equiv \partial_\mu\psi_L - i\sum_A t_A\, V^A_\mu\, \psi_L\,, \qquad D_\mu\phi \equiv \partial_\mu\phi - i\sum_A t_A\, V^A_\mu\, \phi\,. \tag{27.1.28}
$$

Eq. (27.1.27) is thus a suitable gauge-invariant kinematic Lagrangian for the scalar and spinor components of a left-chiral superfield, now supplemented with Yukawa couplings of gaugino fields to the scalar and spinor components of chiral superfields as well as terms involving auxiliary fields $\mathscr{F}_n$ and D_A.

27.2 Gauge-Invariant Action for Abelian Gauge Superfields

We must now consider how to construct a gauge-invariant supersymmetric action for the gauge superfields $V^A(x,\theta)$ that contain the gauge fields $V^A_\mu(x)$. In order to motivate this construction we will first consider the case of a single Abelian gauge field (dropping the superscript A), and then return to the general case in the next section.

In an Abelian gauge theory like quantum electrodynamics, the gauge-invariant field constructed from $V_\mu(x)$ is the familiar field-strength tensor

$$
f_{\mu\nu}(x) = \partial_\mu V_\nu(x) - \partial_\nu V_\mu(x)\,. \tag{27.2.1}
$$

The supersymmetry transformation rule for $f_{\mu\nu}(x)$ is then given by the transformation rule (26.2.15) for $V_\mu(x)$ as

$$\delta f_{\mu\nu} = \left(\bar{\alpha}\,(\partial_\mu \gamma_\nu - \partial_\nu \gamma_\mu)\lambda\right) . \tag{27.2.2}$$

Eq. (26.2.16) gives the transformation rule for $\lambda(x)$ as

$$\delta\lambda = \left(-\tfrac{1}{4} f_{\mu\nu}[\gamma^\mu , \gamma^\nu] + i\gamma_5\, D\right)\alpha\,, \tag{27.2.3}$$

while Eq. (26.2.17) gives the transformation rule for $D(x)$:

$$\delta D = i\left(\bar{\alpha}\,\gamma_5\, \not{\partial}\lambda\right) . \tag{27.2.4}$$

None of this depends on whether or not the superfield $V^A(x,\theta)$ is taken to be in Wess–Zumino gauge. We see that the fields $f_{\mu\nu}(x)$, $\lambda(x)$, and $D(x)$ form a complete supersymmetry multiplet.

It is not hard to construct a suitable kinematic Lagrangian density for the fields of this supermultiplet. The only Lorentz-invariant, parity conserving, and gauge-invariant functions of these fields with dimensionality four are $f_{\mu\nu}f^{\mu\nu}$, $\bar{\lambda}\,\not{\partial}\lambda$, and D^2. We can make V^μ a conventionally normalized vector field by taking the coefficient of $f_{\mu\nu}f^{\mu\nu}$ to be $-\frac{1}{4}$, so we may tentatively take the kinematic Lagrangian density as

$$\mathscr{L}_{\text{gauge}} = -\tfrac{1}{4} f_{\mu\nu}f^{\mu\nu} - c_\lambda\left(\bar{\lambda}\,\not{\partial}\lambda\right) - c_D D^2\,,$$

with coefficients c_λ and c_D to be determined from the condition that $\int \mathscr{L}_{\text{gauge}} d^4x$ is supersymmetric. Using Eqs. (27.2.2)–(27.2.4), an infinitesimal supersymmetry transformation changes the operators in the Lagrangian by

$$\delta\left(f_{\mu\nu}f^{\mu\nu}\right) = 2f^{\mu\nu}\left(\bar{\alpha}\,(\gamma_\nu\partial_\mu - \gamma_\mu\partial_\nu)\lambda\right) ,$$

$$\delta\left(\bar{\lambda}\,\not{\partial}\lambda\right) = 2\left(\bar{\alpha}\left[+\tfrac{1}{4} f_{\mu\nu}[\gamma^\mu , \gamma^\nu] + i\gamma_5\, D\right]\not{\partial}\lambda\right) ,$$

$$\delta D^2 = 2i\,D\left(\bar{\alpha}\,\gamma_5\, \not{\partial}\lambda\right) ,$$

in which we drop derivative terms that do not contribute to the variation of the action. In order to see how these terms cancel, it is necessary to use an identity for gamma matrices*

$$[\gamma^\mu , \gamma^\nu]\,\gamma^\rho = -2\eta^{\mu\rho}\gamma^\nu + 2\eta^{\nu\rho}\gamma^\mu - 2i\epsilon^{\mu\nu\rho\sigma}\gamma_\sigma\gamma_5\,. \tag{27.2.5}$$

The term $-i\epsilon^{\mu\nu\rho\sigma} f_{\mu\nu}(\bar{\alpha}\,\gamma_\sigma\gamma_5\partial_\rho\lambda)$ does not contribute to $\int d^4x\,\delta\mathscr{L}$, because integrating by parts yields a contribution proportional to $\epsilon^{\mu\nu\rho\sigma}\partial_\rho f_{\mu\nu}$, which

* To derive this, use the fact that any 4×4 matrix may be expressed as a linear combination of the 16 independent covariant matrices described in Section 5.4, which in our case is limited by Lorentz invariance and space inversion invariance to the terms shown here. The coefficients of these terms can be calculated by giving $\mu\nu\rho$ the values 121 and 123.

vanishes for an $f_{\mu\nu}$ of the form (27.2.1). This identity then allows us to rewrite the variation of the λ-term as

$$\delta\left(\bar{\lambda}\,\partial\!\!\!/\lambda\right) = -f^{\mu\nu}\left(\bar{\alpha}(\gamma_\nu\partial_\mu - \gamma_\mu\partial_\nu)\lambda\right) + 2iD\left(\bar{\alpha}\gamma_5\,\partial\!\!\!/\lambda\right).$$

The cancellation of terms proportional to $f^{\mu\nu}\lambda$ requires that $c_\lambda = 1/2$, while the cancellation of terms proportional to $D\lambda$ requires that $c_D = -c_\lambda$, so the supersymmetric Lagrangian density takes the form

$$\mathscr{L}_{\text{gauge}} = -\tfrac{1}{4}f_{\mu\nu}f^{\mu\nu} - \tfrac{1}{2}\left(\bar{\lambda}\,\partial\!\!\!/\lambda\right) + \tfrac{1}{2}D^2\,. \tag{27.2.6}$$

This shows that with V^μ canonically normalized, the field λ related to V^μ by the transformation rules (27.2.2) and (27.2.3) is also canonically normalized.

In addition, for Abelian gauge theories there is a superrenormalizable term, known as a *Fayet–Iliopoulos term*:[2]

$$\mathscr{L}_{\text{FI}} = \xi\,D\,, \tag{27.2.7}$$

with ξ an arbitrary constant. Its variation under a supersymmetry transformation is shown by Eq. (27.2.4) to be a derivative, so that it yields another supersymmetric term in the action. As we will see in Section 27.5, the presence of such a term can provide a mechanism for the spontaneous breakdown of supersymmetry.

Both for its own interest and as a tool in constructing supersymmetric interactions involving the fields $f_{\mu\nu}$, λ, and D, it is interesting to ask what sort of superfield has these as component fields. Somewhat surprisingly, it turns out to be a *spinor* superfield $W_\alpha(x)$, with component fields (in the notation of Eq. (26.2.10)) given by

$$\begin{aligned}
&C_{(\alpha)}(x) = \lambda_\alpha(x)\,,\\
&\omega_{(\alpha)\beta}(x) = \tfrac{1}{2}i\left(\gamma^\mu\gamma^\nu\epsilon\right)_{\alpha\beta}f_{\mu\nu}(x) + (\gamma_5\epsilon)_{\alpha\beta}D(x)\,,\\
&V_{(\alpha)\mu}(x) = -i\partial_\mu\left(\gamma_5\lambda(x)\right)_\alpha\,,\\
&M_{(\alpha)}(x) = -i\left(\partial\!\!\!/\gamma_5\lambda(x)\right)_\alpha\,, \qquad N_{(\alpha)}(x) = -\left(\partial\!\!\!/\lambda(x)\right)_\alpha\,,\\
&\lambda_{(\alpha)\beta}(x) = D_{(\alpha)}(x) = 0\,.
\end{aligned} \tag{27.2.8}$$

(The subscript α on these component fields is put inside parentheses to emphasize that it labels the whole superfield.) It is straightforward to use Eqs. (27.2.2)–(27.2.4) to check directly that the superfield components given by Eq. (27.2.8) do transform as in Eqs. (26.2.11)–(26.2.17).

Inserting the component fields (27.2.8) in Eq. (26.2.10) and using

Eq. (26.A.5), we find that the superfield W_α takes the form

$$W_\alpha(x,\theta) = \Big[\lambda(x) + \tfrac{1}{2}\gamma^\mu\gamma^\nu\theta\, f_{\mu\nu}(x) - i\gamma_5\theta\, D(x) - \tfrac{1}{2}\Big(\theta^{\rm T}\epsilon\theta\Big)\, \partial\!\!\!/\gamma_5\lambda(x)$$
$$+\tfrac{1}{2}\Big(\theta^{\rm T}\epsilon\gamma_5\theta\Big)\, \partial\!\!\!/\lambda(x) + \tfrac{1}{2}\Big(\theta^{\rm T}\epsilon\gamma^\mu\theta\Big)\gamma_5\partial_\mu\lambda(x)$$
$$-\tfrac{1}{4}\Big(\theta^{\rm T}\epsilon\theta\Big)\gamma_5\gamma^\mu\gamma^\nu\gamma^\sigma\theta\,\partial_\sigma f_{\mu\nu}(x)$$
$$+\tfrac{1}{2}i\Big(\theta^{\rm T}\epsilon\theta\Big)\gamma^\sigma\theta\,\partial_\sigma D(x) - \tfrac{1}{8}\Big(\theta^{\rm T}\epsilon\theta\Big)^2\Box\lambda(x)\Big]_\alpha . \tag{27.2.9}$$

As we showed in Section 26.3, a superfield like this with zero λ- and D-components is *chiral* — that is, it is the sum of left-chiral and right-chiral superfields

$$W(x,\theta) = W_L(x,\theta) + W_R(x,\theta) . \tag{27.2.10}$$

Here the left- and right-chiral superfields are simply the projections of W on the subspaces with $\gamma_5 = +1$ and $\gamma_5 = -1$, respectively:

$$W_L(x,\theta) = \tfrac{1}{2}(1+\gamma_5)W(x,\theta)$$
$$= \lambda_L(x_+) + \tfrac{1}{2}\gamma^\mu\gamma^\nu\theta_L\, f_{\mu\nu}(x_+) + \Big(\theta_L^{\rm T}\epsilon\theta_L\Big)\, \partial\!\!\!/\lambda_R(x_+) - i\theta_L D(x_+) , \tag{27.2.11}$$

$$W_R(x,\theta) = \tfrac{1}{2}(1-\gamma_5)W(x,\theta)$$
$$= \lambda_R(x_-) + \tfrac{1}{2}\gamma^\mu\gamma^\nu\theta_R\, f_{\mu\nu}(x_-) - \Big(\theta_R^{\rm T}\epsilon\theta_R\Big)\, \partial\!\!\!/\lambda_L(x_-) - i\theta_R D(x_-) , \tag{27.2.12}$$

with $x_\pm^\mu$ given by Eq. (26.3.23).

As we saw in Section 26.3, we can construct suitable Lagrangian densities from the $\mathscr{F}$-term of any scalar function of a left-chiral superfield, plus its Hermitian adjoint. The simplest scalar function of the left-chiral superfield (27.2.11) is $\sum_{\alpha\beta}\epsilon_{\alpha\beta}W_{L\alpha}W_{L\beta}$. To calculate the $\mathscr{F}$-term, we note that, when expressed as a function of θ_L and x_+, the term in $\sum_{\alpha\beta}\epsilon_{\alpha\beta}W_{L\alpha}W_{L\beta}$ of second order in θ_L is

$$-\Big[\sum_{\alpha\beta}\epsilon_{\alpha\beta}W_{L\alpha}W_{L\beta}\Big]_{\theta_L^2} = \Big(\theta_L^{\rm T}\epsilon\theta_L\Big)\Big[-2\Big(\lambda_L^{\rm T}(x)\,\epsilon\,\partial\!\!\!/\lambda_R(x)\Big) + D^2(x)\Big]$$
$$+\frac{1}{16}\Big(\overline{\theta_L}[\gamma^\mu,\gamma^\nu]\,[\gamma^\rho,\gamma^\sigma]\theta_L\Big)\, f_{\mu\nu}(x)f_{\rho\sigma}(x) .$$

(The argument of the fields here can be taken as x^μ rather than x_+^μ because the difference would yield terms that contain at least three factors of θ_L, which therefore vanish.) Lorentz invariance together with the fact that $(\bar{s}[\gamma_\mu,\gamma_\nu]s)$ and $(\bar{s}[\gamma_\mu,\gamma_\nu]\gamma_5 s)$ vanish for any Majorana fermion s tell us that the bilinear $(\overline{\theta_L}[\gamma^\mu,\gamma^\nu][\gamma^\rho,\gamma^\sigma]\theta_L)$ must be proportional to a linear combination of $(\overline{\theta_L}\theta_L)(\eta^{\mu\rho}\eta^{\nu\sigma}-\eta^{\mu\sigma}\eta^{\nu\rho})$ and $(\overline{\theta_L}\theta_L)\epsilon^{\mu\nu\rho\sigma}$. We can find the coefficients by giving $\mu\nu\rho\sigma$ the values 1212 or 1230, and in this way we find that

$$\left(\overline{\theta_L}\,[\gamma^\mu,\gamma^\nu]\,[\gamma^\rho,\gamma^\sigma]\theta_L\right) = 4\left(\overline{\theta_L}\theta_L\right)\left[-\eta^{\mu\rho}\eta^{\nu\sigma}+\eta^{\mu\sigma}\eta^{\nu\rho}+i\epsilon^{\mu\nu\rho\sigma}\right].$$

The $\mathscr{F}$-term is the coefficient of $(\overline{\theta_L}\theta_L)$, so

$$-\Big[\sum_{\alpha\beta}\epsilon_{\alpha\beta}W_{L\alpha}W_{L\beta}\Big]_{\mathscr{F}} = -2\left(\overline{\lambda_R}\,\partial\!\!\!/\lambda_R\right)-\frac{1}{2}f_{\mu\nu}f^{\mu\nu}+\frac{i}{4}\epsilon^{\mu\nu\rho\sigma}f_{\mu\nu}f_{\rho\sigma}+D^2\,. \tag{27.2.13}$$

Eq. (26.A.21) shows that $(\bar\lambda\,\partial\!\!\!/\lambda)$ is real, while $(\bar\lambda\,\partial\!\!\!/\gamma_5\lambda)$ is imaginary, so the real part of Eq. (27.2.13) yields the Lagrangian (27.2.6) for the gauge and gaugino fields

$$-\frac{1}{2}\,\mathrm{Re}\Big[\sum_{\alpha\beta}\epsilon_{\alpha\beta}W_{L\alpha}W_{L\beta}\Big]_{\mathscr{F}} = -\frac{1}{2}\left(\bar\lambda\,\partial\!\!\!/\lambda\right)-\frac{1}{4}f_{\mu\nu}f^{\mu\nu}+\frac{1}{2}D^2\,. \tag{27.2.14}$$

The physical significance of the imaginary part will be discussed in a more general context in the next section.

There is another way to derive the form of the spinor superfield, which will turn out to provide a more convenient way of deriving the components of the gauge superfield in non-Abelian gauge theories. A tedious but straightforward calculation shows that the gauge-invariant superfield (27.2.9) may be expressed in terms of the gauge superfield (27.1.16) as

$$W_\alpha(x,\theta) = \frac{i}{4}\left(\mathscr{D}^{\mathrm{T}}\epsilon\mathscr{D}\right)\mathscr{D}_\alpha\,V(x,\theta)\,, \tag{27.2.15}$$

where $\mathscr{D}_\alpha$ is the superderivative introduced in Eq. (26.2.26):

$$\mathscr{D}_\alpha \equiv \sum_\beta(\gamma_5\epsilon)_{\alpha\beta}\frac{\partial}{\partial\theta_\beta}-(\gamma^\mu\theta)_\alpha\frac{\partial}{\partial x^\mu} = -\frac{\partial}{\partial\bar\theta_\alpha}-(\gamma^\mu\theta)_\alpha\frac{\partial}{\partial x^\mu}\,.$$

This result might have been obtained (aside from the normalization factor) by noting that the function (27.2.15) has the desired property of being a gauge-invariant chiral spinor superfield. First, note that Eq. (27.2.15) *is* a superfield, because it is formed by acting on the superfield V with superderivatives. Also, it follows from the anticommutation of the $\mathscr{D}$s that the product of any three or more $\mathscr{D}_L$s or three or more $\mathscr{D}_R$s vanishes, so that

$$\left(\mathscr{D}^{\mathrm{T}}\epsilon\mathscr{D}\right)\mathscr{D} = \left(\mathscr{D}_L^{\mathrm{T}}\epsilon\mathscr{D}_L\right)\mathscr{D}_R+\left(\mathscr{D}_R^{\mathrm{T}}\epsilon\mathscr{D}_R\right)\mathscr{D}_L\,. \tag{27.2.16}$$

Because $\mathscr{D}_L(\mathscr{D}_L^{\mathrm{T}}\epsilon\mathscr{D}_L) = \mathscr{D}_R(\mathscr{D}_R^{\mathrm{T}}\epsilon\mathscr{D}_R) = 0$, the superfield (27.2.15) is chiral, with

$$W_{L\alpha}(x,\theta) = \frac{i}{4}\left(\mathscr{D}_R^{\mathrm{T}}\epsilon\mathscr{D}_R\right)\mathscr{D}_{L\alpha}\, V(x,\theta)\,, \quad W_{R\alpha}(x,\theta) = \frac{i}{4}\left(\mathscr{D}_L^{\mathrm{T}}\epsilon\mathscr{D}_L\right)\mathscr{D}_{R\alpha}\, V(x,\theta)\,. \tag{27.2.17}$$

Finally, we can show that (27.2.15) is invariant under the generalized gauge transformation (27.1.13), which for a single Abelian gauge superfield is simply

$$V(x,\theta) \to V(x,\theta) + \frac{i}{2}\Big[\Omega(x,\theta) - \Omega^*(x,\theta)\Big]\,, \tag{27.2.18}$$

where $\Omega(x,\theta)$ is an arbitrary left-chiral superfield. Since $\mathscr{D}_L\Omega^* = 0$, the change in $W_{L\alpha}$ is proportional to $(\mathscr{D}_R^{\mathrm{T}}\epsilon\mathscr{D}_R)\mathscr{D}_{L\alpha}\Omega$. But $\mathscr{D}_R\Omega = 0$ and

$$\Big[(\mathscr{D}_R^{\mathrm{T}}\epsilon\mathscr{D}_R),\, \mathscr{D}_{L\alpha}\Big] = -2\Big[(1+\gamma_5)\,\not\partial\mathscr{D}_R\Big]_\alpha\,,$$

so the change in $W_{L\alpha}$ vanishes. A similar argument shows that $W_{R\alpha}$ is also gauge-invariant. (The work of checking Eq. (27.2.15) is greatly reduced by using this gauge-invariance property to put $V(x,\theta)$ in Wess–Zumino gauge.)

The chiral superfields (27.2.11) and (27.2.12) are evidently not of the most general form for left- and right-chiral superfields. To put the constraints satisfied by these superfields in a manifestly supersymmetric form, we note by using the anticommutation relation (26.2.30) that

$$\begin{aligned}\epsilon_{\alpha\beta}\mathscr{D}_{L\alpha}\left(\mathscr{D}_R^{\mathrm{T}}\epsilon\mathscr{D}_R\right)\mathscr{D}_{L\beta} &= -2\mathscr{D}_{R\alpha}\mathscr{D}_{L\beta}\left(\epsilon(1+\gamma_5)\,\not\partial\right)_{\beta\alpha} + \left(\mathscr{D}_R^{\mathrm{T}}\epsilon\mathscr{D}_R\right)\left(\mathscr{D}_L^{\mathrm{T}}\epsilon\mathscr{D}_L\right)\\ &= \epsilon_{\alpha\beta}\mathscr{D}_{R\alpha}\left(\mathscr{D}_L^{\mathrm{T}}\epsilon\mathscr{D}_L\right)\mathscr{D}_{R\beta}\,.\end{aligned} \tag{27.2.19}$$

From Eq. (27.2.17) it follows then that W_L and W_R are related by the constraint

$$\epsilon_{\alpha\beta}\mathscr{D}_{L\alpha}W_{L\beta} = \epsilon_{\alpha\beta}\mathscr{D}_{R\alpha}W_{R\beta}\,. \tag{27.2.20}$$

It is straightforward to show that the most general chiral spinor superfields satisfying Eq. (27.2.20) are of the form (27.2.11) and (27.2.12), with $f_{\mu\nu}$ constrained to satisfy the 'Bianchi' identities $\epsilon^{\mu\nu\rho\sigma}\partial_\rho f_{\mu\nu} = 0$.

27.3 Gauge-Invariant Action for General Gauge Superfields

Our experience in the previous section with supersymmetric Abelian gauge theories suggests immediately that in a general non-Abelian gauge theory the kinematic Lagrangian for the fields $V_\mu^A(x)$, $\lambda^A(x)$, and $D^A(x)$ should appear as part of the gauge-invariant generalization of Eq. (27.2.6):

$$\mathscr{L}_{\text{gauge}} = -\tfrac{1}{4}\sum_A f_{A\mu\nu}f_A^{\mu\nu} - \tfrac{1}{2}\sum_A\left(\overline{\lambda_A}\,(\not{D}\lambda)_A\right) + \tfrac{1}{2}\sum_A D_A D_A\,. \tag{27.3.1}$$

We are now using a basis for the Lie algebra with totally antisymmetric structure constants, and we are consequently not preserving the distinction between upper and lower group indices, writing all indices A, B, etc. as subscripts. Also, $f_{A\mu\nu}$ is the gauge-covariant field-strength tensor

$$f_{A\mu\nu} = \partial_\mu V_{A\nu} - \partial_\nu V_{A\mu} + \sum_{BC} C_{ABC} V_{B\mu} V_{C\nu} , \tag{27.3.2}$$

and $D_\mu \lambda$ is the gauge-covariant derivative of the gaugino field, which in the adjoint representation is

$$(D_\mu \lambda)_A = \partial_\mu \lambda_A + \sum_{BC} C_{ABC} V_{B\mu} \lambda_C . \tag{27.3.3}$$

The question is: does Eq. (27.3.1) yield a supersymmetric action?

Since the Lagrangian density (27.3.1) is manifestly gauge-invariant, we can test whether the action is supersymmetric in any convenient gauge. To find out whether $\delta \mathscr{L}_{\text{gauge}}$ is a derivative at some point X^μ, it is convenient to adopt a specific version of Wess–Zumino gauge in which $V_A^\mu(X) = 0$. Then at X the changes in the component fields are given by Eqs. (26.2.15)–(26.2.17) at $x = X$ as

$$\delta V_{A\mu} = \left(\bar{\alpha} \gamma_\mu \lambda_A\right) , \tag{27.3.4}$$

$$\delta \lambda_A = \left(\frac{1}{4} f_{A\mu\nu} \, [\gamma^\nu , \gamma^\mu] + i\gamma_5 D_A\right)\alpha , \tag{27.3.5}$$

$$\delta D_A = i\left(\bar{\alpha} \gamma_5 \not{\partial} \lambda_A\right) . \tag{27.3.6}$$

(We must set x^μ in these expressions equal to X^μ *after* calculating the change under a supersymmetry transformation, not before.) Also, the non-linear terms in $f_A^{\mu\nu}$ are quadratic in the Vs and therefore at $x = X$ have zero variation, so at $x = X$

$$\delta f_{A\mu\nu} = \left(\bar{\alpha}\,(\gamma_\nu \partial_\mu - \gamma_\mu \partial_\nu) \lambda_A\right) . \tag{27.3.7}$$

With one exception, the terms in Eq. (27.3.1) and their transformations under supersymmetry transformations are thus just a number of copies (labelled by A) of the Abelian theory discussed in the previous section, and therefore give a supersymmetric action. The one exception that might disturb the supersymmetry of the action arises from the second term in the gauge-covariant derivative (27.3.3) of the gaugino field:

$$\mathscr{L}_{\lambda\lambda V} = -\tfrac{1}{2} \sum_{ABC} C_{ABC} \left(\overline{\lambda_A} \, \not{V}_B \lambda_C\right) , \tag{27.3.8}$$

whose variation at $x = X$ is

$$\delta\mathscr{L}_{\lambda\lambda V} = -\tfrac{1}{2}\sum_{ABC} C_{ABC}\left(\overline{\lambda_A}(\delta \not{V}_B)\lambda_C\right) = -\tfrac{1}{2}\sum_{ABC} C_{ABC}\left(\overline{\lambda_A}\gamma_\mu\lambda_C\right)\left(\bar{\alpha}\gamma^\mu\lambda_B\right). \tag{27.3.9}$$

We can write the product of bilinears on the right-hand side as the sum of two terms

$$\left(\overline{\lambda_A}\gamma_\mu\lambda_C\right)\left(\bar{\alpha}\gamma^\mu\lambda_B\right) = X_{ABC} + Y_{ABC}\,,$$

with

$$X_{ABC} \equiv \tfrac{1}{4}\sum_{\pm}\left(\overline{\lambda_A}(1\pm\gamma_5)\gamma_\mu\lambda_C\right)\left(\bar{\alpha}\gamma^\mu(1\pm\gamma_5)\lambda_B\right),$$

$$Y_{ABC} \equiv \tfrac{1}{4}\sum_{\pm}\left(\overline{\lambda_A}(1\pm\gamma_5)\gamma_\mu\lambda_C\right)\left(\bar{\alpha}\gamma^\mu(1\mp\gamma_5)\lambda_B\right).$$

By using standard Fierz identities and the anticommutativity of the spinor fields, we have

$$\left(\overline{\lambda_A}(1\pm\gamma_5)\gamma_\mu\lambda_B\right)\left(\bar{\alpha}(1\pm\gamma_5)\gamma^\mu\lambda_C\right) = \left(\overline{\lambda_A}(1\pm\gamma_5)\gamma_\mu\lambda_C\right)\left(\bar{\alpha}(1\pm\gamma_5)\gamma^\mu\lambda_B\right),$$

$$\left(\overline{\lambda_A}(1\pm\gamma_5)\gamma_\mu\lambda_B\right)\left(\bar{\alpha}(1\mp\gamma_5)\gamma^\mu\lambda_C\right) = \left(\overline{\lambda_C}(1\pm\gamma_5)\gamma_\mu\lambda_B\right)\left(\bar{\alpha}(1\mp\gamma_5)\gamma^\mu\lambda_A\right).$$

(To derive the first of these relations, we note that $[(1\pm\gamma_5)\gamma_\mu]_{\alpha\gamma}[(1\pm\gamma_5)\gamma^\mu]_{\delta\beta}$ may be thought of as the $\alpha\beta$ matrix element of a matrix depending on δ and γ, and may therefore be expanded in a linear combination of $1_{\alpha\beta}$, $\gamma^\mu_{\alpha\beta}$, $[\gamma^\mu,\gamma^\kappa]_{\alpha\beta}$, $(\gamma_5\gamma^\mu)_{\alpha\beta}$, and $(\gamma_5)_{\alpha\beta}$. Because of the factors $(1\pm\gamma_5)$, the only term in the expansion is proportional to $[(1\pm\gamma_5)\gamma^\nu]_{\alpha\beta}$. Lorentz invariance and the presence of the other $1\pm\gamma_5$ factor tell us that this expansion takes the form

$$[(1\pm\gamma_5)\gamma_\mu]_{\alpha\gamma}[(1\pm\gamma_5)\gamma^\mu]_{\delta\beta} = k[(1\pm\gamma_5)\gamma_\mu]_{\alpha\beta}[(1\pm\gamma_5)\gamma^\mu]_{\delta\gamma}\,.$$

To determine the proportionality constant k, we may contract both sides with $(\gamma_\nu)_{\gamma\alpha}$ and find $k = -1$. The minus sign is cancelled by a minus sign arising from the anticommutation of λ_C and $\bar{\alpha}$. The other Fierz identity is proved in the same way, except that we also need to use the symmetry property (26.A.7) for Majorana bilinears.) Hence X_{ABC} is symmetric under interchange of B and C, while Y_{ABC} is symmetric under interchange of A and B. Since C_{ABC} is totally antisymmetric, both X_{ABC} and Y_{ABC} make a vanishing contribution to the sum in Eq. (27.3.9), leaving us $\delta\mathscr{L}_{\lambda\lambda V} = 0$, so that Eq. (27.3.1) gives a supersymmetric action, as was to be shown.

We can understand *why* Eq. (27.3.1) gives a supersymmetric action by identifying the superfield that has $f_{A\mu\nu}$, λ_A, and D_A as component fields. Recall that under a generalized gauge transformation, the vector superfield

$V_A(x,\theta)$ has the transformation property (27.1.12):

$$\exp\left(-2\sum_A t_A V_A(x,\theta)\right) \to \exp\left(-i\sum_A t_A\Omega_A(x,\theta)\right) \times \exp\left(-2\sum_A t_A V_A(x,\theta)\right) \exp\left(+i\sum_A t_A\Omega_A^*(x,\theta)\right) , \quad (27.3.10)$$

where $\Omega_A(x,\theta)$ is a general left-chiral superfield. This is not a gauge-covariant transformation rule, because $\Omega_A^* \neq \Omega_A$. To eliminate the factor involving Ω_A^*, we note that Ω_A^* is a right-chiral superfield, so that $\mathscr{D}_{L\alpha}\Omega_A^* = 0$, and therefore

$$\exp\left(-2\sum_A t_A V_A(x,\theta)\right) \mathscr{D}_{L\alpha} \exp\left(+2\sum_A t_A V_A(x,\theta)\right)$$
$$\to \exp\left(-i\sum_A t_A\Omega_A(x,\theta)\right) \exp\left(-2\sum_A t_A V_A(x,\theta)\right) \times \mathscr{D}_{L\alpha}\left[\exp\left(+2\sum_A t_A V_A(x,\theta)\right) \exp\left(+i\sum_A t_A\Omega_A(x,\theta)\right)\right] . \quad (27.3.11)$$

This is still not gauge-covariant, because the left-superderivative $\mathscr{D}_{L\alpha}$ acts on $\exp(+i\sum_A t_A\Omega_A(x,\theta))$ as well as on $\exp(+2\sum_A t_A V_A(x,\theta))$. This is eliminated if we follow the lead of the Abelian theory discussed in the previous section, and define a spinor superfield

$$2\sum_A t_A W_{AL\alpha}(x,\theta) \equiv \sum_{\beta\gamma} \epsilon_{\beta\gamma}\mathscr{D}_{R\beta}\mathscr{D}_{R\gamma}\left[\exp\left(-2\sum_A t_A V_A(x,\theta)\right) \times \mathscr{D}_{L\alpha} \exp\left(+2\sum_A t_A V_A(x,\theta)\right)\right] . \quad (27.3.12)$$

Because the product of any three $\mathscr{D}_R$s vanishes, $W_{AL\alpha}$ is left-chiral

$$\mathscr{D}_{R\beta} W_{AL\alpha}(x,\theta) = 0 , \quad (27.3.13)$$

and because $\mathscr{D}_{R\beta}\mathscr{D}_{R\gamma}\mathscr{D}_{L\alpha}\Omega_A \propto \mathscr{D}_{R\delta}\Omega_A = 0$, $W_{AL\alpha}$ is gauge-covariant in the sense that, for a generalized gauge transformation,

$$\sum_A t_A W_{AL\alpha}(x,\theta) \to \exp\left(-i\sum_A t_A\Omega_A(x,\theta)\right) \sum_A t_A W_{AL\alpha}(x,\theta) \times \exp\left(+i\sum_A t_A\Omega_A(x,\theta)\right) . \quad (27.3.14)$$

To calculate the spinor superfield at a point $x^\mu = X^\mu$, we can again adopt a version of Wess–Zumino gauge in which $V_A(X) = 0$, and after a straightforward calculation find that in this gauge

$$W_{AL}(X,\theta) = \lambda_{AL}(X_+) + \tfrac{1}{2}\gamma^\mu\gamma^\nu\theta_L\left(\partial_\mu V_{A\nu}(X_+) - \partial_\nu V_{A\mu}(X_+)\right) + \left(\theta_L^{\mathrm{T}}\epsilon\theta_L\right)\not{\partial}\lambda_{RA}(X_+) - i\theta_L D_A(X_+)\,.$$

Since W_{AL} is gauge-covariant, at a general point in a general gauge it must have the value

$$W_{AL}(x,\theta) = \lambda_{AL}(x_+) + \tfrac{1}{2}\gamma^\mu\gamma^\nu\theta_L f_{A\mu\nu}(x_+) + \left(\theta_L^{\mathrm{T}}\epsilon\theta_L\right)\not{D}\lambda_{RA}(x_+) - i\theta_L D_A(x_+)\,. \tag{27.3.15}$$

From this, we can construct a Lorentz- and gauge-invariant $\mathscr{F}$-term bilinear in W

$$-\Big[\sum_{A\alpha\beta}\epsilon_{\alpha\beta}W_{AL\alpha}W_{AL\beta}\Big]_{\mathscr{F}} = \sum_A\Big[-\left(\overline{\lambda_A}\,\not{D}(1-\gamma_5)\lambda_A\right) - \frac{1}{2}f_{A\mu\nu}f_A^{\mu\nu} + \frac{i}{4}\epsilon_{\mu\nu\rho\sigma}f_A^{\mu\nu}f_A^{\rho\sigma} + D_A^2\Big]\,. \tag{27.3.16}$$

Just as in the previous section, the gauge-invariant Lagrangian (27.3.1) is obtained from the real part of this F-term:

$$-\frac{1}{2}\,\mathrm{Re}\Big[\sum_{A\alpha\beta}\epsilon_{\alpha\beta}W_{AL\alpha}W_{AL\beta}\Big]_{\mathscr{F}} = \mathscr{L}_{\text{gauge}}\,. \tag{27.3.17}$$

What about the imaginary part? This is given by

$$-\,\mathrm{Im}\Big[\sum_{A\alpha\beta}\epsilon_{\alpha\beta}W_{AL\alpha}W_{AL\beta}\Big]_{\mathscr{F}} = -i\sum_A\left(\overline{\lambda_A}\,\not{D}\gamma_5\lambda_A\right) + \frac{1}{4}\epsilon_{\mu\nu\rho\sigma}\sum_A f_A^{\mu\nu}f_A^{\rho\sigma}\,. \tag{27.3.18}$$

Eq. (26.A.7) and the antisymmetry of the structure constants show that $(\overline{\lambda_A}\,\not{D}\gamma_5\lambda_A) = \frac{1}{2}\partial_\mu(\overline{\lambda_A}\gamma^\mu\gamma_5\lambda_A)$, so the first term is a total derivative, while Eq. (23.5.4) tells us that the second term is a total derivative also. In Abelian gauge theories, this means that a term like (27.3.18) would have no effect. But as discussed in Sections 23.5 and 23.6, in non-Abelian gauge theories the existence of instanton solutions allows the density (27.3.18) to have a non-vanishing integral over spacetime. We therefore must allow for the possibility of a new term in the Lagrangian density

$$\mathscr{L}_\theta = -\frac{g^2\theta}{16\pi^2}\,\mathrm{Im}\Big[\sum_{A\alpha\beta}\epsilon_{\alpha\beta}W_{AL\alpha}W_{AL\beta}\Big]_{\mathscr{F}}\,, \tag{27.3.19}$$

where θ is a new real parameter, and g is a gauge coupling, which can be conveniently defined for a simple gauge group so that if t_A, t_B, and t_C are in the 'standard' $SU(2)$ subalgebra of the gauge algebra used in calculating instanton effects, we have $C_{ABC} = g\,\epsilon_{ABC}$. With this definition of the gauge coupling, Eq. (23.5.20) gives for simple gauge groups

$$\int d^4x\, \epsilon_{\mu\nu\rho\sigma} \sum_A f_A^{\mu\nu} f_A^{\rho\sigma} = 64\pi^2 \nu / g^2 \,, \tag{27.3.20}$$

where $\nu = 0, \pm 1, \pm 2, \ldots$ is an integer, the winding number, which characterizes the topological class of the gauge field configuration. Thus for instantons of winding number ν, the Lagrangian density $\mathscr{L}_\theta$ contributes a phase to path integrals, given by

$$\left[\exp\left(i \int d^4x\, \mathscr{L}_\theta\right)\right]_\nu = \exp(i\nu\theta) \,, \tag{27.3.21}$$

so the effects of $\mathscr{L}_\theta$ are periodic in θ, with period 2π.

It is often convenient to absorb a factor g into the gauge field, so that the structure constants do not depend on g, and instead the Lagrangian density for the gauge field is multiplied by an over-all factor $1/g^2$. In this notation, the complete Lagrangian density for the gauge field may be written in terms of the rescaled gauge fields and structure constants as

$$\mathscr{L}_{\text{gauge}} + \mathscr{L}_\theta = -\text{Re}\left[\frac{\tau}{8\pi i} \sum_{A\alpha\beta} \epsilon_{\alpha\beta} W_{AL\alpha} W_{AL\beta}\right]_{\mathscr{F}} \,, \tag{27.3.22}$$

where τ is the complex coupling parameter

$$\tau \equiv \frac{4\pi i}{g^2} + \frac{\theta}{2\pi} \,. \tag{27.3.23}$$

According to Eq. (23.5.19), the contribution of instantons of winding number ν to path integrals is suppressed by a factor $\exp(-8\pi^2 |\nu| / g^2)$, which together with the factor (27.3.21) yields an over-all factor

$$\exp\left[i\nu\theta - \frac{8\pi^2|\nu|}{g^2}\right] = \begin{cases} \exp(2\pi i \nu \tau) & \nu \geq 0 \\ \exp(2\pi i \nu \tau^*) & \nu \leq 0 \end{cases} \,. \tag{27.3.24}$$

27.4 Renormalizable Gauge Theories with Chiral Superfields

We will now put together the pieces assembled in the previous three sections, to construct the most general renormalizable action for chiral superfields interacting with general gauge fields. Adding the terms (27.1.27), (27.2.7), and (27.3.1) and the superpotential terms in Eq. (26.4.5) gives the Lagrangian density

$$\mathscr{L} = \frac{1}{2}\left[\Phi^\dagger \exp\left(-2\sum_A t_A V_A\right)\Phi\right]_D - \frac{1}{2}\text{Re}\sum_A \left(W_{AL}^{\text{T}} \epsilon W_{AL}\right)_{\mathscr{F}}$$
$$- \frac{g^2\theta}{16\pi^2} \sum_A \text{Im}\left(W_{AL}^{\text{T}} \epsilon W_{AL}\right)_{\mathscr{F}} - \sum_A \xi_A [V_A]_D + 2\,\text{Re}\,[f]_{\mathscr{F}}$$

$$= -\sum_n (D_\mu\phi)_n^*(D^\mu\phi)_n - \frac{1}{2}\sum_n \left(\overline{\psi_n}\gamma^\mu(D_\mu\psi)_n\right) + \sum_n \mathscr{F}_n^*\mathscr{F}_n$$
$$-\text{Re}\sum_{nm}\frac{\partial^2 f(\phi)}{\partial\phi_n\partial\phi_m}\left(\psi_{nL}^{\rm T}\epsilon\psi_{mL}\right) + 2\text{Re}\sum_n \frac{\partial f(\phi)}{\partial\phi_n}\mathscr{F}_n$$
$$-2\sqrt{2}\,\text{Im}\sum_{Anm}(t_A)_{nm}\left(\overline{\psi_{nL}}\lambda_A\right)\phi_m + 2\sqrt{2}\,\text{Im}\sum_{Anm}(t_A)_{mn}\left(\overline{\psi_{nR}}\lambda_A\right)\phi_m^*$$
$$-\sum_{Anm}\phi_n^*(t_A)_{nm}\phi_m D_A - \sum_A \xi_A D_A + \frac{1}{2}\sum_A D_A D_A$$
$$-\frac{1}{4}\sum_A f_{A\mu\nu}f_A^{\mu\nu} - \frac{1}{2}\sum_A\left(\overline{\lambda_A}(\not{D}\lambda)_A\right) + \frac{g^2\theta}{64\pi^2}\epsilon_{\mu\nu\rho\sigma}\sum_A f_A^{\mu\nu}f_A^{\rho\sigma}\,. \quad (27.4.1)$$

Here $f(\phi)$ is the superpotential, a gauge-invariant complex function of ϕ_n (but not of ϕ_n^*) which the condition of renormalizability requires to be a cubic polynomial; the ξ_A are constants which gauge invariance requires to vanish except where t_A is a $U(1)$ generator; the gauge-covariant derivatives are

$$D_\mu\psi_L \equiv \partial_\mu\psi_L - i\sum_A t_A V_{A\mu}\psi_L\,, \quad (27.4.2)$$

$$D_\mu\phi \equiv \partial_\mu\phi - i\sum_A t_A V_{A\mu}\phi\,, \quad (27.4.3)$$

$$(D_\mu\lambda)_A = \partial_\mu\lambda_A + \sum_{BC} C_{ABC}V_{B\mu}\lambda_C\,, \quad (27.4.4)$$

and $f_{A\mu\nu}$ is the gauge-covariant gauge field-strength tensor

$$f_{A\mu\nu} = \partial_\mu V_{A\nu} - \partial_\nu V_{A\mu} + \sum_{BC} C_{ABC}V_{B\mu}V_{C\nu}\,. \quad (27.4.5)$$

The auxiliary fields enter quadratically, with field-independent constants as the coefficients of the second-order term, so they can be eliminated by setting them equal to the values at which the Lagrangian density is stationary:

$$\mathscr{F}_n = -\left(\partial f(\phi)/\partial\phi_n\right)^*\,, \quad (27.4.6)$$

$$D_A = \xi_A + \sum_{nm}\phi_n^*(t_A)_{nm}\phi_m\,. \quad (27.4.7)$$

Using these back in Eq. (27.4.1), the Lagrangian density becomes

$$
\begin{aligned}
\mathscr{L} = & -\sum_n (D_\mu\phi)_n^* (D^\mu\phi)_n \\
& -\frac{1}{2}\sum_n \left(\overline{\psi_{nL}}\, \gamma^\mu (D_\mu \psi_L)_n\right) + \frac{1}{2}\sum_n \left(\overline{(D_\mu\psi_L)_n}\, \gamma^\mu\, \psi_{nL}\right) \\
& -\frac{1}{2}\sum_{nm} \frac{\partial^2 f(\phi)}{\partial\phi_n\,\partial\phi_m}\left(\psi_{nL}^{\rm T}\epsilon\,\psi_{mL}\right) - \frac{1}{2}\sum_{nm}\left(\frac{\partial^2 f(\phi)}{\partial\phi_n\,\partial\phi_m}\right)^* \left(\psi_{nL}^{\rm T}\epsilon\,\psi_{mL}\right)^* \\
& -\sum_n \left|\frac{\partial f(\phi)}{\partial\phi_n}\right|^2 \\
& + i\sqrt{2}\sum_{Anm}\left(\overline{\psi_{nL}}\,(t_A)_{nm}\,\lambda_A\right)\phi_m - i\sqrt{2}\sum_{Anm}\phi_n^*\left(\overline{\lambda_A}\,(t_A)_{nm}\,\psi_{mL}\right) \\
& -\frac{1}{2}\sum_A\left(\xi_A + \sum_{nm}\phi_n^*\,(t_A)_{nm}\,\phi_m\right)^2 - \frac{1}{4}\sum_A f_{A\mu\nu} f_A^{\mu\nu} \\
& -\frac{1}{2}\sum_A\left(\overline{\lambda_A}\,(\not{D}\lambda)_A\right) + \frac{g^2\theta}{64\pi^2}\epsilon_{\mu\nu\rho\sigma}\sum_A f_A^{\mu\nu} f_A^{\rho\sigma}\,.
\end{aligned}
\tag{27.4.8}
$$

Lorentz invariance requires the fields ψ_{nL}, λ_A, and $f_{A\mu\nu}$ to have vanishing vacuum expectation values, while the tree-value vacuum expectation values of the ϕ_n are at the minimum of the potential

$$
V(\phi) = \sum_n \left|\frac{\partial f(\phi)}{\partial\phi_n}\right|^2 + \frac{1}{2}\sum_A\left(\xi_A + \sum_{nm}\phi_n^*\,(t_A)_{nm}\,\phi_m\right)^2 . \tag{27.4.9}
$$

This potential is positive, so *if* there is a set of field values at which $V(\phi)$ vanishes, then this is automatically also a minimum of the potential. For $V(\phi)$ to vanish at some field value $\phi_n = \phi_{n0}$, it is necessary and sufficient that

$$
\mathscr{F}_{n0} = -\left[\frac{\partial f(\phi)}{\partial\phi_n}\right]^*_{\phi=\phi_0} = 0 \tag{27.4.10}
$$

and

$$
D_{A0} = \xi_A + \sum_{nm}\phi_{n0}^*\,(t_A)_{nm}\,\phi_{m0} = 0\,. \tag{27.4.11}
$$

This in turn is the necessary and sufficient condition for supersymmetry not to be spontaneously broken, since Eq. (26.3.15) gives $\langle\delta\psi_{nL}\rangle_{\rm VAC} = \sqrt{2}\langle\mathscr{F}_n\rangle_{\rm VAC}\,\alpha_L$, and Eq. (26.2.16) gives $\langle\delta\lambda_A\rangle_{\rm VAC} = i\langle D_A\rangle_{\rm VAC}\,\gamma_5\alpha$.

It is worth stressing here that spontaneous symmetry breaking is more difficult for supersymmetry than for other symmetries. For most symmetries of the action there will be field configurations at which the symmetry is unbroken and the potential is stationary, but the symmetry will nevertheless be spontaneously broken if none of these configurations are minima of the potential. In contrast, any supersymmetric field configuration gives the potential a value of zero, which is necessarily lower than the value of the potential for any non-supersymmetric configuration, so the existence of *any* supersymmetric field configuration insures that supersymmetry is unbroken. As we will see in Section 27.6, this conclusion goes beyond the tree approximation used in this section; it is unaffected by corrections of any finite order in perturbation theory.

It may appear that Eqs. (27.4.10) and (27.4.11) impose too many conditions on the scalar fields to expect a solution, without some fine-tuning of the superpotential. However, for a gauge group of dimensionality D, the superpotential $f(\phi)$ is subject to the D constraints

$$\sum_m \frac{\partial f(\phi)}{\partial \phi_m}\Big(t_A\phi\Big)_m = 0\,, \tag{27.4.12}$$

for all A and all ϕ. Hence if ϕ has N independent components, then the number of *independent* conditions (27.4.10) is $N-D$, while the number of conditions (27.4.11) is D, so there are just N conditions altogether. With the number of conditions equal to the number of free variables, it is likely to find solutions for generic superpotentials. In fact, it is more usual to find solutions than not. For instance, for chiral scalar superfields in a non-trivial representation of a semi-simple gauge group we have $\xi_A = 0$, while $f(\phi)$ can have no terms linear in the ϕ_n, so both Eqs. (27.4.10) and (27.4.11) are satisfied for $\phi_{n0} = 0$. There may be other solutions of Eqs. (27.4.10) and (27.4.11) which break the gauge symmetries, but in such a theory supersymmetry cannot be broken, at least not in the tree approximation, and, as we shall see in Section 27.6, not in any order of perturbation theory.

More generally, it is easy to see that, even if the gauge group has $U(1)$ factors and even if the superpotential involves gauge-invariant superfields, if there exists a set of scalar field values ϕ_{n0} that satisfy Eq. (27.4.10), then there is another that satisfies both Eq. (27.4.10) *and* Eq. (27.4.11), provided only that the Fayet–Iliopoulos constants ξ_A all vanish. To show this, we note that since the superpotential $f(\phi)$ does not involve ϕ^*, it is invariant not only under ordinary gauge transformations $\phi \to \exp(i\sum_A \Lambda_A t_A)\phi$ with Λ_A arbitrary real numbers, but also under transformations with Λ_A arbitrary complex numbers. Under all these transformations the $\mathscr{F}$-terms in Eq. (27.4.10) transform linearly, so if ϕ_0 satisfies Eq. (27.4.10), then so does $\phi^\Lambda \equiv \exp(i\sum_A \Lambda_A t_A)\phi_0$. On the other hand, the scalar product $[\phi^\dagger\phi]$ is not invariant under transformations with Λ_A complex, but $[\phi^{\Lambda\dagger}\phi^\Lambda]$ does remain real and positive for complex Λ_A, so it is bounded below, and therefore has a minimum. For $\xi_A = 0$, the condition that $[\phi^{\Lambda\dagger}\phi^\Lambda]$ be stationary at this minimum is just that ϕ^Λ should satisfy Eq. (27.4.11). We see then that in the absence of Fayet–Ilioupoulos D-terms, the question of whether supersymmetry is unbroken in gauge theories is entirely a matter of whether the superpotential allows solutions of Eq. (27.4.10). The same result applies even in non-renormalizable theories.[3]

Now let us assume that there is a set of values ϕ_{n0} at which $V(\phi_0) = 0$, so that supersymmetry is unbroken. The spin 0 degrees of freedom are described by a shifted field

$$\varphi_n = \phi_n - \phi_{n0} \,. \tag{27.4.13}$$

There is then a cross-term between φ and the gauge fields, arising from the first term in Eq. (27.4.1):

$$2\sum_{nA} \mathrm{Im}\left(\partial_\mu \varphi_n \, (t_A\phi_0)_n^*\right) V_A^\mu \,.$$

As shown in Section 21.1, it is always possible to eliminate this term by adopting a 'unitarity gauge,' in which ϕ_n satisfies a constraint that makes this term vanish:

$$\sum_n \mathrm{Im}\left(\phi_n \, (t_A\phi_0)_n^*\right) = 0 \,. \tag{27.4.14}$$

This will have the effect of eliminating the Goldstone bosons associated with broken gauge symmetries.

We shall now work out the masses of the particles of spin 0, spin 1/2, and spin 1 that arise in this theory if supersymmetry is unbroken, taking into account the possibility that the gauge symmetries may be spontaneously broken.

Spin 0

Because $\partial f(\phi)/\partial\phi_n$ and $\xi_A + \sum_{nm} \phi_n^* (t_A)_{nm} \phi_m$ must both vanish at $\phi_n = \phi_{n0}$, the terms in $V(\phi)$ of second order in $\varphi_n \equiv \phi_n - \phi_{n0}$ and/or φ_n^* are of the form

$$V_{\rm quad}(\phi) = \sum_{nm} (\mathscr{M}^*\mathscr{M})_{nm} \varphi_n^* \varphi_m + \sum_{Anm} \left(t_A\phi_0\right)_n \left(t_A\phi_0\right)_m^* \varphi_n^* \varphi_m$$
$$+\frac{1}{2}\sum_{Anm} \left(t_A\phi_0\right)_n^* \left(t_A\phi_0\right)_m^* \varphi_n \varphi_m + \frac{1}{2}\sum_{Anm} \left(t_A\phi_0\right)_n \left(t_A\phi_0\right)_m \varphi_n^* \varphi_m^* \,, \tag{27.4.15}$$

where $\mathscr{M}$ is the complex symmetric matrix (26.4.11):

$$\mathscr{M}_{nm} \equiv \left(\frac{\partial^2 f(\phi)}{\partial\phi_n \partial\phi_m}\right)_{\phi=\phi_0} .$$

This can be written as

$$V_{\text{quad}} = \frac{1}{2}\begin{bmatrix} \varphi \\ \varphi^* \end{bmatrix}^\dagger M_0^2 \begin{bmatrix} \varphi \\ \varphi^* \end{bmatrix}, \tag{27.4.16}$$

where M_0^2 is the block matrix

$$M_0^2 = \begin{bmatrix} \mathcal{M}^*\mathcal{M} + \sum_A (t_A\phi_0)(t_A\phi_0)^\dagger & \sum_A (t_A\phi_0)(t_A\phi_0)^{\rm T} \\ \sum_A (t_A\phi_0)^*(t_A\phi_0)^\dagger & \mathcal{M}\mathcal{M}^* + \sum_A (t_A\phi_0)^*(t_A\phi_0)^{\rm T} \end{bmatrix}. \tag{27.4.17}$$

Now we must look for the eigenvalues of this mass-squared matrix. Differentiating Eq. (27.4.12) with respect to ϕ_n gives

$$\sum_m \frac{\partial^2 f(\phi)}{\partial\phi_n \partial\phi_m}\Big(t_A\phi\Big)_m + \sum_m \frac{\partial f(\phi)}{\partial\phi_m}\Big(t_A\Big)_{mn} = 0\,. \tag{27.4.18}$$

But as we have seen, $\partial f(\phi)/\partial\phi_m$ vanishes at $\phi = \phi_0$, so by setting ϕ at this value in Eq. (27.4.18), we find

$$\sum_m \mathcal{M}_{nm}\,(t_A\phi_0)_m = 0\,. \tag{27.4.19}$$

It follows that

$$M_0^2 \begin{bmatrix} t_B\phi_0 \\ \pm(t_B\phi_0)^* \end{bmatrix} = \sum_A \Big(\phi_0^\dagger[t_A t_B \pm t_B t_A]\phi_0\Big) \begin{bmatrix} t_B\phi_0 \\ \pm(t_B\phi_0)^* \end{bmatrix}.$$

But the vanishing of D_A at $\phi = \phi_0$ and the global gauge invariance of ξ_A tell us that

$$\Big(\phi_0^\dagger[t_A\,,\,t_B]\phi_0\Big) = i\sum_C C_{ABC}\Big(\phi_0^\dagger t_C\phi_0\Big) = -i\Big(\phi_0^\dagger\phi_0\Big)\sum_C C_{ABC}\xi_C = 0\,. \tag{27.4.20}$$

The matrix (27.4.17) therefore has a pair of eigenvectors for each gauge symmetry

$$u = \begin{bmatrix} \sum_B c_B\, t_B\phi_0 \\ \sum_B c_B\,(t_B\phi_0)^* \end{bmatrix}, \qquad v = \begin{bmatrix} \sum_B c_B\, t_B\phi_0 \\ -\sum_B c_B\,(t_B\phi_0)^* \end{bmatrix}, \tag{27.4.21}$$

for which

$$M_0^2 u = \mu^2 u\,, \qquad M_0^2 v = 0\,, \tag{27.4.22}$$

where μ^2 and c_A are any real solutions of the eigenvalue problem*

$$\sum_B \Big(\phi_0^\dagger\{t_A\,,\,t_B\}\phi_0\Big)c_B = \mu^2 c_A\,, \tag{27.4.23}$$

---

* The presence of a factor 1/2 in Eq. (21.1.17) which does not appear in Eq. (27.4.23) is due to a difference in the way that the scalar fields are normalized.

with the exception that if the eigenvalue μ^2 vanishes then $\sum_B c_B t_B \phi_0 = 0$, so that the eigenvectors u and v are absent. The massless particles associated with the v eigenvectors are Goldstone bosons, which are eliminated from the physical spectrum by the unitarity gauge condition (27.4.14). In addition to these mass eigenstates, there is another set that are orthogonal to all the us and vs and that therefore take the form

$$w_\pm = \begin{bmatrix} \zeta \\ \pm\zeta^* \end{bmatrix} , \tag{27.4.24}$$

where

$$\sum_n (t_A \phi_0)_n^* \zeta_n = 0 . \tag{27.4.25}$$

Eq. (27.4.19) shows that the space of ζs satisfying Eq. (27.4.25) is invariant under multiplication with the Hermitian matrix $\mathscr{M}^\dagger \mathscr{M}$, so it is spanned by the eigenvectors of this matrix, satisfying

$$\mathscr{M}^\dagger \mathscr{M} \zeta = m^2 \zeta \tag{27.4.26}$$

with m^2 a set of real positive (or zero) eigenvalues. Eq. (27.4.26) and its complex conjugate together with Eq. (27.4.25) show that $w_\pm$ are eigenvectors of M_0^2 with eigenvalues m^2:

$$M_0^2 w_\pm = m^2 w_\pm . \tag{27.4.27}$$

We thus have *two* self-charge-conjugate spinless bosons of each mass m satisfying Eq. (27.4.27), and one self-charge-conjugate spinless boson of each non-zero mass μ satisfying Eq. (27.4.23).

Spin 1/2

The fermion masses arise from the non-derivative terms in Eq. (27.4.8) that are of second order in the fermion fields ψ_n and λ_A:

$$\mathscr{L}_{1/2} = -\frac{1}{2} \sum_{nm} \mathscr{M}_{nm} \left(\psi_{nL}^{\mathrm{T}} \epsilon \psi_{mL}\right) - i\sqrt{2} \sum_{Am} (t_A \phi_0)_m^* \left(\lambda_{LA}^{\mathrm{T}} \epsilon\, \psi_{mL}\right) + \text{H.c.} \tag{27.4.28}$$

We saw in Section 26.4 that if the fermion mass term in the Lagrangian for a column χ of Majorana spinor fields is put in the form

$$\mathscr{L}_{1/2} = -\frac{1}{2}\left(\chi_L^{\mathrm{T}} \epsilon M \chi_L\right) + \text{H.c.} , \tag{27.4.29}$$

then the fermion squared masses are the eigenvalues of the Hermitian matrix $M^\dagger M$. Eq. (27.4.28) gives the elements of the matrix M here as

$$M_{nm} = \mathscr{M}_{nm} , \quad M_{nA} = M_{An} = i\sqrt{2}(t_A \phi_0)_n^* , \quad M_{AB} = 0 , \tag{27.4.30}$$

for which, using Eqs. (27.4.19) and (27.4.20),

$$(M^\dagger M)_{nm} = (\mathcal{M}^\dagger \mathcal{M})_{nm} + 2\sum_A (t_A\phi_0)_n (t_A\phi_0)^*_m ,$$
$$(M^\dagger M)_{nA} = (M^\dagger M)_{An} = 0 , \qquad (27.4.31)$$
$$(M^\dagger M)_{AB} = 2(\phi_0^\dagger t_B t_A \phi_0) = (\phi_0^\dagger \{t_B , t_A\}\phi_0) .$$

The eigenvectors of the matrix (27.4.30) are of three types. First are those of the form

$$z = \begin{bmatrix} \zeta \\ 0 \end{bmatrix} , \qquad (27.4.32)$$

with eigenvalues m^2, where ζ_n and m^2 are any eigenvectors and corresponding eigenvalues of $\mathcal{M}^\dagger\mathcal{M}$. Next are those of the form

$$g = \begin{bmatrix} 0 \\ c \end{bmatrix} , \qquad (27.4.33)$$

with eigenvalues μ^2, where c_B and μ^2 are any eigenvectors and eigenvalues of the matrix $(\phi_0^\dagger\{t_B , t_A\}\phi_0)$. Finally, there are those of the form

$$h = \begin{bmatrix} \sum_B c_B t_B \phi_0 \\ 0 \end{bmatrix} , \qquad (27.4.34)$$

with eigenvalues μ^2, where c_B and μ^2 are again any eigenvectors and eigenvalues of the matrix $(\phi_0^\dagger\{t_B , t_A\}\phi_0)$. The only exception is that the eigenvectors c of this matrix with eigenvalue zero have $\sum_A c_A t_A \phi_0 = 0$, corresponding to unbroken symmetries, so that in this case the vector (27.4.34) vanishes and we have only the eigenvector (27.4.33). Thus there is one Majorana fermion of each mass m satisfying Eq. (27.4.26), two Majorana fermions of each non-zero mass μ satisfying Eq. (27.4.22), and one Majorana fermion of zero mass for each unbroken gauge symmetry.

Spin 1

The mass terms in the Lagrangian for the gauge fields arise from the part of the first term in Eq. (27.4.1) that is of second order in the gauge field V_A^μ:

$$\mathcal{L}_V = -\sum_{nAB} (t_A\phi_0)^*_n (t_B\phi_0)_n V_{A\mu} V_B^\mu . \qquad (27.4.35)$$

Since the fields $V_{A\mu}$ are real, their mass-squared matrix is the matrix in Eq. (27.4.23):

$$(\mu^2)_{AB} = \left(\phi_0^\dagger \{t_B , t_A\}\phi_0\right) . \qquad (27.4.36)$$

There is one spin 1 particle of mass μ for each eigenvalue μ^2 of the matrix (27.4.36).

Putting this all together, we see that for each eigenvalue m^2 of the matrix $\mathscr{M}^*\mathscr{M}$ there are two self-charge-conjugate spinless particles and one Majorana fermion of mass m; for each non-zero eigenvalue of the matrix μ^2_{AB} there are one self-charge-conjugate spinless boson, two Majorana fermions, and one self-charge-conjugate spin 1 boson of mass μ; and for each zero eigenvalue of this matrix there is one Majorana fermion and one self-charge-conjugate spin 1 boson of zero mass. It is not surprising that the particle multiplets for each zero or non-zero mass are just the same as we found by direct use of the supersymmetry algebra in Sections 25.4 and 25.5. What *is* a bit surprising is that the masses of the gauge and chiral particles are unaffected by each other. The masses m that are given by eigenvalues of $(\mathscr{M}^*\mathscr{M})_{nm}$ and the particles with these masses are just what they would be in a theory of chiral superfields without gauge superfields, and the masses μ that are given by eigenvalues of μ^2_{AB} and the particles with these masses are just what they would be in a theory of gauge superfields with no chiral superfields.

For future use in Section 27.9 we will now apply the method described in Section 26.7 to construct the supersymmetry current for the supersymmetric gauge Lagrangian (27.4.1). In the gauge used earlier, an infinitesimal supersymmetry transformation changes V_A, λ_a, and D_A by the amounts (27.3.4)–(27.3.6). Adding the Noether supersymmetry current given by Eq. (26.7.2) for these fields to the Noether current for ϕ_n, ψ_n, and $\mathscr{F}_n$ already given in Eq. (26.7.8), with derivatives replaced by gauge-invariant derivatives, gives the total Noether supersymmetry current:

$$N^\mu = \sum_A f^{\mu\nu}_A \gamma_\nu \lambda_A - \frac{1}{8}\sum_A f_{A\rho\sigma}[\gamma^\rho,\gamma^\sigma]\gamma^\mu\lambda_A - \frac{1}{2}i\sum_A D_A\gamma_5\gamma^\mu\lambda_A$$
$$+\frac{1}{\sqrt{2}}\sum_n \Big[2\,(D^\mu\phi)^*_n\,\psi_{nL} + 2\,(D^\mu\phi)_n\,\psi_{nR} + (\not{D}\phi)_n\,\gamma^\mu\psi_{nR}$$
$$+(\not{D}\phi)^*_n\,\gamma^\mu\psi_{nL} - \mathscr{F}_n\,\gamma^\mu\psi_{nR} - \mathscr{F}^*_n\,\gamma^\mu\psi_{nL}\Big]\,. \qquad (27.4.37)$$

This is not the supersymmetry current, because the Lagrangian density is not invariant under supersymmetry; instead, its change is the derivative

$$\delta\mathscr{L} = \partial_\mu\left(\bar{\alpha}K^\mu\right)\,, \qquad (27.4.38)$$

where**

$$K^\mu = \frac{1}{2}i\sum_A \epsilon^{\rho\sigma\mu\nu} f_{A\rho\sigma}\gamma_\nu\gamma_5\lambda_A + \frac{1}{8}\sum_A [\gamma^\rho,\gamma^\sigma]\gamma^\mu\lambda_A f_{A\rho\sigma} + \frac{1}{2}i\sum_A D_A\,\gamma_5\gamma^\mu\lambda_A$$
$$-i\sum_{Anm}(t_A)_{nm}\gamma_5\gamma^\mu\lambda_A\phi_n^*\phi_m$$
$$+\frac{1}{\sqrt{2}}\sum_n \gamma^\mu\Big[-(\not{D}\phi)_n\,\psi_{nR} - (\not{D}\phi)_n^*\,\psi_{nL} + \mathscr{F}_n^*\,\psi_{nL} + \mathscr{F}_n\,\psi_{nR}$$
$$+2\left(\frac{\partial f(\phi)}{\partial\phi_n}\right)\psi_{nL} + 2\left(\frac{\partial f(\phi)}{\partial\phi_n}\right)^*\psi_{nR}\Big]\,. \tag{27.4.39}$$

The first two terms are derived using the identity (27.2.5). Using the same identity again with Eq. (26.7.4) gives the total supersymmetry current:

$$S^\mu = N^\mu + K^\mu$$
$$= -\frac{1}{4}\sum_A f_{A\rho\sigma}[\gamma^\rho,\gamma^\sigma]\gamma^\mu\lambda_A - i\sum_{Anm}(t_A)_{nm}\gamma_5\gamma^\mu\lambda_A\phi_n^*\phi_m$$
$$+\frac{1}{\sqrt{2}}\sum_n\Big[(\not{D}\phi)_n\,\gamma^\mu\psi_{nR} + (\not{D}\phi^*)_n\,\gamma^\mu\psi_{nL}$$
$$+2\left(\frac{\partial f(\phi)}{\partial\phi_n}\right)\gamma^\mu\psi_{nL} + 2\left(\frac{\partial f(\phi)}{\partial\phi_n}\right)^*\gamma^\mu\psi_{nR}\Big]\,. \tag{27.4.40}$$

* * *

In Section 26.8 we considered a class of supersymmetric theories with a superpotential $f(\Phi)$ that has an arbitrary dependence on a set of left-chiral scalar superfields Φ_n but not their derivatives, and with a Kahler potential $K(\Phi,\Phi^*)$ that has an arbitrary dependence on the Φ_n and Φ_n^*, but not their derivatives. We can extend the same considerations to gauge theories, again with the dependence of the Lagrangian on the chiral superfields limited only by supersymmetry, but without introducing new superderivatives or spacetime derivatives. The renormalizable Lagrangian density is then replaced with

$$\mathscr{L} = \frac{1}{2}\Big[K\Big(\Phi,\,\Phi^\dagger\exp(-2\sum_A t_A V_A)\Big)\Big]_D + 2\,\mathrm{Re}\Big[f(\Phi)\Big]_{\mathscr{F}}$$
$$-\frac{1}{2}\mathrm{Re}\sum_{AB}\Big[h_{AB}(\Phi)\Big(W_{AL}^{\mathrm{T}}\epsilon W_{BL}\Big)\Big]_{\mathscr{F}}\,, \tag{27.4.41}$$

** The easiest way to calculate the change in the term $[\Phi^\dagger\exp(-2\sum_A t_A V_A)\Phi]_D$ is to calculate the λ-component of $\Phi^\dagger\exp(-2\sum_A t_A V_A)\Phi$ and use Eq. (26.2.17). In this way of doing the calculation, the important term on the second line of the right-hand side of Eq. (27.4.39) arises from the λ-component of $\exp(-2\sum_A t_A V_A)$.

where $h_{AB}(\Phi)$ is a new function of the Φ_n, but not of the Φ_n^* or derivatives.

The chiral gauge and scalar superfields are given by the expansions (26.3.21) and (27.3.15):

$$W_{AL}(x,\theta) = \lambda_{AL}(x_+) + \frac{1}{2}\gamma^\mu\gamma^\nu\theta_L f_{A\mu\nu}(x_+) + \left(\theta_L^{\rm T}\epsilon\theta_L\right) \not{\partial}\lambda_{AR}(x_+) \\ -i\theta_L D_A(x_+)\,,$$

$$\Phi_n(x,\theta) = \phi_n(x_+) - \sqrt{2}\left(\theta_L^{\rm T}\epsilon\psi_{nL}(x_+)\right) + \mathscr{F}_n(x_+)\left(\theta_L^{\rm T}\epsilon\theta_L\right)\,,$$

where x_+^μ is the shifted coordinate (26.3.23). The terms of second order in θ_L (and independent of θ_R) in $\sum_{AB} h_{AB}(\Phi)(W_{AL}^{\rm T}\epsilon W_{BL})$ are then

$$-\left[\sum_{AB} h_{AB}(\Phi)\left(W_{AL}^{\rm T}\epsilon W_{BL}\right)\right]_{\theta_L^2} =$$

$$\left(\theta_L^{\rm T}\epsilon\theta_L\right)\sum_{AB}\left(\lambda_{AL}^{\rm T}\epsilon\lambda_{BL}\right)\left[\frac{1}{2}\sum_{nm}\left(\psi_{nL}^{\rm T}\epsilon\psi_{mL}\right)\frac{\partial^2 h_{AB}(\phi)}{\partial\phi_n\partial\phi_m} - \sum_n \mathscr{F}_n\frac{\partial h_{AB}(\phi)}{\partial\phi_n}\right]$$
$$+\left(\theta_L^{\rm T}\epsilon\theta_L\right)\sum_{AB} h_{AB}(\phi)\Bigg[-\left(\overline{\lambda_A}\,\not{\partial}(1-\gamma_5)\lambda_B\right) - \frac{1}{2}f_{A\mu\nu}f_B^{\mu\nu}$$
$$+\frac{i}{4}\epsilon_{\mu\nu\rho\sigma}f_A^{\mu\nu}f_B^{\rho\sigma} + D_A D_B\Bigg]$$
$$+\sqrt{2}\sum_{ABn}\frac{\partial h_{AB}(\phi)}{\partial\phi_n}\left(\theta_L^{\rm T}\epsilon\psi_{nL}\right)\left[-\left(\lambda_{BL}^{\rm T}\epsilon\gamma^\mu\gamma^\nu\theta_L\right)f_{A\mu\nu} + 2i\left(\lambda_{BL}^{\rm T}\epsilon\theta_L\right)\right]\,,$$

with all fields now understood to be evaluated at x^μ rather than x_+^μ. (The first and second terms on the right-hand side are taken from Eqs. (26.4.4) and (27.3.16), respectively.) Also, by writing $\theta_{L\alpha}\theta_{L\beta}$ as $\frac{1}{2}\epsilon_{\alpha\beta}(\theta_L^{\rm T}\epsilon\theta_L)$, the third term on the right-hand side can also be expressed as proportional to $(\theta_L^{\rm T}\epsilon\theta_L)$:

$$\left(\theta_L^{\rm T}\epsilon\psi_{nL}\right)\left[\left(\overline{\psi_B}\gamma^\mu\gamma^\nu\theta_L\right)f_{A\mu\nu} - 2i\left(\overline{\psi_B}\theta_L\right)\right] =$$
$$\frac{1}{2}\left(\theta_L^{\rm T}\epsilon\theta_L\right)\left[\left(\overline{\psi_B}\gamma^\mu\gamma^\nu\psi_{nL}\right) - 2i\left(\overline{\psi_B}\psi_{nL}\right)D_A\right]\,.$$

The $\mathscr{F}$-term is the coefficient of $(\theta_L^{\rm T}\epsilon\theta_L)$, so

$$-\left[\sum_{AB} h_{AB}(\Phi)\left(W_{AL}^{\rm T}\epsilon W_{BL}\right)\right]_{\mathscr{F}} =$$

$$\sum_{AB}\left(\lambda_{AL}^{\rm T}\epsilon\lambda_{BL}\right)\left[\frac{1}{2}\sum_{nm}\left(\psi_{nL}^{\rm T}\epsilon\psi_{mL}\right)\frac{\partial^2 h_{AB}(\phi)}{\partial\phi_n\partial\phi_m} - \sum_n \mathscr{F}_n\frac{\partial h_{AB}(\phi)}{\partial\phi_n}\right]$$

$$+\sum_{AB} h_{AB}(\phi)\Big[-\left(\overline{\lambda_A}\,\not{D}(1-\gamma_5)\lambda_B\right)-\frac{1}{2}f_{A\mu\nu}f_B^{\mu\nu}+\frac{i}{4}\epsilon_{\mu\nu\rho\sigma}f_A^{\mu\nu}f_B^{\rho\sigma}$$
$$+D_AD_B\Big]$$
$$+\frac{\sqrt{2}}{2}\sum_{ABn}\frac{\partial h_{AB}(\phi)}{\partial\phi_n}\left[-\left(\overline{\psi_B}\gamma^\mu\gamma^\nu\psi_{nL}\right)f_{A\mu\nu}+2i\left(\overline{\psi_B}\psi_{nL}\right)D_A\right].$$

The other terms in Eq. (27.4.41) are just given by the gauge-invariant version of the Lagrangian density (26.8.6). Putting this together gives the Lagrangian density

$$\mathscr{L}=\mathrm{Re}\sum_{nm}\mathscr{G}_{nm}(\phi,\phi^*)\Big[-\frac{1}{2}\left(\overline{\psi_m}\,\not{D}(1+\gamma_5)\psi_n\right)+\mathscr{F}_n\mathscr{F}^*_m-D_\mu\phi_n\,D^\mu\phi^*_m\Big]$$
$$-2\mathrm{Re}\sum_i\frac{\partial K(\phi,\phi^*)}{\partial\phi^*_i}D_A(\phi^*t_A)_i$$
$$+i\sqrt{2}\frac{\partial^2K(\phi,\phi^*)}{\partial\phi_i\partial\phi^*_j}\left[(t_A\phi)_i\overline{\psi_j}\lambda_{AR}-(\phi^*t_A)_j\overline{\psi_i}\lambda_{AL}\right]$$
$$-\mathrm{Re}\sum_{nml}\frac{\partial^3K(\phi,\phi^*)}{\partial\phi_n\,\partial\phi_m\,\partial\phi^*_l}\left(\overline{\psi_n}\psi_{mL}\right)\mathscr{F}^*_\ell$$
$$+\mathrm{Re}\sum_{nml}\frac{\partial^3K(\phi,\phi^*)}{\partial\phi_n\,\partial\phi_m\,\partial\phi^*_l}\left(\overline{\psi_m}\gamma^\mu\psi_{\ell R}\right)D_\mu\phi_n$$
$$+\frac{1}{4}\sum_{nmlk}\frac{\partial^4K(\phi,\phi^*)}{\partial\phi_n\,\partial\phi_m\,\partial\phi^*_l\,\partial\phi^*_k}\left(\overline{\psi_n}\psi_{mL}\right)\left(\overline{\psi_k}\psi_{lR}\right)$$
$$-\mathrm{Re}\sum_{nm}\frac{\partial^2f(\phi)}{\partial\phi_n\,\partial\phi_m}\left(\overline{\psi_n}\,\psi_{mL}\right)+2\,\mathrm{Re}\sum_n\mathscr{F}_n\frac{\partial f(\phi)}{\partial\phi_n}$$
$$+\frac{1}{4}\mathrm{Re}\sum_{ABnm}\left(\overline{\lambda_A}\lambda_{BL}\right)\left(\overline{\psi_n}\psi_{mL}\right)\frac{\partial^2h_{AB}(\phi)}{\partial\phi_n\partial\phi_m}-\frac{1}{2}\mathrm{Re}\sum_{ABn}\left(\overline{\lambda_A}\lambda_{BL}\right)\mathscr{F}_n\frac{\partial h_{AB}(\phi)}{\partial\phi_n}$$
$$+\mathrm{Re}\sum_{AB}h_{AB}(\phi)\Big[-\left(\overline{\lambda_A}\,\not{D}\lambda_{BR}\right)-\frac{1}{4}f_{A\mu\nu}f_B^{\mu\nu}+\frac{1}{8}i\,\epsilon_{\mu\nu\rho\sigma}f_A^{\mu\nu}f_B^{\rho\sigma}+\frac{1}{2}D_AD_B\Big]$$
$$+\frac{\sqrt{2}}{4}\mathrm{Re}\sum_{ABn}\frac{\partial h_{AB}(\phi)}{\partial\phi_n}\left[-\left(\overline{\lambda_B}\gamma^\mu\gamma^\nu\psi_{nL}\right)f_{A\mu\nu}+2i\left(\overline{\lambda_B}\psi_{nL}\right)D_A\right].$$
$$(27.4.42)$$

One interesting feature of this result is the appearance of a gaugino mass in theories with ϕ_n-dependent functions $h_{AB}(\phi)$, when supersymmetry is broken by a non-vanishing value of $\mathscr{F}_n$. This mechanism is used to generate gaugino masses in some theories of gravitationally mediated supersymmetry breaking, discussed in Section 31.7.

27.5 Supersymmetry Breaking in the Tree Approximation Resumed

We saw in the previous section that if the Fayet–Iliopoulos constants ξ_A all vanish and if there exists a solution of the equations $\partial f(\phi)/\partial\phi_n = 0$, then there is also some solution of these equations where the D-components of the gauge superfields all vanish, so that supersymmetry is unbroken. It follows that there are only two (non-exclusive) ways that supersymmetry can be spontaneously broken in the tree approximation in renormalizable theories of gauge and chiral superfields: the superpotential $f(\phi)$ may be arranged so that there are no solutions of all the equations $\partial f(\phi)/\partial\phi_n = 0$, or for gauge groups with $U(1)$ factors there may be Fayet–Iliopoulos terms in the action.

We have already seen in Section 26.5 how it can happen that there might not be any value of ϕ for which $\partial f(\phi)/\partial\phi_n = 0$. No change in that discussion is needed when the chiral superfields interact with gauge fields, so let's turn to the other possibility: spontaneous supersymmetry breaking produced by Fayet–Iliopoulos terms. Since this can only arise for gauge groups with $U(1)$ factors, the simplest case is a theory with a single $U(1)$ gauge group. As discussed in Section 22.4, to avoid $U(1)$-$U(1)$-$U(1)$ and $U(1)$-graviton-graviton anomalies it is necessary that the sum of the $U(1)$ quantum numbers of all left-chiral superfields and the sum of their cubes should vanish. We will consider the simplest possibility: two left-chiral superfields $\Phi_\pm$, with $U(1)$ quantum numbers $\pm e$. (This is a supersymmetric version of quantum electrodynamics, with the spinor components ψ_{-L} and ψ_{+L} of the two superfields providing the left-handed parts of the electron field and of its charge conjugate.) The most general $U(1)$-invariant superpotential in a renormalizable theory is just $f(\Phi) = m\Phi_+\Phi_-$. The scalar potential (27.4.9) for the scalar components $\phi_\pm$ of these superfields is then

$$V(\phi_+, \phi_-) = m^2|\phi_+|^2 + m^2|\phi_-|^2 + \left(\xi + e^2|\phi_+|^2 - e^2|\phi_-|^2\right)^2 . \quad (27.5.1)$$

Unless the Fayet–Iliopoulos constant ξ vanishes, it is evidently not possible to find a supersymmetric vacuum with $V = 0$. For $\xi > m^2/2e^2$ or $\xi < -m^2/2e^2$ the potential (27.5.1) has a minimum with either $\phi_+ = 0$ and $|\phi_-|^2 = (2e^2\xi - m^2)/2e^4$ or $\phi_- = 0$ and $|\phi_+|^2 = (-2e^2\xi - m^2)/2e^4$, so the $U(1)$ gauge symmetry is broken along with supersymmetry. For $|\xi| < m^2/2e^2$ the minimum of the potential is at $\phi_+ = \phi_- = 0$, so here the gauge symmetry is unbroken. There is in general no necessary connection between the possible breakdown of supersymmetry and of gauge symmetries.

Whether supersymmetry is spontaneously broken by the Fayet–Iliopoulos mechanism discussed here or by the O'Raifeartaigh mechanism of Section 26.5 or by some combination of the two, supersymmetry leaves a remnant in the pattern of tree-approximation masses. Inspection of the Lagrangian (27.4.8) for a general renormalizable supersymmetric theory of gauge and chiral superfields shows that the spontaneous breakdown of supersymmetry in this theory produces the following corrections to the masses calculated in Section 27.4.

Spin 0 Masses

If the $\mathscr{F}$-terms $\mathscr{F}_n = -(\partial f(\phi)/\partial\phi_n)^*$ do not vanish at the minimum ϕ_0 of the potential, then the terms in the potential of second order in $\varphi_n \equiv \phi_n - \phi_{n0}$ have additional terms beyond those listed in Eq. (27.4.15):

$$\begin{aligned}V_{\text{quad}}(\varphi) = &\sum_{nm}(\mathscr{M}^*\mathscr{M})_{nm}\varphi_n^*\varphi_m + \sum_{Anm}\left(t_A\phi_0\right)_n\left(t_A\phi_0\right)_m^*\varphi_n^*\varphi_m \\ &+\frac{1}{2}\sum_{Anm}\left(t_A\phi_0\right)_n^*\left(t_A\phi_0\right)_m^*\varphi_n\varphi_m + \frac{1}{2}\sum_{Anm}\left(t_A\phi_0\right)_n\left(t_A\phi_0\right)_m\varphi_n^*\varphi_m^*\,, \\ &+\frac{1}{2}\sum_{nm}\mathscr{N}_{nm}\varphi_n\varphi_m + \frac{1}{2}\sum_{nm}\mathscr{N}_{nm}^*\varphi_n^*\varphi_m^* \\ &+\sum_{Anm}D_{A0}(t_A)_{nm}\varphi_n^*\varphi_m\,, \end{aligned} \tag{27.5.2}$$

where $\mathscr{M}$ is again the complex symmetric matrix (26.4.11):

$$\mathscr{M}_{nm} \equiv \left(\frac{\partial^2 f(\phi)}{\partial\phi_n\partial\phi_m}\right)_{\phi=\phi_0}\,,$$

$\mathscr{N}_{nm}$ is a new ingredient

$$\mathscr{N}_{nm} \equiv -\sum_\ell \mathscr{F}_{\ell 0}\left(\frac{\partial^3 f(\phi)}{\partial\phi_n\partial\phi_m\partial\phi_\ell}\right)_{\phi=\phi_0}\,, \tag{27.5.3}$$

and $\mathscr{F}_0$ and D_{A0} are again the $\mathscr{F}$-terms and D-terms of the chiral scalar and gauge superfields at the minimum of the potential:

$$\mathscr{F}_{n0} = -\left[\frac{\partial f(\phi)}{\partial\phi_n}\right]^*_{\phi=\phi_0}\,, \qquad D_{A0} = \xi_A + \sum_{nm}\phi_{n0}^*\,(t_A)_{nm}\,\phi_{m0}\,.$$

If we write the quadratic part (27.5.2) of the potential in the form (27.4.16):

$$V_{\text{quad}} = \frac{1}{2}\begin{bmatrix} \varphi \\ \varphi^* \end{bmatrix}^\dagger M_0^2 \begin{bmatrix} \varphi \\ \varphi^* \end{bmatrix},$$

then, instead of Eq. (27.4.17), we now have the scalar mass matrix

$$M_0^2 = \begin{bmatrix} \mathscr{M}^*\mathscr{M} + \mathscr{A} + \sum_A D_{A0}\, t_A & \mathscr{B} + \mathscr{N}^* \\ \mathscr{B}^* + \mathscr{N} & \mathscr{M}\mathscr{M}^* + \mathscr{A}^* + \sum_A D_{A0}\, t_A^{\mathrm{T}} \end{bmatrix}, \tag{27.5.4}$$

where

$$\mathscr{A} \equiv \sum_A (t_A\phi_0)(t_A\phi_0)^\dagger, \qquad \mathscr{B} \equiv \sum_A (t_A\phi_0)(t_A\phi_0)^{\mathrm{T}}.$$

Spin 1/2 Masses

The fermion mass matrix M here is given again by Eq. (27.4.30):

$$M_{nm} = \mathscr{M}_{nm}, \qquad M_{nA} = M_{An} = i\sqrt{2}(t_A\phi_0)^*_n, \qquad M_{AB} = 0.$$

However, now instead of Eq. (27.4.19), the gauge invariance condition (27.4.18) yields

$$\sum_m \mathscr{M}_{nm}\,(t_A\phi_0)_m = \sum_m \mathscr{F}_{m0}(t_A)_{mn}. \tag{27.5.5}$$

Thus the Hermitian positive matrix whose eigenvalues are the squared masses of the fermions is given by

$$(M^\dagger M)_{nm} = (\mathscr{M}^\dagger\mathscr{M})_{nm} + 2\sum_A (t_A\phi_0)_n (t_A\phi_0)^*_m,$$

$$(M^\dagger M)_{AB} = 2(\phi_0^\dagger t_B t_A \phi_0), \tag{27.5.6}$$

$$(M^\dagger M)_{An} = (M^\dagger M)^*_{nA} = i\sqrt{2}\sum_m \mathscr{F}_{m0}(t_A)_{mn}.$$

Spin 1 Masses

The squared masses of the vector bosons are again given by the eigenvalues of the matrix (27.4.36):

$$(\mu^2)_{AB} = \left(\phi_0^\dagger, \{t_B, t_A\}\phi_0\right). \tag{27.5.7}$$

With the exception of the D-terms in Eq. (27.5.4), all of the changes in the mass-squared matrices are in their off-diagonal components. Therefore

Eqs. (27.5.4), (27.5.6), and (27.5.7) yield particularly simple results for the traces of these matrices: for spin 0

$$\mathrm{Tr}\, M_0^2 = 2\mathrm{Tr}\,(\mathscr{M}^*\mathscr{M}) + \mathrm{Tr}\,\mu^2 + 2\sum_A D_{A0}\mathrm{Tr}\,t_A\,, \tag{27.5.8}$$

and for spin 1/2

$$\mathrm{Tr}\,(M^\dagger M) = \mathrm{Tr}\,(\mathscr{M}^*\mathscr{M}) + 2\mathrm{Tr}\,\mu^2\,. \tag{27.5.9}$$

Since the trace is the sum of the eigenvalues, we obtain from this a *mass sum rule*:

$$\sum_{\text{spin }0}\text{mass}^2 - 2\sum_{\text{spin }1/2}\text{mass}^2 + 3\sum_{\text{spin }1}\text{mass}^2 = -2\sum_A D_{A0}\mathrm{Tr}\,t_A\,. \tag{27.5.10}$$

The trace of t_A automatically vanishes unless t_A is a $U(1)$ generator, and, as mentioned in Section 22.4, the trace must also vanish (when taken over all left-handed fermions) for $U(1)$ gauge generators to avoid gravitational contributions to an anomaly that would violate conservation of the $U(1)$ current. Thus (27.5.10) leads to the simpler result[4]

$$\sum_{\text{spin }0}\text{mass}^2 - 2\sum_{\text{spin }1/2}\text{mass}^2 + 3\sum_{\text{spin }1}\text{mass}^2 = 0\,. \tag{27.5.11}$$

Of course, the unbroken conservation of charge, color, and baryon and lepton numbers prevents the mass matrices from having elements linking particles with different values of these quantum numbers, so all these results hold separately for each set of conserved quantum numbers.

The sum rule (27.5.11) is often quoted as providing evidence against models in which supersymmetry is spontaneously broken in the tree approximation within the minimum supersymmetric extension of the standard model. We will discuss this along with other arguments in Section 28.3.

As already observed in Section 26.5 (and discussed in greater generality in Sections 29.1 and 29.2), the spontaneous breakdown of supersymmetry necessarily entails the existence of a massless fermion, the goldstino. For renormalizable gauge theories in the tree approximation, the goldstino field g appears as a term in the spinor components ψ_n and λ_A of the chiral and gauge superfields, with coefficients given by

$$\psi_{nL} = i\sqrt{2}\,\mathscr{F}_{n0}\,g_L + \cdots\,, \qquad \lambda_{AL} = D_{A0}\,g_L + \cdots\,, \tag{27.5.12}$$

where the dots denote terms involving spinor fields of definite non-zero mass. To check this we have to confirm that $(i\sqrt{2}\mathscr{F}_{n0}\,, D_{A0})$ is an eigenvector of the fermion mass-squared matrix $M^\dagger M$ with eigenvalue zero. For this purpose we will need to use the condition that the potential (27.4.9) is stationary at $\phi = \phi_0$:

$$0 = \left.\frac{\partial V}{\partial \phi_n}\right|_{\phi=\phi_0} = -\sum_m \mathscr{M}_{nm} \mathscr{F}_{m0} + \sum_A D_{A0}(\phi_0^\dagger t_A)_n \,. \tag{27.5.13}$$

We also need the gauge invariance condition (27.4.12), which at $\phi = \phi_0$ reads

$$\sum_n \mathscr{F}_{n0}\,(t_A \phi_0)_n = 0\,. \tag{27.5.14}$$

Combining Eqs. (27.5.13) and (27.5.14) with Eqs. (27.5.5) and (27.5.6) then yields

$$i\sqrt{2}\sum_m (M^\dagger M)_{nm}\mathscr{F}_{m0} = i\sqrt{2}\sum_A D_A(t_A \mathscr{F}_0^*)_n = -\sum_A (M^\dagger M)_{nA} D_{A0} \tag{27.5.15}$$

and

$$i\sqrt{2}\sum_m (M^\dagger M)_{Am}\mathscr{F}_{m0} = -2\sum_{nm} \mathscr{F}_{n0}\,(t_A)_{nm}\,\mathscr{F}_{m0} = -\sum_B (M^\dagger M)_{AB}\, D_{B0}\,. \tag{27.5.16}$$

That is,

$$M^\dagger M \begin{pmatrix} i\sqrt{2}\mathscr{F}_0 \\ D_0 \end{pmatrix} = 0\,, \tag{27.5.17}$$

as was to be shown.

27.6 Perturbative Non-Renormalization Theorems

From the beginning, several of the ultraviolet divergences in ordinary renormalizable quantum field theories were found to be absent in the supersymmetric versions of these theories. With the development in 1975 of supergraph techniques, in which all particles in each supermultiplet are considered together, it became possible to show that some radiative corrections are not only finite, but are absent altogether in perturbation theory.[5] Supergraphs will be described in detail in Chapter 30, but as it happens they are not needed to prove the most important non-renormalization theorems. This section will give a version of a method developed by Seiberg[6] in 1993, which showed how the non-renormalization theorems may be easily obtained from simple considerations of symmetry and analyticity.

Consider a general renormalizable supersymmetric gauge theory with a number of left-chiral superfields Φ_n and/or gauge superfields V_A. As mentioned in Section 27.3, if we remove the factor g in the t_A and C_{ABC} and include it instead in the gauge superfields, then the Lagrangian density will be of the form

$$\mathscr{L} = \left[\Phi^\dagger e^{-V}\Phi\right]_D + 2\,\mathrm{Re}\left[f(\Phi)\right]_{\mathscr{F}} + \frac{1}{2g^2}\mathrm{Re}\left[\sum_{A\alpha\beta}\epsilon_{\alpha\beta}W_{A\alpha L}W_{A\beta L}\right]_{\mathscr{F}}, \tag{27.6.1}$$

where the superpotential $f(\Phi)$ is a gauge-invariant cubic polynomial in the left-chiral superfields. (We are ignoring a possible θ-term, which has no effect in perturbation theory.)

Suppose we impose an ultraviolet cut-off λ on the momenta circulating in loop graphs. As discussed in Section 12.4, we can find a *local* 'Wilsonian' effective Lagrangian density $\mathscr{L}_\lambda$ that, with this cut-off, gives precisely the same results as the original Lagrangian density for S-matrix elements of processes at momenta below λ. The effective Lagrangian density has masses and coupling parameters that now depend on λ, and usually there will be an infinite number of coupling terms in the effective Lagrangian density, all possible terms allowed by the symmetries of the theory. But things are much simpler in supersymmetric theories. The non-renormalization theorems tell us that, as long as the cut-off preserves supersymmetry and gauge invariance, to all orders in perturbation theory the effective Lagrangian will have the structure

$$\mathscr{L}_\lambda = \left[\mathscr{A}_\lambda(\Phi, \Phi^\dagger, V, \mathscr{D}\cdots)\right]_D + 2\,\mathrm{Re}\left[f(\Phi)\right]_{\mathscr{F}} + \frac{1}{2g_\lambda^2}\mathrm{Re}\left[\sum_{A\alpha\beta}\epsilon_{\alpha\beta}W_{A\alpha L}W_{A\beta L}\right]_{\mathscr{F}}, \tag{27.6.2}$$

where $\mathscr{A}_\lambda$ is a general Lorentz- and gauge-invariant function; '$\mathscr{D}\cdots$' denotes terms involving superderivatives or spacetime derivatives of the preceding arguments; and g_λ is the *one-loop* effective gauge coupling, given by the same formula as the one-loop renormalized gauge coupling constant

$$g_\lambda^{-2} = \text{constant} - 2b\ln\lambda\,, \tag{27.6.3}$$

where b is the coefficient of g^3 in the Gell-Mann–Low function $\beta(g)$, discussed in Chapter 18. This is for a simple gauge group, with a single gauge coupling, but the extension to a direct product of simple and $U(1)$ gauge groups is trivial. Note in particular that the effective superpotential is not only finite in the limit $\lambda \to \infty$, but at least in perturbation theory it contains no terms beyond those in the original superpotential, and there is no change in the coefficients of the terms it does contain.

To prove this theorem, we shall interpret this theory as a special case of one with two additional external gauge-invariant left-chiral superfields

X and Y, with Lagrangian density

$$\mathscr{L}^\sharp = \frac{1}{2}\Big[\Phi^\dagger e^{-V}\Phi\Big]_D + 2\mathrm{Re}\Big[Y\,f(\Phi)\Big]_{\mathscr{F}} + \frac{1}{2}\mathrm{Re}\Big[X\sum_{A\alpha\beta}\epsilon_{\alpha\beta}W_{A\alpha L}\,W_{A\beta L}\Big]_{\mathscr{F}}\,. \tag{27.6.4}$$

This Lagrangian density becomes equal to the original one when the scalar components x and y of X and Y are given the values $x = 1/g^2$ and $y = 1$, and the spinor and auxiliary components of X and Y are set equal to zero. Since supersymmetry and gauge invariance are assumed to be preserved in the cut-off procedure, the effective Lagrangian density in the presence of these external superfields must be the sum of the D-term of a general superfield and the real part of the $\mathscr{F}$-term of a left-chiral superfield:

$$\mathscr{L}^\sharp_\lambda = \Big[\mathscr{A}_\lambda(\Phi,\Phi^\dagger,V,X,X^\dagger,Y,Y^\dagger,\mathscr{D}\cdots)\Big]_D + 2\,\mathrm{Re}\Big[\mathscr{B}_\lambda(\Phi,\,W_L,X,Y)\Big]_{\mathscr{F}}\,, \tag{27.6.5}$$

with $\mathscr{A}_\lambda$ and $\mathscr{B}_\lambda$ both gauge-invariant functions of the displayed arguments. We do not include any superderivatives or spacetime derivatives in the $\mathscr{F}$-term because, as in Section 26.3, terms involving derivatives of any of the left-chiral superfields or their adjoints may be rewritten as contributions to $[\mathscr{A}_\lambda]_D$. (It is true that Eq. (27.3.12) gives W_L itself in terms of two $\mathscr{D}_R$s acting on a superfield $\exp(-2V)\mathscr{D}_L\exp(2V)$, but this superfield is not gauge-invariant, and we are requiring that $\mathscr{A}_\lambda$ be gauge-invariant.)

The dependence of $\mathscr{B}_\lambda$ on X and Y is severely limited by two additional symmetries of the action obtained from the Lagrangian density (27.6.4). (Both of these symmetries are broken by non-perturbative effects, which will be considered in Chapter 29.) The first symmetry is a perturbative $U(1)$ R-symmetry, of the sort discussed in Section 26.3, for which θ_L and θ_R are given the R values $+1$ and -1, the superfields Φ, V, and X are R-neutral, and Y has the R value $+2$. (Recall that $f_{\mathscr{F}}$ is the coefficient of θ_L^2 in f, so in order for $f_{\mathscr{F}}$ to have the R value 0, f must have the R value 2.) Because W_L is given by two $\mathscr{D}_R$s and one $\mathscr{D}_L$ acting on R-neutral superfields, it has the R value $+1$. Now, R invariance requires $\mathscr{B}_\lambda$, like the superpotential, to have the R value $+2$. It cannot depend on any superfields with negative R values, such as adjoints of left-chiral superfields, because it is *holomorphic*, so the terms in $\mathscr{B}_\lambda$ can only be of first order in Y or of second order in W_L, with coefficients depending only on the R-neutral superfields Φ and/or X:

$$\mathscr{B}_\lambda(\Phi,\,W_L,X,Y) = Y\,f_\lambda(\Phi,X) + \sum_{\alpha\beta AB}\epsilon_{\alpha\beta}W_{A\alpha L}W_{B\beta L}\,h_{\lambda AB}(\Phi,X)\,. \tag{27.6.6}$$

(Lorentz invariance requires the spinor indices on the W_Ls to be contracted with $\epsilon_{\alpha\beta}$.) The other symmetry is translation of X by an imaginary numerical constant, $X \to X + i\xi$, with ξ real. This changes the Lagrangian density (27.6.4) by an amount proportional to Im $\sum_{A\alpha\beta} W_{A\alpha L} W_{A\beta L}$, which as we saw in Section 27.3 is a spacetime derivative, and therefore can have no effect in perturbation theory. This translation symmetry prevents X from appearing anywhere in the effective Lagrangian density (27.6.5) except where it appears in the original Lagrangian density (27.6.4). We therefore conclude that f_λ is independent of X, while $h_{\lambda AB}$ consists of a Φ-independent term proportional to $X\delta_{AB}$, plus a term that is independent of X. That is,

$$\mathscr{B}_\lambda(\Phi, W_L, X, Y) = Y\, f_\lambda(\Phi) + \sum_{\alpha\beta AB} \epsilon_{\alpha\beta} W_{A\alpha L} W_{B\beta L} \left[c_\lambda \delta_{AB} X + \ell_{\lambda AB}(\Phi) \right] , \tag{27.6.7}$$

where c_λ is a real cut-off-dependent constant.

The point of introducing the external auxiliary superfields X and Y is that, by giving them suitable values, we can make use of weak-coupling approximations to determine the coefficients in Eq. (27.6.7). If we set the spinor and auxiliary components of X and Y equal to zero, and take their scalar components x and y to approach infinity and zero, respectively, then the gauge coupling constant vanishes as $1/\sqrt{x}$, and all Yukawa and scalar couplings derived from the superpotential vanish as y. In this limit, the only graph that contributes to the term in (27.6.7) proportional to Y has a single vertex arising from the term $2\,\mathrm{Re}\,[Y f(\Phi)]_{\mathscr{F}}$ in Eq. (27.6.4), so

$$f_\lambda(\Phi) = f(\Phi) . \tag{27.6.8}$$

Also, with $Y = 0$ there is a conservation law which requires every term in $\mathscr{L}^\sharp_\lambda$ to have equal numbers of Φs and $\Phi^\dagger$s, so since $\Phi^\dagger$ cannot occur in $\ell_{\lambda AB}$, neither can Φ. Gauge invariance then requires the constant $\ell_{\lambda AB}$ to be proportional to δ_{AB} for a simple group:

$$\ell_{\lambda AB} = \delta_{AB} L_\lambda . \tag{27.6.9}$$

Now, since gauge propagators go as $1/x$ while pure gauge interactions go as x and scalar propagators and interactions are x-independent, with $y = 0$ the number of powers of x in a diagram with V_W pure gauge boson vertices, I_W internal gauge boson lines, and any number of scalar-gauge boson vertices and scalar propagators is

$$N_x = V_W - I_W . \tag{27.6.10}$$

The number of loops is given by

$$L = I_W + I_\Phi - V_W - V_\Phi + 1 , \tag{27.6.11}$$

where I_Φ is the number of internal Φ lines and V_Φ is the number of Φ–V interaction vertices. All the Φ–V vertices have two Φ lines attached, so with no external Φ lines I_Φ and V_Φ are equal, and therefore cancel in

Eq. (27.6.11), so that Eq. (27.6.10) may be written

$$N_x = 1 - L\,. \tag{27.6.12}$$

Thus the coefficient c_λ of X in Eq. (27.6.7) is correctly given by the tree approximation and is therefore what it was in the original Lagrangian, simply $c_\lambda = 1$, while the coefficient L_λ of the X-independent term is given by one-loop diagrams only. Putting this all together, we have

$$\mathscr{L}^\sharp_\lambda = \Big[\mathscr{A}_\lambda(\Phi,\Phi^\dagger,V,X,X^\dagger,Y,Y^\dagger,\mathscr{D}\cdots)\Big]_D + 2\,\mathrm{Re}\Big[Y f(\Phi)\Big]_{\mathscr{F}}$$
$$+\frac{1}{2}\mathrm{Re}\left[\Big(X+L_\lambda\Big)\sum_{A\alpha\beta}\epsilon_{\alpha\beta}W_{A\alpha L}W_{A\beta L}\right]_{\mathscr{F}}\,, \tag{27.6.13}$$

where L_λ is the one-loop contribution. Setting $Y = 1$ and $X = 1/g^2$ then gives Eq. (27.6.2), with $g_\lambda^{-2} = g^{-2} + L_\lambda$. As shown in Section 18.3, the leading order contribution to $\lambda dg_\lambda/d\lambda$ is the same function of g_λ whatever renormalization scheme is used to define this coupling, so to one-loop order we must have

$$\lambda\, dg_\lambda/d\lambda = b\, g_\lambda^3\,, \tag{27.6.14}$$

where b is the same coefficient of g^3 as in the renormalization group equation of Gell-Mann and Low. The solution is Eq. (27.6.3), completing the proof.

In theories with a $U(1)$ gauge superfield V_1, the Lagrangian may contain a Fayet–Iliopoulos term (27.2.7):

$$\mathscr{L}_{\rm FI} = \xi\,\Big[V_1\Big]_D\,. \tag{27.6.15}$$

It is easy to see that the coefficient ξ of such a term is not renormalized.[7] If the corresponding coefficient ξ_λ in the Wilsonian Lagrangian density did depend on the gauge couplings or the couplings in the superpotential, then when we replace the original Lagrangian (27.6.1) with a Lagrangian (27.6.4) involving the external superfields X and Y, this term in the Wilsonian Lagrangian would be required by supersymmetry to take the form

$$\mathscr{L}^\sharp_{\rm FI\,\lambda} = \Big[\xi_\lambda(X,Y,X^*,Y^*)\,V_1\Big]_D\,, \tag{27.6.16}$$

with ξ_λ a function with a non-trivial dependence on X and/or Y and/or their adjoints. But such a term would not be gauge-invariant, because, according to Eq. (27.2.18), a gauge transformation shifts V_1 by a chiral superfield $i(\Omega - \Omega^*)/2$, and although the D-term of a chiral superfield vanishes, the product of $i(\Omega-\Omega^*)/2$ and ξ_λ is not chiral for general gauge transformations if ξ_λ has any dependence on other superfields. There actually are diagrams that make contributions to ξ_λ that are independent

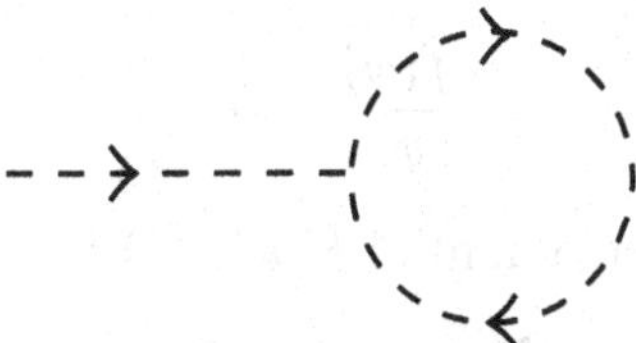

Figure 27.1. A one-loop diagram that could be quadratically divergent in theories with supersymmetry broken by trilinear couplings among scalar fields and their adjoints. The lines all represent complex scalar fields.

of all coupling constants. For the Lagrangian (27.6.1) there are no factors of the gauge coupling g at vertices at which the gauge superfield interacts with chiral matter, but instead a factor g^{-2} for each gauge propagator, so a graph with no internal gauge lines and no self-couplings of the chiral superfield will have no dependence on coupling constants. The only such graphs that contribute to ξ_λ are those in which a single external gauge line is attached to a chiral loop. (See Figure 27.1.) The contribution of all such graphs is proportional to the sum of the gauge couplings of all chiral superfields — that is, to the trace of the $U(1)$ generator. But as discussed in Section 22.4, this trace must vanish (if the $U(1)$ symmetry is unbroken) in order to avoid gravitational anomalies that violate the conservation of the $U(1)$ current.

The most important application of these theorems is a corollary, which tells us that if there is no Fayet–Iliopoulos term and if the superpotential $f(\Phi)$ allows solutions of the equations $\partial f(\phi)/\partial\phi_n = 0$, then supersymmetry is not broken in any finite order of perturbation theory.

To test this we must examine Lorentz-invariant field configurations, in which the Φ_n have only constant scalar components ϕ_n and constant auxiliary components $\mathscr{F}_n$, while (in Wess–Zumino gauge) the coefficients V_A of the gauge generators t_A in the matrix gauge superfield V have only auxiliary components D_A. Supersymmetry is unbroken if there are values of ϕ_n for which $\mathscr{L}_\lambda$ has no terms of first order in $\mathscr{F}_n$ or D_A, in which case there is sure to be an equilibrium solution with $\mathscr{F}_n = D_A = 0$. (In Section 29.2 we will see that this is the sufficient as well as the necessary condition for supersymmetry to be unbroken.) In the absence of Fayet–Iliopoulos terms, this will be the case if for all A

$$\sum_{nm} \frac{\partial K_\lambda(\phi, \phi^*)}{\partial \phi_n^*}(t_A)_{mn}\phi_m^* = 0 \tag{27.6.17}$$

and for all n

$$\frac{\partial f(\phi)}{\partial \phi_n} = 0\,, \tag{27.6.18}$$

where the effective Kahler potential $K_\lambda(\phi, \phi^*)$ is

$$K_\lambda(\phi, \phi^*) = \mathscr{A}_\lambda(\phi, \phi^*, 0, 0 \cdots)\,, \tag{27.6.19}$$

with $\mathscr{A}_\lambda(\phi, \phi^*, 0, 0 \cdots)$ obtained from $\mathscr{A}_\lambda$ by setting the gauge superfield and all superderivatives equal to zero. (With superderivatives required to vanish by Lorentz invariance, the only dependence of $\mathscr{A}_\lambda$ on V is a factor $\exp(-V)$ following every factor $\Phi^\dagger$.) We now use a trick that we have already employed in Section 27.4. If there is any solution $\phi^{(0)}$ of Eq. (27.6.18), then the gauge symmetry tells us that there is a continuum of such solutions, with ϕ_n replaced with

$$\phi_n(z) = \Big[\exp(i \sum_A t_A\, z_A)\Big]_{nm} \phi_m^{(0)}\,, \tag{27.6.20}$$

where (since f depends only on ϕ, not ϕ^*) the z_A are an arbitrary set of *complex* parameters. If $K_\lambda(\phi, \phi^*)$ has a stationary point anywhere on the surface $\phi = \phi(z)$, then at that point

$$0 = \sum_{nmA} \frac{\partial K_\lambda(\phi, \phi^*)}{\partial \phi_n} (t_A)_{nm} \phi_m\, \delta z_A - \sum_{nmA} \frac{\partial K_\lambda(\phi, \phi^*)}{\partial \phi_n^*} (t_A)_{mn} \phi_m^*\, \delta z_A^*\,. \tag{27.6.21}$$

Since this must be satisfied for all infinitesimal *complex* δz_A, the coefficients of both δz_A and δz_A^* must both vanish, and therefore Eq. (27.6.17) as well as Eq. (27.6.18) is satisfied at this point. Thus the existence of a stationary point of $K_\lambda(\phi, \phi^*)$ on the surface $\phi = \phi(z)$ would imply that supersymmetry is unbroken to all orders of perturbation theory. The zeroth-order Kahler potential $(\phi^\dagger \phi)$ is bounded below and goes to infinity as $\phi \to \infty$, so it certainly has a minimum on the surface $\phi = \phi(z)$, where of course it is stationary. If there were no flat directions in which K_λ is constant at this minimum then any sufficiently small perturbation to the Kahler potential might shift the minimum, but would not destroy it. At the minimum of the Kahler potential on the surface $\phi = \phi(z)$ there are flat directions: ordinary global gauge transformations $\delta\phi = i \sum_A \delta z_A t_A \phi$ with z_A real. But these are also flat directions for the perturbation $K_\lambda(\phi, \phi^*) - (\phi^\dagger \phi)$, so there is still a local minimum of K_λ on the surface $\phi = \phi(z)$ for any perturbation in at least a finite range, and thus to all orders in whatever couplings appear in $K_\lambda(\phi, \phi^*)$. As we have seen, this is a set of scalar field values at which $\mathscr{F}_n = 0$ and $D_A = 0$ for all n and A, which means that supersymmetry is unbroken.

* * *

These results may be extended to non-renormalizable theories.[3] In such theories the first term $[\Phi^\dagger e^{-V}\Phi]_D$ in Eq. (27.6.1) is replaced with the D-term of an arbitrary real gauge-invariant scalar function of $\Phi^\dagger$, Φ, V, and their superderivatives and spacetime derivatives, while the second and third terms in Eq. (27.6.1) are replaced with the $\mathscr{F}$-term of an arbitrary globally gauge-invariant scalar function $f(\Phi, W)$ of Φ_n and W_α. It has been shown that, to all orders of perturbation theory, the function $f_\lambda(\Phi, W)$ appearing in the $\mathscr{F}$-term of the Wilsonian Lagrangian is the same as $f(\Phi, W)$, except for the one-loop renormalization of the term quadratic in W.

27.7 Soft Supersymmetry Breaking*

We will see in the next chapter that, even if supersymmetry is an exact symmetry of the action, the spontaneous breakdown of supersymmetry at very high energy can produce superrenormalizable terms that violate supersymmetry conservation in the effective action that describes physics at lower energies. These superrenormalizable terms may explain the lack of supersymmetry observed in phenomena at accessible energies. In this section we will consider the radiative corrections that can be produced by such supersymmetry-breaking superrenormalizable terms, in part to see whether this provides a criterion for including or rejecting such terms in supersymmetric versions of the standard model.

The sign of supersymmetry breaking is the appearance of expectation values of D-terms of general superfields or $\mathscr{F}$-terms of chiral superfields. Any operator $\epsilon\mathscr{O}$ in the Lagrangian density that breaks supersymmetry can be written in a supersymmetric form, as a D-term

$$\epsilon\,\mathscr{O} = \Big[Z\,S\Big]_D\,, \tag{27.7.1}$$

where S is a non-chiral superfield that has $\mathscr{O}$ as its C-term, and Z is a non-chiral external superfield whose only non-vanishing component is $[Z]_D = \epsilon$. Some but not all operators $\epsilon\mathscr{O}$ that break supersymmetry can also be written as $\mathscr{F}$-terms,

$$\epsilon\,\mathscr{O} = \Big[\Omega\,O\Big]_{\mathscr{F}}\,, \tag{27.7.2}$$

or their adjoints, where O is the left-chiral superfield whose $\mathscr{F}$-term is $\mathscr{O}$ and Ω is an external left-chiral superfield whose only non-vanishing component is $[\Omega]_{\mathscr{F}} = \epsilon$. We can count the order in ϵ in which a given

* This section lies somewhat out of the book's main line of development and may be omitted in a first reading.

correction to the effective Lagrangian will occur by counting the powers of Z or Ω needed to construct this correction in a supersymmetric way. We will find interesting limitations on the radiative corrections that can be produced by those interactions that can be written in the form (27.7.2) as well as (27.7.1).

According to the results of the previous section, there are no radiative corrections to $\mathscr{F}$-terms, so all supersymmetry-breaking radiative corrections to the Wilsonian Lagrangian density must take the form of D-terms. This theorem does not prevent any given operator from appearing in the Wilsonian Lagrangian density, because even if an operator $\epsilon\Delta\mathscr{L}$ cannot be expressed in the form $[Z\Lambda]_D$, where Λ is the general superfield whose C-term is $\Delta\mathscr{L}$, yet $\epsilon^2\Delta\mathscr{L}$ *can* be expressed in the form

$$\epsilon^2\Delta\mathscr{L} = 2\Big[\Omega^*\Omega\Lambda\Big]_D\,. \tag{27.7.3}$$

But not all operators can be produced by radiative corrections of *first order* in Ω or Ω^*. In particular, a function only of the ϕ-term of a left-chiral superfield Φ, but not of ϕ^*, cannot be written as the D-term of a superfield linear in Ω. (Note that $[\Omega h(\Phi)]_D$ is a derivative, while $[\Omega^* h(\Phi)]_D = 2[\Phi]_{\mathscr{F}}\partial h(\phi)/\partial\phi$ is not a function of ϕ alone.) We conclude then that *the supersymmetry-breaking terms in the Wilsonian Lagrangian that depend on ϕ alone cannot be produced by radiative corrections that are of first order in supersymmetry-breaking interactions of the form (27.7.2).*

This result is significant because the most divergent radiative corrections are those that are of lowest order in superrenormalizable couplings. To be more specific, the coefficient of an interaction of dimensionality $\mathscr{D}$ has dimensionality (in powers of energy) $4-\mathscr{D}$, so dimensional analysis indicates that the contribution of a set of interactions of dimensionality d_1, d_2, etc. to the coefficient of an interaction of dimensionality d can contain the ultraviolet cut-off to at most a power

$$p = 4 - d - (4 - d_1) - (4 - d_2) - \cdots \tag{27.7.4}$$

and is therefore finite if $p < 0$. (This argument ignores possible ultraviolet divergences in subintegrations; for a thorough treatment of this topic, see Reference 8.) Superrenormalizable interactions are 'soft,' in the sense that they reduce the degree of divergence of the graphs in which they appear. In particular, in a renormalizable theory where all interactions have $d_i \leq 4$, and the strictly renormalizable interactions with $d_i = 4$ are supersymmetric, the contribution of one or more superrenormalizable interactions to the coefficient of an interaction with $d = 4$ will always have $p < 0$, so even if they are not supersymmetric the superrenormalizable interactions will not produce supersymmetry-violating ultraviolet-divergent corrections to the coefficients of the supersymmetric $d = 4$ interactions.

On the other hand, in such a theory there may be divergent radiative corrections to the superrenormalizable interactions themselves.[9] The most worrisome are quadratic (or higher) divergences, which if cut off at some high energy scale M_X may require a fine-tuning of the bare coupling constants to preserve supersymmetry as a good approximate symmetry at energies below M_X. According to Eq. (27.7.4), in renormalizable theories where all interactions with $d_i = 4$ are supersymmetric, radiative corrections can produce quadratic or more highly divergent ($p \geq 2$) supersymmetry-violating operators of dimensionality d only if they involve insertion of a superrenormalizable supersymmetry-violating interaction of dimensionality $d_1 \geq 2 + d$. This allows either $d = 0$ and $d_1 \geq 2$, which arises only when we calculate the cosmological constant, or $d = 1$ and $d_1 = 3$, which arises only when we calculate the 'tadpole' graphs in which a scalar field line disappears into the vacuum. The cosmological constant raises fine-tuning problems for all known theories[10] and will not be considered further here. The tadpole graphs represent operators linear in ϕ or ϕ^* and, as we have seen, can not be produced to first order in supersymmetry-violating interactions that can be put in the form (27.7.2). Such superrenormalizable interactions are therefore 'soft' in the sense that they do not induce quadratic or higher divergences. Along with the superrenormalizable interactions with $d \leq 2$, including arbitrary quadratic polynomials in ϕ and ϕ^*, the supersymmetry-breaking interactions that are soft in this sense include terms of third order in the ϕ, which can be expressed as $\phi^3 = [\Omega\Phi^3]_{\mathscr{F}}$, and likewise terms of third order in the ϕ^*, and also $d = 3$ gaugino mass terms, which can be written as $[\Omega\, \epsilon_{\alpha\beta} W_\alpha W_\beta]_{\mathscr{F}}$, but not terms like $\phi^2\phi^*$ or $\phi\phi^{2*}$, which in general *can* produce quadratically divergent tadpole graphs.[9]

Nevertheless, tadpoles can only arise for scalar fields that are neutral with respect to all exact symmetries. In theories without such neutral scalars, like the supersymmetric standard model discussed in the next chapter, *all* superrenormalizable interactions may be considered to be soft.

27.8 Another Approach: Gauge-Invariant Supersymmetry Transformations*

It is somewhat disturbing that the supersymmetry transformation rules discussed so far involve ordinary spacetime derivatives, not gauge-invariant

* This section lies somewhat out of the book's main line of development and may be omitted in a first reading.

derivatives. For instance, in a $U(1)$ gauge theory, the transformation of the component fields of a chiral scalar superfield is given by Eqs. (26.3.15)–(26.3.17) as

$$\begin{aligned} \delta\psi_L &= \sqrt{2}\partial_\mu\phi\,\gamma^\mu\,\alpha_R\,\phi + \sqrt{2}\mathscr{F}\,\alpha_L\,, \\ \delta\mathscr{F} &= \sqrt{2}\left(\overline{\alpha_L}\,\partial\!\!\!/\psi_L\right), \\ \delta\phi &= \sqrt{2}\left(\overline{\alpha_R}\psi_L\right). \end{aligned} \tag{27.8.1}$$

One might have thought that in the transformation of a chiral superfield that carries $U(1)$ charge q, the ordinary spacetime derivatives in Eq. (27.8.1) should be replaced with gauge-covariant derivatives, given in terms of the $U(1)$ gauge field V_μ by

$$D_\mu = \partial_\mu - iqV_\mu\,. \tag{27.8.2}$$

With such gauge-invariant supersymmetry transformations of the chiral superfields, one would still attempt to formulate supersymmetry transformations of the gauge supermultiplet that involve only the physical and auxiliary fields V_μ, λ, and D:

$$\begin{aligned} \tilde{\delta}V_\mu &= \left(\bar{\alpha}\gamma_\mu\lambda\right), \\ \tilde{\delta}\lambda &= iD\gamma_5\alpha + \frac{1}{2}\left[\partial_\mu V\!\!\!/\,,\,\gamma^\mu\right]\alpha\,, \\ \tilde{\delta}D &= i\left(\bar{\alpha}\gamma_5\,\partial\!\!\!/\lambda\right), \end{aligned} \tag{27.8.3}$$

in which ordinary spacetime derivatives appear because the gauge superfield carries no $U(1)$ charge.

This doesn't work. The algebra of these transformations does not close: the commutator of two of the modified supersymmetry transformations is not a linear combination of bosonic symmetry transformations, such as spacetime translations and gauge transformations. It follows that it is not possible to construct a Lagrangian for the chiral and gauge superfields that would be invariant under these modified supersymmetry transformations, because if there were such a Lagrangian then it would have to be also invariant under the commutators of these transformations, so that these commutators would have to be bosonic symmetries of the Lagrangian.

In 1973 de Wit and Freedman[11] showed that the supersymmetry algebra could be made to close by modifying the supersymmetry transformation properties of the chiral superfields not only by changing ordinary derivatives to gauge-invariant derivatives, but by also adding an extra term in the transformation of the $\mathscr{F}$-component, so that for $U(1)$ gauge theories the modified supersymmetry transformation rules read

$$\begin{aligned}
\tilde{\delta}\psi_L &= \sqrt{2}D_\mu\phi\,\gamma^\mu\,\alpha_R\,\phi + \sqrt{2}\mathscr{F}\,\alpha_L\,, \\
\tilde{\delta}\mathscr{F} &= \sqrt{2}\Big(\overline{\alpha_L}\,\not{D}\psi_L\Big) - 2iq\,\phi\,\Big(\overline{\alpha_L}\lambda_R\Big)\,, \\
\tilde{\delta}\phi &= \sqrt{2}\Big(\overline{\alpha_R}\psi_L\Big)\,.
\end{aligned} \tag{27.8.4}$$

With this change, they were also able to construct a Lagrangian that is invariant under the transformations (27.8.3)–(27.8.4) and that turned out to be just the one we have found in Sections 27.1 and 27.2.

There is nothing wrong with continuing to use the conventional transformation rules (27.8.1), so we do not need the de Wit–Freedman formalism to deal with supersymmetric gauge theories. Nevertheless this formalism is of some interest, because in supergravity theories the analog of the conventional formalism is very cumbersome. As described in Chapter 31, the formalism that has been chiefly used to derive physically interesting results in supergravity theories follows an approach like that of de Wit and Freedman, with supersymmetry transformation rules involving covariant derivatives in place of ordinary derivatives, rather than an approach based on conventional supersymmetry transformations like those of Eq. (27.8.1). It is therefore of some interest to understand the relation between the de Wit–Freedman formalism and the conventional approach in the relatively simple context of $U(1)$ gauge theory, and in particular to explain the origin of the extra term in the transformation rule for $\mathscr{F}$.

In writing supersymmetry transformations (27.8.3) that did not involve the components C, M, N, or ω of the gauge superfield V, de Wit and Freedman were implicitly adopting the Wess–Zumino gauge discussed in Section 27.1. But the choice of Wess–Zumino gauge is not invariant under either the conventional supersymmetry transformations (26.2.11)–(26.2.17) or the extended gauge transformations (27.1.17), so once we adopt this gauge both symmetries are lost. We can, however, define a *combined* transformation, acting on fields in Wess–Zumino gauge, which consists of a conventional supersymmetry transformation followed by an extended gauge transformation that takes us back to Wess–Zumino gauge. *This is the de Wit–Freedman transformation* $\tilde{\delta}$.**

To construct the de Wit–Freedman transformations in this way, note that for a gauge superfield that satisfies the Wess–Zumino gauge conditions

** This was not shown explicitly by de Wit and Freedman. But, in fact, although the point of their paper was to emphasize that the details of the transformations (27.8.3) and (27.8.4) could be inferred from the requirement of a closed supersymmetry algebra (also for non-Abelian gauge theories), they remarked that they had actually found these transformations by identifying the fermionic transformations that survive in Wess–Zumino gauge.

$C = M = N = \omega = 0$, the transformation rules (26.2.11)–(26.2.14) give

$$\delta C = 0\,, \quad \delta\omega = \not{V}\alpha\,, \quad \delta M = -\left(\bar{\alpha}\lambda\right), \quad \delta N = i\left(\bar{\alpha}\gamma_5\lambda\right). \tag{27.8.5}$$

According to Eqs. (27.1.17), we can get back to Wess–Zumino gauge by performing an infinitesimal extended gauge transformation (27.1.13):

$$V \to V + \frac{i}{2}\left[\Omega - \Omega^*\right], \tag{27.8.6}$$

where Ω is a left-chiral superfield with components

$$\phi^\Omega = 0\,, \quad \psi_L^\Omega = -\sqrt{2}\,\not{V}\alpha_R\,, \quad \mathscr{F}^\Omega = -\left(\bar{\alpha}(1-\gamma_5)\lambda\right). \tag{27.8.7}$$

According to Eq. (27.1.11), this extended gauge transformation induces on a chiral superfield of charge q the transformation

$$\delta'\Phi = iq\Omega\Phi\,. \tag{27.8.8}$$

Using the multiplication rules (26.3.27)–(26.3.29), the transformation of the components of Φ is

$$\begin{aligned} \delta'\psi_L &= -i\sqrt{2}q\phi\,\not{V}\alpha_R\,, \\ \delta'\mathscr{F} &= -2iq\phi\left(\overline{\alpha_L}\lambda_R\right) - i\sqrt{2}q\left(\overline{\alpha_L}\,\not{V}\psi_L\right), \\ \delta'\phi &= 0\,. \end{aligned} \tag{27.8.9}$$

Adding this to Eq. (27.8.1) and comparing with Eq. (27.8.4) shows that the de Wit–Freedman transformation is indeed the combination of a conventional supersymmetry transformation and the corresponding extended gauge transformation (27.8.8):

$$\tilde{\delta}\Phi = \delta\Phi + \delta'\Phi\,. \tag{27.8.10}$$

27.9 Gauge Theories with Extended Supersymmetry*

Theories with unbroken extended supersymmetry are not considered to be good candidates for realistic extensions of the standard model, because of the non-chirality of particle multiplets discussed in Section 25.4. Nevertheless gauge theories with extended supersymmetry are worth some consideration here because they have provided paradigms for the use of powerful mathematical methods to solve dynamical problems.

There are a number of special formalisms that have been proposed to construct Lagrangians with $N = 2$ extended supersymmetry,[12] but

* This section lies somewhat out of the book's main line of development and may be omitted in a first reading.

fortunately we can get by with the tools already at hand. Any theory with $N = 2$ supersymmetry also has $N = 1$ supersymmetry, so its Lagrangian must be a special case of the Lagrangians already considered in this chapter. To construct a Lagrangian with $N = 2$ supersymmetry for some set of the $N = 2$ supermultiplets of particles constructed in Sections 25.4 and 25.5, we need only write down the most general Lagrangian with $N = 1$ supersymmetry whose $N = 1$ supermultiplets contain physical fields for the particles in the $N = 2$ supermultiplets, and then impose a discrete R-symmetry on the Lagrangian: a symmetry that acts differently on different components of $N = 2$ supermultiplets. The Lagrangian density will then be invariant under a second supersymmetry, whose supermultiplets are given by acting on the supermultiplets of ordinary $N = 1$ supersymmetry with the R-symmetry.

It will be convenient to choose the discrete R-transformation so that

$$Q_1 \to Q_2 , \qquad Q_2 \to -Q_1 . \tag{27.9.1}$$

If the central charge were zero then the supersymmetry algebra would be invariant under an $SU(2)$ R-symmetry group, which has the transformation (27.9.1) as one finite element $\exp(i\pi\tau_2/2)$, but symmetry under the discrete symmetry is sufficient for our purposes, so we do not need to assume a zero central charge. In fact, it will turn out that the Lagrangians we construct by this method will have an $SU(2)$ R-symmetry, not just symmetry under the discrete transformation (27.9.1).

Let us first consider the renormalizable theory of the gauge bosons of a general gauge group, together with the superpartners required by $N = 2$ extended supersymmetry. We saw in Section 25.4 that in $N = 2$ global supersymmetry theories a massless gauge boson can only belong to a multiplet also containing a pair of massless fermions of each helicity $\pm 1/2$ that transform as a doublet under the $SU(2)$ R-symmetry and a pair of $SU(2)$-singlet spinless bosons. Since $N = 2$ supersymmetry includes $N = 1$ supersymmetry, the renormalizable Lagrangian for this theory must be a special case of the general renormalizable Lagrangian density (27.4.1). One feature of this special case is that since the gauge boson belongs to the adjoint representation of the gauge group, so must also the fermions and scalar field. To furnish $N = 2$ supermultiplets of fields with the correct particle content, we must have one $N = 1$ chiral superfield Φ_A with component fields ϕ_A, ψ_A, $\mathscr{F}_A$ (with ψ_A Majorana and ϕ_A and $\mathscr{F}_A$ both complex) for each $N = 1$ gauge multiplet V^μ_A, λ_A, D_A. We impose a discrete R-symmetry under the transformation

$$\psi_A \to \lambda_A , \qquad \lambda_A \to -\psi_A , \tag{27.9.2}$$

(with all other fields invariant) because this is the effect of the transformation (27.9.1). Since a non-trivial superpotential would give ψ_A interactions or mass terms that are absent for λ_A, the superpotential must vanish. The Lagrangian (27.4.1) therefore takes the special form

$$\begin{aligned}\mathscr{L} = & -\sum_A (D_\mu\phi)^*_A (D^\mu\phi)_A - \frac{1}{2}\sum_A \left(\overline{\psi_A}\,(\not{D}\psi)_A\right) + \sum_A \mathscr{F}^*_A \mathscr{F}_A \\ & -2\sqrt{2}\,\mathrm{Re}\sum_{ABC} C_{ABC}\left(\lambda^{\mathrm{T}}_{AL}\,\psi_{CL}\right)\phi^*_B \\ & +i\sum_{ABC} C_{ABC}\,\phi^*_B\phi_C D_A - \sum_A \xi_A D_A + \frac{1}{2}\sum_A D_A D_A \\ & -\frac{1}{4}\sum_A f_{A\mu\nu}f^{\mu\nu}_A - \frac{1}{2}\sum_A\left(\overline{\lambda_A}\,(\not{D}\lambda)_A\right) + \frac{g^2\theta}{64\pi^2}\epsilon_{\mu\nu\rho\sigma}\sum_A f^{\mu\nu}_A f^{\rho\sigma}_A\,,\end{aligned} \tag{27.9.3}$$

where

$$(D_\mu\psi)_A = \partial_\mu\psi_A + \sum_{BC} C_{ABC} V_{B\mu}\psi_C\,, \tag{27.9.4}$$

$$(D_\mu\lambda)_A = \partial_\mu\lambda_A + \sum_{BC} C_{ABC} V_{B\mu}\lambda_C\,, \tag{27.9.5}$$

$$(D_\mu\phi)_A = \partial_\mu\phi_A + \sum_{BC} C_{ABC} V_{B\mu}\phi_C\,, \tag{27.9.6}$$

and

$$f_{A\mu\nu} = \partial_\mu V_{A\nu} - \partial_\nu V_{A\mu} + \sum_{BC} C_{ABC} V_{B\mu} V_{C\nu}\,. \tag{27.9.7}$$

(Recall that in the adjoint representation $(t_A)_{BC} = -iC_{ABC}$, where C_{ABC} is the real structure constant, defined as usual in this book to include factors of gauge couplings, and taken in a basis in which it is totally antisymmetric.) The Lagrangian density (27.9.3) has $N = 1$ supersymmetry with multiplets ϕ_A, ψ_A, $\mathscr{F}_A$ and V^μ_A, λ_A, D_A, because it is a special case of Eq. (27.4.1), and it has an $SU(2)$ symmetry relating ψ_A and λ_A, including invariance under the finite $SU(2)$ transformation (27.9.2), so it also has a *second* independent $N = 1$ supersymmetry with multiplets ϕ_A, λ_A, $\mathscr{F}_A$ and V^μ_A, $-\psi_A$, D_A. It therefore satisfies the conditions imposed by $N = 2$ supersymmetry.

We can eliminate the auxiliary fields by setting them equal to the values at which the Lagrangian density (27.9.3) is stationary:

$$\mathscr{F}_A = 0\,, \qquad D_A = -i\sum_{BC} C_{ABC}\phi^*_B\phi_C\,. \tag{27.9.8}$$

(We are now assuming that the Fayet–Iliopoulos constants ξ_A all vanish.) Inserting these values back in Eq. (27.9.3) gives an equivalent Lagrangian density

$$\begin{aligned}\mathscr{L} = &-\sum_A (D_\mu\phi)^*_A (D^\mu\phi)_A - \frac{1}{2}\sum_A \left(\overline{\psi_A}(\not{D}\psi)_A\right)\\ &+\sqrt{2}\sum_{ABC} C_{ABC}\left(\overline{\psi_B}\left(\frac{1-\gamma_5}{2}\right)\lambda_A\right)\phi_C\\ &-\sqrt{2}\sum_{ABC} C_{ABC}\left(\overline{\lambda_A}\left(\frac{1+\gamma_5}{2}\right)\psi_C\right)\phi^*_B - V(\phi,\phi^*)\\ &-\frac{1}{4}\sum_A f_{A\mu\nu}f^{\mu\nu}_A - \frac{1}{2}\sum_A\left(\overline{\lambda_A}(\not{D}\lambda)_A\right) + \frac{g^2\theta}{64\pi^2}\epsilon_{\mu\nu\rho\sigma}\sum_A f^{\mu\nu}_A f^{\rho\sigma}_A\,,\end{aligned} \tag{27.9.9}$$

where the potential is

$$V(\phi,\phi^*) = -\frac{1}{2}\sum_A\left[\sum_{BC} C_{ABC}\,\phi^*_B\phi_C\right]^2 = 2\sum_A\left[\sum_{BC} C_{ABC}\,\mathrm{Re}\,\phi_B\,\mathrm{Im}\,\phi_C\right]^2 . \tag{27.9.10}$$

This potential has a minimum value of zero, which is reached not only for $\phi_A = 0$ but also for any set of ϕs for which $\sum_{BC} C_{ABC}\,\phi^*_B\phi_C = 0$ for all A or, in other words, for which

$$[t\cdot\mathrm{Re}\,\phi\,,\,t\cdot\mathrm{Im}\,\phi] = 0\,, \qquad \text{where } t\cdot v \equiv \sum_B t_B v_B\,. \tag{27.9.11}$$

That is, the minimum of the potential is reached for those scalar fields for which all generators $t\cdot\mathrm{Re}\,\phi$ and $t\cdot\mathrm{Im}\,\phi$ belong to a Cartan subalgebra of the full gauge algebra, all of whose generators commute with one another. Though all such values of ϕ give zero potential, and hence unbroken $N = 2$ supersymmetry, they are not physically equivalent, as shown for instance by the different masses they give the gauge bosons associated with the broken gauge symmetries.

One remarkable feature of extended supersymmetry is that the central charges of the supersymmetry algebra in any state can be calculated in terms of 'charges' in that state to which bosonic fields are coupled.[13] The easiest way to do this calculation is to use the transformation properties under ordinary $N = 1$ supersymmetry of the extended supersymmetry currents $S^\mu_r(x)$ with $r = 2, 3, \ldots, N$ to calculate the anticommutators $\{Q_{1\alpha}\,,\,S^\mu_{r\beta}(x)\}$. We can then calculate the central charges from the anticommutators

$$\{Q_{1\alpha}\,,\,Q_{r\beta}\} = \int d^3x\,\{Q_{1\alpha}\,,\,S^0_{r\beta}(x)\}\,. \tag{27.9.12}$$

It turns out that the integrand on the right-hand side is a derivative with respect to space coordinates, but its integral does not vanish if the states have fields that do not vanish rapidly as $\mathbf{x} \to \infty$.

To see how this works in detail, let's consider the case of $N = 2$ supersymmetry with an $SU(2)$ gauge symmetry and a single $N = 2$ gauge supermultiplet, with no additional matter supermultiplets. The Lagrangian here is given by Eq. (27.9.3), with A, B, and C running over the values 1, 2, 3, and

$$C_{ABC} = e\epsilon_{ABC}\,, \qquad \xi_A = 0\,. \tag{27.9.13}$$

(The coupling constant here is denoted e because this is the charge with which the massless gauge field of the unbroken $U(1)$ gauge symmetry interacts.) The usual $N = 1$ supersymmetry current (distinguished now with a subscript 1) is given by Eq. (27.4.40) as

$$S_1^\mu = -\frac{1}{4}\sum_A f_{A\rho\sigma}[\gamma^\rho, \gamma^\sigma]\gamma^\mu\lambda_A - e\sum_{ABC}\epsilon_{ABC}\,\gamma_5\gamma^\mu\lambda_A\,\phi_B^*\,\phi_C$$
$$+\frac{1}{\sqrt{2}}\sum_A\left[(\not{D}\phi)_A\,\gamma^\mu\psi_{AR} + (\not{D}\phi^*)_A\,\gamma^\mu\psi_{AL}\right]. \tag{27.9.14}$$

We can calculate the second supersymmetry current by subjecting S_1^μ to the finite $SU(2)$ R-symmetry used above, which simply amounts to making the replacements $\psi_A \to \lambda_A$, $\lambda_A \to -\psi_A$. This gives

$$S_2^\mu = \frac{1}{4}\sum_A f_{A\rho\sigma}[\gamma^\rho, \gamma^\sigma]\gamma^\mu\psi_A + e\sum_{ABC}\epsilon_{ABC}\,\gamma_5\gamma^\mu\psi_A\,\phi_B^*\,\phi_C$$
$$+\frac{1}{\sqrt{2}}\sum_A\left[(\not{D}\phi)_A\,\gamma^\mu\lambda_{AR} + (\not{D}\phi^*)_A\,\gamma^\mu\lambda_{AL}\right]. \tag{27.9.15}$$

It will be enough for our purposes (and somewhat easier) to calculate only the change in the right-handed part of this current under an $N = 1$ supersymmetry transformation. After setting the auxiliary fields equal to their equilibrium values

$$\mathscr{F}_A = 0\,, \qquad D_A = -ie\sum_{BC}\epsilon_{ABC}\phi_B^*\phi_C\,,$$

we find

$$\delta S_{2R}^\mu = \frac{\sqrt{2}}{4}\sum_A f_{A\rho\sigma}[\gamma^\rho\,, \gamma^\sigma]\gamma^\mu(\not{D}\phi)_A\alpha_R$$
$$-\sqrt{2}\,e\sum_{ABC}\epsilon_{ABC}\gamma^\mu(\not{D}\phi)_A\alpha_R\phi_B^*\phi_C - \frac{\sqrt{2}}{4}f_{A\rho\sigma}(\not{D}\phi)_A\gamma^\mu[\gamma^\rho\,, \gamma^\sigma]\alpha_R$$
$$-\sqrt{2}\,e\sum_{ABC}\epsilon_{ABC}\phi_B^*\phi_C(\not{D}\phi)_A\gamma^\mu\alpha_R + \cdots\,,$$

where the dots denote terms bilinear in fermion fields, which do not concern us here because we are interested in effects of long range boson fields. We can combine terms by using the Dirac anticommutation relations and the identity

$$[\gamma^\rho, \gamma^\sigma]\,[\gamma^\mu, \gamma^\nu] + [\gamma^\mu, \gamma^\nu]\,[\gamma^\rho, \gamma^\sigma] = -8\eta^{\mu\rho}\eta^{\nu\sigma} + 8\eta^{\sigma\mu}\eta^{\rho\nu} + 8\,i\,\epsilon^{\mu\nu\rho\sigma}\gamma_5$$

and find

$$\delta S^\mu_{2R} = -2\sqrt{2}\sum_A f_A^{\mu\nu}(D_\nu\phi)_A\alpha_R - i\sqrt{2}\sum_A \epsilon^{\mu\nu\rho\sigma} f_{A\rho\sigma}(D_\nu\phi)_A\alpha_R$$
$$-2\sqrt{2}\,e\sum_{ABC}\epsilon_{ABC}\phi^*_B\phi_C\,(D^\mu\phi)_A\alpha_R + \cdots .$$

In order to write this as a derivative, we need to use the Yang–Mills field equations given by Eqs. (15.3.6), (15.3.7), and (15.3.9):

$$D_\nu f_A^{\mu\nu} = J_A^\mu = e\sum_{BC}\epsilon_{ABC}\Big((D^\mu\phi)^*_B\phi_C - \phi^*_B(D^\mu\phi)_C\Big)\,,$$
$$\epsilon_{\mu\nu\rho\sigma}(D^\nu f^{\rho\sigma})_A = 0\,.$$

These allow us to write δS^μ_{2R} as a total derivative

$$\delta S^\mu_{2R} = D_\nu X^{\mu\nu}\alpha_R\,, \tag{27.9.16}$$

where

$$X^{\mu\nu} = -2\sqrt{2}\sum_A f_A^{\mu\nu}\phi_A - i\sqrt{2}\sum_A \epsilon^{\mu\nu\rho\sigma} f_{A\rho\sigma}\phi_A + \cdots\,, \tag{27.9.17}$$

with the dots again indicating irrelevant terms involving fermion fields. Eq. (26.1.18) allows us to write Eq. (27.9.16) as an anticommutation relation

$$\Big\{Q_{R\alpha}\,, S^\mu_{R\beta}\Big\} = i\left[\epsilon\left(\frac{1-\gamma_5}{2}\right)\right]_{\alpha\beta} D_\nu X^{\mu\nu}\,. \tag{27.9.18}$$

Since $X^{\mu\nu}$ is a gauge-invariant quantity, its gauge-covariant derivative is the same as its ordinary derivative. Also, $X^{\mu\nu}$ is antisymmetric, so $D_\nu X^{0\nu} = \partial_i X^{0i}$. From Eqs. (27.9.12) and (27.9.18), we have at last

$$\Big\{Q_{R\alpha}\,, Q_{R\beta}\Big\} = i\left[\epsilon\left(\frac{1-\gamma_5}{2}\right)\right]_{\alpha\beta}\int dS_i\, X^{0i}\,, \tag{27.9.19}$$

with the integral taken over a large closed surface enclosing the system in question, with surface area differential **dS** taken normal to the surface. Comparing this with Eq. (25.2.38) gives the central charge

$$Z_{12} = -i\int dS_i\, X^{0i}\,. \tag{27.9.20}$$

If we choose a gauge in which ϕ_A (almost everywhere) has only the constant non-vanishing component $\phi_3 \equiv v$, then

$$\sum_A f_A^{0i}\phi_A = -vE^i\,, \qquad \frac{1}{2}\sum_A \epsilon^{0i\rho\sigma} f_{A\rho\sigma}\phi_A = vB^i\,, \tag{27.9.21}$$

where **E** and **B** are the electric and magnetic fields associated with the unbroken $U(1)$ subgroup of the $SU(2)$ gauge group. Therefore the central charge (27.9.20) here is

$$Z_{12} = 2\sqrt{2}\,v\left[iq - \mathscr{M}\right]\,, \tag{27.9.22}$$

where q and $\mathscr{M}$ are the electric charge and magnetic monopole moment, defined by

$$q = \int dS_i\,E^i\,, \qquad \mathscr{M} = \int dS_i\,B^i\,. \tag{27.9.23}$$

As discussed in Section 23.3, this theory, with $SU(2)$ gauge symmetry spontaneously broken by the expectation value of an $SU(2)$ triplet of scalars, is one in which magnetic monopoles actually do occur.

The application of the results of Section 27.4 to the Lagrangian density (27.9.3) shows that, after the spontaneous breaking of the $SU(2)$ gauge symmetry, this theory will contain elementary particles of charge $\pm e$, zero magnetic monopole moment, and tree-approximation mass $M = \sqrt{2}\,|e\,v|$. Specifically, for each sign of the charge there is one such particle of spin 1, two of them with spin 1/2, and one with spin 0. It is a striking consequence of the results obtained here that *the mass value $\sqrt{2}\,|e\,v|$ is exact, being unaffected by either radiative corrections or non-perturbative effects, provided that the quantity v is defined by Eq. (27.9.22) for the central charge.*[13]

To see this, note that the massive one-particle states for each sign of the charge are 'short' $N = 2$ supermultiplets, which, as shown at the end of Section 25.5, have masses that saturate the lower bound (25.5.24):

$$M = |Z_{12}|/2\,. \tag{27.9.24}$$

Even if we did not trust the tree approximation to give us the precise value of the particle masses, we would not expect corrections to this approximation to turn short multiplets into the full multiplets, with many more states, that could have larger masses, so we can be confident that Eq. (27.9.24) is exactly valid. For particles with electric charge $q = \pm e$ and zero magnetic monopole moment, Eq. (27.9.20) gives $Z_{12} = \pm 2\sqrt{2}\,i\,v\,e$, so Eq. (27.9.24) tells us that their masses are

$$M = \sqrt{2}|e\,v|\,. \tag{27.9.25}$$

This is the result found in the tree approximation, but now we see that it is exact.

The semi-classical calculations described in Section 23.3 show that electrically neutral magnetic monopoles in this theory have monopole strengths**

$$\mathscr{M} = \frac{4\pi\nu}{e}\,, \tag{27.9.26}$$

with ν the winding number, a positive or negative integer. The formula (27.9.22) for the central charge together with the inequality (25.5.24) thus give a lower bound on the monopole masses

$$M \geq \frac{4\pi\sqrt{2}\,|\nu\, v|}{|e|}\,. \tag{27.9.27}$$

Interestingly, this is the same as the Bogomol'nyi lower bound[14] on monopole energies derived in Section 23.3.[†] In fact, the monopole solution for $\nu = 1$ that was described in Section 23.3 saturates this bound. More generally, the 'dyons' of this theory,[15] particles with both charge and magnetic moment, have masses given by[16]

$$M = 2\,|v|\,\sqrt{q^2 + \mathscr{M}^2}\,, \tag{27.9.28}$$

which again is the minimum value allowed by Eqs. (25.5.24) and (27.9.20). Indeed, *all* of the known particles in this theory have masses given in the semi-classical limit by Eq. (27.9.28).[17]

Returning now to general $N = 2$ gauge theories, we may also include additional 'matter' fields in the Lagrangian. For simplicity, we will restrict ourselves to 'short' massive hypermultiplets (with central charge $\mathscr{Z}$ saturating the inequality (25.5.24)), each consisting of a single fermion of spin 1/2 and an $SU(2)$ doublet of spin 0 particles, together with distinct antiparticles. This is the same spin content as is given under $N = 1$ supersymmetry by pairs of left-chiral scalar superfields Φ'_n and Φ''_n, together with their right-chiral adjoints, with the complex scalar field components ϕ'_n and ϕ''_n and their adjoints forming pairs of $SU(2)$ doublets, and the spinor fields all $SU(2)$ singlets. (We are using primes and double primes to distinguish these superfields and their components from Φ_A and its components.) If some of these hypermultiplets Φ'_n and Φ''_n are non-neutral

** Note that the magnetic moment $\mathscr{M}$ defined by Eq. (27.9.23) is related to the magnetic moment g defined in Section 23.3 by $\mathscr{M} = 4\pi g$.

† The canonically normalized field with a non-vanishing vacuum expectation value is $\sqrt{2}\,\mathrm{Re}\,\phi_3$ (for real v), so the quantity $\langle\phi\rangle$ appearing in the Bogomol'nyi inequality (23.3.19) is $\sqrt{2}v$.

under the gauge group, then a superpotential is allowed, of the form:[††]

$$f(\Phi,\Phi',\Phi'') = \frac{1}{2}\sum_{Anm}(s_A)_{nm}\Phi'_n\Phi''_m\Phi_A + \frac{1}{2}\sum_{nm}\mu_{nm}\Phi'_n\Phi''_m\,. \tag{27.9.29}$$

To the Lagrangian density (27.9.3) we then must add a Lagrangian density for these hypermultiplets, given by the first eight terms on the right-hand side of Eq. (27.4.1), and find a total Lagrangian density

$$\begin{aligned}
\mathscr{L} = &-\sum_n (D_\mu\phi')^*_n(D^\mu\phi')_n - \sum_n (D_\mu\phi'')^*_n(D^\mu\phi'')_n - \sum_A (D_\mu\phi)^*_A(D^\mu\phi)_A \\
&-\frac{1}{2}\sum_n\left(\overline{\psi'_n}(\not{D}\psi')_n\right) - \frac{1}{2}\sum_n\left(\overline{\psi''_n}(\not{D}\psi'')_n\right) \\
&-\frac{1}{2}\sum_A\left(\overline{\psi_A}(\not{D}\psi)_A\right) - \frac{1}{2}\sum_A\left(\overline{\lambda_A}(\not{D}\lambda)_A\right) \\
&+\sum_n \mathscr{F}'^*_n\mathscr{F}'_n + \sum_n \mathscr{F}''^*_n\mathscr{F}''_n + \sum_A \mathscr{F}^*_A\mathscr{F}_A \\
&-\mathrm{Re}\sum_{Anm}(s_A)_{nm}\phi_A\left(\psi'^{\mathrm{T}}_{nL}\epsilon\psi''_{mL}\right) - 2\sqrt{2}\,\mathrm{Re}\sum_{ABC}C_{ABC}\left(\lambda^{\mathrm{T}}_{AL}\epsilon\psi_{CL}\right)\phi^*_B \\
&-\mathrm{Re}\sum_{Anm}(s_A)_{nm}\phi'_n\left(\psi''^{\mathrm{T}}_{mL}\epsilon\psi_{AL}\right) - \mathrm{Re}\sum_{Anm}(s_A)_{nm}\phi''_m\left(\psi'^{\mathrm{T}}_{nL}\epsilon\psi_{AL}\right) \\
&+2\sqrt{2}\,\mathrm{Im}\sum_{Anm}(t'_A)_{mn}\left(\psi'^{\mathrm{T}}_{nL}\epsilon\lambda_{AL}\right)\phi'^*_m + 2\sqrt{2}\,\mathrm{Im}\sum_{Anm}(t''_A)_{mn}\left(\psi''^{\mathrm{T}}_{nL}\epsilon\lambda_{AL}\right)\phi''^*_m \\
&+\mathrm{Re}\sum_{Anm}(s_A)_{nm}\phi_A\phi'_n\mathscr{F}''_m + \mathrm{Re}\sum_{Anm}(s_A)_{nm}\phi_A\phi''_m\mathscr{F}'_n \\
&+\mathrm{Re}\sum_{Anm}(s_A)_{nm}\phi'_n\phi''_m\mathscr{F}_A \\
&+\mathrm{Re}\sum_{Anm}\mu_{nm}\phi'_n\mathscr{F}''_m + \mathrm{Re}\sum_{Anm}\mu_{nm}\phi''_m\mathscr{F}'_n - \mathrm{Re}\sum_{nm}\mu_{nm}\left(\psi'^{\mathrm{T}}_{nL}\epsilon\psi''_{mL}\right)
\end{aligned}$$

[††] There still cannot be any terms in the superpotential that are of second or higher order in the Φ_A for the same reason as before: such terms would lead to scalar couplings or masses for the ψ_A, with no corresponding couplings or masses for their $SU(2)$ partners λ_A. Also, there cannot be any term that is trilinear in the Φ'_n and/or Φ''_n, because then the term in Eq. (27.4.1) involving the product of fermion bilinears with second derivatives of the superpotential would lead to a coupling of the $SU(2)$-singlet fermions to $SU(2)$ doublet fields ϕ'_n or ϕ''_n. Thus the only trilinear interaction terms in the superpotential must involve one factor of Φ_A and two factors of Φ'_n and/or Φ''_n. There cannot be any trilinear terms involving a Φ_A and two Φ'_ns or two Φ''_ns, because that would give the $SU(2)$ singlet auxiliary fields $\mathscr{F}_A$ an interaction with $SU(2)$-triplet products $\phi'_n\phi'_m$ or $\phi''_n\phi''_m$, and there cannot be any bilinear terms involving two Φ'_ns or two Φ''_ns, because that would yield $SU(2)$-triplet mass terms $(\psi'^{\mathrm{T}}_n\epsilon\psi'_m)$ or $(\psi''^{\mathrm{T}}_n\epsilon\psi''_m)$. The only remaining allowed bilinear or trilinear terms are of the form (27.9.29).

$$-\sum_{Anm}(t'_A)_{nm}\phi'^*_n\phi'_m D_A - \sum_{Anm}(t''_A)_{nm}\phi''^*_n\phi''_m D_A + i\sum_{ABC} C_{ABC}\,\phi^*_B\phi_C D_A$$
$$-\sum_A \xi_A D_A + \frac{1}{2}\sum_A D_A D_A$$
$$-\frac{1}{4}\sum_A f_{A\mu\nu}f_A^{\mu\nu} + \frac{g^2\theta}{64\pi^2}\epsilon_{\mu\nu\rho\sigma}\sum_A f_A^{\mu\nu}f_A^{\rho\sigma}\,, \tag{27.9.30}$$

where the $(t'_A)_{nm}$ and $(t''_A)_{nm}$ are the matrices (including coupling constant factors) representing the gauge group on the left-chiral scalar superfields Φ'_n and Φ''_n, respectively. The Yukawa couplings between fermions and scalars have a discrete R-symmetry under the transformation

$$\lambda_{AL} \to -\psi_{AL}\,, \qquad \psi_{AL} \to +\lambda_{AL}\,, \qquad \phi''_n \to -\phi'^*_n\,, \qquad \phi'_n \to \phi''^*_n\,, \tag{27.9.31}$$

provided that

$$s_A = -2\sqrt{2}\,i\,t'^{\,\mathrm{T}}_A = +2\sqrt{2}\,i\,t''_A\,. \tag{27.9.32}$$

(Note in particular that Eq. (27.9.32) requires the representations of the gauge group furnished by Φ'_n and Φ''_n to be complex conjugates.) This is also a symmetry of all the other terms in the Lagrangian density (27.9.30), except for those involving the auxiliary fields.

It is not possible to extend the symmetry under the transformation (27.9.31) to the auxiliary fields, but the symmetry appears after the auxiliary fields are eliminated.‡ After setting D_A, $\mathscr{F}'_n$, and $\mathscr{F}''_n$ equal to values at which the Lagrangian density is stationary, and combining D- and $\mathscr{F}$-terms, the Lagrangian density (with s_A and t''_A given by Eq. (27.9.32), and ξ_A taken to vanish) takes the form

$$\mathscr{L} = -\sum_n (D_\mu\phi')^*_n(D^\mu\phi')_n - \sum_n (D_\mu\phi'')^*_n(D^\mu\phi'')_n - \sum_A (D_\mu\phi)^*_A\,(D^\mu\phi)_A$$
$$-\frac{1}{2}\sum_n\left(\overline{\psi'_n}(\not{D}\psi')_n\right) - \frac{1}{2}\sum_n\left(\overline{\psi''_n}(\not{D}\psi'')_n\right)$$
$$-\frac{1}{2}\sum_A\left(\overline{\psi_A}\,(\not{D}\psi)_A\right) - \frac{1}{2}\sum_A\left(\overline{\lambda_A}\,(\not{D}\lambda)_A\right)$$
$$-2\sqrt{2}\,\mathrm{Im}\sum_{Anm}(t'_A)_{mn}\phi_A\left(\psi'^{\mathrm{T}}_{nL}\epsilon\psi''_{mL}\right) - 2\sqrt{2}\,\mathrm{Re}\sum_{ABC} C_{ABC}\left(\lambda^{\mathrm{T}}_{AL}\epsilon\psi_{CL}\right)\phi^*_B$$

‡ After elimination of the auxiliary fields, the resulting action is invariant under the original $N = 2$ supersymmetry transformation only 'on-shell' — that is, only up to terms that vanish when the fields satisfy the interacting field equations. This doesn't hurt, because there still are two conserved supersymmetry currents whose integrated time-components satisfy the $N = 2$ supersymmetry anticommutation relations when the fields of which they are composed are required to satisfy the field equations of the Heisenberg picture. 'Off-shell' formulations of $N = 2$ supersymmetry exist, but are subject to various complications.[18]

$$
\begin{aligned}
&-2\sqrt{2}\,\mathrm{Im}\sum_{Anm}(t'_A)_{mn}\phi'_n\left(\psi''^{\mathrm{T}}_{mL}\epsilon\psi_{AL}\right)-2\sqrt{2}\,\mathrm{Im}\sum_{Anm}(t'_A)_{mn}\phi''_m\left(\psi'^{\mathrm{T}}_{nL}\epsilon\psi_{AL}\right)\\
&+2\sqrt{2}\,\mathrm{Im}\sum_{Anm}(t'_A)_{mn}\left(\psi'^{\mathrm{T}}_{nL}\epsilon\lambda_{AL}\right)\phi'^*_m-2\sqrt{2}\,\mathrm{Im}\sum_{Anm}(t'_A)_{nm}\left(\psi''^{\mathrm{T}}_{nL}\epsilon\lambda_{AL}\right)\phi''^*_m\\
&-\frac{1}{4}\sum_A f_{A\mu\nu}f_A^{\mu\nu}+\frac{g^2\theta}{64\pi^2}\epsilon_{\mu\nu\rho\sigma}\sum_A f_A^{\mu\nu}f_A^{\rho\sigma}\\
&-\sum_{ABnm}\{t'_A\,,\,t'_B\}_{mn}\phi_A\phi_B^*\left(\phi'_n\phi'^*_m+\phi''^*_n\phi''_m\right)\\
&-\frac{1}{2}\sum_A\left[\sum_{nm}(t'_A)_{nm}\left(\phi'^*_n\phi'_m-\phi''_n\phi''^*_m\right)\right]^2\\
&+\frac{1}{2}\sum_{ABCDE}C_{ABC}C_{ADE}\phi_B^*\phi_C\phi_D^*\phi_E-2\sum_A\left|\sum_{nm}(t'_A)_{nm}\phi'_n\phi''_m\right|^2\\
&-4\,\mathrm{Re}\sum_{nm}(t'_A\mu)_{nm}\phi'^*_n\phi'_m-4\,\mathrm{Re}\sum_{nm}(\mu\,t'_A)_{nm}\phi''_n\phi''^*_m\\
&-2\sum_{nm}(\mu^\dagger\mu)_{nm}\phi'^*_n\phi'_m-2\sum_{nm}(\mu\,\mu^\dagger)_{nm}\phi''_n\phi''^*_m\,.
\end{aligned}
\tag{27.9.33}
$$

The last five lines on the right-hand side come from the terms in Eq. (27.9.30) involving auxiliary fields, and now these too are invariant under the discrete transformation (27.9.31), provided that

$$[t'_A\,,\,\mu]=[\mu^\dagger\,,\,\mu]=0\,. \tag{27.9.34}$$

We can now go a step further and consider the case of $N=4$ extended global supersymmetry. (As remarked in Section 25.4, $N=3$ supersymmetry is the same as $N=4$ supersymmetry.) The only massless multiplets of $N=4$ supersymmetry that do not contain gravitons or gravitinos consist of a single particle of helicity 1, an $SU(4)$ quartet of particles of helicity 1/2, and an $SU(4)$ sextet of particles of helicity 0, together with their CPT-conjugates of opposite helicity. There is one such supermultiplet for each generator t_A of the gauge group. These particles can be grouped into supermultiplets of $N=2$ supersymmetry: for each t_A there is one gauge supermultiplet consisting of one particle of helicity 1, two particles of helicity $\pm 1/2$, and one particle of helicity 0, together with their CPT-conjugates of opposite helicity, plus two hypermultiplets, each consisting of one particle of each helicity $\pm 1/2$ and two particles of helicity 0. The $N=2$ gauge superfield consists of an $N=1$ gauge superfield V_A and a left-chiral scalar superfield Φ_A and its complex conjugate, while the two $N=2$ hypermultiplets consist of two additional left-chiral scalar superfields Φ'_A and Φ''_A and their complex conjugates.

Since $N=4$ supersymmetry includes $N=2$ supersymmetry, the

Lagrangian density after elimination of the auxiliary fields of $N = 1$ supersymmetry[18] must be a special case of Eq. (27.9.33), but with the labels n, m, etc. running over the indices A, B, C, etc. of the adjoint representation. Also, the coefficient μ_{nm} in the superpotential (27.9.29) must vanish here, because otherwise Eq. (27.9.33) would contain terms quadratic in the fermion fields ψ'_A and ψ''_A, with no counterpart for their $N = 4$ superpartners λ_A and ψ_A. Also setting $(t'_A)_{BC}$ equal to the generators $-iC_{ABC}$ in the adjoint representation, we find that the Lagrangian density must take the form

$$\begin{aligned}
\mathscr{L} = & -\sum_A (D_\mu\phi')^*_A(D^\mu\phi')_A - \sum_A (D_\mu\phi'')^*_A(D^\mu\phi'')_A - \sum_A (D_\mu\phi)^*_A\,(D^\mu\phi)_A \\
& -\frac{1}{2}\sum_A \left(\overline{\psi'_A}(\not{D}\psi')_A)\right) - \frac{1}{2}\sum_A \left(\overline{\psi''_A}(\not{D}\psi'')_A)\right) \\
& -\frac{1}{2}\sum_A \left(\overline{\psi_A}\,(\not{D}\psi)_A\right) - \frac{1}{2}\sum_A \left(\overline{\lambda_A}\,(\not{D}\lambda)_A\right) \\
& -2\sqrt{2}\,\mathrm{Re}\sum_{ABC} C_{ABC}\phi_A\left(\psi'^{\mathrm{T}}_{BL}\epsilon\psi''_{CL}\right) - 2\sqrt{2}\,\mathrm{Re}\sum_{ABC} C_{ABC}\left(\lambda^{\mathrm{T}}_{AL}\epsilon\psi_{CL}\right)\phi^*_B \\
& -2\sqrt{2}\,\mathrm{Re}\sum_{ABC} C_{ABC}\phi'_B\left(\psi''^{\mathrm{T}}_{CL}\epsilon\psi_{AL}\right) - 2\sqrt{2}\,\mathrm{Re}\sum_{ABC} C_{ABC}\phi''_C\left(\psi'^{\mathrm{T}}_{BL}\epsilon\psi_{AL}\right) \\
& +2\sqrt{2}\,\mathrm{Re}\sum_{ABC} C_{ABC}\left(\psi'^{\mathrm{T}}_{BL}\epsilon\lambda_{AL}\right)\phi'^*_C + 2\sqrt{2}\,\mathrm{Re}\sum_{ABC} C_{ABC}\left(\psi''^{\mathrm{T}}_{BL}\epsilon\lambda_{AL}\right)\phi''^*_C \\
& -\frac{1}{4}\sum_A f_{A\mu\nu}f^{\mu\nu}_A + \frac{g^2\theta}{64\pi^2}\epsilon_{\mu\nu\rho\sigma}\sum_A f^{\mu\nu}_A f^{\rho\sigma}_A - V\,,
\end{aligned} \tag{27.9.35}$$

where the potential is

$$\begin{aligned}
V = & \sum_{ABCDE} C_{ADE}C_{BCE}\left(\phi_A\phi^*_B + \phi_B\phi^*_A\right)\left(\phi'_C\phi'^*_D + \phi''^*_C\phi''_D\right) \\
& +\frac{1}{2}\sum_A\left|\sum_{BC} C_{ABC}\left(\phi'^*_B\phi'_C - \phi''_B\phi''^*_C\right)\right|^2 \\
& -\frac{1}{2}\sum_{ABCDE} C_{ABC}C_{ADE}\phi^*_B\phi_C\phi^*_D\phi_E + 2\sum_A\left|\sum_{BC} C_{ABC}\phi'_B\phi''_C\right|^2 .
\end{aligned} \tag{27.9.36}$$

With no further constraints needed, this Lagrangian has an $SU(4)$ R-symmetry, which implies that it is invariant under $N = 4$ supersymmetry. To see this, we need to use the Jacobi identity to write the cross-term in the second line of the right-hand side of Eq. (27.9.36) in the form

$$\sum_{ABCDE} C_{ABC}C_{ADE}\phi_B'^*\phi_C'\phi_D''^*\phi_E'' = -\sum_{ABCDE} C_{ABC}C_{ADE}\phi_B'^*\phi_D'\phi_E''^*\phi_C'' - \sum_{ABCDE} C_{ABC}C_{ADE}\phi_B'^*\phi_E'\phi_C''^*\phi_D'' \,,$$

which allows us to write the potential (27.9.36) in a form symmetric among the scalars and their adjoints

$$\begin{aligned} V = &\sum_A \Big|\sum_{BC} C_{ABC}\phi_B^*\phi_C'\Big|^2 + \sum_A \Big|\sum_{BC} C_{ABC}\phi_B^*\phi_C''^*\Big|^2 + \sum_A \Big|\sum_{BC} C_{ABC}\phi_B\phi_C'\Big|^2 \\ &+ \sum_A \Big|\sum_{BC} C_{ABC}\phi_B\phi_C''^*\Big|^2 + \sum_A \Big|\sum_{BC} C_{ABC}\phi_B'^*\phi_C''\Big|^2 \\ &+ \sum_A \Big|\sum_{BC} C_{ABC}\phi_B'\phi_C''\Big|^2 + \frac{1}{2}\sum_A \Big|\sum_{BC} C_{ABC}\phi_B'\phi_C'^*\Big|^2 \\ &+ \frac{1}{2}\sum_A \Big|\sum_{BC} C_{ABC}\phi_B''\phi_C''^*\Big|^2 + \frac{1}{2}\sum_A \Big|\sum_{BC} C_{ABC}\phi_B\phi_C^*\Big|^2 \,. \end{aligned} \tag{27.9.37}$$

Now to make the $SU(4)$ symmetry apparent, we introduce an $SU(4)$ notation for the fields. We assemble the left-handed fermion fields into an $SU(4)$ vector:

$$\psi_{1AL} \equiv \psi_{AL}\,, \quad \psi_{2AL} \equiv \lambda_{AL}\,, \quad \psi_{3AL} \equiv \psi'_{AL}\,, \quad \psi_{4AL} \equiv \psi''_{AL}\,. \tag{27.9.38}$$

In order for the fermion kinematic terms in the Lagrangian density to be $SU(4)$-invariant, we then must assemble the right-handed fermion fields into a contragredient vector:

$$\psi^1_{AR} \equiv \psi_{AR}\,, \quad \psi^2_{AR} \equiv \lambda_{AR}\,, \quad \psi^3_{AR} \equiv \psi'_{AR}\,, \quad \psi^4_{AR} \equiv \psi''_{AR}\,. \tag{27.9.39}$$

The Majorana condition on the fermion fields then takes the $SU(4)$-invariant form

$$(\psi_{iAL})^* = -\beta\epsilon\psi^i_{AR}\,, \tag{27.9.40}$$

with indices i, j, etc. running over the values 1, 2, 3, 4. In order for the Yukawa couplings between the fermion and scalar fields to be $SU(4)$-invariant, we must give the scalars the transformation properties of an antisymmetric $SU(4)$ tensor

$$\begin{aligned} &\phi_A^{12} \equiv \phi_A^*\,, \quad \phi_A^{13} \equiv \phi_A''\,, \quad \phi_A^{14} \equiv -\phi_A'\,, \\ &\phi_A^{23} \equiv -\phi_A'^*\,, \quad \phi_A^{24} \equiv -\phi_A''^*\,, \quad \phi_A^{34} \equiv \phi_A\,, \end{aligned} \tag{27.9.41}$$

which also obeys an $SU(4)$-invariant reality condition

$$\left(\phi_A^{ij}\right)^* = \frac{1}{2}\sum_{kl} \epsilon_{ijkl}\,\phi_A^{kl}\,. \qquad (27.9.42)$$

The whole Lagrangian density (27.9.35) can then be written in the manifestly $SU(4)$-invariant form

$$\begin{aligned}\mathscr{L} = &-\frac{1}{2}\sum_{Aij}(D_\mu\phi^{ij})_A(D^\mu\phi^{ij})_A^* \\ &-\frac{1}{2}\sum_{Ai}\left(\psi_{iAL}^{\mathrm{T}}\epsilon(\not{D}\psi_R^i)_A\right) + \frac{1}{2}\sum_{Ai}\left(\psi_{AR}^{i\mathrm{T}}\epsilon(\not{D}\psi_{iL})_A\right) \\ &-\sqrt{2}\,\mathrm{Re}\sum_{ABCij} C_{ABC}\phi_A^{ij}\left(\psi_{iBL}^{\mathrm{T}}\epsilon\psi_{jCL}\right) - V \\ &-\frac{1}{4}\sum_A f_{A\mu\nu}f_A^{\mu\nu} + \frac{g^2\theta}{64\pi^2}\epsilon_{\mu\nu\rho\sigma}\sum_A f_A^{\mu\nu}f_A^{\rho\sigma}\,,\end{aligned} \qquad (27.9.43)$$

where the potential is

$$V = \frac{1}{8}\sum_{Aijkl}\left|\sum_{BC} C_{ABC}\phi_B^{ij}\phi_C^{kl}\right|^2\,. \qquad (27.9.44)$$

The potential has a minimum value zero, so that supersymmetry is not broken in this theory. The minimum is reached when the generators $\sum_A t_A\phi_A^{ij}$ all commute with one another.

For zero theta angle, the gauge theories with a simple gauge group and either $N = 2$ or $N = 4$ supersymmetry have just a single coupling constant, the gauge coupling constant g. Since these theories have $N = 1$ supersymmetry, they share the property discussed in Section 27.6, that the only infinity in higher orders of perturbation theory is in a one-loop correction to this coupling.‡‡ The function $\beta(g)$ in the renormalization group equation $\mu dg/d\mu = \beta(g)$ is then given to all orders of perturbation theory by the one-loop formula (18.7.2), with a suitable correction for the presence of scalar fields:

$$\beta(g) = -\frac{g^3}{4\pi^2}\left(\frac{11}{12}C_1 - \frac{1}{6}C_2^f - \frac{1}{12}C_2^s\right)\,, \qquad (27.9.45)$$

‡‡ The trilinear term in the superpotential (27.9.29) is proportional to the gauge coupling, and is therefore renormalized, despite the no-renormalization theorem of Section 27.6. This is because here we are renormalizing the left-chiral scalar superfields Φ_A, Φ'_n, and Φ''_n as well as the gauge superfield V_A so as to keep them canonically normalized. The bilinear term in Eq. (27.9.29) is renormalized for the same reason.

where

$$\sum_{AB} C_{ABC}C_{ABD} = g^2 C_1 \delta_{CD} ,$$
$$\Big[\mathrm{Tr}\,(t_C t_D)\Big]_{\text{Majorana fermions}} = g^2 C_2^f \delta_{CD} , \tag{27.9.46}$$
$$\Big[\mathrm{Tr}\,(t_C t_D)\Big]_{\text{complex scalars}} = g^2 C_2^s \delta_{CD} .$$

In general theories with $N = 2$ supersymmetry we have two Majorana fermions λ_A and ψ_A in the adjoint representation and H pairs of Majorana fermions ψ'_n and ψ''_n whose left- and right-handed parts are in representations with generators either t'_A or $-t'^{\mathrm{T}}_A$, so

$$C_2^f = 2C_1 + 2HC'_2 , \tag{27.9.47}$$

where C'_2 is defined by

$$\mathrm{Tr}\, t'_C t'_D = g^2 C'_2 \delta_{CD} . \tag{27.9.48}$$

Also, we have one complex scalar ϕ_A in the adjoint representation and H pairs of complex scalars ϕ'_n and ϕ''_n in representations with generators t'_A or $-t'^{\mathrm{T}}_A$, so

$$C_2^s = C_1 + 2HC'_2 . \tag{27.9.49}$$

The beta function (27.9.45) is therefore

$$\beta(g) = -\frac{g^2}{8\pi^2}\Big(C_1 - HC'_2\Big) . \tag{27.9.50}$$

The case of $N = 4$ supersymmetry is just the special case with $H = 1$ pairs of $N = 2$ hypermultiplets in the adjoint representation, with $C'_2 = C_1$, so in this case the beta function vanishes. This is therefore a finite theory, with no renormalizations at all.[19]

Gauge theories with $N = 4$ supersymmetry have another remarkable property, known as *duality*. This was first conjectured by Montonen and Olive[17] for purely bosonic theories in which a simple gauge group is spontaneously broken to a $U(1)$ electromagnetic gauge group. They noticed that semi-classical calculations (of the sort described in Section 23.3) give the mass of particles with charge $q = ne$ and magnetic monopole moment $\mathscr{M} = 4\pi m/e$ (with n and m integers of any sign) as

$$M = \sqrt{2}\left|v\left(ne + \frac{4\pi i m}{e}\right)\right| , \tag{27.9.51}$$

which is invariant under the transformations

$$m \to n , \qquad n \to -m , \qquad e \leftrightarrow 4\pi/e . \tag{27.9.52}$$

On this basis, they suggested that the theory with a weak gauge coupling e is fully equivalent to one with a strong gauge coupling $4\pi/e$. Neither the purely bosonic theory nor the simplest versions of the $N = 1$ and $N = 2$ extended supersymmetry theories really have this property;[20] for one thing, the massive charged elementary vector bosons of the broken gauge symmetries have spin 1, while all of the monopoles and dyons have spins 1/2 or 0. (We will see in Section 29.5 that $N = 2$ theories do have a duality property of a more subtle sort.) But for $N = 4$ supersymmetry the monopole states form multiplets with one particle of spin 1, four particles of spin 1/2, and two particles of spin 0, just like the elementary particles.[20] Evidence has accumulated[21] that $N = 4$ supersymmetric gauge theories are indeed invariant under the interchange of electric and magnetic quantum numbers and of e with $4\pi/e$. The equivalence of theories with large and small coupling constants has become an increasingly important theme in string theory, but this is beyond the scope of this book.

Problems

1. To second order in gauge coupling constants, calculate the components of the superfields Ω^A that are needed to put the gauge superfield V^A in Wess–Zumino gauge by the transformation (27.1.12).

2. Show that the most general chiral spinor superfield W_α satisfying the condition (27.2.20) has components $f_{\mu\nu}$ satisfying the homogeneous Maxwell equations $\epsilon^{\mu\nu\rho\sigma}\partial_\rho f_{\mu\nu} = 0$. What conditions on the other components of W_α are imposed by Eq. (27.2.20)?

3. Consider a general renormalizable $N = 1$ supersymmetric gauge theory with an $SU(2)$ gauge group and a single chiral superfield belonging to the 3-vector representation of $SU(2)$. What is the most general superpotential for this theory? Construct the Lagrangian density of the whole theory explicitly. Eliminate the auxiliary fields. Show that supersymmetry is not broken in this theory. What are the masses of the particles of this theory?

4. Express the gaugino and chiral fermion fields in the supersymmetric version of quantum electrodynamics described in Section 27.5 in terms of the goldstino field and other spinor fields of definite mass.

5. Consider the renormalizable $N = 2$ supersymmetric theory with an $SU(3)$ gauge symmetry and no hypermultiplets. What are the values of the scalar fields for which the potential vanishes? What are the massless gauge fields for non-zero values of these scalars? Calculate

the central charge in terms of the quantities to which these massless gauge fields are coupled.

References

1. Supersymmetry was applied first to Abelian gauge theories without using the superfield formalism by J. Wess and B. Zumino, *Nucl. Phys.* **B78**, 1 (1974). It was then extended to non-Abelian gauge theories by S. Ferrara and B. Zumino, *Nucl. Phys.* **B79**, 413 (1974); A. Salam and J. Strathdee, *Phys. Lett.* **51B**, 353 (1974). These references are reprinted in *Supersymmetry*, S. Ferrara, ed. (North Holland/World Scientific, Amsterdam/Singapore, 1987).
2. P. Fayet and J. Iliopoulos, *Phys. Lett.* **51B**, 461 (1974). This reference is reprinted in *Supersymmetry*, Ref. 1.
3. S. Weinberg, *Phys. Rev. Lett.* **80**, 3702 (1998).
4. S. Ferrara, L. Girardello, and F. Palumbo, *Phys. Rev.* **D20**, 403 (1979). This article is reprinted in *Supersymmetry*, Ref. 1. Special cases of this sum rule had been given by P. Fayet, *Phys. Lett.* **84B**, 416 (1979).
5. M. T. Grisaru, W. Siegel, and M. Roček, *Nucl. Phys.* **B159**, 429 (1979).
6. N. Seiberg, *Phys. Lett.* **B318**, 469 (1993).
7. This was first proved in the supergraph formalism by W. Fischler, H. P. Nilles, J. Polchinski, S. Raby, and L. Susskind, *Phys. Rev. Lett.* **47**, 757 (1981). The proof presented here was given by M. Dine, in *Fields, Strings, and Duality: TASI 96*, C. Efthimiou and B. Greene, eds. (World Scientific, Singapore, 1997); S. Weinberg, Ref. 3.
8. A detailed proof of the statement that superrenormalizable terms that violate some global symmetry do not introduce infinite symmetry breaking radiative corrections to the coefficients of renormalizable interactions is given by K. Symanzik, in *Cargèse Lectures in Physics*, Vol. 5, D. Bessis, ed. (Gordon and Breach, New York, 1972). This is discussed briefly in the footnote on p. 507 of Volume I.
9. L. Girardello and M. T. Grisaru, *Nucl. Phys.* **B194**, 65 (1982), reprinted in *Supersymmetry*, Ref. 1; K. Harada and N. Sakai, *Prob. Theor. Phys.* **67**, 67 (1982).
10. For a review, see S. Weinberg, *Rev. Mod. Phys.* **61**, 1–23 (1989).

11. B. de Wit and D. Z. Freedman, *Phys. Rev.* **D12**, 2286 (1975). This article is reprinted in *Supersymmetry*, Ref. 1.

12. The first examples of gauge theories with $N = 2$ extended supersymmetry were given by P. Fayet, *Nucl. Phys.* **B113**, 135 (1976); reprinted in *Supersymmetry*, Ref. 1. The approach presented here is similar to that of Fayet. A superfield formalism was subsequently given by R. Grimm, M. Sohnius, and J. Scherk, *Nucl. Phys.* **B113**, 77 (1977). Gauge theories with $N = 2$ and $N = 4$ supersymmetry in four spacetime dimensions were constructed by dimensional reduction of higher-dimensional theories with simple supersymmetry by L. Brink, J. H. Schwarz, and J. Scherk, *Nucl. Phys.* **B113**, 77 (1977); M. F. Sohnius, K. S. Stelle, and P. C. West, *Nucl. Phys.* **B113**, 127 (1980). For other approaches, see M. F. Sohnius, *Nucl. Phys.* **B138**, 109 (1979); A. Halperin, E. A. Ivanov, and V. I. Ogievetsky, *Prima JETP* **33**, 176 (1981); P. Breitenlohner and M. F. Sohnius, *Nucl. Phys.* **B178**, 151 (1981); P. Howe, K. S. Stelle, and P. K. Townsend, *Nucl. Phys.* **B214**. 519 (1983).

13. E. Witten and D. Olive, *Phys. Lett.* **78B**, 97 (1978). Also see H. Osborn, *Phys. Lett.* **83B**, 321 (1979).

14. E. B. Bogomol'nyi, *Sov. J. Nucl. Phys.* **24**, 449 (1976).

15. D. Zwanziger, *Phys. Rev.* **176**, 1480, 1489 (1968); J. Schwinger, *Phys. Rev.* **144**, 1087 (1966); **173**, 1536 (1968); B. Julia and A. Zee, *Phys. Rev.* **D11**, 2227 (1974); F. A. Bais and J. R. Primack, *Phys. Rev.* **D13**, 819 (1975). (The last article was incorrectly attributed to Julia and Zee in Chapter 23 of the first printings of Volume II.)

16. M. K. Prasad and C. M. Sommerfield, *Phys. Rev. Lett.* **35**. 760 (1975); E. B. Bogomol'nyi, Ref. 14; S. Coleman, S. Parke, A. Neveu, and C. M. Sommerfield, *Phys. Rev.* **D15**, 544 (1977).

17. This was noted by C. Montonen and D. Olive, *Phys. Lett.* **72B**, 117 (1977). The absence of one-loop corrections to the masses was shown by A. D'Adda, R. Horsley, and P. Di Vecchia, *Phys. Lett.* **76B**, 298 (1978).

18. Obstacles to formulating theories with auxiliary fields for $N = 4$ supersymmetry have been analyzed by W. Siegel and M. Roček, *Phys. Lett.* **105B**, 275 (1981).

19. The finiteness of the $N = 4$ theory was shown by M. F. Sohnius and P. C. West, *Nucl. Phys.* **B100**, 245 (1981); P. S. Howe, K. S. Stelle, and P. Townsend, *Nucl. Phys.* **B214**, 519 (1983); S. Mandelstam, *Nucl. Phys.* **B213**, 149 (1983); L. Brink, O. Lindgren, and B. E. W. Nilsson, *Nucl. Phys.* **B212**, 401 (1983); *Phys. Lett.* **123B**, 328 (1983). The proof

of finiteness was extended to non-perturbative effects by N. Seiberg, *Phys. Lett.* **B206**, 75 (1988). Also see S. Kovacs, hep-th/9902047, to be published. There is a wide class of ultraviolet-finite N = 2 theories; see P. S. Howe, K. S. Stelle, and P. C. West, *Phys. Lett.* **B124**, 55 (1983).

20. H. Osborn, Ref. 13.

21. A. Sen, *Phys. Lett.* **B329**, 217 (1994); C. Vafa and E. Witten, *Nucl. Phys.* **B 431**, 3 (1994); L. Girardello, A. Giveon, M. Porrati, and A. Zaffaroni, *Phys. Lett.* **B334**, 331 (1994).

28
Supersymmetric Versions of the Standard Model

Physical phenomena at energies accessible in today's accelerator laboratories are accurately described by the standard model, the renormalizable theory of quarks, leptons, and gauge bosons, governed by the gauge group $SU(3) \times SU(2) \times U(1)$, described in Sections 18.7 and 21.3. The standard model is today usually[1] understood as a low-energy approximation to some as-yet-unknown fundamental theory in which gravitation appears unified with the strong and electroweak forces at an energy somewhere in the range of 10^{16} to 10^{18} GeV. This raises the *hierarchy problem*: what accounts for the enormous ratio of this fundamental energy scale and the energy scale ≈ 300 GeV that characterizes the standard model?

The strongest theoretical motivation for supersymmetry is that it offers a hope of solving the hierarchy problem. Quarks, leptons, and gauge bosons are required by the $SU(3) \times SU(2) \times U(1)$ gauge symmetry to appear with zero masses in the Lagrangian of the standard model, so that the physical masses of these particles are proportional to the electroweak breaking scale, which in turn is proportional to the mass of the scalar fields responsible for the electroweak symmetry breakdown. The crux of the hierarchy problem[1a] is that the scalar fields, unlike the fermion and gauge boson fields, are not protected from acquiring large bare masses by any symmetry of the standard model, so it is difficult to see why their masses, and hence all other masses, are not in the neighborhood of 10^{16} to 10^{18} GeV.

It has been hoped that this problem could be solved by embedding the standard model in a supersymmetric theory. If the scalar fields appear in supermultiplets along with fermions in a chiral representation of some gauge group, then supersymmetry would require vanishing bare masses for the scalars as well as the fermions. All the masses of the standard model would then be tied to the energy scale at which supersymmetry is broken. The hope of a solution of the hierarchy problem along these lines has been the single strongest motivation for trying to incorporate supersymmetry in a realistic theory.

Unfortunately, none of the new particles required by supersymmetric theories have been detected, and no entirely satisfactory supersymmetric version of the standard model has emerged so far. This chapter will describe the attempts that have been made in this direction.

28.1 Superfields, Anomalies, and Conservation Laws

In this section we shall attempt to decide at least tentatively what ingredients should appear in a supersymmetric version of the standard model.

None of the quark and lepton fields of the standard model belong to the adjoint representation of the $SU(3) \times SU(2) \times U(1)$ gauge group, so they cannot be the superpartners of known gauge bosons and must therefore be included in chiral scalar superfields. We will define U_i, D_i, $\bar{U}_i$, $\bar{D}_i$, N_i, E_i, and $\bar{E}_i$ as the left-chiral superfields whose ψ_L components are the left-handed fields of the quarks of charge $2e/3$ and $-e/3$, of the antiquarks of charge $-2e/3$ and $+e/3$, of the leptons of charge 0 and $-e$, and of the antileptons of charge $+e$, respectively, with i a generation label running over values 1, 2, and 3. (For instance, the spinor components of U_1, U_2, and U_3 are the left-handed fields of the u, c, and t quarks, respectively.) Of these superfields, U_i and D_i form $SU(2)$ doublets, N_i and E_i also form $SU(2)$ doublets, and the others are $SU(2)$ singlets. The quark superfields form $SU(3)$ triplets and the antiquark superfields form $SU(3)$ antitriplets, with color indices suppressed, and the leptons and antilepton superfields are $SU(3)$ singlets. As mentioned earlier, the particles described by the scalar components of these superfields are known as squarks, antisquarks, sleptons, and antisleptons. There are also the gauginos, the spin 1/2 superpartners of the gauge bosons of $SU(3)$, $SU(2)$, and $U(1)$, respectively known as the gluino, wino, and bino.*

We must also add some mechanism that produces a spontaneous breakdown of $SU(2) \times U(1)$ and gives mass to all the quarks and leptons as well as to the $W^\pm$ and Z^0. The simplest possibility is to suppose the existence

* As discussed in Section 28.3, the energy scale that characterizes supersymmetry breaking is expected to be considerably higher than the $\approx$ 300 GeV that characterizes the breaking of $SU(2) \times U(1)$, so there is a substantial range of energies in which supersymmetry but not $SU(2) \times U(1)$ may be considered to be broken. In this range, the gauginos have masses that are governed by $SU(2) \times U(1)$ symmetry, so the neutral electroweak gauginos of definite mass are superpartners of the $SU(2)$ triplet W^0 and the $SU(2)$ singlet B, known as the neutral wino and the bino, rather than superpartners of the Z^0 and the photon. When $SU(2) \times U(1)$ breaking is taken into account there is a small mixing of the neutral wino and bino.

of just two more $SU(2)$ doublets of left-chiral superfields:

$$H_1 = \begin{pmatrix} H_1^0 \\ H_1^- \end{pmatrix}, \qquad H_2 = \begin{pmatrix} H_2^+ \\ H_2^0 \end{pmatrix}, \tag{28.1.1}$$

which appear in the Lagrangian density in linear combinations of the $SU(3) \times SU(2) \times U(1)$-invariant $\mathscr{F}$-terms:

$$\left[\left(D_i H_1^0 - U_i H_1^-\right)\bar{D}_j\right]_{\mathscr{F}}, \qquad \left[\left(E_i H_1^0 - N_i H_1^-\right)\bar{E}_j\right]_{\mathscr{F}}, \tag{28.1.2}$$

and

$$\left[\left(D_i H_2^+ - U_i H_2^0\right)\bar{U}_j\right]_{\mathscr{F}}, \tag{28.1.3}$$

with obvious contractions of color indices. According to Eq. (26.4.11), a non-vanishing expectation value of the scalar component of H_1^0 gives mass to the charged leptons and charge $-e/3$ quarks, while a non-vanishing expectation value of the scalar component of H_2^0 gives mass to the charge $+2e/3$ quarks. These expectation values of course also give mass to the $W^\pm$ and Z^0 vector bosons and, since the H_1 and H_2 are $SU(2)$ doublets, we automatically get the same successful results for these masses as found in Section 21.3. Note that supersymmetry does not allow the complex conjugates of the H_1 and H_2 left-chiral superfields to appear in the superpotential, so a vacuum expectation value of the scalar component of H_1^0 cannot give masses to the charge $+2e/3$ quarks, and a vacuum expectation value of the scalar component of H_2^0 cannot give mass to the charge $-e/3$ quarks or the charged leptons, which is why both H_1 and H_2 are needed to give masses to all the quarks and leptons.

Of course, there might be more than one of the H_1 and/or H_2 doublets. Their numbers are partly constrained by the condition of anomaly cancellation. We saw in Section 22.4 that the gauge symmetries of the non-supersymmetric standard model are anomaly-free, as they must be for quantum mechanical consistency, but now there are additional spinor fields in the Lagrangian. The gaugino fields don't create any problems, because their left-handed components belong to the adjoint representation of the gauge group, which is real for all gauge groups. The only problem can arise from the higgsinos — the spin 1/2 components of the superfields (H_1^0, H_1^-) and (H_2^+, H_2^0). The spinor components of each (H_1^0, H_1^-) doublet of superfields produces an $SU(2)$-$SU(2)$-$U(1)$ anomaly proportional to $\sum t_3^2 y = (\frac{1}{2}g)^2(\frac{1}{2}g') + (-\frac{1}{2}g)^2(\frac{1}{2}g') = \frac{1}{2}g^2 g'$, while the spinor components of each (H_2^+, H_2^0) doublet of superfields produces an $SU(2)$-$SU(2)$-$U(1)$ anomaly proportional to $\sum t_3^2 y = (\frac{1}{2}g)^2(-\frac{1}{2}g') + (-\frac{1}{2}g)^2(-\frac{1}{2}g') = -\frac{1}{2}g^2 g'$. *The cancellation of anomalies thus requires an equal number of (H_1^0, H_1^-) and (H_2^+, H_2^0) doublets.* In this case, all anomalies cancel, including the $U(1)^3$ and $U(1)$-graviton-graviton anomalies. The next section will give an argument that there is in fact just one of each type of doublet.

In a theory constructed along these lines, we have to give up one of the attractive features of the non-supersymmetric standard model, that is, that it *automatically* excludes any renormalizable interactions that violate the conservation of baryon or lepton number. There are several renormalizable supersymmetric $SU(3) \times SU(2) \times U(1)$-invariant $\mathscr{F}$-terms that could be included in the Lagrangian density that would violate baryon and/or lepton number conservation without violating the $SU(3) \times SU(2) \times U(1)$ gauge symmetries:

$$\left[\left(D_iN_j - U_iE_j\right)\bar{D}_k\right]_{\mathscr{F}}, \qquad \left[\left(E_iN_j - N_iE_j\right)\bar{E}_k\right]_{\mathscr{F}}, \tag{28.1.4}$$

and also

$$\left[\bar{D}_i\bar{D}_j\bar{U}_k\right]_{\mathscr{F}}, \tag{28.1.5}$$

with the three suppressed color indices in Eq. (28.1.5) understood to be contracted with an antisymmetric ϵ-symbol to give a color singlet. With all of these interactions present, there would be no clever way to assign baryon and lepton numbers to the squarks and sleptons that would avoid an unsuppressed violation of baryon and lepton number conservation. For instance, the exchange of the scalar boson of the $\bar{D}$ superfield between vertices for interactions (28.1.4) and (28.1.5) would lead to the process $u_Ld_Ru_R \to \overline{e_R}$, observed for instance as $p \to \pi^0 + e^+$, at a catastrophic rate that is only suppressed by factors of coupling constants. To avoid this, it is necessary to make an independent assumption that would rule out some or all of the interactions (28.1.4)–(28.1.5).

Note that it is not necessary to rule out *all* of the interactions (28.1.4) and (28.1.5). For instance, suppose that we assume only that baryon number is conserved, with conventional baryon number assignments: the U_i and D_i left-chiral superfields are assigned baryon number $+1/3$; the $\bar{U}_i$ and $\bar{D}_i$ are assigned baryon number $-1/3$; and the L_i, $\bar{E}_i$, H_1, and H_2 all are assigned baryon number 0. This would allow the interactions (28.1.4) while forbidding the interactions (28.1.5). Despite appearances, the interactions (28.1.4) alone do not violate the conservation of lepton number, provided the scalar components of the superfields are assigned appropriate lepton numbers. This can be done by assigning lepton number 0 to the N_i and E_i superfields, lepton number -1 to the U_i, D_i, $\bar{U}_i$, and $\bar{D}_i$ superfields, lepton number -2 to the $\bar{E}_i$ superfields, lepton number 0 to the H_1 and H_2 superfields, and lepton numbers -1 and $+1$ to θ_L and θ_R, respectively. (Recall that such symmetries, under which θ transforms non-trivially, are known as *R-symmetries.*) Then all quarks and leptons have conventional lepton numbers: the fermion components ν_{iL} and e_{iL}, which are the coefficients of θ_L in N_i and E_i, have the lepton numbers $0+1 = +1$; the fermion components $\overline{e_{iR}}$ of the $\bar{E}_i$ superfields have lepton numbers $-2+1=-1$, and the quarks and antiquarks have lepton numbers $-1+1=0$. The higgsinos (the fermion components of H_1 and H_2) have lepton numbers $0+1=+1$. On the other hand, the scalar components of the superfields have the same lepton numbers as the superfields themselves, which are unconventional. Further, the $\mathscr{F}$-term of a left-chiral superfield is the coefficient of θ_L^2, so the interactions in Eq. (28.1.4) have lepton numbers $-1+0-1+2=0$ and $0+0-2+2=0$; the H_1 interactions (28.1.2) have lepton number $-1+0-1+2=0$ and $0+0-2+2=0$, respectively; and the H_2 interaction (28.1.3) has lepton number $-1+0-1+2=0$; so none of these interactions violate the conservation of lepton number. Also, the scalar components of H_1 and H_2 have lepton number 0, so their vacuum expectation values also do not violate the conservation of lepton number. With this assignment of lepton numbers, lepton number conservation rules out any renormalizable interactions that would violate baryon number conservation: the interaction (28.1.5) has lepton number $-1-1-1+2=-1$, and so is forbidden.

The interactions (28.1.4) would allow an alternative mechanism for breaking $SU(2)\times U(1)$ and giving mass to the charged leptons and charge $-e/3$ quarks: the scalar components of the neutrino superfields N_i might have non-vanishing vacuum expectation values. (With the lepton number assignments of the previous paragraph this expectation value would not violate lepton number conservation, because these scalar components have the lepton number of the N_i superfields, which is zero.) But we cannot rely on this mechanism to let us do without the H_1 superfields altogether, because we still need the H_2 interactions (28.1.3) to give mass to the charge $+2e/3$ quarks and, as we have seen, the cancellation of anomalies requires equal numbers of H_1 and H_2 superfields.

It is usually assumed instead that some symmetries forbid both interactions (28.1.4) and (28.1.5). Obviously, these symmetries could be baryon and lepton number conservation, with conventional assignments of these numbers: U_i, D_i having baryon number $B=1/3$ and lepton number $L=0$, $\bar{U}_i$ and $\bar{D}_i$ having baryon number $B=-1/3$ and lepton number 0, N_i and E_i having lepton number $L=+1$ and baryon number 0, $\bar{E}_i$ having lepton number -1 and baryon number 0, and H_1^0, H_1^- and H_2^+, H_2^0 *and* θ_L and θ_R all having baryon and lepton numbers 0. The same results apply if we only assume the conservation of certain linear combinations of baryon and lepton number, such as the anomaly-free combination $B-L$ discussed in Section 22.4.

There are widespread doubts about whether it is possible to have exact continuous global symmetries, because in string theory the existence of any exact continuous symmetry would imply the existence of a massless spin 1 particle coupled to the symmetry current, so that the symmetry would have to be local, not global.[1b] But the interactions (28.1.4) and

(28.1.5) may also be banned by assuming a *discrete* global symmetry, known as the conservation of *R parity*.[2] The R parity is defined to be +1 for quarks, leptons, gauge bosons, and Higgs scalars, and −1 for their superpartners. This R parity is equal to

$$\Pi_R = (-1)^F \, (-1)^{3(B-L)} \, , \tag{28.1.6}$$

where $(-1)^F$ is the fermion parity, which is +1 for all bosons and −1 for all fermions. Fermion parity is the same sign as is produced by a 2π rotation, and is therefore always conserved, so if $B-L$ is conserved then so is R parity.** It is possible that R parity may be conserved even if $B-L$ is not, but in fact the interactions (28.1.4) and (28.1.5) are forbidden by R parity conservation, so as far as renormalizable interactions are concerned R parity conservation implies the conservation of both baryon and lepton number. This is *not* true of the non-renormalizable supersymmetric interactions that are presumably produced by physical processes at very high energies. The baryon- and lepton-non-conserving processes produced by such interactions are discussed in Section 28.7.

All of the new 'sparticles' (squarks, sleptons, gauginos, and higgsinos) that are required by supersymmetry theories have negative R parity, so if R parity is exact and unbroken then the lightest of the new particles required by supersymmetry must be absolutely stable. All of the other new particles will then undergo a chain of decays, ultimately yielding ordinary particles and the lightest new particle. Much of the phenomenology of various supersymmetry models is governed by the choice of which of the new particles is the lightest.

With supersymmetry and either R parity or $B-L$ conserved, the most general renormalizable Lagrangian for the superfields discussed above consists of the usual gauge-invariant kinematic part for the chiral superfields, given by a sum of terms of the form $(\Phi^* \exp(-V)\Phi)_D$ for each of the quark, lepton, and Higgs chiral superfields, plus the usual gauge-invariant kinematic term for the gauge superfields, given by a sum of terms of the form $\epsilon_{\alpha\beta}(W_\alpha W_\beta)_{\mathscr{F}}$ for each of the $SU(3)$, $SU(2)$, and $U(1)$ field strength superfields, plus the supersymmetric Yukawa couplings, given by a linear combination of the interactions (28.1.2), (28.1.3), and a new $\mathscr{F}$-term

** The value of $(-1)^{3(B-L)}$ is −1 for quark and lepton superfields and +1 for all other superfields, so the conservation of R parity is equivalent to invariance under a transformation in which all quark and lepton superfields change sign with no change in other superfields. This invariance principle was introduced in Reference 3 in order to rule out the interactions (28.1.4) and (28.1.5).

coupling H_1 and H_2:

$$\mathscr{L}_Y = \sum_{ij} h^D_{ij}\Big[\Big(D_i H^0_1 - U_i H^-_1\Big)\bar{D}_j\Big]_{\mathscr{F}} + \sum_{ij} h^E_{ij}\Big[\Big(E_i H^0_1 - N_i H^-_1\Big)\bar{E}_j\Big]_{\mathscr{F}}$$
$$+ \sum_{ij} h^U_{ij}\Big[\Big(D_i H^+_2 - U_i H^0_2\Big)\bar{U}_j\Big]_{\mathscr{F}} + \mu\Big[H^+_2 H^-_1 - H^0_2 H^0_1\Big]_{\mathscr{F}} + \text{H.c.} \tag{28.1.7}$$

As we will see in Section 28.3, more terms will have to be added to the Lagrangian in order to account for supersymmetry breaking.

The coefficient μ in Eq. (28.1.7) has the dimensions of mass, and is the only dimensional parameter that enters in the supersymmetric version of the standard model Lagrangian. It is somewhat disappointing to find that this term is still allowed, because it revives the hierarchy problem: why is μ not of order 10^{16} to 10^{18} GeV? The μ-term in Eq. (28.1.7) can be avoided if we assume that lepton number is conserved, with the unconventional lepton number assignments discussed above that would allow the interactions (28.1.4) but not the interactions (28.1.5). In this case, the μ-term carries lepton number +2 and is therefore also forbidden. This term can also be forbidden if we assume a $U(1)$ 'Peccei–Quinn symmetry,'[4] for which the superfields H_1 and H_2 carry equal quantum numbers, say +1, while θ_L and θ_R are neutral. The interactions (28.1.2) and (28.1.3) which give mass to the quarks and leptons are then allowed if, for instance, we give Peccei–Quinn quantum numbers −1 to the left-chiral superfields of antisquarks and antisleptons while the left-chiral superfields of squarks and sleptons are taken to be neutral. This choice then also forbids the dangerous interactions (28.1.4) and (28.1.5). Unfortunately, as we will see in Section 28.4, the μ-term in Eq. (28.1.7) seems to be needed for phenomenological reasons. The theories of gravity-mediated supersymmetry breaking discussed in Section 31.7 provide a natural mechanism for producing a μ-term of an acceptable magnitude.

We can obtain a crude upper bound on the masses of the new particles by assuming that supersymmetry does solve the hierarchy problem discussed at the beginning of this chapter. In accord with the theorem of Section 27.6, if supersymmetry were unbroken the contribution to the mass of the scalar component of H_1 or H_2 from one-loop diagrams with an intermediate quark, lepton, W or Z loop would be cancelled by the corresponding one-loop diagram with an intermediate squark, slepton, wino, or bino. Therefore with supersymmetry broken the contribution δm^2_H of such diagrams to the squared masses of the H_1 and H_2 scalars is a sum of terms of order $(\mathscr{G}^2_s/8\pi^2)\Delta m^2_s$, where $\mathscr{G}_s$ is the Yukawa or gauge coupling of the Higgs scalar to supermultiplet s, and Δm^2_s is the mass-squared splitting within the supermultiplet. To avoid having to fine-tune these corrections, we need δm_H^2 not much larger than the coefficient of order (300 GeV)^2 of the term in the standard model Lagrangian density that gives the observed $SU(2) \times U(1)$ breaking in the tree approximation, so we will assume that $\delta m_H^2 < (1\ \text{TeV})^2$. For instance, the top quark and squark have couplings to H_2 of order unity, so we expect that the splittings Δm^2 should be less than about $8\pi^2$ TeV^2, and hence the masses of the top squarks should be less than about 10 TeV. We will see in Section 28.4 that the rates of flavor-changing processes can be brought within experimental upper bounds by taking the masses of the squarks to be nearly equal, in which case this can be taken as a rough upper limit on the masses of all the squarks. (However it is possible that the rates of these processes may be suppressed instead by very large masses of the first two generations of squarks, while the mass of the top squark is below the 10 TeV naturalness bound.[4a]) The limits set by this sort of argument on the masses of other particles with $R = -1$ are somewhat weaker, but at least in the popular class of models discussed in Section 28.6, none of the masses of these particles are expected to be much greater than the squark masses, so 10 TeV can be taken as an upper bound on all of them. On the other hand, the fact that none of these particles have been observed only indicates that their masses are probably greater than about 100 GeV, so there is an ample mass range in which they may yet be found.

* * *

If R parity conservation or some other conservation law makes the lightest of the new particles predicted by supersymmetry stable, then some of these particles may be left from the early universe. The number density of these relics can be estimated using techniques that were originally applied to the cosmic density of massive neutrinos.[4b] To give one example of this sort of calculation, we shall show that for a broad range of plausible masses, the new stable particle of supersymmetry theories cannot be a charged and uncolored particle, like a charged slepton, wino, or higgsino.[4c]

Once the cosmic temperature T (in energy units, with the Boltzmann constant set equal to unity) drops below the mass m of any stable charged untrapped particle, their number nR^3 in a volume R^3 that expands with the universe is decreased by annihilation at a rate per particle equal to $\overline{v\sigma}n$, where $\overline{v\sigma}$ is the mean value of the product of the relative velocity and the annihilation cross-section. That is,

$$\frac{d(nR^3)}{dt} = -\overline{v\sigma}n^2R^3 ,$$

so that

$$\frac{1}{nR^3} = \left(\frac{1}{nR^3}\right)_0 + \int_{t_0}^{t} \frac{\overline{v\sigma}}{R^3}\, dt , \tag{28.1.8}$$

where 0 labels the epoch at which $T \simeq m$. The annihilation process is exothermic, so $\overline{v\sigma}$ approaches a constant for $v \ll 1$. Also, in a radiation-dominated phase of the cosmic expansion $R \propto t^{1/2}$, so the integral converges, and gives

$$\begin{aligned}\left(\frac{1}{nR^3}\right)_{t\to\infty} &= \left(\frac{1}{nR^3}\right)_0 + \overline{v\sigma}\int_{t_0}^{\infty}\frac{dt}{R_0^3\,(t/t_0)^{3/2}} \\ &= \left(\frac{1}{nR^3}\right)_0 + \frac{2\,\overline{v\sigma}\,t_0}{R_0^3}\,. \end{aligned} \tag{28.1.9}$$

The density n_B of baryon number (baryons minus antibaryons) goes as R^{-3}, so this can be rewritten as a formula for the present ratio of new particles to baryons:

$$(n/n_B)_\infty = \left[(n_B/n)_0 + 2\,\overline{v\sigma}\,n_{B0}\,t_0\right]^{-1}\,. \tag{28.1.10}$$

We expect that the ratio $(n/n_B)_0$ at the time that T drops to a value $\approx m$ is roughly of order unity, and since in any realistic theory the present ratio $(n/n_B)_\infty$ must be much less than unity, we can neglect the first term in the denominator on the right-hand side of Eq. (28.1.10), and write instead

$$(n/n_B)_\infty \simeq \frac{1}{\overline{v\sigma}\,n_{B0}\,t_0}\,. \tag{28.1.11}$$

The precise value of $\overline{v\sigma}$ depends on the particle spin and its interactions; keeping track only of factors of 2π, the particle mass m, and the electric charge, we can estimate it generally to be of order

$$\overline{v\sigma} \approx \frac{e^4\mathcal{N}}{2\pi m^2} \approx 10^{-3}\frac{\mathcal{N}}{m^2}\,, \tag{28.1.12}$$

where $\mathcal{N}$ is the number of charged particle spin states with mass less than m, into which this particle may annihilate. Also, the age of the universe at a temperature $T_0 \simeq m$ is $t_0 \approx m^4/m_{PL}$, where $m_{PL} \simeq 10^{18}$ GeV, and the density of baryon number is about 10^{-9} times the photon number density, which is of order T^3, so that $n_{B0} \approx 10^{-9}m^3$. Putting this together, we find the present ratio of the new charged particles to baryons:

$$(n/n_B)_\infty \approx 10^{12}\,\frac{m}{m_{PL}\mathcal{N}} \approx 10^{-6}\,\frac{m\,(\mathrm{GeV})}{\mathcal{N}}\,. \tag{28.1.13}$$

These new charged particles would experience the same condensations into galaxies, stars, and planets as ordinary baryons, so this would be the ratio observed today on earth. But experiments[4d] that apply mass spectroscopy to samples of water that have been strongly enriched in heavy water-like molecules by electrolysis have set limits of about $10^{-21}n_B$ on the number density of new charged particles with 6 GeV $< m <$ 330 GeV in terrestrial matter. Thus, even if $\mathcal{N}$ is as large as 1000, these measurements decisively rule out the existence of any new charged untrapped particles in this mass range in the numbers that would have been left over from the early universe.

On the other hand, neutral untrapped particles would be left in intergalactic space. Such particles might well provide the 'missing mass,' that seems to be necessary to account for the gravitational field that governs the motion of galaxies in clusters of galaxies. One of these possible neutral particles is the gravitino, whose cosmological abundance is discussed in Section 28.3. Ellis *et al.*[4c] have extended cosmological considerations to all the new particles required by supersymmetry.

28.2 Supersymmetry and Strong–Electroweak Unification

We shall have to defer a detailed assessment of supersymmetric models of particle physics until we are ready to consider how supersymmetry is broken. In this section we will consider the quantitative application of supersymmetry in one context in which the mechanism for the breakdown of supersymmetry is relatively unimportant, and in which supersymmetry has scored what so far is its greatest empirical success.

If the $SU(3)\times SU(2)\times U(1)$ gauge group of the strong and electroweak interactions is embedded in a simple group G that has the known quarks and leptons (plus perhaps some $SU(3)\times SU(2)\times U(1)$-neutral fermions) as a representation, then, as described in Section 21.5, at energies at or above the scale M_X at which G is spontaneously broken, the $SU(3)\times SU(2)\times U(1)$ coupling constants will be related by

$$g_s^2 = g^2 = \frac{5g'^2}{3} \quad \text{at energies } \geq M_X\,. \tag{28.2.1}$$

At energies far below M_X, these couplings are seriously affected by renormalization corrections. If measured at a scale $\mu < M_X$, the couplings will have values $g_s^2(\mu)$, $g^2(\mu)$, $g'^2(\mu)$, governed by one-loop renormalization group equations

$$\mu\frac{d}{d\mu}g'(\mu) = \beta_1\Big(g'(\mu)\Big)\,, \quad \mu\frac{d}{d\mu}g(\mu) = \beta_2\Big(g(\mu)\Big)\,, \quad \mu\frac{d}{d\mu}g_s(\mu) = \beta_3\Big(g_s(\mu)\Big)\,, \tag{28.2.2}$$

with initial conditions at M_X satisfying Eq. (28.2.1). In the original use[5] of these renormalization group equations, discussed in Section 21.5, the beta functions were calculated in one-loop order to be

$$\beta_1 = \frac{5n_g g'^3}{36\pi^2}\,, \tag{28.2.3}$$

$$\beta_2 = \frac{g^3}{4\pi^2}\left(-\frac{11}{6} + \frac{n_g}{3}\right)\,, \tag{28.2.4}$$

$$\beta_3 = \frac{g_s^3}{4\pi^2}\left(-\frac{11}{4}+\frac{n_g}{3}\right), \tag{28.2.5}$$

where n_g is the number of generations of quarks and leptons and the relatively small contributions of scalar fields are here neglected. Since M_X will turn out to be many orders of magnitude larger than the energies accessible with today's accelerators, it seems reasonable to suppose that supersymmetry is unbroken over most of the range below M_X, in which case all of the new fields discussed in the previous section need to be included in calculations of the beta functions in Eq. (28.2.1). These new fields introduce three major changes in the calculations of the beta functions:

1. For every gauge boson, there is a Majorana gaugino with the same $SU(3) \times SU(2) \times U(1)$ quantum numbers. Eq. (17.5.41) shows that the ratio of the contribution to the beta function for any gauge coupling of a Dirac fermion that furnishes a contribution of the gauge group with generators t_A to the contribution of the corresponding gauge boson is $-4C_2/11C_1$, where according to Eqs. (17.5.33) and (17.5.34) the ratio of C_1 and C_2 is given by:

$$\sum_{AB} C_{CAB}C_{DBA} = -(C_1/C_2)\mathrm{Tr}\,(t_C t_D)\,. \tag{28.2.6}$$

For the adjoint representation, $(t_C)_{AB} = iC_{ABC}$, so $C_1 = C_2$, and so a Dirac fermion in the adjoint representation makes a contribution that is $-4/11$ that of the gauge bosons. But the gauginos are Majorana fermions, so their contribution is $-2/11$ that of the gauge boson. Thus the term $11/6$ and $11/4$ in Eqs. (28.2.4) and (28.2.5) are reduced by a factor $9/11$ to $9/6$ and $9/4$, respectively.

2. For every left-handed quark, lepton, antiquark, or antilepton field, there is a complex scalar field with the same $SU(3) \times SU(2) \times U(1)$ quantum numbers. Following the same method as in Section 17.5, it is not hard to calculate that the contribution of a complex scalar field belonging to a representation of a gauge group with generators t_A to the beta function for the gauge coupling g_i is

$$[\beta_i(g_i)]_{\text{scalar}} = \frac{g_i^3 C_{2i}}{48\pi^2}\,, \tag{28.2.7}$$

where $\mathrm{Tr}\,(t_A t_B) = g_i^2 C_{2i}\delta_{AB}$. This is $1/4$ the contribution of a Dirac spinor field in the same representation, given by Eq. (18.7.2), and hence $1/2$ the contribution of each left-handed spinor field (including the complex conjugates of the right-handed components of the Dirac fields). Thus the coefficients of n_g in Eqs. (28.2.3)–(28.2.5) should be increased by a factor $3/2$.

3. The decrease by a factor 9/11 of the negative gauge boson terms in the beta function and the increase by a factor 3/2 of the positive squark and slepton terms both lead to a general decrease in the rate at which the three gauge coupling constants diverge below M_X from the ratios (28.2.1). This will increase our estimate of M_X, but, as we shall see, in itself it would have no affect on the prediction of the electroweak mixing parameter $\sin^2\theta$. But these changes do enhance the relative contribution of the Higgs scalars, which was neglected in Eqs. (28.2.3)–(28.2.5), and which is now also accompanied with the larger contribution of the accompanying higgsinos. With n_s of the superfields (H_1^0, H_1^-) or (H_2^+, H_2^0) discussed in the previous section, the constant C_{2i} in Eq. (28.2.7) is $[(1/2)^2 + (-1/2)^2]n_s = n_s/2$ for $SU(2)$, and is $2n_s(\pm 1/2)^2 = n_s/2$ for $U(1)$. According to Eq. (28.2.7), the scalar components of these superfields make a contribution to β_1 equal to $n_s g'^3/96\pi^2$ and a contribution to β_2 also equal to $n_s g^3/96\pi^2$. As we have seen, the Majorana higgsinos contribute twice as much to the beta functions as complex scalars with the same quantum numbers, so the superfields (H_1^0, H_1^-) or (H_2^+, H_2^0) make a total contribution to β_1 and β_2 that is 3/2 the contribution of the Higgs scalars, and hence equal to $n_s g'^3/32\pi^2$ and $n_s g^3/32\pi^2$, respectively.

Making all these changes in the beta functions, we now have

$$\beta_1 = \frac{g'^3}{4\pi^2}\left(\frac{5n_g}{6} + \frac{n_s}{8}\right), \tag{28.2.8}$$

$$\beta_2 = \frac{g^3}{4\pi^2}\left(-\frac{9}{6} + \frac{n_g}{2} + \frac{n_s}{8}\right), \tag{28.2.9}$$

$$\beta_3 = \frac{g_s^3}{4\pi^2}\left(-\frac{9}{4} + \frac{n_g}{2}\right). \tag{28.2.10}$$

The solutions of the renormalization group equations (28.2.2) are then

$$\frac{1}{g'^2(\mu)} = \frac{1}{g'^2(M_X)} + \frac{1}{2\pi^2}\left(\frac{5n_g}{6} + \frac{n_s}{8}\right)\ln\left(\frac{M_X}{\mu}\right), \tag{28.2.11}$$

$$\frac{1}{g^2(\mu)} = \frac{1}{g^2(M_X)} + \frac{1}{2\pi^2}\left(-\frac{3}{2} + \frac{n_g}{2} + \frac{n_s}{8}\right)\ln\left(\frac{M_X}{\mu}\right), \tag{28.2.12}$$

$$\frac{1}{g_s^2(\mu)} = \frac{1}{g_s^2(M_X)} + \frac{1}{2\pi^2}\left(-\frac{9}{4} + \frac{n_g}{2}\right)\ln\left(\frac{M_X}{\mu}\right). \tag{28.2.13}$$

It is convenient to take $\mu = m_Z$, so that $SU(2) \times U(1)$ can be regarded as unbroken over almost all of the range of energies in which we use the formulas (28.2.11)–(28.2.13). Using Eq. (28.2.1), the difference between Eqs. (28.2.12) and (28.2.13) gives

$$\frac{1}{g^2(m_Z)} - \frac{1}{g_s^2(m_Z)} = \frac{1}{2\pi^2}\left(\frac{3}{4} + \frac{n_s}{8}\right)\ln\left(\frac{M_X}{m_Z}\right), \tag{28.2.14}$$

Table 28.1. Values of the electroweak mixing parameter $\sin^2\theta$ and unification mass M_X given by Eqs. (28.2.17) and (28.2.18), as functions of the number n_s of left-chiral superfield doublets (H_1^0, H_1^-) or (H_2^+, H_2^0).

n_s	$\sin^2\theta$	M_X (GeV)
0	0.203	8.7×10^{17}
2	0.231	2.2×10^{16}
4	0.253	1.1×10^{15}

while the difference between Eq. (28.2.12) and 3/5 of Eq. (28.2.11) gives

$$\frac{1}{g^2(m_Z)}-\frac{3}{5g'^2(m_Z)}=\frac{1}{2\pi^2}\left(-\frac{3}{2}+\frac{n_s}{20}\right)\ln\left(\frac{M_X}{m_Z}\right). \tag{28.2.15}$$

Eq. (21.3.19) allows us to express the electroweak couplings in terms of the electroweak mixing angle θ and the positron charge e:

$$g(m_Z)=-e(m_Z)/\sin\theta\,,\qquad g'(m_Z)=-e(m_Z)/\cos\theta\,. \tag{28.2.16}$$

We can then solve for the unknowns $\ln(M_X/m_Z)$ and $\sin^2\theta$ in terms of input parameters $e(m_Z)$ and $g_s(m_Z)$:

$$\sin^2\theta=\frac{18+3n_s+(e^2(m_Z)/g_s^2(m_Z))(60-2n_s)}{108+6n_s}\,, \tag{28.2.17}$$

$$\ln\left(\frac{M_X}{m_Z}\right)=\left(\frac{8\pi^2}{e^2(m_Z)}\right)\left(\frac{1-(8e^2(m_Z)/3g_s^2(m_Z))}{18+n_s}\right). \tag{28.2.18}$$

For $n_s=0$ Eq. (28.2.17) gives the same result (21.5.15) for $\sin^2\theta$ as was originally calculated (ignoring the small contribution of Higgs scalars) in non-supersymmetric theories, but the value (28.2.18) of $\ln(M_X/m_Z)$ is larger than the original result (21.5.16) by a factor 11/9, which as we have seen arises from gaugino contributions to the beta functions.

Using the same input parameters $e^2(m_Z)/4\pi=(128)^{-1}$, $g_s^2(m_Z)/4\pi=0.118$, $m_Z=91.19$ GeV as in Section 21.5 now gives the numerical results shown in Table 28.1. As discussed in the previous section, the necessity of cancelling anomalies in the electroweak currents requires equal numbers of (H_1^0, H_1^-) and (H_2^+, H_2^0) doublets, so we consider only even values of the number n_s of these superfields.

Remarkably, the value $n_s=2$ for the simplest plausible theory yields a value[6] $\sin^2\theta=0.231$ which is in perfect agreement with the experimentally observed value, $\sin^2\theta = 0.23$. The value of M_X is 20 times greater[7] than calculated in this way in non-supersymmetric theories, leading to a decrease by a factor 20^{-4} in the rate for proton decay processes like $p \to \pi^0 + e^+$, thus removing a conflict with the experimental non-observation of such processes. (Proton decay is discussed in more detail in Section 28.7.) This increase in the value of M_X brings it closer to the energy scale $\approx 10^{18}$ GeV at which gravitation has the same strength as other interactions. It may be that this remaining gap may be filled by a change in gravitational interactions at very high energies.[7a]

A value $n_s = 4$ would give a value for $\sin^2\theta$ in serious disagreement with experiment, and a value of M_X low enough to revive the conflict with expectations for proton decay. This makes a strong case for having just one of each superfield (H_1^0, H_1^-) and (H_2^+, H_2^0).

Unlike the calculated values of $\sin^2\theta$ and M_X, the calculated value of the common gauge coupling (28.2.1) at M_X does depend on the number of generations as well as the number of scalar doublets. With $n_g = 3$ and $n_s = 2$ and our previous input parameters, Eq. (28.2.13) gives

$$\frac{g^2(M_X)}{4\pi} = \frac{g_s^2(M_X)}{4\pi} = \frac{1}{17.5} . \tag{28.2.19}$$

28.3 Where is Supersymmetry Broken?

Supersymmetry if valid at all is certainly not apparent in the menu of known particles, so any consideration of the implications of supersymmetry at ordinary energies requires us to make some assumption about the mechanism of supersymmetry breaking. It would be simplest to suppose that supersymmetry is broken like $SU(2) \times U(1)$, by effects occurring in the tree approximation of the supersymmetric standard model. This possibility may be definitely ruled out.

One argument against a tree-approximation breakdown of supersymmetry is based on the mass sum rule (27.5.11), which holds separately for each value of the unbroken conserved quantities color and electric charge. In the color-triplet sector with electric charge $-e/3$ the only known fermions are the d, s, and b quarks, for which

$$m_d^2 + m_s^2 + m_b^2 \simeq (5\text{ GeV})^2 . \tag{28.3.1}$$

According to the sum rule, if there are no other fermions with this color and charge, then the sum of all squared masses for bosons (counting each spin state separately) with the same color and charge must equal about $2(5\text{ GeV})^2$. In particular, each of the squarks with this color and charge must have a mass no greater than 7 GeV. The existence of such light squarks is definitely ruled out experimentally; they would have shown up, for instance, as a contribution to the rate for electron–positron annihilation into hadrons at energies where this process has been studied very thoroughly.

This argument could be invalidated if there were a heavy fourth generation of quarks. There is another argument, due to Dimopoulos and Georgi,[3] which would apply however many heavy quarks there are and which yields an even stronger upper bound on the mass of the lightest squark. The unbroken conservation of charge and color tells us that the only non-zero D_{A0}-terms in the supersymmetric standard model are for the generators y of $U(1)$ and t_3 of $SU(2)$, which we shall call D_1 and D_2, respectively. The values of these generators are $y = -g'/6$ and $t_3 = +g/2$ for the left-handed quarks of charge $2e/3$; $y = -g'/6$ and $t_3 = -g/2$ for the left-handed quarks of charge $-e/3$; $y = 2g'/3$ and $t_3 = 0$ for the right-handed quarks of charge $2e/3$; and $y = -g'/3$ and $t_3 = 0$ for the right-handed quarks of charge $-e/3$. Also, the squark fields are color-triplets, and therefore cannot have vacuum expectation values. According to Eq. (27.5.4), the mass-squared matrix of the charge $2e/3$ color-triplet (not antitriplet) squarks is

$$M_{0U}^2 = \begin{bmatrix} \mathscr{M}_U^*\mathscr{M}_U - g'D_1/6 + gD_2/2 & \mathscr{F}_U^* \\ \mathscr{F}_U & \mathscr{M}_U\mathscr{M}_U^* + 2g'D_1/3 \end{bmatrix}, \qquad (28.3.2)$$

while the mass-squared matrix of the charge $-e/3$ color-triplet squarks is

$$M_{0D}^2 = \begin{bmatrix} \mathscr{M}_D^*\mathscr{M}_D - g'D_1/6 - gD_2/2 & \mathscr{F}_D^* \\ \mathscr{F}_D & \mathscr{M}_D\mathscr{M}_D^* - g'D_1/3 \end{bmatrix}. \qquad (28.3.3)$$

Also, Eq. (27.5.6) gives the mass-squared matrices of the quarks of charge $2e/3$ and $-e/3$ here as just $\mathscr{M}_U^*\mathscr{M}_U$ and $\mathscr{M}_D^*\mathscr{M}_D$, respectively, with no mixing with the gauginos.

Now let v_u and v_d be the normalized eigenvectors of the quark mass-squared matrices $\mathscr{M}_U^*\mathscr{M}_U$ and $\mathscr{M}_D^*\mathscr{M}_D$ corresponding to the quarks u and d of lowest mass, and consider the expectation values of the corresponding squark mass-squared matrices

$$\begin{bmatrix} 0 \\ v_u^* \end{bmatrix}^\dagger M_{0U}^2 \begin{bmatrix} 0 \\ v_u^* \end{bmatrix} = m_u^2 + \frac{2g'D_1}{3}, \qquad (28.3.4)$$

$$\begin{bmatrix} 0 \\ v_d^* \end{bmatrix}^\dagger M_{0D}^2 \begin{bmatrix} 0 \\ v_d^* \end{bmatrix} = m_d^2 - \frac{g'D_1}{3}. \qquad (28.3.5)$$

These expectation values are weighted averages of the squared masses of the squarks of charge $2e/3$ and $-e/3$, respectively, so at least one squark of charge $2e/3$ must have a squared mass less than $m_u^2 + 2g'D_1/3$, and at least one squark of charge $-e/3$ must have a squared mass less than $m_d^2 - g'D_1/3$. Thus, depending on the sign of D_1, *there must be either a squark of charge $2e/3$ lighter than the u quark, or a squark of charge $-e/3$ lighter than the d quark.*

Needless to say, the existence of a charged color-triplet scalar this light would radically change strong-interaction phenomenology. Like the u and d quarks, this colored scalar would appear as an ingredient of hadrons with a 'constituent' mass of a few hundred MeV, which is certainly not seen. Since this scalar is electrically charged, it would also be created in pairs in e^+–e^- annihilation at energies above a few hundred MeV, making a contribution to the annihilation cross-section that would destroy the excellent agreement between theory and experiment for this cross-section. Even worse, since the u and d quarks are so light, and D_1 is expected to be of the order of the supersymmetry-breaking scale, Eqs. (28.3.4) and (28.3.5) indicate that one of the squarks would have a negative squared mass, meaning that this squark field would have to develop a non-vanishing expectation value, breaking both color and charge conservation. We are forced to reject the simple picture of supersymmetry broken spontaneously in the tree approximation in a supersymmetric version of the standard model.

One way out of this conclusion would be to add another $U(1)$ gauge superfield to the theory. If all the quark superfields carry the same value $\tilde{g}$ of this new $U(1)$ generator, then the corresponding D-term $\tilde{D}$ would make an additive contribution $\tilde{g}\tilde{D}$ to the right-hand sides of both Eqs. (28.3.4) and (28.3.5). If this term were sufficiently large, then it could give a large positive value to all of the squark squared masses, avoiding all the problems mentioned above. But there is no sign of such a new neutral gauge boson at accessible energies, and in any case we would still have an upper bound of 7 GeV on the masses of all the squarks of charge $-e/3$.

It is not necessarily a bad thing that we have to look for the breaking of supersymmetry elsewhere than in the tree approximation of the supersymmetric standard model. If supersymmetry were broken in this approximation, then the characteristic mass that sets the scale of supersymmetry breaking would be some mass parameter in the Lagrangian, which would in turn set the scale of all other masses in the standard model. We would then still be confronted with the hierarchy problem: why is this mass scale so much less than 10^{16}–10^{18} GeV?

There is one known way to explain such large mass ratios. If supersymmetry is not spontaneously broken in the tree approximation in whatever field theory unifies all the interactions at some high mass scale M_X, then as shown in Section 27.6, it will not be broken in any order of perturbation theory. But it can be broken by non-perturbative effects. In particular, if there is some gauge field with an asymptotically free gauge coupling

$\mathcal{G}(\mu)$ at renormalization scale μ, and if $\mathcal{G}^2(\mu)/8\pi^2$ is substantially less than unity for $\mu \approx M_X$, then as discussed in Section 18.3, this gauge interaction will become strong at an energy of order $M_S = M_X \exp(-8\pi^2 b/\mathcal{G}^2(M_X))$, where b is a number of order unity. It is not necessary for $\mathcal{G}^2(M_X)/8\pi^2$ to be very small in order to have M_S many orders of magnitude less than M_X. We will see in Section 29.4 that supersymmetry can indeed be broken in just this way, by a gauge coupling that has become strong at some energy $M_S \ll M_X$. Indeed, this is just what happens to chiral symmetry in quantum chromodynamics; there is no mystery why the proton mass (or at least its main part, due to the dynamical breakdown of chiral symmetry, and not to the tiny masses of the u and d quarks) is so much smaller than the unification scale M_X. Alternatively, the forces that are strong at energy M_S may produce a potential for scalar fields, whose vacuum expectation value then breaks supersymmetry.

There are no signs of any new strong interaction of the known quarks and leptons, so we have to assume that the observed particles of the standard model are neutral with respect to the strong force that breaks supersymmetry. Supersymmetry breaking therefore occurs in a 'hidden sector' of particles that do feel this new strong force. The remaining question, then, is what is the mechanism by which supersymmetry breaking in this hidden sector is communicated to the known particles of the standard model? As we shall see, most of our expectations for the phenomenological implications of supersymmetry depend on the answer to this question, rather than on the details of the breakdown of supersymmetry itself.

Of course, the mechanism for communicating supersymmetry breaking to observed particles must be some sort of interaction that is felt by these particles. There are two leading candidates. One mechanism is provided by the $SU(3) \times SU(2) \times U(1)$ gauge interactions themselves, to be discussed in Section 28.6. The other is gravitation, or rather the auxiliary fields that are superpartners of the gravitational field, to be discussed in Sections 31.4 and 31.7.

Without going into details here, we can make a crude estimate of the supersymmetry-breaking scale M_S for these two possibilities. For gauge-mediated supersymmetry breaking, we expect that the mass splitting between the observed quarks, leptons, and gauge bosons and their superpartners would be of order $g_s^2/16\pi^2$ or $g'^2/16\pi^2$ or $g^2/16\pi^2$ (where g_s, g, and g' are the $SU(3)$, $SU(2)$, and $U(1)$ gauge couplings), depending on which quantum numbers are carried by the supermultiplet in question. (This guess is verified in Section 28.6.) Hence if the squarks, sleptons, and gauginos have masses in the range of 100 GeV to 10 TeV, as argued at the end of Section 28.1, then the supersymmetry-breaking scale M_S would be higher by two or three orders of magnitude — say, of order 100 TeV. On the other hand, if it is gravitation that serves as the mediator of supersymmetry breaking, then on dimensional grounds we would expect the mass splittings Δm between the observed particles and their superpartners to be of order $\sqrt{G}M_S^2$, or perhaps of order GM_S^3. (Results of both sorts will be encountered in models described in Section 31.7.) If the squarks, sleptons, and gauginos have masses in the range of 100 GeV to 10 TeV, then M_S would be of order 10^{11} GeV for $\Delta m \approx \sqrt{G}M_S^2$, or 10^{13} GeV for $\Delta m \approx GM_S^3$.

The large difference in estimates of the supersymmetry-breaking scale M_S for gauge- and gravitation-mediated supersymmetry breaking makes an important difference in particle phenomenology and cosmology. As already mentioned several times, supersymmetry dictates that the graviton must have a partner of spin 3/2, the gravitino. When supersymmetry is spontaneously broken at a scale M_S, the gravitino acquires a mass m_g of order $\sqrt{G}M_S^2$. (A precise formula will be given in Section 31.3.) For gauge-mediated supersymmetry breaking, this is very small; if $M_S \approx 100$ TeV then $m_g \approx 1$ eV, so the gravitino would be by far the lightest of the new particles required by supersymmetry — that is, the lightest particle with negative R parity (28.1.6). On the other hand, for gravitationally mediated supersymmetry breaking the gravitino mass is just of the same order of magnitude $\sqrt{G}M_S^2$ as the mass splitting between known particles and their superpartners, so the gravitino would have roughly the same mass as the squarks, sleptons, and gauginos. The gravitino then might or might not be the lightest particle with negative R parity, but its interactions with known particles and their superpartners in this case are of gravitational strength, so that gravitinos would play no direct role in experiments on elementary particles.

* * *

There are limits on the number of gravitinos that could survive from the big bang, which set useful constraints on the scale M_S of supersymmetry breaking. At some point in the distant past the temperature T was presumably high enough so that even purely gravitational interactions would have kept gravitinos in thermal equilibrium with other particles, in which case the number density of gravitinos would have been of the order of T^3, roughly the same as the number density of photons. (We are using units in which the Boltzmann constant k_B as well as $\hbar$ and c are equal to unity.) If gravitinos do not annihilate or decay, then the expansion of the universe will lower their number density in the same way as the number density of photons, so even after the gravitinos go out of equilibrium they would be present in numbers comparable to those of photons. More precisely, since the photons but not the gravitinos are heated by the annihilation of other particles, the number density n_{g0} of gravitinos at present would be one or two orders of magnitude less than the number density $n_{\gamma 0}$ of photons in the cosmic microwave radiation background. In order for the mass density $m_g n_{g0}$ of the gravitinos not to exceed the upper bound on the cosmic mass density set by the observed value of the Hubble constant, m_g would have to be less[8] than about 1 keV. As we have seen, this limit is well satisfied in theories of gauge-mediated supersymmetry breaking, where the gravitino is too light for cosmic gravitinos to contribute appreciably to the mass density of the universe. Since some of the fields that break supersymmetry in these theories must interact at least indirectly with the known quark, lepton, and gauge fields in order for the known particles to show the effects of supersymmetry breaking, the interactions of the gravitino with the known particles and their superpartners are suppressed only by powers of gauge and Yukawa coupling constants, so all the superpartners of the quarks, leptons, and gauge bosons would decay quickly into these known particles and gravitinos. Thus these particles also do not provide candidates in these models for the 'missing mass' sought by cosmologists. (It is possible that conservation laws could keep some particles of the supersymmetry-breaking sector stable, in which case they could conceivably serve as the missing mass.)

On the other hand, for supersymmetry breaking that is gravitationally mediated, gravitinos are heavy enough to be unstable (though gravitino annihilation is still negligible), so the above limit need not apply.[9] We will see in Section 31.3 that the coupling of the gravitino to other fields is proportional to $\sqrt{G}$, so on dimensional grounds the decay rate Γ_g of a gravitino at rest is roughly of the order of Gm_g^3. This is to be compared with the rate of expansion of the universe, which at temperature T is of order $\sqrt{GT^4}$. (We are here ignoring factors of order 10–100, including those involving non-gravitational coupling constants and the number of particle species.) When the cosmic temperature drops to the value $T \approx m_g$ at which gravitinos become non-relativistic, the ratio of their decay rate to the expansion rate is of order $\sqrt{G}m_g = m_g/m_{\text{Planck}} \ll 1$, so gravitino decay becomes significant only after this time, when the gravitinos are highly non-relativistic. As we have seen, their number density will be of order T^3, so their energy density will then be of order $m_g T^3$, which is greater than the energy density of order T^4 of the photons and other particles in thermal equilibrium at temperature T, and therefore makes the dominant contribution to the cosmic gravitational field that governs the rate of expansion of the universe. The expansion rate under these conditions is therefore of order $\sqrt{Gm_g T^3}$, and gravitino decay becomes significant when this equals the gravitino decay rate of order Gm_g^3, and therefore at a temperature

$$T_g \approx G^{1/3} m_g^{5/3} .$$

As we have seen, if these gravitinos did not decay before the present then their mass had better be less than 1 keV, but they can lead to cosmological difficulties even if they did decay before now. After they decay, their energy must go into the energy of photons and other relativistic particles, so the temperature T'_g after decay is related to the temperature T_g calculated above by the energy conservation condition $m_g T_g^3 \approx T_g'^4$, and hence

$$T'_g \approx G^{1/4} m_g^{3/2} .$$

In particular, since $T_g \ll m_g$, we have $T'_g \gg T_g$. If T_g were less than the temperature $T_n \simeq 0.1$ MeV at which cosmological nucleosynthesis can occur, then gravitinos would still be abundant before nucleosynthesis, giving a higher energy density and hence a faster expansion, so that there would be less time for free neutrons to decay before being incorporated into complex nuclei, and hence more helium would be produced when nucleosynthesis occurs. Also, the ratio of the photon and baryon densities would have been subsequently increased by gravitino decay, so this ratio at the time of nucleosynthesis would have been considerably less than is usually estimated from the present cosmic microwave background temperature, and so nuclear reactions would have incorporated neutrons more completely into helium, and less deuterium would be left today. The present agreement between theory and observation for the cosmic helium and deuterium abundances would thus be destroyed. This problem is avoided if $T_g > 0.1$ MeV, but it can also be avoided under the much weaker condition that $T'_g > 0.4$ MeV, because then after the gravitinos decay the temperature would have been high enough to break up the excess helium and give cosmological nucleosynthesis a fresh start as the universe recools. This condition requires that $m_g > 10$ TeV, which is just barely consistent with the upper bound derived in Section 28.1 on the masses of the superpartners of the known quarks, leptons, and gauge bosons, which for gravitationally mediated supersymmetry breaking are of order m_g. This limit on m_g corresponds to a supersymmetry-breaking scale $M_S > 10^{11}$ GeV for $m_g \approx \sqrt{G}M_S^2$ or $M_S > 10^{13}$ GeV for $m_g \approx GM_S^3$.

28.4 The Minimal Supersymmetric Standard Model

In the previous section we identified two different ways that the breakdown of supersymmetry at a high energy scale M_S could be communicated to the known quarks and leptons: through gauge or gravitational superfields. The supersymmetry-breaking terms in the resulting low-energy effective Lagrangian will then be suppressed by powers of gauge couplings or the Newton constant. Most of these terms will therefore be rather small, with the exception that, along with factors of gauge couplings or the Newton constant, on dimensional grounds the mass terms and other superrenormalizable terms in the effective Lagrangian will be proportional to one or more factors of the supersymmetry-breaking scale M_S, which is quite large compared with known particle masses. We can conclude then that, to a fair approximation for gauge-mediated supersymmetry breaking and a very good approximation for gravitationally mediated supersymmetry breaking, the main effect of supersymmetry breaking will be in the superrenormalizable terms of the effective Lagrangian of the supersymmetric standard model. This version of the standard model,[10] which is supersymmetric except for superrenormalizable terms, is usually known as the *minimal supersymmetric standard model.*

With R parity or $B-L$ conserved, the most general superrenormalizable Lagrangian density allowed by $SU(3)\times SU(2)\times U(1)$ gauge symmetry takes the form

$$\begin{aligned}
\mathscr{L}_{SR} = &-\sum_{ij} M_{ij}^{2\,Q}\left(\mathscr{Q}_i^\dagger \mathscr{Q}_j\right) - \sum_{ij} M_{ij}^{2\,\bar{U}}\left(\bar{\mathscr{U}}_i^\dagger \bar{\mathscr{U}}_j\right) - \sum_{ij} M_{ij}^{2\,\bar{D}}\left(\bar{\mathscr{D}}_i^\dagger \bar{\mathscr{D}}_j\right) \\
&-\sum_{ij} M_{ij}^{2\,L}\left(\mathscr{L}_i^\dagger \mathscr{L}_j\right) - \sum_{ij} M_{ij}^{2\,\bar{E}}\left(\bar{\mathscr{E}}_i^\dagger \bar{\mathscr{E}}_j\right) \\
&-\left(\overline{\lambda_3}\, m_{\text{gluino}}\, \lambda_3\right) - \left(\overline{\lambda_2}\, m_{\text{wino}}\, \lambda_2\right) - \left(\overline{\lambda_1}\, m_{\text{bino}}\, \lambda_1\right) \\
&-\sum_{ij} A_{ij}^D h_{ij}^D \left(\mathscr{Q}_i^{\rm T} e \mathscr{H}_1\right)\bar{\mathscr{D}}_j - \sum_{ij} A_{ij}^E h_{ij}^E \left(\mathscr{L}_i^{\rm T} e \mathscr{H}_1\right)\bar{\mathscr{E}}_j \\
&-\sum_{ij} A_{ij}^U h_{ij}^U \left(\mathscr{Q}_i^{\rm T} e \mathscr{H}_2\right)\bar{\mathscr{U}}_j - \sum_{ij} C_{ij}^D h_{ij}^D \left(\mathscr{Q}_i^{\rm T} \mathscr{H}_2^*\right)\bar{\mathscr{D}}_j \\
&-\sum_{ij} C_{ij}^E h_{ij}^E \left(\mathscr{L}_i^{\rm T} \mathscr{H}_2^*\right)\bar{\mathscr{E}}_j - \sum_{ij} C_{ij}^U h_{ij}^U \left(\mathscr{Q}_i^{\rm T} \mathscr{H}_1^*\right)\bar{\mathscr{U}}_j \\
&-B\mu\left(\mathscr{H}_2^{\rm T} e \mathscr{H}_1\right) + \text{H.c.}
\end{aligned} \tag{28.4.1}$$

Script letters are used here to denote the scalar components of left-chiral superfields. Sums over $SU(2)$ and color indices are understood, as necessary for invariance under $SU(3)\times SU(2)\times U(1)$, with e the usual antisymmetric 2×2 matrix $i\sigma_2$. All coefficients may be complex, and the gaugino masses may involve terms proportional to the γ_5 as well as the unit matrix.

We follow the custom here of writing the coefficients of the terms involving scalar fields but not their adjoints as equal to the coefficients of the corresponding supersymmetric $\mathscr{F}$-terms in Eq. (28.1.7) times factors A_{ij}^D, A_{ij}^E, A_{ij}^U, and B. This is motivated by the consideration that the smallness of the Yukawa couplings of the light quarks in Eq. (28.1.7) reflects a number of approximate chiral symmetries that if extended to the whole supermultiplet would also make the corresponding trilinear terms in Eq. (28.4.1) small, while the appearance of the μ-term in Eq. (28.1.7) violates a possible Peccei–Quinn[4] symmetry that if approximately valid would make both μ and $B\mu$ small. Similar considerations suggest the form in which we write the coefficients of the terms that involve both scalars and their complex conjugates. Also, in Section 31.4 we will describe contributions to Ah and $B\mu$ that really are proportional to h and μ, respectively. However, here we are leaving it an open question whether the Ah, Ch, and $B\mu$ coefficients in Eq. (28.4.1) are necessarily small when the corresponding h and μ coefficients in Eq. (28.1.7) are small.

The Ch-terms in Eq. (28.4.1) have generally been omitted in discussions of the minimal supersymmetric standard model. This is partly because, as discussed in Section 27.7, terms like these that involve the ϕ-components of left-chiral scalar superfields and also their complex conjugates can potentially produce quadratic divergences and thereby raise fine-tuning problems. But we saw in Section 27.7 that the quadratic divergences occur only in 'tadpole' graphs in which a scalar field line disappears into the vacuum, and in the minimum supersymmetric standard model there are no scalars that are neutral under all gauge symmetries, and hence no scalar tadpoles. The Ch-terms are absent in the theories of gravitationally mediated supersymmetry breaking discussed in Section 31.6, and small in the theories of gauge-mediated supersymmetry breaking described in Section 28.6, but there is no reason to suppose that this will always be the case.

Even though they are not supersymmetric, superrenormalizable interactions like those in Eq. (28.4.1) are shown in Section 27.7 to not produce supersymmetry-violating ultraviolet-divergent corrections to the coefficients of the supersymmetric $d = 4$ interactions. The condition of supersymmetry that is imposed on the dimensionless couplings of the minimal supersymmetric standard model therefore does not get in the way of the cancellation of ultraviolet divergences by renormalization of coupling constants. It was this property, rather than any theory of supersymmetry breaking in a high energy hidden sector, that motivated the introduction of the minimal supersymmetric standard model in Reference 10.

The best reason today for studying the implications of the supersymmetric standard model is that, as already mentioned, theories in which supersymmetry is spontaneously broken at a high energy scale are naturally described by the minimal supersymmetric standard model at much lower energies. We can explore the phenomenological implications of the supersymmetric standard model, and be reasonably confident that the results will be relevant whatever detailed model of supersymmetry breaking and its mediation turns out to be correct.

Even without the Ch-terms, if all of the other coefficients in the Lagrangian are constrained only by gauge symmetries and R parity conservation, the minimal supersymmetric standard model will contain over 100 free parameters.[11] Here 'minimal' means nothing more than that the theory contains only a minimal menu of superfields. Sometimes the term 'minimal supersymmetric standard model' is reserved for models that also satisfy restrictions on the coefficients of the superrenormalizable terms, motivated either by some underlying theory or by empirical constraints. For instance, the minimal supersymmetric standard model is sometimes optimistically assumed to satisfy the universality conditions

$$\begin{aligned} &M^{2\,Q}_{ij} = M^{2\,\bar{D}}_{ij} = M^{2\,\bar{U}}_{ij} = M^{2\,L}_{ij} = M^{2\,\bar{E}}_{ij} = M^2\delta_{ij}\,, \\ &m_{\text{gluino}} = m_{\text{wino}} = m_{\text{bino}}\,, \\ &A^D_{ij} = A^E_{ij} = A^U_{ij} = A\,, \qquad C^D_{ij} = C^E_{ij} = C^U_{ij} = 0\,. \end{aligned} \tag{28.4.2}$$

Often these conditions are imposed at the scale $M_X \approx 10^{16}$ GeV of coupling constant unification, with corrections produced only by the renormalization group flow to lower energies. We will not be making such assumptions here.

In analyzing the phenomenological implications of the minimal supersymmetric standard model, we must deal not only with the search for new particles, but also with two classes of severe empirical constraint on processes involving known particles: the experimental upper bounds on various flavor non-conserving processes, and on various modes of CP non-conservation.

Flavor Changing Processes

We saw in Section 21.3 that there is an automatic suppression of flavor-changing processes like K^0–$\bar{K}^0$ oscillations and $K^0 \to \mu^+\mu^-$ in the non-supersymmetric standard model. This is due to the feature of this theory, that it is only the mass splittings of the quarks that prevent them from being defined so that each flavor is separately conserved, so the amplitude of these flavor-changing process must be proportional to several factors of small quark masses. Also, in this theory lepton flavor is automatically conserved, so that processes like $\mu \to e\gamma$ are absolutely forbidden. These satisfactory results are put at risk by the presence of squarks and sleptons in supersymmetric extensions of the standard model, because there is in general no reason to expect that the squark and slepton mass matrices will be diagonal in the same basis as the quark and lepton mass matrices. This does not introduce flavor changing in the interaction of these particles with gauge bosons, which are flavor-independent, but it can produce flavor-changing transitions in which squarks or sleptons turn into quarks or leptons with the emission or absorption of gauginos. Of course, there

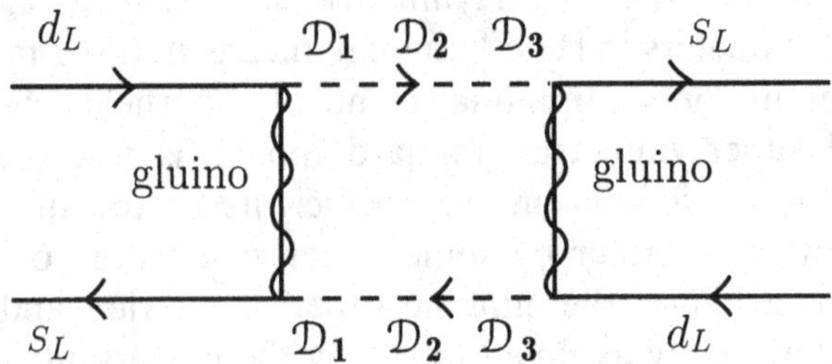

Figure 28.1. A one-loop diagram that can contribute to the $\Delta S = 2$ effective interaction $(\bar{s}_L \gamma^\mu d_L)(\bar{s}_L \gamma_\mu d_L)$ in the supersymmetric standard model. Here solid lines are quarks; dashed lines are squarks; and combined solid and wavy lines are gluinos.

is no problem if the squarks and sleptons are degenerate, in which case their mass matrices are diagonal in any basis.

The most stringent limits on squark mass splittings and/or mixing angles are set by measurements of K^0–$\bar{K}^0$ transitions.[12] These transitions are produced by operators in the effective low-energy Lagrangian density like $(\overline{s_L}\gamma^\mu d_L)(\overline{s_L}\gamma_\mu d_L)$, which can be produced by diagrams like Figure 28.1. The superpartners of the quarks d_L and s_L are in general linear combinations $\sum_i V_{di}\mathscr{D}_i$ and $\sum_i V_{si}\mathscr{D}_i$ of squarks $\mathscr{D}_i$ of definite mass, where V_{ji} is a 3×3 unitary matrix, so the two squark propagators in this diagram contribute a factor

$$\sum_i \frac{V_{di}V_{si}^*}{k^2 + M_i^2 - i\epsilon} \times \sum_j \frac{V_{dj}V_{sj}^*}{k^2 + M_j^2 - i\epsilon} ,$$

where k is the four-momentum circulating in the loop. Because V_{ji} is unitary, this vanishes if the three squark masses M_i are all equal. If the squark square masses differ from some common value M^2_{squark} by relatively small amounts ΔM_i^2, then this becomes

$$\left(\frac{1}{k^2 + M^2_{\text{squark}} - i\epsilon}\right)^4 \left(\sum_i V_{di}V_{si}^*\Delta M_i^2\right)^2 .$$

The amplitude for $d_L\overline{s_L} \to s_L\overline{d_L}$ has dimensionality mass^{-2}, so after multiplying by the gluino propagators and four factors of the strong coupling g_s, and integrating over k, we must get an amplitude proportional to

$$\frac{g_s^4}{\tilde{M}^6}\left(\sum_i V_{di}V_{si}^*\Delta M_i^2\right)^2 , \tag{28.4.3}$$

where $\tilde{M}$ is the larger of M_{squark} and m_{gluino}. This may be compared with the result for this amplitude in the non-supersymmetric standard model, which is produced by W exchange, as shown in Figure 28.2. Ignoring the third generation of quarks, which has only small transition amplitudes to the first two generations, the amplitudes for $d \to u$, $d \to c$, $s \to u$ and $s \to c$ by W^- emission are, respectively, $\cos\theta_c$, $-\sin\theta_c$, $\sin\theta_c$, and $\cos\theta_c$, where θ_c is the Cabibbo angle defined in Section 21.3. Hence in place of the squark propagators here we have the quark propagators

$$\sin\theta_c \cos\theta_c \left(\frac{i\not{k} + m_u}{k^2 + m_u^2 - i\epsilon} - \frac{i\not{k} + m_c}{k^2 + m_c^2 - i\epsilon} \right),$$

and in place of the strong coupling g_s here we have the $SU(2)$ coupling g. Thus in the non-supersymmetric standard model the amplitude for $d_L\overline{s_L} \to s_L\overline{d_L}$ is proportional to

$$\frac{g^4 \sin^2\theta_c \cos^2\theta_c}{m_W^4} \left(m_c - m_u \right)^2, \tag{28.4.4}$$

with a proportionality coefficient of the same order as that in Eq. (28.4.3). With a plausible guess on how to calculate the K^0–$\bar{K}^0$ transition amplitude from the amplitude for $d_L\overline{s_L} \to s_L\overline{d_L}$, the amplitude corresponding to Figure 28.2 is known to give a result in good agreement with experiment. (Indeed, Gaillard and Lee[13] used this calculation to predict that $m_c \approx 1.5$ GeV, before the c quark was discovered.) It therefore seems reasonable to require that the squark exchange result (28.4.3) should be less than the quark exchange result (28.4.4). This yields the condition

$$\left| \sum_i V_{di} V_{si}^* \frac{\Delta M_i^2}{\tilde{M}^2} \right| < \frac{g^2 \sin\theta_c \cos\theta_c}{g_s^2} \frac{(m_c - m_u)\tilde{M}}{m_W^2}. \tag{28.4.5}$$

Taking $g^2/4\pi = 0.036$, $g_s^2/4\pi = 0.118$, $\sin\theta_c = 0.22$, $m_W = 80.4$ GeV, $m_c = 1.5$ GeV, and $m_u \ll m_c$, we find

$$\left| \sum_i V_{di} V_{si}^* \frac{\Delta M_i^2}{\tilde{M}^2} \right| < 1.5 \times 10^{-3} \times (\tilde{M}/100\,\text{GeV}). \tag{28.4.6}$$

The squark masses are unlikely to be much less than m_{gluino}, so we can conclude that either the squark masses are split by no more than about one part in 10^3, or the non-diagonal terms of the mixing matrix V_{ji} are less than about 10^{-3}, or the squarks are heavier than about 10 TeV, or we have some combination of nearly degenerate squarks, nearly zero mixing angles, and heavy squarks. In itself, this result only constrains the superpartners $\mathscr{D}_i$ of the left-handed quarks of charge $-e/3$, but similar limits on the masses and mixing angles of the $\bar{\mathscr{D}}_i$ squarks can be obtained by considering the amplitude for $d_R\overline{s_R} \to s_R\overline{d_R}$. We can also obtain somewhat weaker

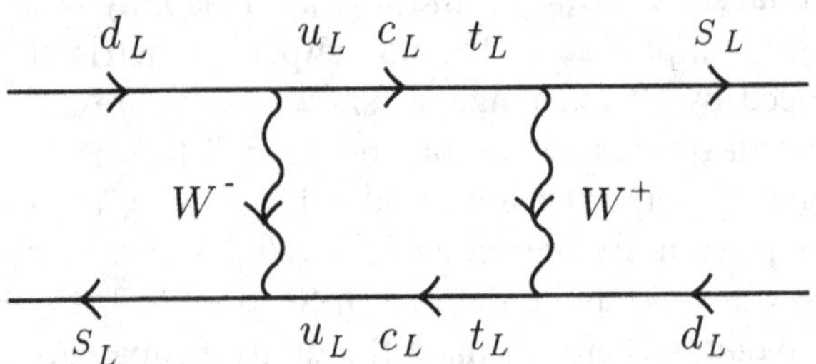

Figure 28.2. A one-loop diagram that can contribute to the $\Delta S = 2$ effective interaction $(\bar{s}_L\gamma^\mu d_L)(\bar{d}_L\gamma_\mu s_L)$ in both the supersymmetric and non-supersymmetric standard models. Here solid lines are quarks and wavy lines are $W^\pm$ bosons.

limits on the masses and mixing angles of the $\mathscr{U}_i$ squarks by considering the amplitudes produced by wino exchange rather than gluino exchange. It should be noted, however, that these arguments put no constraints on the differences between the masses of squarks of different charge, or on the differences between the masses of the superpartners $\mathscr{Q}_i$ and $\bar{\mathscr{Q}}_i$ of left-handed quarks and antiquarks.

Just as for the squarks, the sleptons of definite mass are expected to be non-diagonal linear combinations of the superpartners of the leptons. This leads to the decay process $\mu \to e + \gamma$ through diagrams like Figure 28.3. The experimental upper bound 4.9×10^{-11} on the branching ratio of this process then sets a limit of about 10^{-3} on the fractional mass splitting of sleptons of the same charge but different generations for generic mixing angles, or on the mixing angles for non-degenerate sleptons.[14]

Attempts have been made to explain the degeneracy of the squarks and sleptons in terms of a gauge symmetry connecting the different generations.[14a] In Section 28.6 we will describe an approach to supersymmetry breaking in which this degeneracy appears without needing to impose such symmetries.

CP violation

The second important class of constraint provided by experimental information about known particles has to do with CP-violating effects, such as the electric dipole moments of the neutron and electron.[15] In Section 21.3 we saw that these effects are rather weak in the non-supersymmetric standard model with only one scalar doublet, apart from a potential problem with the parameter θ of quantum chromodynamics, discussed in

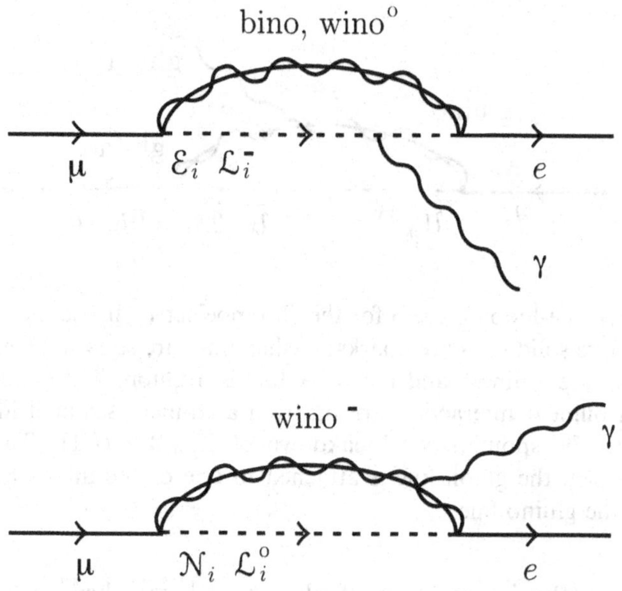

Figure 28.3. One-loop diagrams for the process $\mu \to e + \gamma$. Here solid lines are leptons; dashed lines are sleptons; combined solid and wavy lines are gauginos; and wavy lines are photons.

Section 23.6. This is because all CP-violating phases in the mass matrix of quarks and leptons and their interaction with gauge bosons could be absorbed into the definition of the quark and lepton fields if there were only two generations of quarks and leptons, and although there is a third generation, its mixing with the first two generations is (for mysterious reasons) quite weak. (This argument does not apply to processes that directly involve quarks of the third generation, such as B^0–$\bar{B}^0$ mixing, to be measured in the planned 'B factories.') The electric dipole moment of the neutron in this simple non-supersymmetric version of the standard model is consequently expected[16] to be less than about $10^{-30}\, e$ cm, well below the experimental upper bound, $6.3 \times 10^{-26}\, e$ cm.[16a]

In contrast, the over 100 parameters of the minimum supersymmetric standard model in its most general form involve dozens of CP-violating relative phases. After integrating out the heavy superpartners of the known particles, these phases produce a number of CP-violating effective interactions to be added to the Lagrangian of the standard model. Those of minimum dimensionality, which on dimensional grounds are likely to be the most important, include electric dipole moments of the quarks and leptons,[17] similar CP-violating 'chromoelectric' dipole moments

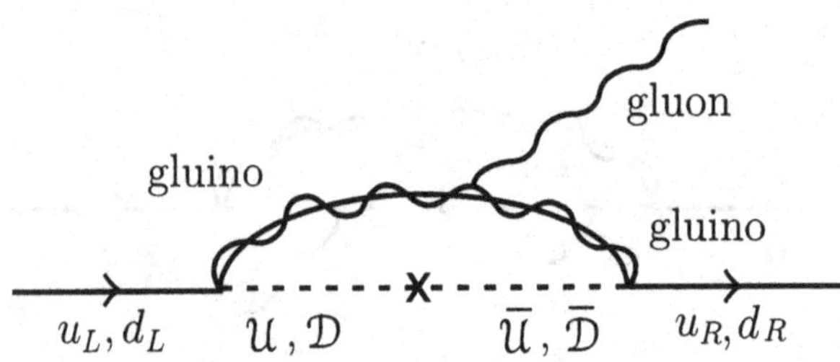

Figure 28.4. A one-loop diagram for the chromoelectric dipole moment of the *u* or *d* quarks. Here solid lines are quarks; dashed lines are squarks; combined solid and wavy lines are gluinos; and the wavy line is a gluon. The X represents the insertion of a bilinear interaction arising from a trilinear scalar field interaction combined with the spontaneous breakdown of $SU(2) \times U(1)$. There are also diagrams in which the gluon line is attached to one of the internal squark lines instead of to the gluino line.

contributing to the interactions of gluons with quarks,[18] a CP-violating purely gluonic interaction,[19] and a CP-violating interaction of the lightest Higgs scalar with leptons.[20]

To take one example, consider the quark chromoelectric dipole moments, which in some models make the largest contribution to the electric dipole moment of the neutron. The CP-violating chromoelectric dipole moment operator is $(\bar{q}\gamma_5[\gamma_\mu, \gamma_\nu]\lambda_a q)f_a^{\mu\nu}$ (where q is a u or d color-triplet quark field, $f_a^{\mu\nu}$ is the $SU(3)$ field-strength tensor, and λ_a are the 3×3 generators of $SU(3)$). Since $\gamma_5[\gamma_\mu, \gamma_\nu]$ has matrix elements only between $\bar{q}_L$ and q_R or between $\bar{q}_R$ and q_L, in order for a one-loop graphs to contribute to the chromoelectric dipole moment an external left-handed u or d quark line must emit an internal gluino line and turn into a $\mathcal{U}$ or $\mathcal{D}$ squark line, which next turns into a $\bar{\mathcal{U}}^*$ or $\bar{\mathcal{D}}^*$ squark line, and then into a right-handed u or d quark line by absorbing the internal gluino line, with the external gluon line attached either to the internal gluino line or to one of the internal squark lines. (See Figure 28.4.)

To calculate this, we need to know the mixing of the scalar components $\mathcal{U}_i$ (or $\mathcal{D}_i$) of the left-chiral quark superfields Q_i with the complex conjugates $\bar{\mathcal{U}}_j^*$ (or $\bar{\mathcal{D}}_j^*$) of the scalar components of the left-chiral antiquark superfields $\bar{U}_j$ (or $\bar{D}_j$), produced by the spontaneous breakdown of $SU(2) \times U(1)$, and represented by the X in Figure 28.4. Part of this mixing arises from a contribution of the supersymmetric $\mathcal{F}$-term interaction (28.1.7) to the last term in Eq. (26.4.7):

$$\mathcal{L}_{\mathcal{Q}\bar{\mathcal{Q}}\mathcal{H}} = -\Big|\sum_{ij} h_{ij}^U \mathcal{U}_i \bar{\mathcal{U}}_j + \mu \mathcal{H}_1^0\Big|^2 - \Big|\sum_{ij} h_{ij}^D \mathcal{D}_i \bar{\mathcal{D}}_j + \mu \mathcal{H}_2^0\Big|^2 . \quad (28.4.7)$$

There is also a contribution from the A- and C-terms in Eq. (28.4.1):

$$\mathcal{L}'_{\mathcal{D}\bar{\mathcal{D}}\mathcal{H}} = -\sum_{ij} h^D_{ij}\mathcal{D}_i\bar{\mathcal{D}}_j\Big[-A^D_{ij}\,\mathcal{H}^0_1 + C^D_{ij}\,\mathcal{H}^{0*}_2\Big]$$
$$-\sum_{ij} h^U_{ij}\mathcal{U}_i\bar{\mathcal{U}}_j\Big[A^U_{ij}\,\mathcal{H}^0_2 + C^U_{ij}\,\mathcal{H}^{0*}_1\Big] - \text{H.c.} \qquad (28.4.8)$$

Replacing the neutral Higgs scalar fields with their expectation values gives the quadratic terms

$$\mathcal{L}_{\mathcal{Q}\bar{\mathcal{Q}}} = -2\,\text{Re}\sum_{ij} m^U_{ij}\mathcal{U}_i\bar{\mathcal{U}}_j\left(\mu^*\cot\beta + A^U_{ij} + C^U_{ij}\cot\beta\right)$$
$$-2\,\text{Re}\sum_{ij} m^D_{ij}\mathcal{D}_i\bar{\mathcal{D}}_j\left(\mu^*(\tan\beta)^* + A^D_{ij} - C^D_{ij}(\tan\beta)^*\right), \qquad (28.4.9)$$

where $m^U_{ij} = \langle\mathcal{H}^0_2\rangle h^U_{ij}$ and $m^D_{ij} = -\langle\mathcal{H}^0_1\rangle h^D_{ij}$ are the mass matrices of the quarks of charge $2e/3$ and $-e/3$, and

$$\tan\beta \equiv \langle\mathcal{H}^0_2\rangle/\langle\mathcal{H}^0_1\rangle^*\,. \qquad (28.4.10)$$

Neglecting Cabibbo mixing, and for definiteness taking the As and Cs diagonal, Figure 28.4 makes contributions to the chromoelectric dipole moment of the u and d quarks of the form

$$d^{ce}_u = \frac{g_s^3}{16\pi^2}\text{Im}\left[m_u\,A'_u I(m_{\mathcal{U}}, m_{\bar{\mathcal{U}}}, m_{\text{gluino}})\right], \qquad (28.4.11)$$

$$d^{ce}_d = \frac{g_s^3}{16\pi^2}\text{Im}\left[m_d\,A'_d I(m_{\mathcal{D}}, m_{\bar{\mathcal{D}}}, m_{\text{gluino}})\right], \qquad (28.4.12)$$

where

$$A'_u = (\mu^* + C_u)\cot\beta + A_u\,, \quad A'_d = (\mu^* - C_d)(\tan\beta)^* + A_d\,, \qquad (28.4.13)$$

and I is a complicated dimensionless function of its arguments arising from the integration over the virtual four-momentum. For $m_{\mathcal{Q}} \simeq m_{\bar{\mathcal{Q}}}$ and the gluino field defined to make m_{gluino} as well as $m_{\mathcal{Q}}$ real, the function I takes the form

$$I(m_{\mathcal{Q}}, m_{\mathcal{Q}}, m_{\text{gluino}}) = m^{-3}_{\text{gluino}}\, J\left(\frac{m^2_{\text{gluino}}}{m^2_{\mathcal{Q}} - m^2_{\text{gluino}}}\right), \qquad (28.4.14)$$

where[21]

$$J(z) = 2\left(-z^4 + \frac{4}{3}z^3 + z^2\right)\ln\left(\frac{1+z}{z}\right) + 2z^3 - \frac{11}{3}z^2\,. \qquad (28.4.15)$$

The hard part of this sort of calculation always lies in estimating the contribution of an operator like the chromoelectric dipole interaction to hadronic matrix elements such as the electric dipole moment of the neutron. We can anticipate that there will be renormalization group corrections required because this operator is to be used at energies of the order of the neutron mass, rather than the masses of the squarks and gluino. More important is simply getting dimensional factors and factors of 4π right. For this purpose, it is usual to use a counting rule[22] known as 'naive dimensional analysis.' A connected graph with V_i vertices of type i and I internal lines will have a number L of loops given by $L = I - \sum_i V_i + 1$. If there are N_i lines attached to a vertex of type i and N external lines of the whole graph, then $2I + N = \sum_i V_i N_i$, so

$$L = 1 - \frac{N}{2} + \sum_i V_i \left(\frac{N_i}{2} - 1\right) .$$

We expect a factor of order $1/16\pi^2$ for each loop, so the coefficient of an operator $\mathcal{O}$ in the low-energy effective Lagrangian with N field factors will contain an over-all factor

$$(4\pi)^{N-2} \prod_i (4\pi)^{(2-N_i)V_i} .$$

If the operator $\mathcal{O}$ has dimensionality d and the interactions $\mathcal{O}_i$ of type i have dimensionality d_i, then the coefficient of $\mathcal{O}$ will have dimensionality $4-d-\sum_i(4-d_i)$, so this coefficient will also have a factor $M^{4-d}\prod_i M^{d_i-4}$, where M is some scale that is typical of hadronic physics, such as the nucleon mass or the energy $2\pi F_\pi \simeq 1200$ MeV, where the low-energy expansions discussed in Section 19.5 begin to break down. Finally, the contribution to the coefficient of $\mathcal{O}$ from some graph will of course be proportional to the couplings of all the operators $\mathcal{O}_i$ associated with the vertices in the graph. These remarks can be conveniently summarized by defining a 'reduced coupling': the reduced coupling associated with any operator $\mathcal{O}_i$ having N_i field factors, dimensionality d_i, and coupling constant g_i is

$$g_i^{\text{reduced}} \equiv g_i (4\pi)^{2-N_i} M^{\mathscr{D}_i - 4} . \tag{28.4.16}$$

The above estimates suggest the rule of naive dimensional analysis: the reduced coupling of any operator $\mathcal{O}$ in the effective hadronic Lagrangian is roughly equal to the product of the reduced couplings of the interactions that contribute to this effective coupling.

The neutron electric dipole moment is the coefficient of an operator with one photon and two neutron fields and dimensionality 5, so its reduced coupling is $Md_n^e/4\pi$. Likewise, the quark chromoelectric magnetic moment has reduced coupling $Md_q^{ce}/4\pi$. In addition to one factor of this reduced coupling, the reduced coupling of the neutron electric dipole operator must have a factor of the reduced coupling $e/4\pi$ of the electromagnetic coupling and an indeterminate number of factors of the reduced strong coupling $g_s/4\pi$, which *at the low-energy scale M* are not very different from unity and will be ignored. Taking the d quark contribution as representative of the contributions of both u and d quarks, the result then is that

$$d_n^e \approx \frac{e\, d_d^{ce}}{4\pi} \approx e \left(\frac{g_s}{4\pi}\right)^3 \text{Im}\left[m_d A_d'\right] I(m_{\mathscr{D}}, m_{\bar{\mathscr{D}}}, m_{\text{gluino}}) \,. \tag{28.4.17}$$

Further simplifying by setting $m_{\text{gluino}} \simeq m_{\mathscr{D}} \simeq m_{\bar{\mathscr{D}}}$, so that $J = 7/18$, and taking $g_s^2/4\pi$ at the scale of squark and gluino masses to have the same value 0.12 as at m_Z, and $|m_d| \approx 7$ MeV, we have then

$$|d_n^e| \approx 0.5 \times 10^{-23}\, e\, \text{cm}\; \frac{|A_d'|\,|\sin\varphi| \times (100\,\text{GeV})^2}{m_{\text{gluino}}^3} \,, \tag{28.4.18}$$

where φ is the phase of A_d', with the convention that the gluino, quark, and squark masses are taken real. The contribution of the electric dipole moment of the quark is somewhat larger, while the contribution of the purely gluonic CP-odd operator is considerably smaller.[23]

To avoid conflict with the experimental upper bound of $0.97 \times 10^{-25}\, e$ cm, either the phases associated with CP violation in the supersymmetric standard model must be less than about 10^{-2}, or some of the new particles of this model must be heavier than about 1 TeV. Similar conclusions have been reached from calculations of the electric dipole moments of atoms and molecules.[23] Even more stringent conditions on CP-violating phases have been derived[24] by considering the contribution of Figure 28.1 to the one precisely measured CP-violating effect, the imaginary part of the amplitude for K^0–$\bar{K}^0$ oscillation.

28.5 The Sector of Zero Baryon and Lepton Number

Despite the large number of parameters of the supersymmetric standard model, in some contexts it is surprisingly predictive. This is true in particular when we consider the scalar fields whose vacuum expectation values spontaneously break the $SU(2) \times U(1)$ gauge symmetry. In this section we will consider these scalars, along with other fields of zero baryon and lepton number: neutral scalars odd under charge conjugation, charged scalars, and the fermionic superpartners of these scalars and of the $W^\pm$ and Z^0.

It is a crucial requirement for supersymmetric versions of the standard model that they should contain scalar doublet 'Higgs' superfields with the right mass and interaction parameters to account for the breakdown of the $SU(2) \times U(1)$ gauge group of the electromagnetic and weak interactions. We saw in Section 28.1 that at least two left-chiral scalar doublets are needed to give mass to the quarks of both charge $2e/3$ and $-e/3$ and the charged leptons, while we found in Section 28.2 that two doublets are just what is needed to bring the $SU(3)$, $SU(2)$, and $U(1)$ gauge couplings together at some very high energy. We have therefore assumed that there are two left-chiral scalar $SU(2)$ doublets

$$H_1 = \begin{pmatrix} H_1^0 \\ H_1^- \end{pmatrix} , \qquad H_2 = \begin{pmatrix} H_2^+ \\ H_2^0 \end{pmatrix} . \tag{28.5.1}$$

These have $SU(2)$ and $U(1)$ D-terms (27.4.7) given (assuming zero Fayet–Iliopoulos constant $\xi_{U(1)}$) by

$$\vec{D} = \frac{g}{2}\left(\mathscr{H}_1^\dagger \vec{\tau}\, \mathscr{H}_1\right) + \frac{g}{2}\left(\mathscr{H}_2^\dagger \vec{\tau}\, \mathscr{H}_2\right) , \tag{28.5.2}$$

$$D_y = \frac{g'}{2}\left(\mathscr{H}_1^\dagger \mathscr{H}_1\right) - \frac{g'}{2}\left(\mathscr{H}_2^\dagger \mathscr{H}_2\right) , \tag{28.5.3}$$

where $\mathscr{H}_{1,2}$ are the scalar components of the superfield doublets $H_{1,2}$ and τ_r are the Pauli matrices, with $\tau_r^2 = 1$. As shown in Eq. (27.4.9), in renormalizable theories this gives the D-term contribution to the scalar field potential

$$\begin{aligned} V_D &= \frac{1}{2}\vec{D}^2 + \frac{1}{2}D_y^2 \\ &= \frac{g^2}{8}\left[\left(\mathscr{H}_1^\dagger \vec{\tau}\, \mathscr{H}_1\right) + \left(\mathscr{H}_2^\dagger \vec{\tau}\, \mathscr{H}_2\right)\right]^2 + \frac{g'^2}{8}\left[\left(\mathscr{H}_1^\dagger \mathscr{H}_1\right) - \left(\mathscr{H}_2^\dagger \mathscr{H}_2\right)\right]^2 . \end{aligned} \tag{28.5.4}$$

This can be put in a more convenient form by using the relation

$$(\vec{\tau})_{i\ell} \cdot (\vec{\tau})_{kj} = 2\delta_{ij}\delta_{k\ell} - \delta_{i\ell}\delta_{kj} . \tag{28.5.5}$$

(To prove this, use rotational invariance to show that $\delta_{ij}\delta_{k\ell}$ may be expressed as a linear combination of $(\vec{\tau})_{i\ell} \cdot (\vec{\tau})_{kj}$ and $\delta_{i\ell}\delta_{kj}$, and calculate the coefficients by taking traces on the indices i, j and i, ℓ.) In this way, we may rewrite the D-term part of the scalar field potential as

$$V_D = \frac{g^2}{2}\left|\left(\mathscr{H}_1^\dagger \mathscr{H}_2\right)\right|^2 + \frac{g^2 + g'^2}{8}\left[\left(\mathscr{H}_1^\dagger \mathscr{H}_1\right) - \left(\mathscr{H}_2^\dagger \mathscr{H}_2\right)\right]^2 . \tag{28.5.6}$$

As mentioned in Section 28.1, there is just one possible renormalizable term in the superpotential for these two left-chiral doublets, of the form

$$f(H_1, H_2) = \mu\left(H_1^{\mathrm{T}} e H_2\right) , \tag{28.5.7}$$

where μ is a constant with the dimensions of a mass and e is the anti-symmetric matrix $i\tau_2$. According to Eq. (27.4.9), this gives an additional contribution to the scalar field potential

$$V_\mu = \sum_r \left| \frac{\partial f(\mathscr{H}_1, \mathscr{H}_2)}{\partial \mathscr{H}_{1r}} \right|^2 + \sum_r \left| \frac{\partial f(\mathscr{H}_1, \mathscr{H}_2)}{\partial \mathscr{H}_{2r}} \right|^2$$

$$= |\mu|^2 \left[\left(\mathscr{H}_1^\dagger \mathscr{H}_1 \right) + \left(\mathscr{H}_2^\dagger \mathscr{H}_2 \right) \right] . \tag{28.5.8}$$

For $\mu \neq 0$ the potential $V_D + V_\mu$ evidently has a minimum value of zero, reached at the unique point $\mathscr{H}_1 = \mathscr{H}_2 = 0$. With just these terms in the potential, $SU(2) \times U(1)$ as well as supersymmetry is not spontaneously broken. (The case $\mu = 0$ is not much better; there is a continuous infinity of vacuum states with supersymmetry unbroken and with $SU(2) \times U(1)$ broken down to electromagnetic gauge invariance with all possible strengths, including zero.) This is one more example of the general difficulty, already seen in Section 28.3, of formulating realistic theories in which supersymmetry is spontaneously broken within the standard model.

Under the assumption of the previous section that supersymmetry is violated in the effective Lagrangian only by superrenormalizable terms, the most general such supersymmetry-breaking term involving the scalar doublets is of the form

$$V_m = m_1^2 \left(\mathscr{H}_1^\dagger \mathscr{H}_1 \right) + m_2^2 \left(\mathscr{H}_2^\dagger \mathscr{H}_2 \right) + \text{Re} \left\{ B\mu \left(\mathscr{H}_1^{\text{T}} e \mathscr{H}_2 \right) \right\} ,$$

where m_1^2 and m_2^2 are real parameters (not necessarily positive) and $B\mu$ is a parameter of arbitrary phase. We will adjust the *over-all* phase of the superfields H_1 and H_2 so that $B\mu$ is real and positive, and so

$$V_m = m_1^2 \left(\mathscr{H}_1^\dagger \mathscr{H}_1 \right) + m_2^2 \left(\mathscr{H}_2^\dagger \mathscr{H}_2 \right) + B\mu \, \text{Re} \left(\mathscr{H}_1^{\text{T}} e \mathscr{H}_2 \right) . \tag{28.5.9}$$

The total scalar potential in the tree approximation is then

$$V = V_D + V_\mu + V_m$$

$$= \frac{g^2}{2} \left| \left(\mathscr{H}_1^\dagger \mathscr{H}_2 \right) \right|^2 + \frac{g^2 + g'^2}{8} \left[\left(\mathscr{H}_1^\dagger \mathscr{H}_1 \right) - \left(\mathscr{H}_2^\dagger \mathscr{H}_2 \right) \right]^2$$

$$+ (m_1^2 + |\mu|^2) \left(\mathscr{H}_1^\dagger \mathscr{H}_1 \right) + (m_2^2 + |\mu|^2) \left(\mathscr{H}_2^\dagger \mathscr{H}_2 \right)$$

$$+ B\mu \, \text{Re} \left(\mathscr{H}_1^{\text{T}} e \mathscr{H}_2 \right) . \tag{28.5.10}$$

Note in particular that μ^2, m_1^2, and m_2^2 only appear in the combinations $m_1^2 + |\mu|^2$ and $m_2^2 + |\mu|^2$.

There is one condition on the supersymmetry-breaking parameters m_i^2, derived from the requirement that the potential should be bounded below. For scalar fields going to infinity in generic directions, the potential is dominated by the quartic terms V_D, which are positive. There are special directions in which V_D vanishes: those for which (up to an $SU(2) \times U(1)$ gauge transformation)

$$\mathcal{H}_1 = \begin{pmatrix} \phi \\ 0 \end{pmatrix}, \qquad \mathcal{H}_2 = \begin{pmatrix} 0 \\ \phi \end{pmatrix},$$

with ϕ an arbitrary complex quantity. For such directions, $V = (2|\mu|^2 + m_1^2 + m_2^2)|\phi|^2 - B\mu\phi^2$, so (since $B\mu$ has been defined to be positive) in order for this not to go to $-\infty$ as $\phi \to +\infty$, it is necessary that

$$2|\mu|^2 + m_1^2 + m_2^2 \geq B\mu\,. \tag{28.5.11}$$

We wish to look for a minimum of the potential at which electromagnetic gauge invariance is not broken, so let us consider the behavior of the potential as a function of the neutral scalar fields, with the charged scalar fields set equal to zero. In this case Eq. (28.5.10) gives the potential of the neutral scalars as

$$\begin{aligned} V^{\rm N} = \frac{g^2+g'^2}{8}\left[|\mathcal{H}_1^0|^2 - |\mathcal{H}_2^0|^2\right]^2 + (m_1^2 + |\mu|^2)\left|\mathcal{H}_1^0\right|^2 \\ +(m_2^2+|\mu|^2)\left|\mathcal{H}_2^0\right|^2 - B\mu\,{\rm Re}\left(\mathcal{H}_1^0\mathcal{H}_2^0\right). \end{aligned} \tag{28.5.12}$$

To find a stationary point, we expand $V^{\rm N}$ around the constant values $\mathcal{H}_i^0 = v_i$, writing

$$\mathcal{H}_i^0 = v_i + \varphi_i\,. \tag{28.5.13}$$

To second order in the φ_i, Eq. (28.5.12) gives

$$\begin{aligned} V^{\rm N}_{\rm quad} = & \frac{g^2+g'^2}{4}(|v_1|^2 - |v_2|^2)\left[2{\rm Re}\,(v_1^*\varphi_1 - v_2^*\varphi_2) + |\varphi_1|^2 - |\varphi_2|^2\right] \\ & +\frac{g^2+g'^2}{2}\left[{\rm Re}\,(v_1^*\varphi_1 - v_2^*\varphi_2)\right]^2 + (m_1^2+|\mu|^2)\left(2{\rm Re}\,v_1^*\varphi_1 + |\varphi_1|^2\right) \\ & +(m_2^2+|\mu|^2)\left(2{\rm Re}\,v_2^*\varphi_2 + |\varphi_2|^2\right) - B\mu\,{\rm Re}\left(v_1\varphi_2 + v_2\varphi_1 + \varphi_1\varphi_2\right) \\ & + {\rm constant}\,. \end{aligned} \tag{28.5.14}$$

For the v_i to be equilibrium values of the fields, the terms of first order in the φ_i must vanish

$$\left(m_1^2 + |\mu|^2\right)v_1^* + \frac{g^2+g'^2}{4}\left(|v_1|^2 - |v_2|^2\right)v_1^* - \frac{1}{2}B\mu\, v_2 = 0\,, \tag{28.5.15}$$

$$\left(m_2^2 + |\mu|^2\right)v_2^* + \frac{g^2+g'^2}{4}\left(|v_2|^2 - |v_1|^2\right)v_2^* - \frac{1}{2}B\mu\, v_1 = 0\,. \tag{28.5.16}$$

Without changing the over-all phase of the φ_i, we may adjust their relative phases so that v_1 is real. Then Eqs. (28.5.15) and (28.5.16) show that v_2 is also real, so that these equations become

$$\left(m_1^2+|\mu|^2\right)v_1+\frac{g^2+g'^2}{4}\left(v_1^2-v_2^2\right)v_1-\frac{1}{2}B\mu v_2=0\,, \quad (28.5.17)$$

$$\left(m_2^2+|\mu|^2\right)v_2+\frac{g^2+g'^2}{4}\left(v_2^2-v_1^2\right)v_2-\frac{1}{2}B\mu v_1=0\,. \quad (28.5.18)$$

These conditions may be used to express the mass parameters in the potential in terms of the convenient quantities

$$\tan\beta\equiv v_2/v_1\,, \quad (28.5.19)$$

$$m_Z^2=\tfrac{1}{2}(g^2+g'^2)\,(v_1^2+v_2^2)\,, \quad (28.5.20)$$

and

$$m_A^2\equiv 2|\mu|^2+m_1^2+m_2^2\,. \quad (28.5.21)$$

(The parameter m_Z is the mass of the Z vector boson.* We will soon see that m_A is the mass of one of the physical scalars.) Multiplying Eqs. (28.5.17) and (28.5.18), respectively, by v_2 and v_1 and taking the sum and difference gives

$$B\mu=m_A^2\sin 2\beta\,, \quad (28.5.22)$$

and

$$m_1^2-m_2^2=-\,(m_A^2+m_Z^2)\cos 2\beta\,, \quad (28.5.23)$$

which with Eq. (28.5.21) yields

$$m_1^2+|\mu|^2=\tfrac{1}{2}m_A^2-\tfrac{1}{2}(m_A^2+m_Z^2)\cos 2\beta\,,\ \ m_2^2+|\mu|^2=\tfrac{1}{2}m_A^2+\tfrac{1}{2}(m_A^2+m_Z^2)\cos 2\beta\,. \quad (28.5.24)$$

With linear terms cancelling, the quadratic part (28.5.14) of the neutral scalar potential may then be written

$$\begin{aligned}V^{\rm N}_{\rm quad}&=\frac{g^2+g'^2}{4}(v_1^2-v_2^2)\Big[|\varphi_1|^2-|\varphi_2|^2\Big]+\frac{g^2+g'^2}{2}\Big[{\rm Re}\,(v_1\varphi_1-v_2\varphi_2)\Big]^2\\&\quad+(m_1^2+|\mu|^2)\,|\varphi_1|^2+(m_2^2+|\mu|^2)\,|\varphi_2|^2-B\mu\,{\rm Re}\left(\varphi_1\varphi_2\right)+\text{constant}\\&=\tfrac{1}{2}m_Z^2\cos 2\beta\Big[|\varphi_1|^2-|\varphi_2|^2\Big]+m_Z^2\Big[{\rm Re}\,(\cos\beta\varphi_1-\sin\beta\varphi_2)\Big]^2\\&\quad+\tfrac{1}{2}m_A^2\Big(|\varphi_1|^2+|\varphi_2|^2\Big)-\tfrac{1}{2}(m_A^2+m_Z^2)\cos 2\beta\Big[|\varphi_1|^2-|\varphi_2|^2\Big]\\&\quad-m_A^2\sin 2\beta\,{\rm Re}\left(\varphi_1\varphi_2\right)+\text{constant}\,.\end{aligned} \quad (28.5.25)$$

* There is a difference of a factor of 2 between the formulas for m_Z^2 given by Eqs. (28.5.20) and (21.3.30), due to the fact that the scalar fields are normalized differently here and in Section 21.3.

We see from Eq. (28.5.25) that the real and imaginary parts of the φ_i are decoupled. (This is because the potential (28.5.12) is invariant under a charge conjugation or CP transformation $\varphi_i \to \varphi_i^*$.) The mass-squared matrix of the imaginary parts of the φ_i is

$$M^2_{\mathrm{Im}\,\varphi} = \begin{pmatrix} \frac{1}{2}m_A^2\,(1-\cos 2\beta) & \frac{1}{2}m_A^2\,\sin 2\beta \\ \frac{1}{2}m_A^2\,\sin 2\beta & \frac{1}{2}m_A^2\,(1+\cos 2\beta) \end{pmatrix} . \tag{28.5.26}$$

The determinant vanishes, so one eigenvalue is zero, and the other equals the trace, which is just m_A^2. The zero-mass scalar is of course the neutral Goldstone boson associated with the spontaneous breakdown of $SU(2) \times U(1)$ to electromagnetic gauge invariance, and as discussed in Chapter 21 it is eliminated by the Higgs mechanism. As promised, m_A is the mass of one of the physical scalars, the non-Goldstone boson with C negative. This shows that for the field value $\varphi_i = v_i$ to be at least a local minimum of the potential, the parameter m_A^2 defined by Eq. (28.5.21) must be positive. The condition (28.5.11) for good behavior at large field strengths shows that Eq. (28.5.22) has a solution here for β in the range $0 \leq \beta \leq \pi/2$.

In particular, if $B\mu = 0$ and $0 < \beta < \pi/2$, then Eq. (28.5.22) shows that $m_A = 0$. In this case, the particle A is the Goldstone boson of a $U(1)$ Peccei–Quinn symmetry[4] of the potential (28.5.12) under an equal phase change of $\mathscr{H}_1^0$ and $\mathscr{H}_2^0$, which for $v_1 \neq 0$ and $v_2 \neq 0$ is spontaneously broken without leaving any combination of this and the electroweak $U(1)$ symmetry unbroken. This is the original version of the axion,[25] which as we saw in Section 23.6 acquires only a small mass from the Yukawa interactions of the scalars with the quarks and is experimentally ruled out. Thus we can conclude that $B\mu$ definitely does not vanish.

The elements of the mass-squared matrix for the real scalars are given by Eq. (28.5.25) as

$$\begin{aligned} (M^2_{\mathrm{Re}\,\varphi})_{11} &= \tfrac{1}{2}m_A^2(1-\cos 2\beta) + \tfrac{1}{2}m_Z^2(1+\cos 2\beta)\,, \\ (M^2_{\mathrm{Re}\,\varphi})_{12} = (M^2_{\mathrm{Re}\,\varphi})_{21} &= -\tfrac{1}{2}(m_A^2 + m_Z^2)\sin 2\beta\,, \\ (M^2_{\mathrm{Re}\,\varphi})_{22} &= \tfrac{1}{2}m_A^2(1+\cos 2\beta) + \tfrac{1}{2}m_Z^2(1-\cos 2\beta)\,. \end{aligned} \tag{28.5.27}$$

Solving the secular equation gives the eigenvalues

$$m_H^2 = \frac{1}{2}\left[m_A^2 + m_Z^2 + \sqrt{(m_A^2+m_Z^2)^2 - 4m_A^2 m_Z^2\cos^2 2\beta}\right]\,, \tag{28.5.28}$$

$$m_h^2 = \frac{1}{2}\left[m_A^2 + m_Z^2 - \sqrt{(m_A^2+m_Z^2)^2 - 4m_A^2 m_Z^2\cos^2 2\beta}\right]\,. \tag{28.5.29}$$

To calculate the masses of the charged scalars, we evaluate the potential

V with the neutral scalars set equal to their vacuum expectation values:

$$\mathscr{H}_1 = \begin{pmatrix} v_1 \\ \mathscr{H}_1^- \end{pmatrix}, \qquad \mathscr{H}_2 = \begin{pmatrix} \mathscr{H}_2^+ \\ v_2 \end{pmatrix}. \tag{28.5.30}$$

Using this in Eq. (28.5.10) then gives the quadratic part of the charged scalar potential

$$V^{\rm C}_{\rm quad} = \frac{g^2}{2}\left|v_2(\mathscr{H}_1^-)^* + v_1\mathscr{H}_2^+\right|^2 + \frac{g^2+g'^2}{4}(v_1^2 - v_2^2)\Big(|\mathscr{H}_1^-|^2 - |\mathscr{H}_2^+|^2\Big)$$
$$+(m_1^2 + |\mu|^2)|\mathscr{H}_1^-|^2 + (m_2^2 + |\mu|^2)|\mathscr{H}_2^+|^2 + B\mu\mathscr{H}_1^-\mathscr{H}_2^+ \,. \tag{28.5.31}$$

Using Eqs. (28.5.22) and (28.5.24), this may be written

$$V^{\rm C}_{\rm quad} = \frac{1}{2}(m_W^2 + m_A^2)\Big[|\mathscr{H}_1^-|^2(1-\cos 2\beta) + |\mathscr{H}_2^+|^2(1+\cos 2\beta)$$
$$+2\sin 2\beta\,\mathscr{H}_1^-\mathscr{H}_2^+\Big], \tag{28.5.32}$$

where m_W is the charged gauge boson mass:

$$m_W^2 = \frac{1}{2}g^2\Big(|v_1|^2 + |v_2|^2\Big). \tag{28.5.33}$$

The charged scalar mass matrix is then

$$M_{\rm C}^2 = \frac{1}{2}(m_W^2 + m_A^2)\begin{pmatrix} 1-\cos 2\beta & \sin 2\beta \\ \sin 2\beta & 1+\cos 2\beta \end{pmatrix}. \tag{28.5.34}$$

This has determinant zero, so it has one eigenvalue equal to zero and the other eigenvalue equal to the trace

$$m_{\rm C}^2 = m_W^2 + m_A^2 \,. \tag{28.5.35}$$

The zero-mass charged scalar is of course the other Goldstone boson associated with the spontaneous breakdown of $SU(2)\times U(1)$, and like the neutral Goldstone boson found earlier it is eliminated by the Higgs mechanism.

Even without knowing the parameters m_A and β, these results tell us a lot about the relative magnitudes of the scalar boson masses. We can rewrite Eqs. (28.5.28) and (28.5.29) in the form

$$m_H^2 = \frac{1}{2}\Big[m_A^2 + m_Z^2 + \sqrt{(m_A^2 - m_Z^2)^2 + 4m_A^2 m_Z^2 \sin^2 2\beta}\Big], \tag{28.5.36}$$

$$m_h^2 = \frac{1}{2}\Big[m_A^2 + m_Z^2 - \sqrt{(m_A^2 - m_Z^2)^2 + 4m_A^2 m_Z^2 \sin^2 2\beta}\Big]. \tag{28.5.37}$$

We see that *the heavier neutral scalar mass m_H is larger than the larger of m_Z and m_A, while the lighter neutral scalar mass m_h is smaller than the smaller of m_Z and m_A*. If the large ratio of top to bottom quark masses is due to a large ratio $v_2/v_1 = \tan\beta$ of scalar field vacuum expectation values, rather than a large ratio of Yukawa couplings, then we expect

β to be near $\pi/2$, in which case these inequalities become approximate equalities. Further, Eq. (28.5.35) shows that *the charged scalar mass is greater than both m_A and m_W.*

These results are quantitatively modified by various radiative corrections *within* the standard model (as opposed to radiative effects that produce the input parameters m_i^2 in theories of gauge-mediated supersymmetry breaking). The most important corrections arise from the presence in the scalar potential V of terms arising from graphs consisting of a single loop of top or bottom quarks interacting any number of times with external scalar field lines. This is because the top and bottom quarks have by far the strongest couplings to $\mathscr{H}_2$ and $\mathscr{H}_1$, respectively. (It is prudent to include bottom as well as top quark loops here because, as mentioned above, the larger mass of the top quark may be due to a large ratio v_2/v_1, rather than to a large ratio of Yukawa couplings, but even in this case we will see that the dominant corrections are due to top quark loops.)

Let us first consider the neutral scalars, at least one of which would be lighter than the Z boson in the absence of radiative corrections. The effect of these top and bottom loops is to contribute a term of form $U_t(|\mathscr{H}_2^0|^2) + U_b(|\mathscr{H}_1^0|^2)$ to $V^{\rm N}$. We will absorb any terms in U_b or U_t that are linear in $|\mathscr{H}_1^0|^2 - v_1^2$ or $|\mathscr{H}_2^0|^2 - v_2^2$ into the input parameters m_1^2 and m_2^2, so that

$$U_b'(v_1^2) = U_t'(v_2^2) = 0\,. \tag{28.5.38}$$

Our earlier results (28.5.24) and (28.5.22) for $m_1^2 + |\mu|^2$, $m_2^2 + |\mu|^2$ and $B\mu$ are then unchanged. Also, the mass matrix for the C-odd neutral scalars is still given by Eq. (28.5.26). On the other hand, the elements of the mass-squared matrix for the C-even neutral scalars are now given by

$$\begin{aligned}(M^2_{\mathrm{Re}\,\varphi})_{11} &= \tfrac{1}{2}m_A^2(1-\cos 2\beta) + \tfrac{1}{2}m_Z^2(1+\cos 2\beta) + \Delta_b\,,\\ (M^2_{\mathrm{Re}\,\varphi})_{12} = (M^2_{\mathrm{Re}\,\varphi})_{21} &= -\tfrac{1}{2}(m_A^2 + m_Z^2)\sin 2\beta\,,\\ (M^2_{\mathrm{Re}\,\varphi})_{22} &= \tfrac{1}{2}m_A^2(1+\cos 2\beta) + \tfrac{1}{2}m_Z^2(1-\cos 2\beta) + \Delta_t\,,\end{aligned} \tag{28.5.39}$$

where

$$\Delta_b = 2v_1^2\, U_b''(v_1^2)\,, \qquad \Delta_t = 2v_2^2\, U_t''(v_2^2)\,. \tag{28.5.40}$$

The solutions of the secular equation are then

$$m_H^2 = \frac{1}{2}\Bigg[m_A^2 + m_Z^2 + \Delta_t + \Delta_b + \sqrt{\Big((m_A^2 - m_Z^2)\cos 2\beta + \Delta_t - \Delta_b\Big)^2 + \Big(m_A^2 + m_Z^2\Big)^2 \sin^2 2\beta}\;\Bigg]\,, \tag{28.5.41}$$

$$m_h^2 = \frac{1}{2}\Bigg[m_A^2 + m_Z^2 + \Delta_t + \Delta_b$$
$$-\sqrt{\left((m_A^2 - m_Z^2)\cos 2\beta + \Delta_t - \Delta_b\right)^2 + \left(m_A^2 + m_Z^2\right)^2 \sin^2 2\beta}\,\Bigg]\,. \tag{28.5.42}$$

In considering searches for these particles, it is important to note that the light Higgs mass m_h increases as the unknown mass m_A increases, reaching a finite upper bound for $m_A \to \infty$

$$m_h \leq m_h(m_A \to \infty) = m_Z^2 \cos^2 2\beta + \Delta_t \sin^2\beta + \Delta_b \cos^2\beta\,. \tag{28.5.43}$$

To calculate Δ_b and Δ_t, we recall from Section 16.2 that the potentials U_b and U_t are given by

$$U_b(|\mathscr{H}_1^0|^2) = -\frac{3}{16\pi^2}\left|\lambda_b\mathscr{H}_1^0\right|^4\left[\ln\frac{\left|\lambda_b\mathscr{H}_1^0\right|^2}{M_{sb}^2} - \frac{3}{2}\right] + \text{linear terms}\,, \tag{28.5.44}$$

$$U_t(|\mathscr{H}_2^0|^2) = -\frac{3}{16\pi^2}\left|\lambda_t\mathscr{H}_2^0\right|^4\left[\ln\frac{\left|\lambda_t\mathscr{H}_2^0\right|^2}{M_{st}^2} - \frac{3}{2}\right] + \text{linear terms}\,, \tag{28.5.45}$$

where $\lambda_t = m_t/v_2$ and $\lambda_b = m_b/v_1$ are the Yukawa couplings of the top and bottom quarks; M_{st} and M_{sb} are the masses of the stop and sbottom (the scalar superpartners of the top and bottom quarks); these masses and the terms $-3/2$ inside the square brackets are chosen to satisfy the condition that supersymmetry-breaking corrections due to stop and sbottom loops would cancel the corrections due to top and bottom loops if the masses were equal; and the 'linear terms' are linear in $|\mathscr{H}_2^0|^2$ or $|\mathscr{H}_1^0|^2$, with coefficients adjusted to satisfy Eq. (28.5.38). (The factor 3 takes account of the three quark colors.) Then Eq. (28.5.40) gives

$$\Delta_b = -\frac{3}{4\pi^2}|\lambda_b|^4 v_1^2 \ln\left(\frac{\lambda_b v_1^2}{M_{sb}^2}\right) = \frac{3\sqrt{2}\,m_b^4\,G_F}{2\pi^2\cos^2\beta}\ln\left(\frac{M_{sb}^2}{m_b^2}\right)\,, \tag{28.5.46}$$

$$\Delta_t = -\frac{3}{4\pi^2}|\lambda_t|^4 v_2^2 \ln\left(\frac{\lambda_t v_2^2}{M_{st}^2}\right) = \frac{3\sqrt{2}\,m_t^4\,G_F}{2\pi^2\sin^2\beta}\ln\left(\frac{M_{st}^2}{m_t^2}\right)\,, \tag{28.5.47}$$

where $G_F = 1.17 \times 10^{-5}\ \text{GeV}^{-2}$ is the Fermi coupling constant, given by Eq. (21.3.34) as $G_F = g^2/4\sqrt{2}m_W^2$. Taking $m_b = 4.3$ GeV, $m_t = 180$ GeV, $M_{st} \sim M_{sb} \sim 1$ TeV, and $m_Z = 91.2$ GeV gives $\Delta_b \sim 1.1 \times 10^{-6}\, m_Z^2/\cos^2\beta$ and $\Delta_t \sim 1.1\, m_Z^2/\sin^2\beta$. We see that even if $\tan\beta$ is as large as m_t/m_b, the top-quark correction Δ_t will still be much larger than Δ_b.

The effect of Δ_t is to *increase* both m_H and m_h. Taking this and other radiative corrections into account,[26] for $\tan\beta > 10$ the upper bound (28.5.43) on the lightest neutral scalar mass is raised by radiative corrections from just below m_Z to between 100 GeV and 110 GeV for stop masses between 300 GeV and 1 TeV. For comparison, an experimental *lower* bound[27] of 62.5 GeV on m_h, m_H, and m_A is set by the absence of an hA or HA final state in e^+e^- collisions at 130 to 172 GeV. Also, calculations of radiative corrections involving Higgs scalars are consistent with precision measurements of electroweak phenomena for m_h in the range[27a] of 27 to 140 GeV.

Radiative corrections are less important for the charged scalars. Since the attachment of charged scalar lines allows transitions between top and bottom quarks, the correction to the scalar field potential here takes a more general form, constrained by $SU(2) \times U(1)$ to be

$$\Delta V = U(\mathscr{H}_2^\dagger \mathscr{H}_2,\, \mathscr{H}_1^\dagger \mathscr{H}_1,\, \mathscr{H}_2^\dagger \mathscr{H}_1,\, \mathscr{H}_1^\dagger \mathscr{H}_2,\, \mathscr{H}_1^{\rm T} e \mathscr{H}_2)\,. \tag{28.5.48}$$

(Quark loops do not actually produce any dependence on $\mathscr{H}_1^{\rm T} e \mathscr{H}_2$.) Every appearance of a $\mathscr{H}_1$ or $\mathscr{H}_2$ doublet is accompanied with a factor λ_b or λ_t, respectively, so terms involving $\mathscr{H}_1$ will be suppressed, as we have already seen in calculating the neutral scalar masses. To a good approximation, then, the correction to the effective potential is of the form

$$\Delta V \simeq U(\mathscr{H}_2^\dagger \mathscr{H}_2,\, 0,\, 0,\, 0) = U(|v_2 + \varphi_2|^2 + |\mathscr{H}_2^-|^2,\, 0,\, 0,\, 0)\,. \tag{28.5.49}$$

By going back to the case where the charged fields vanish, we see that the function U must be just the same as what was earlier called U_t. Any terms in U of first order in an expansion in powers of $|v_2 + \varphi_2|^2 + |\mathscr{H}_2^-|^2 - v_2^2$ would simply serve to redefine the constant m_2^2, and are eliminated by the convention (28.5.38). Terms in U of second order in $|v_2+\varphi_2|^2+|\mathscr{H}_2^-|^2-v_2^2$ are genuine radiative corrections, but although they contain terms of second order in $|\varphi_2|^2$ which do affect the neutral scalar masses, they contain no terms of second order in $|\mathscr{H}_2^-|^2$ which could shift the non-Goldstone charged scalar mass. Fortunately, radiative corrections are not needed to avoid a conflict with experiment, because in the absence of an upper bound on m_A there is no theoretical upper bound on the charged scalar mass (28.5.35). An experimental lower bound[28] $m_C \geq 59$ GeV is provided by the non-observation of the process $e^+e^- \to \mathscr{H}^+\mathscr{H}^-$ at 181 to 184 GeV. There is a much more stringent lower bound on m_C provided by the rate of the process $b \to s\gamma$ (measured in decays like $B \to K^*\gamma$), which can occur by transitions to an intermediate $\mathscr{H}^- u$ or $\mathscr{H}^- c$ state, with the photon radiated from the virtual quark or $\mathscr{H}^-$. The present agreement between theory and experiment for this process sets a lower bound[29] of about 150 GeV on m_C (and higher for $\tan\beta < 1$). Using Eq. (28.5.35), this gives an important lower bound $m_A > 125$ GeV.

There are two conditions on the masses m_i^2 that must be satisfied by any model of supersymmetry breaking in order to give a successful account of electroweak symmetry breaking. One of them is provided by the requirement that the potential is bounded below, which as we have seen requires that

$$2|\mu|^2 + m_1^2 + m_2^2 > B\mu \,.$$

Since $B\mu$ is defined to be positive, this ensures that the C-odd neutral scalar squared mass (28.5.21) is positive. The other condition is provided by Eqs. (28.5.22) and (28.5.24), which for arbitrary values of β require that

$$4\,(m_1^2 + |\mu|^2)\,(m_2^2 + |\mu|^2) \leq (B\mu)^2 \,. \tag{28.5.50}$$

We can easily see from Eq. (28.5.10) that this condition ensures that the matrix of second derivatives of the potential has a negative eigenvalue at $\mathscr{H}_1 = \mathscr{H}_2 = 0$, so this $SU(2) \times U(1)$-invariant point is one of unstable equilibrium, and therefore $SU(2) \times U(1)$ must be spontaneously broken. If β is very near $\pi/2$, then Eq. (28.5.24) tells us that this condition is satisfied by having $m_1^2 + |\mu|^2$ positive and $m_2^2 + |\mu|^2$ *negative.* As we will see in the next section, the renormalization group flow of the parameters in the scalar field Lagrangian provides a mechanism for driving $m_2^2 + |\mu|^2$ negative.

Even with a minimal set of superfields, in supersymmetry theories there are several pairs of particles with different $SU(2) \times U(1)$ transformation properties, but the same charge, color, and baryon and lepton number, which become mixed when $SU(2) \times U(1)$ is spontaneously broken. We have already seen an example of this in the previous section, in which we had to deal with the mixing of the scalar superpartners of the left-handed quarks with the complex conjugates of the scalar superpartners of the left-handed antiquarks. A similar mixing occurs between the higgsinos and gauginos, both charged and neutral; the particles of definite mass are not higgsinos or gauginos, but mixtures known as *charginos* and *neutralinos.* Let us consider the charginos, which provide a useful bound on μ. According to Eq. (27.4.8), there are off-diagonal supersymmetric mass terms in the Lagrangian density

$$-\mathrm{Re}\,\left[\mu\left(h_{1L}^{-\mathrm{T}}\epsilon h_{2L}^{+}\right) + i\sqrt{2}m_W \cos\beta\left(w_L^{-\mathrm{T}}\epsilon h_{2L}^{+}\right) + i\sqrt{2}m_W \sin\beta\left(w_L^{+\mathrm{T}}\epsilon h_{1L}^{-}\right)\right] \,,$$

To this, we should add a term for the wino mass generated by gauge interactions with the supersymmetry-breaking sector

$$-m_{\mathrm{wino}}\mathrm{Re}\left(w_L^{+\mathrm{T}}\epsilon w_L^{-}\right) .$$

The squared chargino masses are then the eigenvalues of the matrix

$\mathscr{M}_C^\dagger \mathscr{M}_C$, where

$$\mathscr{M}_C = \begin{pmatrix} m_{\rm wino} & i\sqrt{2}m_W \sin\beta \\ i\sqrt{2}m_W \cos\beta & \mu \end{pmatrix} . \qquad (28.5.51)$$

These two eigenvalues are

$$m^2_{\rm chargino} = \frac{1}{2}\Big[m^2_{\rm wino} + 2m^2_W + |\mu|^2 \pm \Big((m^2_{\rm wino} - |\mu|^2)^2 + 4m^4_W \cos^2 2\beta$$
$$+4m^2_W(m^2_{\rm wino} + |\mu|^2 - 2m_{\rm wino}{\rm Re}\,\mu \sin 2\beta)\Big)^{1/2}\Big] . \qquad (28.5.52)$$

We expect the wino mass $m_{\rm wino}$ to be much larger than m_W. If it is also much larger than $|\mu|$, then the heavier chargino is mostly a wino, with mass $m_{\rm wino}$, while the lightest chargino is mostly a higgsino, with mass $|\mu|$. In any case, $|\mu|$ is greater than the lightest chargino mass, which the non-appearance of gauginos in e^+–e^- annihilation tells us is greater than about 60 GeV, and probably greater than m_W. The search for neutralinos in e^+–e^- annihilation (with assumptions like (28.4.2)) has set a lower bound of 27 GeV on the mass of the lightest neutralino.[29a]

28.6 Gauge Mediation of Supersymmetry Breaking

In this section we will consider the possibility that the breakdown of supersymmetry is transmitted to the known particles through interactions of the ordinary $SU(3) \times SU(2) \times U(1)$ gauge bosons and their superpartners.[30] It is assumed here that supersymmetry is dynamically broken in a sector of superfields, *not* including the superfields of the observed quarks and leptons, and that some of the chiral superfields in the symmetry-breaking sector, known as the *messenger superfields*, have non-vanishing $SU(3) \times SU(2) \times U(1)$ quantum numbers. In order for the messenger particles to be able to get large (say, of order 1 TeV) masses without breaking $SU(3) \times SU(2) \times U(1)$, it is necessary for them to furnish a real (or pseudoreal) representation of $SU(3) \times SU(2) \times U(1)$, which automatically also means that they do not introduce any new anomalies. Although most treatments of gauge-mediated supersymmetry breaking in the literature also make specific assumptions about the interactions of the messenger superfields with the other superfields responsible for supersymmetry breaking, the most important predictions of this class of theories in fact do not depend on these assumptions. We shall therefore put off assuming anything about the interactions of the messenger superfields with the other superfields of the symmetry-breaking sector. We will, however, make another assumption about the $SU(3)\times SU(2)\times U(1)$ properties of the

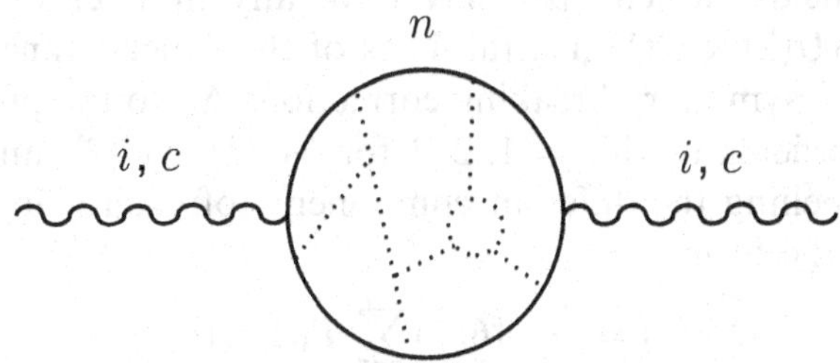

Figure 28.5. A diagram of the sort that introduces a breakdown of supersymmetry into the propagator of the gauge superfields. Here wavy lines are any component fields of the gauge superfields; solid lines are component fields of the messenger superfields; and dotted lines are component fields of the $SU(3) \times SU(2) \times U(1)$-neutral superfields of the supersymmetry-breaking sector.

messenger superfields, which has a strong phenomenological motivation. In order for the messenger particles not to interfere with the unification of couplings discussed in Section 28.2, we assume that they have the same ratios for the total traces of all squared $SU(3) \times SU(2) \times U(1)$ gauge generators as the ordinary quarks and leptons. This condition will automatically be satisfied if the messenger superfields (together perhaps with some $SU(3) \times SU(2) \times U(1)$-neutral chiral superfields) furnish a complete representation of some simple group G that contains $SU(3) \times SU(2) \times U(1)$, of which the quarks and leptons (again, together perhaps with some $SU(3) \times SU(2) \times U(1)$-neutral chiral superfields) also form a complete representation. (For instance, these left-chiral superfields might form N $SU(2)$ singlet $SU(3)$ triplets with charge $e/3$ and N $SU(2)$ doublet $SU(3)$ singlets with charges 0 and $-e$, which together form N representations $\mathbf{5}$ of $SU(5)$, together with an equal number of left-chiral superfields in the complex-conjugate representations of $SU(3) \times SU(2) \times U(1)$, which form N representations $\overline{\mathbf{5}}$.) However, for our present purposes we will neither need to assume that G is an actual symmetry group of the theory, nor adopt any particular choice of G or of the representations furnished by the messenger particles.

This interaction of the messenger superfields both with the other chiral and/or gauge superfields of the supersymmetry-breaking sector and with the $SU(3) \times SU(2) \times U(1)$ gauge superfields can be expected to produce a breakdown of supersymmetry in the propagators of the component fields of the $SU(3) \times SU(2) \times U(1)$ gauge superfields. To lowest order in the $SU(3) \times SU(2) \times U(1)$ couplings, the leading contribution to the propagators comes from the diagram shown in Figure 28.5, in which a pair of gauge, gaugino, or auxiliary D-field lines is attached to a loop of messenger fields, which also may have any number of interactions with $SU(3) \times SU(2) \times U(1)$-neutral fields of the supersymmetry-breaking sector. The supersymmetry-breaking corrections Δ_{ic} to the propagators of the gauge superfields (with $i = 1, 2, 3$ for $SU(3)$, $SU(2)$, and $U(1)$, and $c = V, \lambda, D$ labelling the different components of each gauge superfield) therefore have the forms

$$\begin{aligned}
\Delta_{3c}(q) &= (g_s^2/16\pi^2)\sum_n T_{3n}\Pi_{cn}(q)\,, \\
\Delta_{2c}(q) &= (g^2/16\pi^2)\sum_n T_{2n}\Pi_{cn}(q)\,, \\
\Delta_{1c}(q) &= (g'^2/16\pi^2)\sum_n T_{1n}\Pi_{cn}(q)\,,
\end{aligned} \tag{28.6.1}$$

where n labels the different messenger superfields; $\Pi_{cn}(q)$ are more-or-less complicated functions of the four-momentum q; T_{3n} and T_{2n} are the traces of the squares of any generator of $SU(3)$ and $SU(2)$, respectively, in the representations furnished by the nth messenger superfield (normalized so that in the defining representations $T_3 = T_2 = 1/2$); and T_{1n} is the sum of the squares of the electroweak hypercharges of the nth messenger superfield. One immediate consequence is that the gauginos acquire masses of the same form:*

$$\begin{aligned}
m_{\text{gluino}} &= (g_s^2/16\pi^2)\sum_n T_{3n}M_{gn}\,, \\
m_{\text{wino}} &= (g^2/16\pi^2)\sum_n T_{2n}M_{gn}\,, \\
m_{\text{bino}} &= (g'^2/16\pi^2)\sum_n T_{1n}M_{gn}\,,
\end{aligned} \tag{28.6.2}$$

where the M_{gn} are masses that characterize the different messenger superfields. As already mentioned, in order to preserve the unification of couplings at very high energy we assume that the sums of the T_n have the same ratios as for the observed quarks and leptons:

$$\sum_n T_{3n} = \sum_n T_{2n} = \sum_n 3T_{1n}/5 \equiv T\,. \tag{28.6.3}$$

The breakdown of supersymmetry in these propagators is then communicated to the squarks and sleptons of the supersymmetric standard model through the diagrams shown in Figure 28.6, in which an

* Recall that the bino is the superpartner of the $U(1)$ gauge field B_μ that appears in the Lagrangian of the standard model. We are not yet taking $SU(2) \times U(1)$ breaking into account, so the gaugino, squark, and slepton masses calculated here should be understood as parameters appearing in the $SU(3) \times SU(2) \times U(1)$-invariant effective Lagrangian of the standard mode.

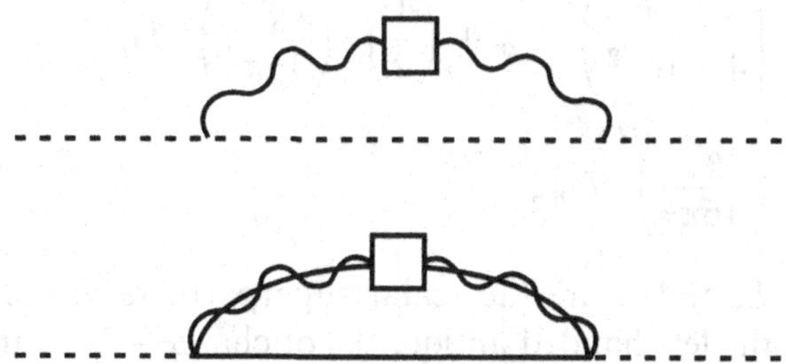

Figure 28.6. Diagrams that communicate supersymmetry breaking to the squarks and sleptons. Here dashed lines are squarks or sleptons; wavy lines are $SU(3) \times SU(2) \times U(1)$ gauge bosons or auxiliary D fields; solid lines are quarks or leptons; combined solid and wavy lines are $SU(3)\times SU(2)\times U(1)$ gauginos; and the squares represent insertions of the supersymmetry-breaking propagator correction shown in Figure 28.5.

$SU(3) \times SU(2) \times U(1)$ gauge boson or gaugino or auxiliary D field is emitted and reabsorbed by the squark or slepton. We are calculating the effective low-energy theory in which $SU(3)\times SU(2)\times U(1)$-breaking is not yet taken into account, so there is no mixing among the $SU(3)$, $SU(2)$, and $U(1)$ propagators, and each propagator acts like a unit matrix on the gauge indices. Thus the squared mass given to any squark or slepton will be proportional to a sum over the squares of all the $SU(3)\times SU(2)\times U(1)$ generators (including coupling constants) in the representation furnished by that squark or slepton. The sums of the squared $SU(2)$ and $SU(3)$ generators in the defining representations are

$$\sum_{a=1}^{3}\left(g\sigma_a/2\right)^2 = \frac{3g^2}{4}\cdot 1\,, \qquad \sum_{\alpha=1}^{8}\left(g_s\lambda_\alpha/2\right)^2 = \frac{4g_s^2}{3}\cdot 1\,,$$

where σ_a are the Pauli isospin matrices (5.4.18) and λ_α are the Gell-Mann matrices (19.7.2). For $U(1)$ the generator is just the weak hypercharge (21.3.7), including a factor g'. The squared squark and slepton masses therefore have the forms

$$M_{\tilde{Q}}^2 = 2\sum_n M_{sn}^2\left[\frac{4}{3}\left(\frac{g_s^2}{16\pi^2}\right)^2 T_{3n} + \frac{3}{4}\left(\frac{g^2}{16\pi^2}\right)^2 T_{2n} + \left(\frac{1}{6}\right)^2\left(\frac{g'^2}{16\pi^2}\right)^2 T_{1n}\right]$$

$$M_{\tilde{U}}^2 = 2\sum_n M_{sn}^2\left[\frac{4}{3}\left(\frac{g_s^2}{16\pi^2}\right)^2 T_{3n} + \left(\frac{2}{3}\right)^2\left(\frac{g'^2}{16\pi^2}\right)^2 T_{1n}\right]\,,$$

$$M_{\tilde{D}}^2 = 2\sum_n M_{sn}^2\left[\frac{4}{3}\left(\frac{g_s^2}{16\pi^2}\right)^2 T_{3n} + \left(-\frac{1}{3}\right)^2\left(\frac{g'^2}{16\pi^2}\right)^2 T_{1n}\right]\,, \qquad (28.6.4)$$

$$M_{\tilde{L}}^2 = 2\sum_n M_{sn}^2 \left[\frac{3}{4}\left(\frac{g^2}{16\pi^2}\right)^2 T_{2n} + \left(\frac{1}{2}\right)^2 \left(\frac{g'^2}{16\pi^2}\right)^2 T_{1n}\right],$$

$$M_{\tilde{\bar{E}}}^2 = 2\sum_n M_{sn}^2 \left(\frac{g'^2}{16\pi^2}\right)^2 T_{1n},$$

where Q, $\bar{U}$, $\bar{D}$, L, and $\bar{E}$ are the scalar superpartners of the left-handed quark doublets, the left-handed antiquarks of charge $-2e/3$ and $+e/3$, the left-handed lepton doublets, and the left-handed charged antileptons, and M_{sn} are some new masses that characterize the nth messenger superfields. (The factor 2 is extracted from M_{sn}^2 for future convenience.) The squark and slepton masses that are produced in this way are automatically the same in all three generations, thus avoiding the problem with flavor-changing processes discussed in Section 28.4.

We expect all M_{gn} and M_{sn} to be roughly of the same order of magnitude, so that the gluino and squarks will have comparable masses, while the wino, bino, and sleptons will be much lighter, with masses suppressed by squares of electroweak coupling constants.

We can go considerably further than this with some reasonable dynamical assumptions. Suppose that the effects of supersymmetry breaking on the messenger superfields may be modeled by including these superfields along with a set of $SU(3)\times SU(2)\times U(1)$-neutral chiral superfields S_n (not necessarily all distinct) in a superpotential

$$f(\Phi,\bar{\Phi},S) = \sum_n \lambda_n S_n \Phi_n \bar{\Phi}_n , \tag{28.6.5}$$

where $\bar{\Phi}_n$ and Φ_n are left-chiral messenger superfields in complex-conjugate representations of $SU(3) \times SU(2) \times U(1)$ and λ_n are a set of coupling coefficients. (Here and below we suppress the $SU(3) \times SU(2)$ indices that are summed over in calculating scalar products like $\Phi_n\bar{\Phi}_n$.) The superfields S_n are supposed to have non-vanishing vacuum expectation values $\mathscr{S}_n$ and $\mathscr{F}_n$ for their scalar and auxiliary components, respectively. It is the non-zero values of $\mathscr{F}_n$ that in these models introduce supersymmetry breaking in the masses of the Φ_n and $\bar{\Phi}_n$ particles. Section 26.4 shows that, with gauge couplings neglected, the squared masses of the spinor components of Φ_n (and $\bar{\Phi}_n$) are the eigenvalues of the matrix $\mathscr{M}_n^\dagger \mathscr{M}_n$, with $\mathscr{M}_n$ defined by Eq. (26.4.11), which gives

$$\mathscr{M}_n = \begin{pmatrix} 0 & \lambda_n \mathscr{S}_n \\ \lambda_n \mathscr{S}_n & 0 \end{pmatrix},$$

so that the messenger fermions have masses $|\lambda_n \mathscr{S}_n|$. To find the mass terms for the scalar components ϕ_n and $\bar{\phi}_n$ of the Φ_n and $\bar{\Phi}_n$ superfields, we note that integrating out the auxiliary fields of the Φ_n and $\bar{\Phi}_n$ yields a potential

$$\sum_n \left|\frac{\partial f(\phi,\bar{\phi},\mathscr{S})}{\partial \phi_n}\right|^2 + \sum_n \left|\frac{\partial f(\phi,\bar{\phi},\mathscr{S})}{\partial \bar{\phi}_n}\right|^2 = \sum_n |\lambda_n \mathscr{S}_n|^2 \Big[|\phi_n|^2 + |\bar{\phi}_n|^2\Big] ,$$

to which we now must add the contribution of the auxiliary component of the S_n, given by the second term of Eq. (26.4.4) as:

$$2\text{Re} \sum_n \left[\lambda_n \mathscr{F}_n \frac{\partial f(\phi,\bar{\phi},\mathscr{S})}{\partial \mathscr{S}_n}\right] = 2\text{Re} \sum_n \left[\mathscr{F}_n \lambda_n \phi_n \bar{\phi}_n\right] .$$

The complex scalar fields of definite mass are then $(\phi_n \pm e^{-i\alpha_n}\bar{\phi}_n)/\sqrt{2}$, where α_n is the phase of $\lambda_n \mathscr{F}_n$, with squared masses $|\lambda_n \mathscr{S}_n|^2 \pm |\lambda_n \mathscr{F}_n|$. (Note that this pattern, of a pair of complex scalars with squared masses equidistant above and below a Majorana fermion squared mass, is just what we would expect from the sum rule (27.5.11).) Since these squared masses must be positive, it follows that

$$|\mathscr{F}_n| \leq |\lambda_n||\mathscr{S}_n|^2 . \tag{28.6.6}$$

The gaugino masses in models based on Eq. (28.6.5) are given by diagrams of the form shown in Figure 28.5, but now with just a single loop, not including what are shown as dotted lines in Figure 28.5. A detailed calculation gives the coefficients M_{gn} in Eq. (28.6.2) as[31]

$$M_{gn} = \frac{|\mathscr{F}_n|}{|\mathscr{S}_n|} g\left(\frac{|\mathscr{F}_n|}{|\lambda_n||\mathscr{S}_n|^2}\right) , \tag{28.6.7}$$

where

$$\begin{aligned} g(x) &= \frac{1}{2x^2}\Big[(1+x)\ln(1+x) + (1-x)\ln(1-x)\Big] \\ &= 1 + \frac{x^2}{6} + \frac{x^4}{15} + \frac{x^6}{28} + \cdots . \end{aligned} \tag{28.6.8}$$

The masses of the squarks and sleptons are given by the diagrams of Figure 28.6, which now involve just two loops. Another detailed calculation gives the mass parameters M^2_{sn} in Eq. (28.6.4) as[31]

$$M^2_{sn} = \frac{|\mathscr{F}_n|^2}{|\mathscr{S}^2_n|} f\left(\frac{|\mathscr{F}_n|}{|\lambda_n||\mathscr{S}_n|^2}\right) , \tag{28.6.9}$$

where

$$\begin{aligned} f(x) &= \frac{1+x}{x^2}\left[\ln(1+x) - 2\,\text{Li}_2\left(\frac{x}{1+x}\right) + \frac{1}{2}\text{Li}_2\left(\frac{2x}{1+x}\right)\right] + x \to -x \\ &= 1 + \frac{1}{36}x^2 - \frac{11}{450}x^4 - \frac{319}{11760}x^6 + \cdots , \end{aligned} \tag{28.6.10}$$

with Li_2 the dilogarithm

$$\text{Li}_2(x) \equiv -\int_0^x \frac{\ln(1-t)}{t}\,dt\,. \tag{28.6.11}$$

In particular, if (as is usually assumed) the various S_n are all the same, and if $|\mathscr{F}| \ll |\lambda_n||\mathscr{S}|^2$ for all n, then f and g in Eqs. (28.6.9) and (28.6.7) may be set equal to 1, so that

$$M_{gn} = M_{sn} = |\mathscr{F}|/|\mathscr{S}| \equiv M\,. \tag{28.6.12}$$

Using Eq. (28.6.3), we can express the gaugino masses (28.6.2) as

$$\begin{aligned} m_{\text{wino}} &= (g^2/16\pi^2)TM\,, \\ m_{\text{bino}} &= (5/3)(g'^2/16\pi^2)TM\,, \\ m_{\text{gluino}} &= (g_s^2/16\pi^2)TM\,, \end{aligned} \tag{28.6.13}$$

while the squared squark and slepton masses (28.6.4) become

$$\begin{aligned} M^2_{\tilde{Q}} &= 2TM^2\left[\frac{4}{3}\left(\frac{g_s^2}{16\pi^2}\right)^2 + \frac{3}{4}\left(\frac{g^2}{16\pi^2}\right)^2 + \frac{5}{3}\left(\frac{1}{6}\right)^2\left(\frac{g'^2}{16\pi^2}\right)^2\right]\,, \\ M^2_{\tilde{\bar{U}}} &= 2TM^2\left[\frac{4}{3}\left(\frac{g_s^2}{16\pi^2}\right)^2 + \frac{5}{3}\left(\frac{2}{3}\right)^2\left(\frac{g'^2}{16\pi^2}\right)^2\right]\,, \\ M^2_{\tilde{\bar{D}}} &= 2TM^2\left[\frac{4}{3}\left(\frac{g_s^2}{16\pi^2}\right)^2 + \frac{5}{3}\left(-\frac{1}{3}\right)^2\left(\frac{g'^2}{16\pi^2}\right)^2\right]\,, \\ M^2_{\tilde{L}} &= 2TM^2\left[\frac{3}{4}\left(\frac{g^2}{16\pi^2}\right)^2 + \frac{5}{3}\left(\frac{1}{2}\right)^2\left(\frac{g'^2}{16\pi^2}\right)^2\right]\,, \\ M^2_{\tilde{\bar{E}}} &= 2TM^2\frac{5}{3}\left(\frac{g'^2}{16\pi^2}\right)^2\,. \end{aligned} \tag{28.6.14}$$

There is no special reason to expect that $|\mathscr{F}| \ll |\lambda_n \mathscr{S}|^2$, but this assumption is not actually very restrictive, because Eq. (28.6.6) already requires that $|\mathscr{F}| \leq |\lambda_n||\mathscr{S}|^2$, and it turns out that the functions $f(x)$ and $g(x)$ do not differ much from unity for $x < 1$ unless x is very close to 1.

The extreme simplicity of the results (28.6.13) and (28.6.14) has been explained by Giudice and Rattazzi[32] by using Seiberg's arguments of holomorphy,[33] described in Section 27.6. Suppose we introduce a superpotential for the messenger superfields like (28.6.5), but with a single external singlet superfield S:

$$f(S,\Phi) = S\sum_n \lambda_n\,\Phi_n\bar{\Phi}_n\,. \tag{28.6.15}$$

Among other effects the kinematic term for the gauge superfield V_i (with $i = 3, 2, 1$ for $SU(3)$, $SU(2)$, and $U(1)$) in the Wilsonian effective Lagrangian at a renormalization scale μ will now take the form

$$\mathscr{L}_{\text{gauge},\mu} = \text{Re}\left[\sum_i N_i(S,\mu)\sum_{\alpha\beta}\left(W_{iL\alpha}\epsilon_{\alpha\beta}W_{iL\beta}\right)\right]_{\mathscr{F}}, \tag{28.6.16}$$

with some functions $N_i(S,\mu)$ replacing the factors $1/2g_i^2(\mu)$ in Eq. (27.3.22). (The θ-term is dropped here because it has no effect in perturbation theory. A sum is implied over indices on $W_{iL\alpha}$ that label different members of the adjoint representations of $SU(3)$ and $SU(2)$, which are not explicitly shown.) The gauge coupling constant is now given by setting the superfield S equal to the expectation value $\mathscr{S}$ of its scalar component

$$\frac{1}{2g_i^2(\mu)} = N_i(\mathscr{S},\mu)\,. \tag{28.6.17}$$

Also, recalling that $W_{iL\alpha} = \lambda_{iL\alpha} + O(\theta)$ and using Eq. (27.2.11), the terms in the Lagrangian density (28.6.16) of second order in the gaugino fields are, up to derivative terms,

$$-2\sum_i \text{Re}\left[N_i(\mathscr{S},\mu)\left(\bar{\lambda}_{iR}\,\partial\!\!\!/\lambda_{iR}\right) + [N_i(S,\mu)]_{\mathscr{F}}\left(\lambda_{iL}^{\text{T}}\epsilon\lambda_{iL}\right)\right]\,.$$

This gives the gaugino masses

$$m_{gi}(\mu) = \left|\frac{[N_i(S,\mu)]_{\mathscr{F}}}{2N_i(\mathscr{S},\mu)}\right| = g_i^2(\mu)\left|[N_i(S,\mu)]_{\mathscr{F}}\right|\,. \tag{28.6.18}$$

Now let us consider the behavior of $N_i(\mathscr{S},\mu)$ as a function of a real positive $\mathscr{S}$, with the phases of the messenger superfields adjusted so that all λ_n are real and positive. Suppose we fix the values of the gauge couplings $g_i(\mu)$ at some scale $\mu = K$ above all the messenger particle masses. Taking account of the change in the constants b_i in the renormalization group equations $\mu dg_i(\mu)/d\mu = b_i g_i^3$ as μ passes through the various messenger masses, the solution of this equation when μ is *below* all the messenger particle masses takes the form

$$\frac{1}{g_i^2(\mu)} = \frac{1}{g_i^2(K)} - 2b_i^{(0)}\ln\left(\frac{M_1}{K}\right) - 2b_i^{(1)}\ln\left(\frac{M_2}{M_1}\right) - \cdots - 2b_i^{(N)}\ln\left(\frac{\mu}{M_N}\right),$$

where we label the messenger particles so that their masses $M_n = \lambda_n\mathscr{S}$ satisfy

$$M_1 > M_2 > \cdots > M_N\,,$$

and $b_i^{(n)}$ is calculated taking account only of particles with masses less than M_n. Since all M_n are proportional to $\mathscr{S}$, we see that $N_i(\mathscr{S},\mu)$ has the

$\mathscr{S}$ dependence

$$N_i(\mathscr{S},\mu) = -b_i^{\text{messenger}} \ln \mathscr{S} + \mathscr{S}\text{-independent terms}\,, \tag{28.6.19}$$

where $b_i^{\text{messenger}} = b_i^{(0)} - b_i^{(N)}$ is the contribution to b_i of all the messenger superfields. According to Eq. (27.9.45) (with $C_{i1} = 0$ and $C_{i2}^f = C_{i2}^s = \sum_n T_{in}$), this is

$$b_i^{\text{messenger}} = \frac{1}{16\pi^2}\sum_n T_{in}\,. \tag{28.6.20}$$

Since $N_i(S,\mu)$ is required by supersymmetry to be a holomorphic function of S, we see that

$$N_i(S,\mu) = -\frac{1}{16\pi^2}\sum_n T_{in} \ln S + S\text{-independent terms}\,. \tag{28.6.21}$$

Expanding around $S = \mathscr{S}$, to first order in $\mathscr{F}$ we have $[\ln \mathscr{S}]_{\mathscr{F}} = \mathscr{F}/\mathscr{S}$ and so Eq. (28.6.18) gives the gaugino masses

$$m_{gi}(\mu) = \frac{g_i^2(\mu)}{16\pi^2}\sum_n T_{in}\left|\frac{\mathscr{F}}{\mathscr{S}}\right|\,. \tag{28.6.22}$$

Using Eq. (28.6.3), we see that this is the same as our previous result (28.6.13). Eq. (28.6.14) for the squark and slepton masses was obtained by Giudice and Rattazzi in a similar way, by studying the kinematic terms for the quark and lepton superfields instead of the gauge superfields.

Incidentally, Eqs. (28.6.13) and (28.6.14) could be obtained (in general with different values of M in Eq. (28.6.13) and Eq. (28.6.14)) without a specific dynamical assumption like Eq. (28.6.5), if we supposed that the supersymmetry-breaking sector respected invariance under some grand unified group G that had both the known quarks and leptons and the messenger fields as complete representations. In this case the coefficients M_{gn} and M_{sn} in Eqs. (28.6.2) and (28.6.4) would have values $M_g(d)$ and $M_s(d)$, respectively, that depend only on the irreducible representation d of G to which the nth messenger fields belong. The sums of T_{in} over n belonging to any irreducible representation d of G have the same ratios as the sums in Eq. (28.6.3), so $\sum_{n\in d} T_{in} = k_i T(d)$, where $k_3 = k_2 = 1$, $k_1 = 5/3$, and therefore

$$\sum_n T_{in} M_{gn} = \sum_d M_g(d) \sum_{n\in d} T_{in} = k_i M_g\,,$$

where $M_g = \sum_d M_g(d) T(d)$. Likewise

$$\sum_n T_{in} M_{sn}^2 = \sum_d M_s^2(d) \sum_{n\in d} T_{in} = k_i M_s^2\,,$$

where $M_s^2 = \sum_d M_s^2(d)T(d)$. Eqs. (28.6.2) and (28.6.4) would then yield Eqs. (28.6.13) and (28.6.14), except with M_g in place of TM and M_s^2 in place of $2TM^2$. But the assumption that M_{gn} and M_{sn} respect invariance under G is implausible if the messenger mass scale is far below the grand unified scale, since whatever we assume about a grand unified gauge group, the $SU(3) \times SU(2) \times U(1)$ gauge interactions would make coupling constants like λ_n run differently for Φ_n with different $SU(3) \times SU(2) \times U(1)$ quantum numbers in the same representation of the grand unified group.

These results are subject to various radiative corrections, of which the most important is that we must use values of g_s, g, and g' renormalized at a scale comparable to the mass being calculated. Indeed, the gaugino mass ratios given in Eq. (28.6.13) could also be derived under quite different assumptions: that all gaugino masses are equal at the grand unification scale where the coupling constants are related by $g_s^2 = g^2 = 5g'^2/3$ and become different at lower energies as described by the renormalization group equations.

For an illustrative example of numerical results, suppose that the messenger superfields form an $SU(2)$ singlet $SU(3)$ triplet with charge $e/3$ and an $SU(2)$ doublet $SU(3)$ singlet with charges 0 and $-e$, together with left-chiral superfields in the complex-conjugate representations of $SU(3) \times SU(2) \times U(1)$. Then, as already mentioned, Eq. (28.6.3) is satisfied with $T = 2 \times 1/2 = 1$, so by using the correct values of the gauge couplings, the masses of the squarks, gluino, L sleptons, wino, E sleptons, and bino are calculated to be in the ratios[34] 11.6 :: 7.0 :: 2.5 :: 2 :: 1.1 :: 1.0. In larger representations of G we could have $T \gg 1$, in which case the gluino would be the heaviest of these particles, and the sleptons the lightest ones.

There are also radiative corrections aside from the running of the gauge couplings. According to one calculation,[34] in the model with an $SU(2)$ singlet $SU(3)$ triplet with charge $e/3$ and an $SU(2)$ doublet $SU(3)$ singlet with charges 0 and $-e$, together with left-chiral superfields in the complex-conjugate representations of $SU(3) \times SU(2) \times U(1)$, radiative corrections give the ratios of the masses of the squarks, gluino, L sleptons, wino, E sleptons, and bino as 9.3 :: 6.4 :: 2.6 :: 1.9 :: 1.35 :: 1.0.

As we saw in the previous section, the wino and bino may mix with the charged and neutral higgsinos, so the wino and bino masses that have been calculated here must be regarded as inputs to a calculation of physical masses of mixtures known as *charginos* and *neutralinos*, rather than physical masses themselves.

Now we shall consider the masses of the Higgs scalars in these models. If we were to consider only the two-loop diagrams by which these scalars get masses through gauge interactions with the supersymmetry-breaking sector, then since (apart from signs) they have the same $SU(3) \times SU(2) \times$

$U(1)$ quantum numbers as the left-handed lepton doublets, their masses would be given by a formula like the fourth of Eqs. (28.6.4)

$$[m_1^2]_{2\text{ loop}} = [m_2^2]_{2\text{ loop}} = M_L^2$$
$$= \sum_n M_{sn}^2 \left[\frac{3}{4} \left(\frac{g^2}{16\pi^2} \right)^2 T_{2n} + \left(\frac{1}{2} \right)^2 \left(\frac{g'^2}{16\pi^2} \right)^2 T_{1n} \right]^2 . \quad (28.6.23)$$

If this were the whole story, then it would be impossible to meet the condition for $SU(2) \times U(1)$ breaking found in the previous section that (unless $\tan\beta$ is very close to unity) one of $m_1^2 + |\mu|^2$ and $m_2^2 + |\mu|^2$ must be negative. Fortunately, the large masses of the top quark and squark produce a negative contribution to m_2^2 that leads naturally to a spontaneous breakdown of electroweak symmetry. The couplings of the Higgs doublets to the third-generation quark superfields are described by the superpotential

$$f_{3\text{rd gen}} = \lambda_b \left(H_1^{\mathrm{T}} e Q \right) \bar{B} + \lambda_t \left(H_2^{\mathrm{T}} e Q \right) \bar{T} , \quad (28.6.24)$$

where Q is the $SU(2)$ quark doublet left-chiral superfield (T, B), while $\bar{T}$ and $\bar{B}$ are the left-chiral superfields of the left-handed top and bottom antiquarks, and λ_t and λ_b are the Yukawa couplings, related to the t and b quark masses by $m_t = \lambda_t v_2$ and $m_b = \lambda_b v_1$. The last term of Eq. (26.4.7) then gives the terms in the potential involving an interaction between the squark fields and Higgs fields as

$$V_{sq\,H} = \left| \lambda_b \mathscr{H}_1^- \bar{\mathscr{B}} + \lambda_t \mathscr{H}_2^0 \bar{\mathscr{T}} \right|^2 + \left| \lambda_b \mathscr{H}_1^0 \bar{\mathscr{B}} + \lambda_t \mathscr{H}_2^+ \bar{\mathscr{T}} \right|^2$$
$$+ \left| \lambda_b \right|^2 \left| \mathscr{H}_1^0 \mathscr{B} - \mathscr{H}_1^- \mathscr{T} \right| + \left| \lambda_t \right|^2 \left| \mathscr{H}_2^+ \mathscr{B} - \mathscr{H}_2^0 \mathscr{T} \right|^2 , \quad (28.6.25)$$

where script letters denote the scalar field components of superfields. The squark loop contributions to the potential of $\mathscr{H}_1$ and $\mathscr{H}_2$ are then

$$V_H^{\text{squark loop}} = 3\langle \mathscr{S} \mathscr{S}^* \rangle \left[2|\lambda_b|^2 \left(\mathscr{H}_1^\dagger \mathscr{H}_1 \right) + 2|\lambda_t|^2 \left(\mathscr{H}_2^\dagger \mathscr{H}_2 \right) \right] , \quad (28.6.26)$$

where $\langle \mathscr{S} \mathscr{S}^* \rangle$ is the vacuum expectation value of the product of any one of the squark fields and its complex conjugate at the same spacetime point. (In taking this to be the same for all squark types, we are here using Eq. (28.6.4), which tells us that the squark masses M_Q do not vary much among $\mathscr{T}$, $\mathscr{B}$, $\bar{\mathscr{T}}$, and $\bar{\mathscr{B}}$ squarks. The factor 3 in Eq. (28.6.26) takes account of the three colors of each squark type.) The vacuum expectation value $\langle \mathscr{S} \mathscr{S}^* \rangle$ is given in lowest order by

$$\langle \mathscr{S} \mathscr{S}^* \rangle \equiv \langle \mathscr{S}(x) \mathscr{S}^*(x) \rangle_{\text{VAC}} = \frac{-i}{(2\pi)^4} \int \frac{d^4 p}{p^2 + M_Q^2 - i\epsilon} . \quad (28.6.27)$$

This of course is divergent, but the contribution to supersymmetry-breaking coefficients would be cancelled by quark loops if the squarks had the zero bare mass of the quarks, so the effect of quark loops is to subtract from (28.6.27) the same expression with M_Q replaced with zero, which with a Wick rotation becomes

$$\langle \mathscr{S}\mathscr{S}^* \rangle \to \frac{M_Q^2\, i}{(2\pi)^4} \int \frac{d^4p}{(p^2 + M_Q^2 - i\epsilon)(p^2 - i\epsilon)}$$
$$= -\frac{M_Q^2}{(2\pi)^4} \int_0^{M^2} \frac{\pi^2 dp^2}{p^2 + M_Q^2} \simeq -\frac{M_Q^2}{16\pi^2} \ln\left(\frac{M^2}{M_Q^2}\right) .$$

We have inserted an ultraviolet cut-off at the messenger mass M, because at momenta above M the squark mass must be replaced with a momentum-dependent mass, which goes to the supersymmetric value zero at very high momenta. Making this substitution in Eq. (28.6.26) gives a net contribution to the potential due to squark and quark loops:

$$(V_m)^{3\text{ loop}} = \frac{3M_Q^2}{16\pi^2} \ln\left(\frac{M^2}{M_Q^2}\right) \left[2|\lambda_b|^2\left(\mathscr{H}_1^\dagger \mathscr{H}_1\right) + 2|\lambda_t|^2\left(\mathscr{H}_2^\dagger \mathscr{H}_2\right)\right] . \tag{28.6.28}$$

(This is a three-loop contribution, because the squared squark masses are given by two-loop diagrams. There are also terms in the potential that are quadratic in both the Higgs and squark superfields, arising from products of the Higgs and squark terms in the squares of the $SU(2) \times U(1)$ gauge field D-components. These do not make a three-loop contribution to the Higgs masses because the sum of each $SU(2) \times U(1)$ quantum number of the squarks vanishes, so that their contributions to the $SU(2) \times U(1)$ gauge field D-components have zero vacuum expectation value.) Comparing Eq. (28.6.28) with Eq. (28.5.9) and adding the two-loop contribution (28.6.23) to the masses, we see then that

$$m_1^2 \simeq M_L^2 - \frac{3M_Q^2|\lambda_b|^2}{8\pi^2} \ln\left(\frac{M^2}{M_Q^2}\right) , \tag{28.6.29}$$

$$m_2^2 \simeq M_L^2 - \frac{3M_Q^2|\lambda_t|^2}{8\pi^2} \ln\left(\frac{M^2}{M_Q^2}\right) . \tag{28.6.30}$$

Using Eq. (28.6.14) and $|\lambda_t| = m_t/v_2 = m_t(2\sqrt{2}G_F)^{1/2}/\sin\beta$, we may write

(28.6.30) as

$$m_2^2 \simeq 2TM^2\left[\frac{3}{4}\left(\frac{g^2}{16\pi^2}\right)^2+\frac{5}{12}\left(\frac{g'^2}{16\pi^2}\right)^2\right.$$
$$\left.-\frac{\sqrt{2}G_F m_t^2}{\pi^2\sin^2\beta}\left(\frac{g_s^2}{16\pi^2}\right)^2\ln\left(\frac{3}{8T(g_s^2/16\pi^2)}\right)\right]. \qquad (28.6.31)$$

For $T=1$, $g_s^2/4\pi=0.118$, $g^2/4\pi=0.0340$, $g'^2/4\pi=0.0101$, and $m_t=180$ GeV, this is

$$m_2^2 \simeq M_L^2\left[1-\frac{3.06}{\sin^2\beta}\right], \qquad (28.6.32)$$

which is negative for all values of β, thus providing a natural mechanism for the spontaneous breaking of the electroweak gauge symmetry. Also, $M_L^2=(0.91\times10^{-4})M^2/8\pi^2$. Unless $\tan\beta$ is huge, we have $|\lambda_b|\ll|\lambda_t|$, so Eq. (28.6.27) gives

$$m_1^2 \simeq M_L^2\,. \qquad (28.6.33)$$

The aspect of electroweak phenomenology for which the predictions of gauge-mediated supersymmetry models are most uncertain and unsatisfactory has to do with the parameter μ in the supersymmetry-preserving term $\mu[(H_1^{\rm T}eH_2)]_{\mathscr{F}}$ and the related supersymmetry-breaking term $B\mu$ in the Lagrangian density. These are related because the interactions of the Higgs superfields with the gauge, lepton, and quark superfields are invariant under a symmetry

$$\begin{array}{ll} H_1\to e^{i\varphi}H_1\,, & H_2\to e^{i\varphi}H_2\,,\\ Q\to e^{-i\varphi}Q\,, & V_i\to V_i\,,\\ \bar D\to\bar D\,, & \bar U\to\bar U\,, \end{array} \qquad (28.6.34)$$

which in the absence of a superpotential term $\mu(H_1^{\rm T}eH_2)$ would forbid radiative corrections from producing a term $B\mu\,{\rm Re}(\mathscr{H}_1^{\rm T}e\mathscr{H}_2)$ in the scalar field potential.

It is not possible for $B\mu$ to vanish because then Eq. (28.5.22) and the fact that (as we have seen) $m_A\neq0$ would imply that $\sin2\beta=0$, or in other words either $v_1=0$ or $v_2=0$, which would imply that either all the charge $-e/3$ quarks and charged leptons are massless or all the charge $+2e/3$ quarks are massless. (If $B\mu=0$ and $\mu=0$ then this problem persists to all orders, because the appearance of non-zero vacuum expectation values for both v_1 and v_2 would imply that symmetry under any combination of the transformation (28.6.34) and electroweak $U(1)$ gauge transformations is spontaneously broken, so the C-odd neutral scalar would be a Goldstone boson with $m_A=0$.) It is natural to try to account for a non-zero value of $B\mu$ as a radiative correction in a theory in which symmetry under the transformation (28.6.34) is explicitly broken by a supersymmetric term $\mu[(H_1^{\mathrm{T}} e H_2)]_{\mathscr{F}}$ in the Lagrangian density. This gives a very small value of $B\mu$ at the messenger scale,[34] although renormalization group effects greatly increase $B\mu$ at lower energies. According to Eq. (28.5.22), a relatively small $B\mu$ would fit well with the idea that the large mass of the top quark arises from a large value of $\tan\beta$. In any case, as already mentioned, the experimental lower bound on the chargino mass tells us that $|\mu|$ is at least about 60 GeV.

The trouble is that the appearance of a non-zero value of μ resurrects the hierarchy problem that supersymmetry was supposed to solve: instead of asking why the Higgs mass terms in the Lagrangian density are so much smaller than the Planck mass or the mass at which gauge couplings are unified, we now have to ask why μ is this small?

The hierarchy problem would be put to rest if the Higgs superfields interact with the supersymmetry-breaking sector in such a way that a term $\mu[(H_1^{\mathrm{T}} e H_2)]_{\mathscr{F}}$ is forbidden by some symmetry, but appears when that symmetry is spontaneously broken. A massless Goldstone boson may be avoided if the symmetry is discrete rather than continuous. The simplest possibility is simply to extend the symmetry transformation (28.6.34) to include a transformation

$$S \to e^{-2i\varphi} S ,$$

which would allow a term in the superpotential of the form

$$\lambda' S(\mathscr{H}_1^{\mathrm{T}} e \mathscr{H}_2) .$$

We can avoid a continuous symmetry by also including a term S^3 in the superpotential, so that the Lagrangian is invariant only under transformations with φ a multiple of $2\pi/3$, which is still enough to forbid a non-zero bare value of μ. In this case the appearance of non-zero vacuum values $\mathscr{S}$ and $\mathscr{F}$ for the scalar and auxiliary components of S yields

$$B\mu = |\lambda' \mathscr{F}| , \qquad \mu = |\lambda' \mathscr{S}| .$$

This would result in B having a very large $|\lambda'|$-independent value M, given by (28.6.12), which is greater by a factor of order $(g_s^2/16\pi)^{-1} \simeq 100$ than the squark or gluino masses. But then, since Eqs. (28.6.32) and (28.6.33) give $m_1^2 + m_2^2 < 0$, the stability condition (28.5.11) would require that $|\mu| \geq M/2$, and hence Eq. (28.5.22) would require that m_A is also much greater than the squark and slepton masses. This is ruled out by the relation (28.5.23) and the estimates (28.6.33) and (28.6.32) of m_1^2 and m_2^2 unless $\tan\beta$ is very close to unity.

We will see in Section 31.6 that theories of gravity-mediated supersymmetry breaking naturally yield acceptable values of $B\mu$ and μ. Such theories are characterized by a very high energy scale of supersymmetry breaking, in various versions of order either 10^{11} GeV or 10^{13} GeV. There have been several suggestions[35] on ways to obtain acceptable values of $B\mu$ and μ in theories in which supersymmetry is broken at relatively low energy, such as theories of gauge-mediated supersymmetry breaking, but none of them is particularly compelling. Also, since we do not know where μ comes from, we do not have any reason to suppose that it is real, so theories of gauge-mediated supersymmetry breaking are at risk of yielding too much CP violation, just as in the more general framework of Section 28.4.

Like the scalar squared masses m_1^2 and m_2^2, the parameters A_{ij} and C_{ij} in Eq. (28.4.1) are given by two-loop diagrams. However, they have the dimensions of mass rather than mass squared, and they are much less than the scalar and gaugino masses, so they make a relatively unimportant contribution to supersymmetry breaking.

As in any model with supersymmetry broken at energies much less than 10^{10} GeV, the lightest R-odd particle in all models based on gauge-mediated supersymmetry breaking is the gravitino. As we will see in Section 31.3, the gravitino mass is of the order of $\sqrt{G}$ times the squared energy F characteristic of supersymmetry breaking, defined so that the vacuum energy is $F^2/2$. Where supersymmetry is broken by the $\mathscr{F}$-terms $\mathscr{F}_{n0}$ of the $SU(3) \times SU(2) \times U(1)$-neutral chiral superfields S_n, we have $F^2 = \sum_n |\mathscr{F}_{n0}|^2$. If there are no large dimensionless quantities in the Lagrangian for the S_n, then squark masses in this model are of the order of $g_s^2\sqrt{F}/16\pi^2 \approx 10^{-2}\sqrt{F}$, so in order for this to be less than the naturalness bound of 10^4 GeV, we must have $\sqrt{F} < 10^6$ GeV, which gives a gravitino mass less than 1 keV. Gravitational couplings are so weak at accessible energies that it is only the helicity $\pm 1/2$ states of the gravitino that can actually be produced, and these behave just like goldstino states. As shown by Eq. (27.5.12), in the models discussed here the goldstino field appears with coefficient $i\sqrt{2}\mathscr{F}_n$ in the fermionic component ψ_n of S_n. Goldstinos are emitted in the decay of the R-odd sparticles into the corresponding R-even particles of the standard model through radiative corrections, with goldstinos emerging from vertices connecting internal ψ_n and $\mathscr{S}_n$ lines. According to Eq. (29.2.10), the goldstino emission amplitudes are inversely proportional to F, which makes these decays comparatively slow, although perhaps fast enough to be observed.

Because the decay of R-odd particles into goldstinos is slow, it is phenomenologically important in these models to identify the next-to-lightest R-odd particle, into which all the heavier R-odd particles will decay before it in turn decays into a goldstino. As we have seen, the next-to-lightest R-odd particle is usually a slepton, wino, or bino. (It is possible in models with a messenger superfield having the same $SU(3) \times SU(2) \times U(1)$

quantum numbers as the Higgs doublets that the mixing between these superfields would lower the mass of the doublet messengers so much that the lightest R-odd particle would be the gluino.[35a]) Detailed calculations including the effects of $SU(2)\times U(1)$ breaking indicate that here is a large region of parameter space in which the next-to-lightest R-odd particle is one of the two tau sleptons.[35b]

28.7 Baryon and Lepton Non-Conservation

The extra particles in supersymmetric models provide several new mechanisms for baryon and lepton non-conservation. We saw in Section 28.1 that there are various baryon- and lepton-number non-conserving supersymmetric operators (28.1.2) and (28.1.3) of dimensionality four that can be included in a renormalizable $SU(3)\times SU(2)\times U(1)$-invariant theory and that would lead to processes like proton decay at a catastrophic rate. These terms can be excluded from the Lagrangian by imposing R parity conservation (or, equivalently, invariance under a change of sign of all quark and lepton chiral superfields) but this does not exclude various $SU(3)\times SU(2)\times U(1)$-invariant but baryon- and lepton-number non-conserving operators of dimensionality $d>4$. As discussed in Section 21.3, if there is an underlying mechanism for baryon and lepton non-conservation characterized by some high mass scale M, then these operators will appear in the effective Lagrangian of the standard model with coefficients proportional to M^{4-d}. With only the fields of the non-supersymmetric standard model, operators that can violate baryon conservation have a minimum dimensionality six,[36] and thus give baryon non-conserving amplitudes proportional to M^{-2}. The new fields required by supersymmetry lead to two important changes in estimates of baryon non-conserving processes like proton and bound neutron decay. As we saw in Section 28.2, the change in the renormalization group equations gives larger estimates of M, which decreases the effect of the dimension six operators. At the same time, these new fields allow the construction of new operators of dimensionality five, which give baryon non-conserving amplitudes proportional to M^{-1} and are therefore likely to make the dominant contribution to proton and bound neutron decay.[37]

The supersymmetric operators of dimensionality five that can be formed out of chiral superfields (generically called Φ) are of the forms $(\Phi^*\Phi\Phi)_D$ and $(\Phi\Phi\Phi\Phi)_{\mathscr{F}}$, and their complex conjugates. (We do not consider operators that include derivatives or gauge fields, because they do not offer any additional possibility of baryon or lepton non-conservation.) In the notation of Section 28.1, the $SU(3)\times SU(2)\times U(1)$-invariant operators

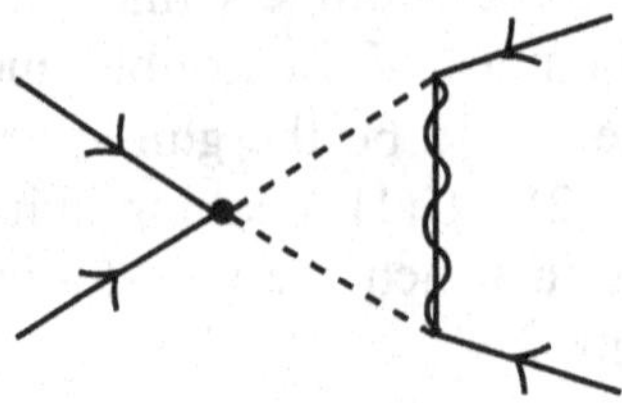

Figure 28.7. A diagram that can produce a four-fermion interaction among quarks and/or leptons that violates baryon and lepton number conservation. Here solid lines are quarks and/or leptons; dashed lines are squarks and/or sleptons; the combined solid and wavy line is a gaugino; and the dot is a vertex arising directly from the $\mathscr{F}$-term interactions (28.7.3).

of dimensionality five that also conserve R parity are

$$(LLH_2H_2)_{\mathscr{F}}\,, \tag{28.7.1}$$

$$(L\bar{E}H_2^*)_D\,, \quad (Q\bar{D}H_2^*)_D\,, \quad (Q\bar{U}H_1^*)_D\,, \quad (QQ\bar{U}\bar{D})_{\mathscr{F}}\,, \quad (Q\bar{U}L\bar{E})_{\mathscr{F}}\,, \tag{28.7.2}$$

and

$$(QQQL)_{\mathscr{F}}\,, \qquad (\bar{U}\bar{U}\bar{D}\bar{E})_{\mathscr{F}}\,, \tag{28.7.3}$$

with obvious contraction of indices as dictated by $SU(3)$ and $SU(2)$ conservation. The interaction (28.7.1) is the supersymmetric version of the dimensionality five operator which would provide small neutrino masses in some theories.[38] The interactions (28.7.2) only provide small corrections to processes that already occur in the renormalizable terms of the supersymmetric standard model. It is the interactions (28.7.3) that provide new mechanisms for baryon as well as lepton non-conservation.

According to Eq. (26.4.4), the quarks and leptons enter in the interactions (28.7.3) through terms involving a pair of quark and/or lepton fields and a pair of squark and/or slepton fields. In order to generate reactions among quarks and leptons alone, it is necessary for the pair of squarks and/or sleptons to be converted into a pair of quarks and/or leptons by exchanging a gaugino in the one-loop diagram shown in Figure 28.7. This will produce effective four-fermion $qqq\ell$ interactions with $d = 6$ among three quarks and a lepton. The coupling g_6 of these interactions will be proportional to the square of the gauge coupling g or g' or g_s of the gaugino, to the supersymmetry-breaking mass of the gaugino, to the inverse square of the larger of the gaugino and squark or slepton masses (which is needed to give a coupling of the right dimensionality), and to a factor of order $1/8\pi^2$ from the loop integration.

It might be thought that the greater strength of the gluino coupling would make gluino exchange the dominant contribution to g_6. (Indeed, in gauge-mediated supersymmetry-breaking theories, for moderate values of the trace (28.6.3) and with $g \approx g'$, Eqs. (28.6.13)–(28.6.14) give

$$m_{\rm gluino} \approx m_{\rm squark} \approx \frac{g_s^2}{16\pi^2} M_* ,$$

$$m_{\rm wino} \approx m_{\rm slepton} \approx m_{\rm bino} \approx \frac{g^2}{16\pi^2} M_* ,$$

where M_* is a mass characterizing the messenger sector. Therefore in such theories individual gluino exchange diagrams give contributions proportional to $g_s^2/m_{\rm gluino}$, while wino or bino exchange (or, more precisely, chargino or neutralino exchange) makes a contribution proportional to $g^2 m_{\rm wino}/m^2_{\rm squark}$, which is smaller by a factor roughly equal to $m_{\rm wino} g^2/m_{\rm gluino} g_s^2 \approx g^4/g_s^4$.) However, there is a cancellation among the diagrams for gluino exchange that strongly suppresses their contribution. This was originally shown using a Fierz identity among four-fermion operators,[39] but the same result can be obtained with no equations at all. To conserve color, the coefficients of the operators $(QQQL)_{\mathcal{F}}$ and $(\bar{U}\bar{U}\bar{D}\bar{E})_{\mathcal{F}}$ must be totally antisymmetric in the colors of the three quark or antiquark superfields, and since these superfields are bosonic, they must then also be antisymmetric in their flavors as well. Gluino interactions are flavor-independent, so if we can neglect the flavor dependence of the squark masses then the coefficients of the four-fermion $d = 6$ operators will also be totally antisymmetric in flavor as well as color. Fermi statistics then requires the coefficients of these operators to be also totally antisymmetric in the spin indices of the quark or antiquark fields. But the three quark or antiquark fields in the $d = 6$ operators derived by gaugino exchange from $QQQL_{\mathcal{F}}$ or $(\bar{U}\bar{U}\bar{D}\bar{E})_{\mathcal{F}}$ are all left-handed and therefore have only two independent spin indices, so no coefficient can be antisymmetric in all three spins. The contribution of gluino exchange to the $d = 6$ operators would therefore vanish if the squark masses were all equal and so this contribution is suppressed by the fractional differences among the masses of the different squarks. According to Eq. (28.6.4), in theories of gauge-mediated supersymmetry breaking the fractional differences between the $\bar{U}$ and $\bar{D}$ squark masses are of order g'^4/g_s^4, so gluino exchange between the antisquarks in the $(\bar{U}\bar{U}\bar{D}\bar{E})_{\mathcal{F}}$ operator generates a dimension six four-fermion operator with a coefficient of the same order as that generated by bino exchange. However, since gluinos conserve flavor this operator like the $(\bar{U}\bar{U}\bar{D}\bar{E})_{\mathcal{F}}$ operator must be totally antisymmetric in antiquark flavors, so that it must involve c or t quarks and therefore cannot contribute directly to proton or bound neutron decay. On the other hand, Eq. (28.6.4) indicates that the fractional mass differences among the

Q quarks of different flavors is much less than of order g^4/g_s^4, so that gluino exchange between the squarks in the $(QQQL)_{\mathscr{F}}$ operator makes a contribution to g_5 that is much smaller than wino or bino exchange. We conclude that at least in the theories of gauge-mediated supersymmetry breaking, gluino exchange makes a smaller contribution to proton or bound neutron decay than wino or bino exchange. In other models gluino exchange may make a contribution comparable to that of other processes.[40]

For $g \approx g'$ and $m_{\rm wino} \approx m_{\rm bino}$, the contribution of wino or bino exchange to the dimension six operators is of order

$$g_6 \approx \frac{g^2\, g_5\, m_{\rm wino}}{8\pi^2 m_{\rm squark}^2}\,, \tag{28.7.4}$$

where g_5 is a typical value of the couplings of the effective $d = 5$ interactions (28.7.3). If the wino and squark masses have the same ratio (g^2/g_s^2) as in theories of gauge-mediated supersymmetry breaking, then this gives

$$g_6 \approx \frac{g^4\, g_5}{8\pi^2 g_s^2 m_{\rm squark}}\,. \tag{28.7.5}$$

The four-fermion $qqq\ell$-terms of dimensionality six in the effective Lagrangian are the same as those that had been supposed to generate processes like proton decay in non-supersymmetric theories.[36] They produce proton and bound neutron decay at a rate which on dimensional grounds must be of the form

$$\Gamma_N = c_N\, m_N^5\, g_6^2\,, \tag{28.7.6}$$

where c_N is a pure number that must be calculated by non-perturbative calculations in quantum chromodynamics. Much work has gone into these calculations, with results[41] generally in the range $c_N \approx 3 \times 10^{-3\pm 0.7}$.

To estimate g_5, we note that it is not possible to produce $\mathscr{F}$-terms like (28.7.3) that involve only left-chiral superfields by the tree-approximation exchange of gauge supermultiplets, which always interact with both left-chiral superfields and their right-chiral complex conjugates. Thus the interactions (28.7.3) arise in the tree approximation only from the exchange of particles of chiral superfields, and therefore g_5 is of the order of g_T^2/M_T^2, where g_T is a typical baryon and lepton non-conserving coupling of some superheavy left-chiral superfield of mass M_T to the quark and lepton superfields. To produce the interactions (28.7.3), these superheavy particles must be color triplets or antitriplets, and $SU(2)$ triplets or singlets. Whatever gauge group unifies the strong and electroweak interactions presumably dictates some relation between the interactions of the superheavy color triplet T and the familiar color singlet H_1 and H_2. Then g_T will be of the same order as the Yukawa coupling constants in the interactions (28.1.2) and (28.1.3) that give mass to the known quarks and leptons, and which are equal to quark or lepton masses divided by the vacuum expectation values of order $G_F^{-1/2} \simeq 300$ GeV of $\mathscr{H}_1^0$ or $\mathscr{H}_2^0$. We therefore take

$$g_5 \approx \frac{G_F m_f^2}{M_T} , \tag{28.7.7}$$

where m_f is some typical quark or lepton mass. We have seen that the dimension five operators are antisymmetric in quark flavors, so as a compromise between the masses of the s quark and the u or d quarks, we will take $m_f = 30$ MeV. Combining Eqs. (28.7.5)–(28.7.7) and taking $M_T = 2 \times 10^{16}$ GeV (as suggested by the results of Section 28.2), $c_N = 0.003$, $g_s^2/4\pi = 0.118$, $g^2/4\pi = 1/(0.23 \times 137)$, and $m_{\text{squark}} = 1$ TeV, we find a proton (or bound neutron) lifetime Γ_N^{-1} of about 2×10^{31} years.[42] This is not very different from experimental lower bounds on the partial lifetimes of what are expected to be the leading modes of proton decay, variously quoted as ranging from 10^{31} to 5×10^{32} years. At the time of writing the most stringent bounds are set by the non-observation of proton decay in the large Super Kamiokande neutrino detector in Japan:[42a] the partial lifetimes for the decays $p \to e^+\pi^0$ and $p \to \bar{\nu}K^+$ are greater than 2.1×10^{33} years and 5.5×10^{32} years, respectively. There is an uncertainty of a factor of at least 100 in the above estimate of the theoretical lifetime from the uncertainty in the squark mass alone, so it is too soon to say that there is any discrepancy between experiment and theoretical expectations. On the other hand, supersymmetry raises the possibility that baryon non-conservation may be discovered soon.

We can also say a little on general grounds about the expected branching ratios for various proton and bound neutron decay modes. As we have mentioned, the dimension five operators (28.7.3) must be totally antisymmetric in the flavor of the quark superfields, so the only operators that concern us here are of the forms $(U_i D_j D_k N_\ell)_\mathscr{F}$, $(D_i U_j U_k E_\ell)_\mathscr{F}$, and $(\bar{D}_i \bar{U}_j \bar{U}_k \bar{E}_\ell)_\mathscr{F}$, where i, j, k, ℓ are generation indices, and in each case $j \neq k$. The exchange of neutral winos or binos then produces $d = 6$ four-fermion operators of the forms $u_i d_j d_k \nu_\ell$, $d_i u_j u_k e_\ell$, and $\bar{d}_i \bar{u}_j \bar{u}_k \bar{e}_\ell$, with $j \neq k$ and i and ℓ arbitrary, while charged wino exchange produces the same four-fermion operators with $i \neq j$ and k and ℓ arbitrary. The only quarks light enough to be involved in proton decay are u, s, and d; ignoring all the others and the small mixing angles in the third generation, we have

$$u_1 = u\,, \quad d_1 = d\cos\theta_c + s\sin\theta_c\,, \quad d_2 = -d\sin\theta_c + s\cos\theta_c\,,$$

where θ_c is the Cabibbo angle, while u_2, u_3, and d_3 can be ignored. The four-fermion operators that can be produced by wino or bino exchange and can contribute to proton or bound neutron decay thus are

$u\,d\,s\,\nu_\ell \cos(2\theta_c)$, $u\,d\,d\,\nu_\ell \sin(2\theta_c)$, $u\,u\,s\,e_\ell \cos\theta_c$, and $u\,u\,d\,e_\ell \sin\theta_c$, plus others with quarks and leptons replaced with antiquarks and antileptons. All other things being equal, the dominant decay modes are therefore $p \to K^+\bar{\nu}$, $n \to K^0\bar{\nu}$, $p \to K^0 e^+$, and $p \to K^0\mu^+$, while the rates for the decay modes $p \to \pi^+\bar{\nu}$, $n \to \pi^0\bar{\nu}$, $p \to \pi^0 e^+$, $p \to \pi^0\mu^+$, and $n \to \pi^- e^+$ are suppressed by factors $\sin^2\theta_c = 0.05$, though also enhanced somewhat by the greater phase space available.

These considerations do not lead to definite predictions of branching ratios, because in addition to all the factors mentioned above, the coefficients generically called g_5 of the operators (28.7.3) may have a strong dependence on the superfield flavors appearing in these operators. To go further, one needs a specific theory for the generation of the dimension five operators. Most of the authors of Reference 42 concluded on the basis of a supersymmetric version of $SU(5)$ theories that proton and bound neutron decay would be dominated by the processes $p \to K^+\bar{\nu}$ and $n \to K^0\bar{\nu}$, but for a model based on $SO(10)$ charged lepton modes can become prominent.[43] Also, in some models higgsino exchange can compete with wino and bino exchange,[44] increasing the rate of $p \to K^+\bar{\nu}$. It seems a good idea in searches for baryon non-conservation to keep an open mind as to the decay modes to be expected in proton or bound neutron decay.

Of course, it is possible that all of these baryon non-conserving processes are prohibited by some sort of conservation law. As mentioned in Section 28.1, string theory argues against baryon conservation being a fundamental global continuous symmetry, but the baryon non-conserving operators (28.7.3) may be forbidden by a $\mathbb{Z}_3$ multiplicative symmetry known as baryon parity,[45] under which the Q superfield is neutral; the H_2 and $\bar{D}$ superfields are multiplied by the phase $\exp(i\pi/3)$, and the L, H_1, $\bar{U}$, and $\bar{E}$ superfields are multiplied by the opposite phase $\exp(-i\pi/3)$. This symmetry is designed to allow the fundamental Yukawa couplings (28.1.2) and (28.1.3) as well as the μ-term (28.5.7) and the lepton non-conserving terms (28.1.4) and (28.7.1), but it rules out the dimension four baryon non-conserving term (28.1.5) and the dimension five baryon non-conserving terms (28.7.3). This symmetry is spontaneously broken by the appearance of vacuum expectation values of $\mathscr{H}_1^0$ and $\mathscr{H}_2^0$ (and perhaps the sneutrino fields $\mathscr{N}$), and with no conservation law for R parity, there would be nothing to keep the lightest supersymmetric particle stable.

Problems

1. Suppose that the interactions (28.1.4) and (28.1.5) were actually present in the Lagrangian of a supersymmetric version of the

standard model. Make a rough estimate of how heavy the squarks and sleptons would have to be to avoid a conflict with experimental bounds on the proton lifetime.

2. Suppose that the typical mass m of the gauginos, higgsinos, squarks and sleptons is very much larger than m_Z. Give the renormalization group equations for the running gauge couplings at energies both above and below m. Use the results, together with the unification assumption employed in Section 28.2, to give formulas for $\sin^2\theta$ and the unification scale M in terms of m, m_Z, $e(m_Z)$, $g_s(m_Z)$, and n_s. How large could m be without violating experimental bounds on $\sin^2\theta$ and M?

3. Give formulas for the couplings of the quarks and leptons to the lightest CP-even neutral scalar particle in the minimum supersymmetric standard model in terms of the parameters m_A, m_Z, β, G_F, and the quark and lepton masses.

4. Use holomorphy arguments to derive the one-loop formula for the gluino mass in a theory of gauge-mediated supersymmetry breaking, in which the messenger superfields Φ_n and $\bar{\Phi}_n$ get their masses from a term $\sum_n \lambda_n S_n(\bar{\Phi}_n\Phi_n)$ in the superpotential, in terms of the expectation values $\mathscr{S}_n$ and $\mathscr{F}_n$ of the ϕ- and $\mathscr{F}$-components of the singlet superfields S_n, in the limit where $|\mathscr{F}_n| \ll |\lambda_n||\mathscr{S}_n|^2$.

5. Taking account of the possibility of a small flavor dependence of squark masses, estimate the gluino exchange contribution to baryon and lepton non-conserving four-fermion interactions among the quarks and leptons. Set an upper bound on these contributions, using the bound on squark mass splittings from the rate of $K^0 \to \overline{K}^0$ conversion.

References

1. We will not consider here the possibility of unification at much lower energies raised by I. Antoniadis, *Phys. Lett.* **B246**, 377 (1990); J. Lykken, *Phys. Rev.* **D54**, 3693 (1996), and revived by N. Arkani-Hamed, S. Dimopoulos, and G. Dvali, *Phys. Lett.* **B429**, 263 (1998); K. R. Dienes, E. Dudea, and T. Gherghetta, *Phys. Lett.* **B436**, 55 (1998); I. Antoniadis, N. Arkani-Hamed, S. Dimopoulos, and G. Dvali, *Phys. Rev. Lett.* **B436**, 257 (1998).

1*a*. S. Weinberg, in *Proceedings of the XVII International Conference on High Energy Physics, London, 1974*, J. R. Smith, ed. (Rutherford Laboratory, Chilton, Didcot, England, 1974); S. Weinberg, *Phys.*

Rev. **D13**, 974 (1976); E. Gildener and S. Weinberg, *Phys. Rev.* **D13**, 3333 (1976).

1*b*. T. Banks and L. Dixon, *Nucl. Phys.* **B307**, 93 (1988). For a detailed treatment, see J. Polchinski, *String Theory* (Cambridge University Press, Cambridge, 1998): Chapter 18.

2. Additive *R*-conservation laws were introduced by A. Salam and J. Strathdee, *Nucl. Phys.* **B87**, 85 (1975); P. Fayet, *Nucl. Phys.* **B90**, 104 (1975), reprinted in *Supersymmetry*, S. Ferrara, ed., (North Holland/World Scientific, Amsterdam/Singapore, 1987). The *R* parity can be defined in terms of an *R* quantum number as $\exp(i\pi R)$, and may be conserved multiplicatively even if *R* is not conserved additively; see G. Farrar and P. Fayet, *Phys. Lett.* **76B**, 575 (1978); P. Fayet, in *Unification of the Fundamental Particle Interactions*, S. Ferrara, J. Ellis, and P. van Nieuwenhuizen, eds. (Plenum, New York, 1980); S. Dimopoulos, S. Raby, and F. Wilczek, *Phys. Lett.* **112B**, 133 (1982); G. Farrar and S. Weinberg, *Phys. Rev.* **D27**, 1731 (1983), reprinted in *Supersymmetry*.

3. S. Dimopoulos and H. Georgi, *Nucl. Phys.* **B193**, 150 (1981), reprinted in *Supersymmetry*, Ref. 2.

4. R. D. Peccei and H. Quinn, *Phys. Rev. Lett.* **38**, 1440 (1977); *Phys. Rev.* **D16**, 1791 (1977).

4*a*. S. Dimopoulos and G. F. Giudice, *Phys. Lett.* **B357**, 573 (1995); A. Pomerol and D. Tommasini, *Nucl. Phys.* **B466**, 3 (1996); G. Dvali and A. Pomerol, *Phys. Rev. Lett.* **77**, 3728 (1996); *Nucl. Phys.* **B522**, 3 (1998); A. G. Cohen, D. B. Kaplan, and A. E. Nelson, *Phys. Lett.* **B388**, 588 (1996); R. N. Mohapatra and A. Riotto, *Phys. Rev.* **D55**, 1 (1997); R.-J. Zhang, *Phys. Lett.* **B402**, 101 (1997); H-P. Nilles and N. Polonsky, *Phys. Lett.* **B412**, 69 (1997); D. E. Kaplan, F. Lepeintre, A. Masiero, A. E. Nelson, and A. Riotto, hep-ph/9806430, to be published; J. Hisano, K. Kurosawa, and Y. Nomura, *Phys. Lett.* **B445**, 316 (1999). This pattern of masses can arise naturally from radiative corrections; see J. L. Feng, C. Kolda, and N. Polonsky, *Nucl. Phys.* **B546**, 3 (1999); J. Bagger, J. L. Feng, and N. Polonsky, hep-ph/9905292, to be published.

4*b*. B. W. Lee and S. Weinberg, *Phys. Rev. Lett.* **39**, 165 (1977); D. A. Dicus, E. W. Kolb, and V. L. Teplitz, *Phys. Rev. Lett.* **39**, 168 (1977).

4*c*. S. Wolfram, *Phys. Lett.* **82B**, 65 (1979); J. Ellis, J. S. Hagelin, D. V. Nanopoulos, K. Olive, and M. Srednicki, *Nucl. Phys.* **B238**, 453 (1984).

4*d*. P. F. Smith and J. R. J. Bennett, *Nucl. Phys.* **B149**, 525 (1979).

5. H. Georgi, H. R. Quinn, and S. Weinberg, *Phys. Rev. Lett.* **33**, 451 (1974).

6. S. Dimopoulos and H. Georgi, Ref. 3; J. Ellis, S. Kelley, and D. V. Nanopoulos, *Phys. Lett.* **B260**, 131 (1991); U. Amaldi, W. de Boer, and H. Furstmann, *Phys. Lett.* **B260**, 447 (1991); C. Giunti, C. W. Kim and U. W. Lee, *Mod. Phys. Lett.* **16**, 1745 (1991); P. Langacker and M.-X. Luo, *Phys. Rev.* **D44**, 817 (1991). For other references and more recent analyses of the data, see P. Langacker and N. Polonsky, *Phys. Rev.* **D47**, 4028 (1993); **D49**, 1454 (1994); L. J. Hall and U. Sarid, *Phys. Rev. Lett.* **70**, 2673 (1993).

7. S. Dimopoulos, S. Raby, and F. Wilczek, *Phys. Rev.* **D24**, 1681 (1981), reprinted in *Supersymmetry*, Ref. 2.

7*a*. P. Hořava and E. Witten, *Nucl. Phys.* **B460**, 506 (1996); *ibid.* **B475**, 94 (1996); E. Witten, *Nucl. Phys.* **B471**, 135 (1996); P. Hořava, *Phys. Rev.* **D54**, 7561 (1996).

8. H. Pagels and J. R. Primack, *Phys. Rev. Lett.* **48**, 223 (1982).

9. S. Weinberg, *Phys. Rev. Lett.* **48**, 1303 (1983).

10. S. Dimopoulos and H. Georgi, Ref. 3; N. Sakai, *Z. Phys. C* **11**, 153 (1981). For reviews, see H. E. Haber and G. L. Kane, *Phys. Reports* **117**, 75 (1985); J. A. Bagger, in *QCD and Beyond: Proceedings of the Theoretical Advanced Study Institute in Elementary Particle Physics, University of Colorado, June 1995*, D. E. Soper, ed. (World Scientific, Singapore, 1996); V. Barger, in *Fundamental Particles and Interactions: Proceedings of the FCP Workshop on Fundamental Particles and Interactions, Vanderbilt University, May 1997*, R. S. Panvini, T. J. Weiler, eds. (American Institute of Physics, Woodbury, NY, 1998); J. F. Gunion, in *Quantum Effects in the MSSM – Proceedings of the International Workshop on Quantum Effects in the MSSM, Barcelona, September 1997*, J. Solà, ed. (World Scientific Publishing, Singapore, 1998); S. Dawson, in *Proceedings of the 1997 Theoretical Advanced Study Institute on Supersymmetry, Supergravity, and Supercolliders*, J. Bagger, ed. (World Scientific, Singapore, 1998); S. P. Martin, in *Perspectives on Supersymmetry*, G. L. Kane, ed. (World Scientific, Singapore, 1998); K. R. Dienes and C. Kolda, in *Perspectives on Supersymmetry*, ibid.

11. S. Dimopoulos and D. Sutter, *Nucl. Phys.* **B194**, 65 (1995); H. Haber, *Nucl. Phys. Proc. Suppl.* **62**, 469 (1998).

12. S. Dimopoulos and H. Georgi, Ref. 3; J. Ellis and D. V. Nanopoulos, *Phys. Lett.* **110B**, 44 (1982); J. F. Donoghue, H-P. Nilles, and D. Wyler, *Phys. Lett.* **128B**, 55 (1983). For strong-interaction corrections to these calculations, see J. A. Bagger, K. T. Matchev, and

R.-J. Zhang, *Phys. Lett.* **B412**, 77 (1997). Conditions under which limits on flavor changing processes do not constrain squark masses are discussed by R. Barbieri and R. Gatto, *Phys. Lett.* **110B**, 211 (1981); Y. Nir and N. Seiberg, *Phys. Lett.* **B309**. 337 (1993). For a detailed review, see F. Gabbiani, E. Gabrielli, A. Masiero, and L. Silvestrini, *Nucl. Phys.* **B477**, 321 (1996).

13. M. K. Gaillard and B. W. Lee, *Phys. Rev.* **D10**, 897 (1974).

14. J. Ellis and D. V. Nanopoulos, Ref. 12. Detailed results are given by F. Gabbiani and A. Masiero, *Nucl. Phys.* **B322**, 235 (1989); J. S. Hagelin, S. Kelley, and T. Tanaka, *Nucl. Phys.* **B415**, 293 (1994). The most complete treatment is by D. Sutter, Stanford University Ph. D. thesis (unpublished) and S. Dimopoulos and D. Sutter, Ref. 11.

14*a*. M. Dine, R. Leigh, and A. Kagan, *Phys. Rev.* **D48**, 4269 (1993).

15. For more recent reviews, see Y. Grossman, Y. Nir, and R. Rattazzi, in *Heavy Flavours II*, A. J. Buras and M. Lindner, eds. (World Scientific, Singapore, 1998); A. Masiero and L. Silvestrini, in *Perspectives on Supersymmetry*, Ref. 10.

16. J. Ellis and M. K. Gaillard, *Nucl. Phys.* **B150**, 141 (1979); D. V. Nanopoulos, A. Yildiz, and P. H. Cox, *Ann. Phys. (N.Y.)* **127**, 126 (1980); M. B. Gavela, A. Le Yaouanc, L. Oliver, O. Pène, J.-C. Raynal, and T. N. Pham, *Phys. Lett.* **109B**, 215 (1982); B. H. J. McKellar, S. R. Choudhury, X-G. He, and S. Pakvasa, *Phys. Lett.* **B197**, 556 (1987).

16*a*. P. G. Harris *et al.*, *Phys. Rev. Lett.* **82**, 904 (1999).

17. J. Ellis, S. Ferrara, and D. V. Nanopoulos, *Phys. Lett.* **114B**, 231 (1982); J. Polchinski and M. B. Wise, *Phys. Lett.* **125B**, 393 (1983); M. Dugan, B. Grinstein, and L. Hall, *Nucl. Phys.* **B255**, 413 (1985).

18. R. Arnowitt, J. Lopez, and D. Nanopoulos, *Phys. Rev.* **D42**, 2423 (1990); R. Arnowitt, M. Duff, and K. Stelle, *Phys. Rev.* **D43**, 3085 (1991); Y. Kizuri and N. Oshimo, *Phys. Rev.* **D45**, 1806 (1992).

19. S. Weinberg, *Phys. Rev. Lett.* **63**, 2333 (1989); D. Dicus, *Phys. Rev.* **D41**, 999 (1990); J. Dai, H. Dykstra, R. G. Leigh, S. Paban, and D. A. Dicus, *Phys. Lett.* **B237**, 216 (1990); E. Braaten, C. S. Li, and T. C. Yuan, *Phys. Rev. Lett.* **64**, 1709 (1990); A. De Rújula, M. B. Gavela, O. Pène, and F. J. Vegas, *Phys. Lett.* **B245**, 640 (1990); R. Arnowitt, M. J. Duff, and K. S. Stelle, Ref. 18; T. Ibrahim and P. Nath, *Phys. Lett.* **148B**, 98 (1998).

20. K. S. Babu, C. Kolda, J. March-Russell, and F. Wilczek, *Phys. Rev.* **D59**, 016004 (1999).

21. R. Arnowitt, M. J. Duff, and K. S. Stelle, Ref. 18. In terms of the functions J_1 and J_2 in this reference, the function J here is taken as $2J_1 + \frac{2}{3}J_2$, under the assumption that the graphs with the gluon attached to the squark or gluino lines add constructively.

22. H. Georgi and L. Randall, *Nucl. Phys.* **B276**, 241 (1980); A. Manohar and H. Georgi, *Nucl. Phys.* **B238**, 189 (1984); S. Weinberg, Ref. 19.

23. W. Fischler, S. Paban, and S. Thomas, *Phys. Lett.* **B 289**, 373 (1992).

24. J. Ellis and D. V. Nanopoulos, Ref. 12; F. Gabbiani and A. Masiero, Ref. 14; F. Dine, A. Kagan, and S. Samuel, *Phys. Lett.* **B243** 250 (1990); F. Gabbiani, E. Gabrielli, A. Masiero, and L. Silvestrini, Ref. 12.

25. S. Weinberg, *Phys. Rev. Lett.* **40**, 223 (1978); F. Wilczek, *Phys. Rev. Lett.* **40**, 279 (1978).

26. A. Brignole, J. Ellis, G. Ridolfi, and F. Zwirner, *Phys. Lett.* **B271**, 123 (1991); M. Carena, M. Quiros, and C. E. M. Wagner, *Nucl. Phys.* **B461**, 407 (1996); S. Heinemayer, W. Hollik, and G. Weiglein, hep-ph/9812472, hep-ph/9903404, hep-ph/9903504, to be published. The numerical results for m_h cited here are taken from calculations quoted by S. Dawson, Ref. 10.

27. R. Barate *et al.* (ALEPH collaboration), *Phys. Lett.* **B412**, 173 (1997).

27*a*. M. Grünewald and D. Karlen, in *Proceedings of the XXIX International Conference on High Energy Nuclear Physics*, A. Astbury, D. Axen, and J. Robinson, eds. (TRIUMF, Vancouver, 1999).

28. R. Barate *et al.* (ALEPH collaboration), 1999 CERN preprint EP-99-011, to be published in *Phys. Lett.* A lower bound of 54.5 GeV had been obtained earlier from experiments on e^+–e^- annihilation at 130 to 172 GeV by P. Abreu *et al.* (DELPHI collaboration), *Phys. Lett.* **B420**, 140 (1998).

29. A. J. Buras, M. Misiak, M. Münz, and S. Pokorski, *Nucl. Phys.* **B424**, 374 (1994).

29*a*. R. Barate *et al.* (ALEPH collaboration), 1999 CERN preprint EP-99-014, to be published in *Eur. Phys. J.*

30. M. Dine, W. Fischler, and M. Srednicki, *Nucl. Phys.* **B189**, 575 (1981); S. Dimopoulos and S. Raby, *Nucl. Phys.* **B192**, 353 (1982); M. Dine and W. Fischler, *Phys. Lett.* **110B**, 227 (1982); *Nucl. Phys.* **B204**, 346 (1982); C. Nappi and B. Ovrut, *Phys. Lett.* **113B**, 175 (1982); L. Alvarez-Gaumé, M. Claudson, and M. Wise, *Nucl. Phys.*

B207, 96 (1982); S. Dimopoulos and S. Raby, *Nucl. Phys.* **B219**, 479 (1983). This class of models was revived by M. Dine and A. E. Nelson, *Phys. Rev.* **D48**, 1277 (1993); **D51**, 1362 (1995); J. Bagger, Ref. 10; M. Dine, A. Nelson, and Y. Shirman, *Phys. Rev.* **D51**, 1362 (1995); M. Dine, A. Nelson, Y. Nir, and Y. Shirman, *Phys. Rev.* **D53**, 2658 (1996). For reviews, see C. Kolda, *Nucl. Phys. Proc. Suppl.* **62**, 266 (1998); G. F. Giudice and R. Rattazzi, hep-ph/9801271, to be published in *Phys. Rep*; S. L. Dubovsky, D. S. Gorbunov, and S. V. Troitsky, hep-ph/9905466, to be published. The phenomenology of these models is described by S. Dimopoulos, S. Thomas, and J. D. Wells, *Nucl. Phys.* **B488**, 39 (1997).

31. S. Dimopoulos, G. F. Giudice, and A. Pomerol, *Phys. Lett.* **389B**, 37 (1997); S. P. Martin, *Phys. Rev.* **D55**, 3177 (1997).

32. G. F. Giudice and R. Rattazzi, *Nucl. Phys.* **B511**, 25 (1998). This work has been extended by N. Arkani-Hamed, G. F. Giudice, M. A. Luty, and R. Rattazzi, *Phys. Rev.* **D58**, 115005 (1998).

33. N. Seiberg, *Phys. Lett.* **B318**, 469 (1993).

34. K. S. Babu, C. Kolda, and F. Wilczek, *Phys. Rev. Lett.* **77**, 3070 (1996).

35. J. E. Kim and H-P. Nilles, *Phys. Lett.* **138B**, 150 (1984); J. Ellis, J. F. Gunion, H. E. Haber, L. Roszkowski, and F. Zwirner, *Phys. Rev.* **D39**, 844 (1989); E. J. Chun, J. E. Kim, and H-P. Nilles, *Nucl. Phys.* **B370**, 105 (1992); M. Dine and A. E. Nelson, Ref. 30; M. Dine, A. E. Nelson, Y. Nir, and Y. Shirman, Ref. 30; G. Dvali, G. F. Giudice, and A. Pomerol, *Nucl. Phys.* **B478**, 31 (1996); S. Dimopoulos, G. Dvali, and R. Rattazzi, *Phys. Lett.* **413B**, 336 (1997); H-P. Nilles and N. Polonsky, *Nucl. Phys.* **B484**, 33 (1997); G. Cleaver, M. Cvetič, J. R. Espinosa, L. Everett, and P. Langacker, *Phys. Rev.* **D57**, 2701 (1998); P. Langacker, N. Polonsky, and J. Wang, hep-ph/9905252, to be published; J. E. Kim, hep-ph/9901204, to be published.

35*a*. S. Raby, *Phys. Lett.* **B422**, 158 (1998).

35*b*. D. A. Dicus, B. Dutta, and S. Nandi, *Phys. Rev. Lett.* **78**, 3055 (1997); *Phys. Rev.* **D56**, 5748 (1997).

36. S. Weinberg, *Phys. Rev. Lett.* **43**, 1566 (1979); F. Wilczek and A. Zee, *Phys. Rev. Lett.* **43**, 1571 (1979).

37. S. Weinberg, *Phys. Rev.* **D26**, 287 (1982); N. Sakai and T. Yanagida, *Nucl. Phys.* **B197**, 533 (1982). These articles are reprinted in *Supersymmetry*, Ref. 2.

38. S. Weinberg, Ref. 36.

39. J. Ellis, J. S. Hagelin, D. V. Nanopoulos, and K. Tamvakis, *Phys. Lett.* **124B**, 484 (1983); V. M. Belyaev and M. I. Vysotsky, *Phys. Lett.* **127B**, 215 (1983).

40. V. Lucas and S. Raby, *Phys. Rev.* **D55**, 6986 (1997).

41. This estimate is taken from a compilation of non-supersymmetric calculations of the total two-body decay proton decay rate in terms of the superheavy gauge boson masses M_X by P. Langacker, in *Proceedings of the 1983 Annual Meeting of the Division of Particles and Fields of the American Physical Society* (American Institute of Physics, New York, 1983): 251. To express the result in terms of g_6, I have assumed that, in effect, the coupling g_6 that was used in these calculations was given by $g_6 = g^2(M_X)/M_X^2$, where $g(M_X)$ has the value appropriate to non-supersymmetric theories, with $g^2(M_X)/4\pi \simeq 1/41$.

42. For more detailed (mostly model-dependent) calculations, including renormalization group corrections to g_5, see S. Dimopoulos, S. Raby, and F. Wilczek, *Phys. Lett.* **112B**, 133 (1982); J. Ellis, D. V. Nanopoulos, and S. Rudaz, *Nucl. Phys.* **B202**, 43 (1982); W. Lang, *Nucl. Phys.* **B203**, 277 (1982); J. Ellis, J. S. Hagelin, D. V. Nanopoulos, and K. Tamvakis, Ref. 39; V. M. Belyaev and M. I. Vysotsky, Ref. 39; L. E. Ibáñez and C. Muñoz, *Nucl. Phys.* **B245**, 425 (1984); P. Nath, A. H. Chamseddine, and R. Arnowitt, *Phys. Rev.* **D32**, 2385 (1985); J. Hisano, H. Murayama, and T. Yanagida, *Nucl. Phys.* **B402**, 46 (1993); V. Lucas and S. Raby, Ref. 40. For a review, see P. Nath and R. Arnowitt, *Phys. Atom. Nucl.* **61**, 975 (1997).

42*a*. M. Takita *et al.*, in *Proceedings of the XXIX International Conference on High Energy Nuclear Physics*, Ref. 27a.

43. K. S. Babu, J. C. Pati, and F. Wilczek, *Phys. Lett.* **423B**, 337 (1998).

44. V. Lucas and S. Raby, Ref. 40; T. Goto and T. Nihei, *Phys. Rev.* **D59**, 115009 (1999).

45. L. Ibáñez and G. Ross, *Nucl. Phys.* **B368**, 3 (1991).

29

Beyond Perturbation Theory

Most of the implications of supersymmetry discussed so far have been inferred with the use of perturbation theory. In this chapter we shall consider some results that apply even when non-perturbative effects are taken into account.

29.1 General Aspects of Supersymmetry Breaking

Supersymmetry is not observed in the spectrum of known particles, so it must be broken. We saw in the previous chapter that supersymmetry breaking in the tree approximation of the standard model is experimentally ruled out, and that the large disparity between the electroweak breaking scale and the Planck or grand unification scales suggests that supersymmetry is likely to be broken when some running gauge coupling becomes strong. It is therefore essential for us to explore spontaneous supersymmetry breaking without using perturbation theory.

We saw in Section 26.7 that the supersymmetry of the action implies the existence of a supersymmetry current $S^\mu(x)$. This current is a Majorana spinor in the sense of Eq. (26.A.2):

$$S^\mu(x)^* = -\beta\gamma_5\epsilon S^\mu(x) \; ; \tag{29.1.1}$$

it is conserved,

$$\partial_\mu S^\mu_\beta(x) = 0 \; ; \tag{29.1.2}$$

and the integral of its time component is the supersymmetry generator

$$\int d^3x \; S^0_\beta(x) = Q_\beta \, , \tag{29.1.3}$$

for which the commutator of $-i(\bar{\alpha}Q)$ with any operator gives the change of that operator under a supersymmetry transformation with infinitesimal Majorana spinor parameter α.

The arguments that led to these results relied on the supersymmetry of the action; nothing depended on whether or not supersymmetry is spontaneously broken, except perhaps for the assumption that the integral (29.1.3) exists. In fact this assumption may be violated in theories with massless fermions, which can have long range effects (or, equivalently, poles at zero four-momentum) that would make this integral not converge. We will see here that such massless fermions are necessary consequences of supersymmetry breaking. In order to avoid the issue of the convergence of this integral even in theories with massless fermions, it is very convenient to work in a space of finite volume V. We can do this while maintaining translation invariance by imposing periodic boundary conditions: all fields are assumed to be unaffected by translations of any spatial coordinate x^i by an amount $V^{1/3}$.

The existence of an operator Q_α that induces supersymmetric transformations on quantum fields would allow us to derive all the consequences of supersymmetry, provided that there exists a supersymmetric vacuum state $|\text{VAC}\rangle$ of zero three-momentum from which multiparticle states may be constructed by acting with field operators. But if $|\text{VAC}\rangle$ is supersymmetric in the sense that $Q_\alpha|\text{VAC}\rangle = 0$, then it follows from the anticommutation relation (25.2.36) that this state has zero energy as well as zero momentum. Conversely, by taking the vacuum expectation value of the positive operator (returning for a moment to two-component notation) $\{\text{Q}_a, \text{Q}_a^*\}$, we see that if the vacuum has zero energy then it must be annihilated by Q_a and Q_a^*, and hence be supersymmetric, while if it is not supersymmetric then its energy must be positive-definite. *The question of whether supersymmetry is or is not spontaneously broken is thus entirely a question of whether the vacuum has a positive-definite or a zero energy.*

The same reasoning led Witten to the conclusion that extended supersymmetry with $N > 1$ two-component spinor generators Q_{ar} and their adjoints cannot be spontaneously broken to extended supersymmetry with fewer generators or to simple supersymmetry, because if any one of the generators does not annihilate the vacuum then the vacuum energy cannot vanish, and it follows from this that none of the generators can annihilate the vacuum.[1] Usually the energy of the vacuum state appears as an ill-defined additive constant in the energies of all states, but here this constant is given a meaning by the appearance of the energy-momentum four-vector in the anticommutation relations of supersymmetry. One of the advantages of working in a finite volume is that it makes it meaningful to talk about the total energy of the vacuum.

Hughes, Liu, and Polchinski have pointed out that there are theories that exhibit a sort of partly broken supersymmetry.[1a] These theories do not have a supersymmetry algebra of the sort described in Chapter 25. Instead, they have an algebra of *currents*, based on anticommutation relations like Eq. (26.7.45):

$$\int d^3x \left\{ S^0_{r\alpha}(x), \bar{S}^\mu_{s\beta}(y) \right\} = -2i\,\delta_{rs}\gamma_\nu \Theta^{\mu\nu}(y) + 2i\gamma^\mu_{\alpha\beta} C_{rs} ,$$

where $\Theta^{\mu\nu}$ is an energy-momentum tensor satisfying the conservation condition $\partial_\mu \Theta^{\mu\nu} = 0$, and C_{rs} is a new ingredient, a constant. For $N = 1$ this constant could be regarded as a term $-\eta^{\mu\nu}C$ in $\Theta^{\mu\nu}$, but this is not possible for extended supersymmetry unless $C_{rs} \propto \delta_{rs}$, which need not be the case. This algebra is not ruled out by the Haag–Lopuszanski–Sohnius theorem proved in Section 25.2 because it could not be a symmetry of the S-matrix. It is true that spontaneously broken symmetries are never symmetries of the S-matrix, but they are usually assumed to be based on algebras or superalgebras that *could* be symmetries of the S-matrix in some phases of some theories. With C_{rs} not proportional to δ_{rs}, the superalgebra of currents is one that could not generate a symmetry of the S-matrix in any phase of any theory. Here we will only consider superalgebras of the sort described in Chapter 25, to which Witten's argument does apply.

Another advantage of working in a finite volume is that all states become discrete and normalizable. An immediate consequence of the fact that Q_α commutes with P_μ is that any state of non-zero energy is paired with another state of the same energy and momentum but opposite statistics. To see this, note that for any three-momentum $\mathbf{p}$ we may find a two-component spinor u_a such that $\sum_{ab} u^*_a \boldsymbol{\sigma}_{ab} \cdot \mathbf{p} u_b = 0$ and $\sum_a |u_a|^2 = 1$. (For $\mathbf{p}$ in the 3-direction, take $u = (1, 1)/\sqrt{2}$. For $\mathbf{p}$ in any other direction, apply to this u the spin 1/2 representation of the rotation that takes the 3-direction into the direction of $\mathbf{p}$.) Then, within the space with four-momentum p^μ, the anticommutation relations (25.2.31) and (25.2.32) give

$$\mathrm{Q}^2(p) = p^0 , \tag{29.1.4}$$

where $\mathrm{Q}(p)$ is the Hermitian linear combination of supersymmetry generators:

$$\mathrm{Q}(p) = \sum_a u_a \mathrm{Q}_a + \sum_a u^*_a \mathrm{Q}^*_a . \tag{29.1.5}$$

Acting with $\mathrm{Q}(p)$ on any normalized state $|X\rangle$ with four-momentum p^μ and $p^0 > 0$, we get another normalized state $|Y\rangle = \mathrm{Q}(p)|X\rangle/\sqrt{p^0}$ with opposite statistics and the same four-momentum. Furthermore, $|X\rangle$ is the only state related to $|Y\rangle$ in this way, because if $|Y\rangle = \mathrm{Q}(p)|X\rangle/\sqrt{p^0}$ then, according to Eq. (29.1.4), $|X\rangle = \mathrm{Q}(p)|Y\rangle/\sqrt{p^0}$. The multiplicity of supersymmetry generators and of spin states will usually cause these pairs of fermionic and bosonic states to be joined by other pairs, all with the same four-momentum, but for the moment it is enough to know that all states of non-zero energy may at least be grouped in these pairs.

When supersymmetry is broken we do not expect states with a definite number of particles to form supermultiplets with other states of opposite statistics and with the same four-momentum and the same number of particles. The pairing of states then requires the existence of a massless fermion, so that an n-particle state can be paired with a state of the same energy and momentum but opposite statistics, consisting of the same n particles together with a massless fermion of zero energy and momentum. This massless fermion is known as a *goldstino.* To be more precise, any n-particle state is accompanied with two states of the same energy and momentum and opposite statistics, containing an additional zero-momentum goldstino of spin up or down, and with another state of the same energy and momentum and the *same* statistics, containing two additional zero-momentum goldstinos of opposite spin. In particular, when supersymmetry is spontaneously broken the vacuum state has non-zero energy, so it must be paired with a fermionic state of the same energy and zero momentum; more precisely, the vacuum and the state containing two zero-momentum goldstinos are paired with the two states of a single zero-momentum goldstino. It is only when supersymmetry is unbroken that there is a state of zero energy, the vacuum, that can be unpaired.

The pairing of states of non-zero energy provides a valuable diagnostic tool that can in some cases tell us that supersymmetry is not spontaneously broken, even where perturbation theory is not adequate to answer this question. When all interactions are weak, we can rely on perturbation theory to give us a *qualitative* picture of the spectrum. If it turns out that in the tree approximation there are n vacuum states with zero energy, and no massless fermions, then we can be confident that for weak coupling there are no zero-energy fermionic states with which we could pair the n vacuum states, so these unpaired states would have to have precisely zero energy. Then, as we increase the strength of the couplings or vary the parameters of the theory in any other way, states may move from positive to zero energy or vice-versa, but they will generally not suddenly appear or disappear. (There is an exception for changes in the parameters that change the asymptotic behavior of the Lagrangian for large fields; as we will shortly see, this *can* produce or destroy states.) Because each state of non-zero energy is always paired with another of opposite statistics, they can only make the transition from zero to non-zero energy or vice-versa in such pairs, so the number of bosonic zero-energy states minus the number of fermionic zero-energy states does not change as the parameters of the theory are varied, as long as the large-field behavior of the Lagrangian is not changed. This difference is known as the *Witten index.*[2] Formally, this index is $\mathrm{Tr}\,(-1)^F$, where F is the fermion number; the pairing of states discussed earlier insures that this trace can receive no contribution from states of non-zero energy. If the Witten index is non-zero then there must be *some* states of zero energy, and so supersymmetry cannot be broken. In particular, in a theory where the tree approximation gives n zero-energy vacuum states and no zero-energy fermions, the Witten index is n for weak coupling, where the tree approximation can be trusted to give a qualitative picture of the spectrum, and the index remains equal to n as the strength of the coupling is increased, so we can be certain that higher-order effects or even non-perturbative effects do not break supersymmetry.

As an example of the use of the Witten index, consider the Wess–Zumino theory of a single chiral superfield with a cubic polynomial superpotential of the form (26.4.16):

$$f(\phi) = \tfrac{1}{2}m^2\phi^2 + \tfrac{1}{6}g\phi^3 ,$$

where ϕ is the complex scalar field component of the superfield. We saw in Section 26.4 that this model does not exhibit supersymmetry breaking in the tree approximation, but what about higher orders of perturbation theory, and what about non-perturbative effects? Perturbation theory gives a good approximation to the energy spectrum when m is large and g is small; it tells us that in this case there are two bosonic states near zero energy, corresponding to the solutions $\phi = 0$ and $\phi = -2m^2/g$ of the equation $\partial f(\phi)/\partial\phi = 0$, and no fermionic states near zero energy; the lowest-energy fermionic state is a zero-momentum one-fermion state of energy near $|m|$. In typical scalar field theories we would not expect the two bosonic states to have precisely zero energy; even though each has zero energy in the tree approximation, higher-order effects (including tunneling through the barrier between $\phi = 0$ and $\phi = -2m^2/g$) would be expected to mix them and shift their energies away from zero. (It is only in the limit of infinite volume that this barrier becomes impassable.) But in supersymmetric theories these states must have precisely zero energy, because there is no low-energy fermion state with which they could be paired. Thus for m large and g small the Witten index is 2. Because the Witten index is invariant under changes in the parameters of the theory, the Witten index remains equal to 2 even when g is large, where perturbation theory breaks down, and even when m vanishes, where the two potential wells merge. (It is not easy to calculate the Witten index directly in this case, because of the presence in the tree approximation of massless bosons as well as massless fermions.) Since the Witten index is not zero, supersymmetry remains strictly unbroken in the Wess–Zumino model, whatever the values of its parameters.

The same arguments can generally be used in theories with several chiral scalar superfields to show that the Witten index is positive, and that supersymmetry is therefore not spontaneously broken. The O'Raifeartaigh models discussed in Section 26.5 are an exception, because there are flat directions in which the potential remains constant as the fields go to infinity, rather than growing as some power of the fields. These models provide a good illustration of the fact that although the Witten index must be zero for supersymmetry to be broken, a zero Witten index does not necessarily imply that supersymmetry *is* broken. For instance, if we write the superpotential used as an illustration in Section 26.5 in terms of canonically normalized superfields, then it takes the form

$$f(X, Y_1, Y_2) = mY_1(X - a) + gY_2X^2 ,$$

with arbitrary parameters m, g, and a. The potential is then

$$U(x, y_1, y_2) = |g|^2|x|^4 + |m|^2|x - a|^2 + |my_1 + 2gxy_2|^2 ,$$

with lower-case letters denoting the scalar components of the left-chiral superfields. For m and a non-zero and g small, perturbation theory gives a good estimate of the spectrum, and tells us that there is a minimum of the potential near $x = a - 2|g|^2|a|^4/|m|^2$ with $my_1 + 2xy_2 = 0$, at which the vacuum energy is approximately equal to $|ga^2|^2V$. Since this energy appears as an additive constant in all states, there are no zero-energy states, and the Witten index vanishes. The matrix $\mathcal{M}$ of second derivatives of the superpotential (with rows and columns labelled in the order x, y_1, y_2) is here

$$\mathcal{M} = \begin{pmatrix} 2gy_2 & m & 2gx \\ m & 0 & 0 \\ 2gx & 0 & 0 \end{pmatrix} .$$

This has an eigenvector $(0, 2gx, -m)$ with eigenvalue zero, so there is a massless fermion here; this is the goldstino associated with the breaking of supersymmetry. The fermionic state degenerate with the vacuum consists of a single goldstino of zero energy and momentum. (Once again, there are two vacuum-energy fermionic states, with opposite orientations for the goldstino spin, and two vacuum-energy bosonic states: the vacuum, and a state containing two goldstinos with opposite spin.) Now, as $a \to 0$ supersymmetry may become unbroken (and we shall see later that it does), but the Witten index must remain 0; in this case the massless fermion is no longer a goldstino, but continuity demands that its mass remains zero, so it remains paired with the vacuum state. This of course is a general feature of theories in which supersymmetry is restored at isolated values of the parameters; continuity demands that the massless fermion that plays the role of the goldstino when supersymmetry is broken remains massless (though no longer a goldstino) at the value of the parameters where supersymmetry is restored, so that the vacuum remains paired with the massless fermion state, and the Witten index remains zero.

This model provides a good illustration of why we had to qualify the statement that the Witten index does not change when we vary the parameters of a supersymmetric theory, with the proviso that the parameters must not be changed in such a way as to alter the asymptotic behavior of the Lagrangian density for large fields. Suppose we unflatten the flat direction in this model by adding a small term to the superpotential, so that it now reads

$$f(X, Y_1, Y_2) = mY_1(X - a) + gY_2X^2 + \tfrac{1}{2}\epsilon(Y_1^2 + Y_2^2)\,,$$

where ϵ is a small mass parameter. Now there are two solutions of the conditions for supersymmetry to be preserved:

$$0 = \frac{\partial f}{\partial x} = \frac{\partial f}{\partial y_1} = \frac{\partial f}{\partial y_2}\,.$$

At these solutions, x is at one of the roots of the quadratic equation $2g^2x^2 + m(x - a) = 0$, while y_1 and y_2 are of order $1/\epsilon$: $y_1 = -m(x-a)/\epsilon$ and $y_2 = -gx^2/\epsilon$. We see that the reason that the Witten index can change from 0 to 2 when we turn on the small parameter ϵ is that two new minima of the potential come in from infinite field values.

In deciding whether or not supersymmetry is broken in theories with a vanishing Witten index, it is often useful to use conservation laws to limit the pairings that may occur, and to define a new sort of index. If K is a quantum operator that commutes with the supersymmetry generators Q_α (and hence also with the Hamiltonian), then all states of non-zero energy that have a definite value for K are paired with states of opposite statistics and the same energy and momentum *and* the same value of K. Also, not only is the Witten index $\mathrm{Tr}\,(-1)^F$ independent of the parameters of the theory (as long as they are not varied in a way that varies the large-field asymptotic behavior of the Lagrangian) — so also is the *weighted Witten index*, given by $\mathrm{Tr}\,g(K)(-1)^F$, where $g(K)$ is an arbitrary function of the conserved quantity. To use the conservation law in this way, it is not necessary that it be unbroken when the volume V becomes infinite; it is only necessary that K commute with the supersymmetry generators.

In diagnosing the possibilities of supersymmetry breaking, it is sometimes helpful to work with a linear combination of weighted Witten indices for a number of different conserved quantities. In particular, consider the quantity

$$W_G = \sum_{h \in G} \mathrm{Tr}\left\{h(-1)^F\right\}, \tag{29.1.6}$$

with the sum running over all elements of some symmetry group G. (For compact continuous groups, this sum should be interpreted as an integral over the group volume, with a suitable invariant measure.) Within any irreducible representation other than the identity, the 'characters' $\text{Tr}\, h$ add up to zero when summed over a finite or compact group, so

$$W_G = \sum_f N(f)(-1)^f , \tag{29.1.7}$$

where $N(f)$ is the number of times that the identity representation of G appears among states with fermion number f. In other words, W_G is just the Witten index, but evaluated using only G-invariant states. As long as G is conserved, W_G remains independent of the parameters of the theory, and if non-zero indicates that supersymmetry is unbroken.

Conservation laws will be used in this way in Section 29.4 to study the spontaneous breaking of supersymmetry in gauge theories, but a simpler (though academic) example is provided by the O'Raifeartaigh-style model discussed earlier, only now with the parameter a set equal to zero. The superpotential here is

$$f(X, Y_1, Y_2) = mY_1X + gY_2X^2 , \tag{29.1.8}$$

yielding a tree-approximation potential

$$U(x, y_1, y_2) = |m|^2|x|^2 + |g|^2|x|^4 + |my_1 + 2gxy_2|^2 . \tag{29.1.9}$$

We know that the Witten index vanishes here, because we have seen that it vanished for $a \neq 0$, but is supersymmetry broken? With $a = 0$, there are now field values with $x = y_1 = 0$ where the potential vanishes in the tree approximation, but how can we tell whether effects of higher order in g or even non-perturbative effects give the corresponding states a small energy? To answer this, we note that this superpotential (and hence the Lagrangian density) is invariant under a discrete symmetry K, under which the superfields are transformed by

$$KXK^{-1} = iX , \quad KY_1K^{-1} = -iY_1 , \quad KY_2K^{-1} = -Y_2 . \tag{29.1.10}$$

(Note that this symmetry is violated by the term $-maY_1$ in the original superpotential, so none of the results obtained using K will apply to that superpotential.) Since the potential vanishes for $x = y_1 = 0$ and y_2 arbitrary, for small g we may use perturbation theory to tell us that for each y_2 there is a bosonic vacuum state *near* zero energy. For $y_2 = 0$ this vacuum is even under K. For any non-zero value of $|y_2|$ we can take linear combinations of the two zero-energy states with $y_2 = \pm|y_2|$, of which one will be even under K and the other odd. As we have seen, there is also a massless fermion here, the fermionic component of Y_2, but this fermion is odd under K, so it cannot be paired with the even vacuum states. The only other fermions in the theory have tree-approximation masses $|m|$, so for small g they can't be paired with the even vacuum states either. We conclude then that for small g the even vacuum states must have precisely zero energy, and supersymmetry is unbroken. It is not easy to calculate the weighted Witten index here, because there are an infinite number of bosonic zero-energy states that are even under K, consisting of either zero or two Y_2-fermions with zero momentum plus any even number of zero-momentum Y_2 bosons, but it is clear that $\mathrm{Tr}\, K(-)^F > 0$, and since this is independent of g (as long as $g \neq 0$), supersymmetry cannot be broken for any finite g.

29.2 Supersymmetry Current Sum Rules

We now turn to sum rules that yield exact quantitative relations between the vacuum energy and parameters describing the strength of supersymmetry breaking.

Let us again start by assuming that the world is placed in a box of volume V, with periodic boundary conditions to preserve translation invariance. The vacuum expectation value of the anticommutation relations (25.2.36) may then be expressed as a sum over discrete states $|X, \text{Box}\rangle$:

$$\sum_X \Big\langle \text{VAC}\Big|Q_\alpha\Big|X, \text{Box}\Big\rangle\Big\langle \text{VAC}\Big|Q_\beta\Big|X, \text{Box}\Big\rangle^* + \sum_X \Big\langle \text{VAC}\Big|Q^*_\beta\Big|X, \text{Box}\Big\rangle\Big\langle \text{VAC}\Big|Q^*_\alpha\Big|X, \text{Box}\Big\rangle^* = -2i\Big(\gamma_\mu\beta\Big)_{\alpha\beta}\Big\langle \text{VAC}\Big|P^\mu\Big|\text{VAC}\Big\rangle\,, \tag{29.2.1}$$

with the label 'Box' indicating that the states are normalized to have Kronecker deltas rather than delta functions as their scalar products. Setting $\beta = \alpha$, summing over α and using Eq. (25.2.37) gives

$$\sum_{X,\alpha} \left|\Big\langle \text{VAC}\Big|Q_\alpha\Big|X, \text{Box}\Big\rangle\right|^2 = 4\Big\langle \text{VAC}\Big|P^0\Big|\text{VAC}\Big\rangle\,. \tag{29.2.2}$$

Because Q_α commutes with the four-momentum, it is only states with zero three-momentum and the same energy as the vacuum that can contribute to this sum. To find the volume dependence of the matrix elements that do contribute in Eq. (29.2.2), we note that a box-normalized state $|X, \text{Box}\rangle$ containing N_X particles is related according to Eq. (3.4.3) to the corresponding continuum-normalized state $|X\rangle$ by

$$\Big|X, \text{Box}\Big\rangle = \Big((2\pi)^3/V\Big)^{N_X/2}\Big|X\Big\rangle . \tag{29.2.3}$$

For states with $\mathbf{p}_X = 0$, the spatial integral of the supersymmetry current time-component S^0_α gives another factor of V, so we conclude that for box-normalized states with $\mathbf{p}_X = 0$

$$\Big\langle \text{VAC}\Big|Q_\alpha\Big|X, \text{Box}\Big\rangle = (2\pi)^{3N_X/2}V^{1-N_X/2}\Big\langle \text{VAC}\Big|S^0_\alpha(0)\Big|X\Big\rangle . \tag{29.2.4}$$

Since invariance under 2π-rotations does not allow X to be a zero-particle state, the dominant terms in Eq. (29.2.2) when $V \to \infty$ will be those from one-particle states. In this limit, Eq. (29.2.2) becomes

$$(2\pi)^3 \sum_{X,\alpha}{}^{(0)}\Big|\Big\langle \text{VAC}\Big|S^0_\alpha(0)\Big|X\Big\rangle\Big|^2 = 4\,\rho_{\text{VAC}} , \tag{29.2.5}$$

where ρ_{VAC} is the vacuum energy density

$$\rho_{\text{VAC}} \equiv \Big\langle \text{VAC}\Big|P^0\Big|\text{VAC}\Big\rangle/V , \tag{29.2.6}$$

and the superscript (0) indicates that the sum in Eq. (29.2.5) runs only over one-particle states with zero four-momentum. These of course are the two helicity states of the goldstino.

We see again from Eq. (29.2.5) that if the vacuum energy density does not vanish then the vacuum is not invariant under supersymmetry transformations, but is rather transformed into one-goldstino states. Conversely, Eq. (29.2.2) shows that if the vacuum is not invariant under supersymmetry then according to Eq. (29.2.2) its energy in a finite box cannot vanish, although it is conceivable that supersymmetry transformations might take the vacuum only into multiparticle states, in which case the vacuum energy density would vanish in the limit of large volume.

To evaluate the one-goldstino contribution in Eq. (29.2.5), we use Lorentz invariance to write the matrix element of the supersymmetry current between the vacuum and a one-goldstino state $|\mathbf{p}, \sigma\rangle$, with momentum

p and helicity λ, in the form*

$$\left\langle \text{VAC}\middle|S^\mu(0)\middle|\mathbf{p},\lambda\right\rangle = (2\pi)^{-3/2}\left[\left(\frac{1+\gamma_5}{2}\right)\left(\gamma^\mu F + ip^\mu F'\right)\right.$$
$$\left. + \left(\frac{1-\gamma_5}{2}\right)\left(\gamma^\mu F^* + ip^\mu F'^*\right)\right] u(\mathbf{p},\lambda)\,, \quad (29.2.7)$$

where $u(\mathbf{p},\lambda)$ are the coefficient functions for a massless Dirac field introduced in Section 5.5, and F and F' are unknown constants. The matrix element (29.2.7) satisfies a conservation condition $p_\mu\left\langle \text{VAC}\middle|S^\mu(0)\middle|\mathbf{p},\lambda\right\rangle = 0$ because $u(\mathbf{p},\lambda)$ satisfies the momentum-space Dirac equation (5.5.42) for zero mass, and p^μ is on the light-cone. The sum over helicities gives

$$\sum_\lambda u(\mathbf{p},\lambda)\,\bar{u}(\mathbf{p},\lambda) = -i\,\not{p}/2p^0\,.$$

(The Dirac spinor $u(\mathbf{p},\lambda)$ for a massless particle of momentum **p** and spin λ is not well defined for $\mathbf{p}\to 0$, but there is no problem if we regard the sum over X in Eq. (29.2.5) as a sum over helicities for a small momentum **p** of fixed direction.) After a straightforward calculation, we then have

$$4\,\rho_{\text{VAC}} = \text{Tr}\left\{\left(F\frac{1+\gamma_5}{2} + F^*\frac{1-\gamma_5}{2}\right)\gamma^0\frac{-i\not{p}\beta}{2p^0}(\gamma^0)^\dagger\left(F^*\frac{1+\gamma_5}{2} + F\frac{1-\gamma_5}{2}\right)\right\},$$

and therefore

$$\rho_{\text{VAC}} = |F|^2/2\,. \quad (29.2.8)$$

An alternative proof of this formula will be given at the end of this section.

---

* Lorentz invariance alone would yield this formula with independent coefficients F_L, F'_L and F_R, F'_R for the matrices proportional to $(1+\gamma_5)/2$ and $(1-\gamma_5)/2$, respectively. It is CPT invariance that imposes the relations $F_R = F^*_L$ and $F'_R = F'^*_L$. To see this, we must use the CPT-transformation properties of the supersymmetry current

$$\text{CPT}\,S^\mu(x)(\text{CPT})^{-1} = -\gamma_5 S^\mu(-x)^* = -\beta\epsilon S^\mu(-x)$$

(see Section 5.8) and of the one-particle states

$$\text{CPT}|\mathbf{p},\lambda\rangle = \chi_\lambda|\mathbf{p},-\lambda\rangle$$

with χ_λ a phase factor that depends on how we define the relative phases of the helicity states. We also need the reality properties of the coefficient functions $u(\mathbf{p},\lambda)$. These are related to those of the one-particle states by the definition

$$\langle\text{VAC}|\psi_{\text{REN}}(x)|\mathbf{p},\lambda\rangle = (2\pi)^{-3/2}\exp(ip\cdot x)\,u(\mathbf{p},\lambda)\,,$$

where $\psi_{\text{REN}}(x)$ is a renormalized Majorana field, with CPT-transformation property

$$\text{CPT}\,\psi_{\text{REN}}(x)\,(\text{CPT})^{-1} = \gamma_5\,\psi_{\text{REN}}(-x)^* = \beta\,\epsilon\,\psi_{\text{REN}}(-x)\,.$$

which yields $u(\mathbf{p},\lambda) = \chi^*_\lambda\,\beta\,\epsilon\,u^*(\mathbf{p},-\lambda)$. Putting this together with the CPT-transformation properties of the supersymmetry current and one-particle states yields the relations $F_R = F^*_L$ and $F'_R = F'^*_L$.

The parameter F plays much the same role in the interactions of soft goldstinos as does the parameter F_π (introduced in Section 19.4) in the interactions of soft pions. The matrix element of the supersymmetry current between any two states X and Y may be split into terms that have a one-goldstino pole at $p^\mu = 0$ in the momentum transfer $p \equiv p_X - p_Y$ and terms that do not have this pole

$$\langle X|S^\mu(0)|Y\rangle = \left\{\left(\frac{1+\gamma_5}{2}\right)[\gamma^\mu F + ip^\mu F'] + \left(\frac{1-\gamma_5}{2}\right)[\gamma^\mu F^* + ip^\mu F'^*]\right\}$$
$$\times\left(\frac{-i\not{p}}{p^2}\right) M(X \to Y + g) + \langle X|S^\mu(0)|Y\rangle_{\text{no pole}}\,, \quad (29.2.9)$$

where $\bar{u}M(X \to Y+g)$ is the amplitude for emitting a goldstino with four-momentum p and Dirac wave function u, and the subscript 'no pole' denotes the terms in the matrix element that do not have the one-goldstino pole in the four-momentum p. The conservation of the current S^μ tells us that this vanishes if contracted with p_μ, so in the limit $p^\mu \to 0$ the goldstino emission amplitude is**

$$M(X \to Y+g) \to -i\left\{\left(\frac{1+\gamma_5}{2F}\right) + \left(\frac{1-\gamma_5}{2F^*}\right)\right\} p_\mu \langle X|S^\mu(0)|Y\rangle_{\text{no pole}}\,. \quad (29.2.10)$$

There is another sum rule that provides an alternative proof of the existence of goldstinos when supersymmetry is spontaneously broken, and that also relates the parameter F and the vacuum energy density to the D-terms and $\mathscr{F}$-terms that characterize the strength of supersymmetry breaking. (This use of a sum rule is analogous to the second proof in Section 19.2 of the existence of Goldstone bosons when ordinary symmetries are spontaneously broken.) To derive this sum rule, we will now give up the device of a finite volume, and instead avoid the question of the convergence of the integral (29.1.3) by working with a local consequence of the supersymmetry of the action

$$\left[\left(\bar{S}^0(\mathbf{x},t)\,\alpha\right),\, \chi(\mathbf{y},t)\right] = \left[\left(\bar{\alpha}\, S^0(\mathbf{x},t)\right),\, \chi(\mathbf{y},t)\right] = i\,\delta^3(\mathbf{x}-\mathbf{y})\,\delta\chi(\mathbf{x},t) + \cdots\,, \quad (29.2.11)$$

** The amplitude $\langle X|S^\mu(0)|Y\rangle_{\text{no pole}}$ may have poles that go as $1/p\cdot k$ as $p^\mu \to 0$; they would not arise from the goldstino propagator, which is explicitly excluded in this matrix element, but rather from other particle propagators produced by the insertion of the supersymmetry current in external lines of momentum k of the process $X \to Y$. In the limit $p^\mu \to 0$, the contribution of these poles in Eq. (29.2.10) would dominate the amplitudes for emission or absorption of soft goldstinos. But for such poles to arise, the goldstino would have to be emitted in a transition between a pair of *degenerate* particles of opposite statistics, which are not likely to appear in theories in which supersymmetry is spontaneously broken. The interactions of soft goldstinos differ in this respect from the interactions of soft pions, photons, or gravitons.

where $\chi(x)$ is an arbitrary fermionic or bosonic field, $\delta\chi(x)$ is the change in $\chi(x)$ induced by the supersymmetry transformation with infinitesimal parameter α that leaves the action invariant, and the dots denote terms involving derivatives of $\delta^3(\mathbf{x}-\mathbf{y})$. Let us consider the vacuum expectation value of the anticommutator of an arbitrary left-handed spinor field $\psi_L(x)$ with the covariant conjugate $\bar{S}^\mu(y)$ of the supersymmetry current. By summing over a complete set $|X\rangle$ of intermediate states (including integration over particle momenta), this vacuum expectation value may be put in the form

$$\left\langle \mathrm{VAC}\middle|\left\{\psi_{L\alpha}(x),\,\bar{S}^\mu_\beta(y)\right\}\middle|\mathrm{VAC}\right\rangle = \int d^4p\; e^{ip\cdot(x-y)}\left[G^\mu_{\alpha\beta}(p)+\tilde{G}^\mu_{\alpha\beta}(-p)\right], \tag{29.2.12}$$

where

$$G^\mu_{\alpha\beta}(p) \equiv \sum_X \delta^4(p-p_X)\left\langle \mathrm{VAC}\middle|\psi_{L\alpha}(0)\middle|X\right\rangle\left\langle X\middle|\bar{S}^\mu_\beta(0)\middle|\mathrm{VAC}\right\rangle, \tag{29.2.13}$$

$$\tilde{G}^\mu_{\alpha\beta}(p) = \sum_X \delta^4(p-p_X)\left\langle \mathrm{VAC}\middle|\bar{S}^\mu_\beta(0)\middle|X\right\rangle\left\langle X\middle|\psi_{L\alpha}(0)\middle|\mathrm{VAC}\right\rangle. \tag{29.2.14}$$

Lorentz invariance requires that the matrices $G^\mu(p)$ and $\tilde{G}^\mu(p)$ must take the forms:

$$G^\mu(p) = \theta(p^0)\left(\frac{1+\gamma_5}{2}\right)\Big[\gamma^\mu G^{(1)}(-p^2)+\not{p}p^\mu G^{(2)}(-p^2) +p^\mu G^{(3)}(-p^2)+\not{p}\gamma^\mu G^{(4)}(-p^2)\Big] \tag{29.2.15}$$

and

$$\tilde{G}^\mu(p) = \theta(p^0)\left(\frac{1+\gamma_5}{2}\right)\Big[\gamma^\mu \tilde{G}^{(1)}(-p^2)+\not{p}p^\mu \tilde{G}^{(2)}(-p^2) +p^\mu \tilde{G}^{(3)}(-p^2)+\not{p}\gamma^\mu \tilde{G}^{(4)}(-p^2)\Big]. \tag{29.2.16}$$

The right-hand sides of Eqs. (29.2.15) and (29.2.16) are unaffected if we replace $-p^2$ with m^2, multiply by a factor $\delta(p^2+m^2)$, and integrate over m^2. In this way, Eq. (29.2.12) becomes

$$(2\pi)^{-3}\left\langle \mathrm{VAC}\middle|\left\{\psi_L(x),\,\bar{S}^\mu(y)\right\}\middle|\mathrm{VAC}\right\rangle = \left(\frac{1+\gamma_5}{2}\right)\int_0^\infty dm^2\,\Big[\gamma^\mu G^{(1)}(m^2) - \not{\partial}\partial^\mu G^{(2)}(m^2) - i\partial^\mu G^{(3)}(m^2) - i\,\not{\partial}\gamma^\mu G^{(4)}(m^2)\Big]\Delta_+(x-y,m)$$

$$+\left(\frac{1+\gamma_5}{2}\right)\int_0^\infty dm^2\left[\gamma^\mu \tilde{G}^{(1)}(m^2) - \not{\partial}\partial^\mu \tilde{G}^{(2)}(m^2) - i\partial^\mu \tilde{G}^{(3)}(m^2)\right.$$
$$\left. -i\not{\partial}\gamma^\mu \tilde{G}^{(4)}(m^2)\right]\Delta_+(y-x,m)\,, \qquad (29.2.17)$$

where $\Delta_+(x,m)$ is the standard function

$$\Delta_+(x,m) \equiv (2\pi)^{-3}\int d^4p\;\theta(p^0)\,\delta(p^2+m^2)\,\exp(ip\cdot x)\,. \qquad (29.2.18)$$

Causality requires that the anticommutator on the left-hand side of Eq. (29.2.17) should vanish for spacelike separations $x-y$. For such separations, $\Delta_+(x-y,m)$ is an even function of $x-y$, so Eq. (29.2.17) vanishes for all spacelike separations if and only if

$$G^{(1)}(m^2) = -\tilde{G}^{(1)}(m^2)\,, \qquad G^{(2)}(m^2) = -\tilde{G}^{(2)}(m^2)\,,$$
$$G^{(3)}(m^2) = +\tilde{G}^{(3)}(m^2)\,, \qquad G^{(4)}(m^2) = +\tilde{G}^{(4)}(m^2)\,, \qquad (29.2.19)$$

so that for general $x-y$, Eq. (29.2.17) now reads

$$(2\pi)^{-3}\Big\langle \text{VAC}\Big|\Big\{\psi_L(x),\,\bar{S}^\mu(y)\Big\}\Big|\text{VAC}\Big\rangle = \int_0^\infty dm^2\left[G^{(1)}(m^2)\gamma^\mu - G^{(2)}(m^2)\,\not{\partial}\partial^\mu\right.$$
$$\left. -iG^{(3)}(m^2)\partial^\mu - iG^{(4)}(m^2)\,\not{\partial}\gamma^\mu\right]\Delta(x-y,m)\left(\frac{1+\gamma_5}{2}\right)\,, \qquad (29.2.20)$$

where, as usual,

$$\Delta(x-y,m) \equiv \Delta_+(x-y,m) - \Delta_+(y-x,m)\,. \qquad (29.2.21)$$

We next impose the supersymmetry current conservation condition (29.1.2), which (because $\Box\Delta = m^2\Delta$) yields

$$G^{(1)}(m^2) = m^2G^{(2)}(m^2)\,, \qquad m^2G^{(3)}(m^2) = -m^2G^{(4)}(m^2)\,. \qquad (29.2.22)$$

Finally, we must relate these spectral functions to the breaking of supersymmetry. Recall that for $x^0 = y^0$,

$$\Delta(\mathbf{x}-\mathbf{y},0,m) = 0\,, \qquad \dot{\Delta}(\mathbf{x}-\mathbf{y},0,m) = -i\delta^3(\mathbf{x}-\mathbf{y})\,.$$

Setting $x^0 = y^0 = t$ and $\mu = 0$ in Eq. (29.2.20) thus yields

$$\Big\langle \text{VAC}\Big|\Big\{\psi_L(\mathbf{x},t),\,\bar{S}^0(\mathbf{y},t)\Big\}\Big|\text{VAC}\Big\rangle = (2\pi)^3\delta^3(\mathbf{x}-\mathbf{y})\left(\frac{1+\gamma_5}{2}\right)$$
$$\times\int_0^\infty dm^2\left[G^{(3)}(m^2) + G^{(4)}(m^2)\right]\,. \qquad (29.2.23)$$

Contracting on the right with an infinitesimal Majorana fermionic supersymmetry transformation parameter α and using Eq. (29.2.11) then gives

$$i\Big\langle\delta\psi_L\Big\rangle_{\text{VAC}} = (2\pi)^3\left(\frac{1+\gamma_5}{2}\right)\alpha\int_0^\infty dm^2\;\left[G^{(3)}(m^2) + G^{(4)}(m^2)\right]\,. \qquad (29.2.24)$$

But Eq. (29.2.22) shows that the integrand of the integral over m^2 in Eq. (29.2.24) vanishes, except perhaps at $m^2 = 0$, so we can conclude that

$$G^{(3)}(m^2) + G^{(4)}(m^2) = \delta(m^2)\mathscr{G}\,, \tag{29.2.25}$$

with a constant coefficient $\mathscr{G}$ given by

$$i\Big\langle \delta\psi_L \Big\rangle_{\rm VAC} = (2\pi)^3 \mathscr{G} \alpha_L\,. \tag{29.2.26}$$

As we saw in Sections 26.4 and 27.4, a breakdown of supersymmetry is signaled by the appearance of vacuum expectation values of the changes $\delta\psi$ under supersymmetry transformations of one or more spinor fields ψ. Eqs. (29.2.25) and (29.2.26) show that for any such spinor field, the spectral function $G^{(3)}(m^2) + G^{(4)}(m^2)$ has a delta function singularity at $m^2 = 0$, which could only arise from the appearance of a massless one-particle state $|g\rangle$ in the sums over states in Eqs. (29.2.13) and/or (29.2.14). For the matrix elements $\langle {\rm VAC}|\psi|g\rangle$ or $\langle g|\psi|{\rm VAC}\rangle$ not to vanish, this massless particle must have spin 1/2. This is the goldstino.

Now let's calculate the contribution of a one-goldstino state $|\mathbf{p}, \lambda\rangle$ with momentum $\mathbf{p}$ and helicity λ to the spectral functions $G^{(i)}(m^2)$. Lorentz invariance tells us that the matrix element of a general fermion field (not just the renormalized goldstino field) in Eq. (29.2.13) takes the form[†]

$$\Big\langle {\rm VAC}\Big|\psi(0)\Big|\mathbf{p},\lambda\Big\rangle = \frac{1}{(2\pi)^{3/2}}\left[N\left(\frac{1+\gamma_5}{2}\right) + N^*\left(\frac{1-\gamma_5}{2}\right)\right] u(\mathbf{p},\lambda)\,, \tag{29.2.27}$$

where N is a constant, characterizing the particular fermion field. Also, the delta function in Eq. (29.2.13) is

$$\delta^4(p - p_g) = 2p^0\delta(p^2)\theta(p^0)\delta^3(\mathbf{p} - \mathbf{p}_g)\,.$$

Together with Eq. (29.2.7), this yields

$$\Big[G^\mu_{\alpha\beta}(p)\Big]_1 = \frac{N}{(2\pi)^3}\left(\frac{1+\gamma_5}{2}\right)\theta(p^0)\delta(p^2)\, \not{p}\Big[\gamma^\mu F + ip^\mu F'\Big]\,,$$

with the subscript 1 indicating the one-goldstino contribution. Comparing this with Eq. (29.2.15) shows that

$$\Big[G^{(2)}(m^2)\Big]_1 = i(2\pi)^{-3}NF'\delta(m^2)\,, \qquad \Big[G^{(4)}(m^2)\Big]_1 = (2\pi)^{-3}NF\delta(m^2)\,,$$
$$\Big[G^{(1)}(m^2)\Big]_1 = \Big[G^{(3)}(m^2)\Big]_1 = 0\,. \tag{29.2.28}$$

[†] Again, Lorentz invariance would allow independent coefficients N_L and N_R for $(1+\gamma_5)$ and $(1-\gamma_5)$, respectively. Using the CPT transformation of the one-particle state and the reality property of the coefficient functions discussed in the first footnote of this section, and the CPT transformation of a general fermion field, ${\sf CPT}\psi(x)({\sf CPT})^{-1} = \beta\epsilon\psi(-x)$, we find that $N_L = N_R^*$.

Thus Eqs. (29.2.25) and (29.2.26) yield

$$i\left\langle\delta\psi_L\right\rangle_{\text{VAC}} = NF\alpha_L\,. \tag{29.2.29}$$

Evidently if the change in any fermion field under a supersymmetry transformation has a non-zero vacuum expectation value then the N-factor for that field cannot vanish, so there must be a one-goldstino state that contributes to these spectral functions.

To be a little more specific, recall that the fermionic components ψ_{Ln} of a left-chiral scalar superfield Φ_{Ln} obey the supersymmetry transformation rule (26.3.15):

$$\delta\psi_{Ln} = \sqrt{2}\partial_\mu\phi_n\gamma^\mu\alpha_R + \sqrt{2}\mathscr{F}_n\alpha_L\,, \tag{29.2.30}$$

while Eq. (27.3.5) gives the supersymmetry transformation of the gaugino fields as

$$\delta\lambda_A = \left(\tfrac{1}{4}f_{A\mu\nu}\,[\gamma^\nu,\gamma^\mu] + i\gamma_5 D_A\right)\alpha\,. \tag{29.2.31}$$

Hence the N-factors in the matrix elements of the ψ_n and λ_A between the vacuum and a one-goldstino state are given by

$$N_n = iF^{-1}\sqrt{2}\left\langle\mathscr{F}_n\right\rangle_{\text{VAC}}\,, \qquad N_A = -F^{-1}\left\langle D_A\right\rangle_{\text{VAC}}\,. \tag{29.2.32}$$

Let us see how the results (29.2.32) arise in the tree approximation. For a renormalizable theory of gauge and chiral superfields, Eq. (27.4.30) gives the left-handed part of the fermion mass matrix as

$$\begin{aligned} M_{nm} &= \left(\frac{\partial^2 f(\phi)}{\partial\psi_n\partial\psi_m}\right)_{\phi=\phi_0}\,, \\ M_{nA} &= M_{An} = i\sqrt{2}(t_A\phi_0)^*_n\,, \qquad M_{AB} = 0\,. \end{aligned} \tag{29.2.33}$$

The vacuum fields ϕ_{n0} are at a minimum of the potential given by Eq. (27.4.9), so

$$0 = \left(\frac{\partial V(\phi)}{\partial\phi_n}\right)_0 = -\sum_m M_{nm}\mathscr{F}_{m0} + \sum_A\left(\phi_0^\dagger t_A\right)_n D_{A0}\,, \tag{29.2.34}$$

where the $\mathscr{F}$s and Ds are given by Eqs. (27.4.6) and (27.4.7)

$$\mathscr{F}_n = -\left(\partial f(\phi)/\partial\phi_n\right)^*\,, \qquad D_A = \xi_A + \sum_{nm}\phi_n^*(t_A)_{nm}\phi_m\,,$$

and the subscript 0 indicates that we are to set $\phi_n = \phi_{n0}$. Furthermore, the gauge invariance of the superpotential requires that

$$\sum_n \mathscr{F}_n\left(t_B\phi\right)^*_n = 0\,, \tag{29.2.35}$$

for all values of ϕ. Therefore the left-handed quark mass matrix M has an eigenvector v with $Mv = 0$, where

$$v_n = \sqrt{2}\mathscr{F}_{n0}\,, \qquad v_A = iD_{A0}\,. \tag{29.2.36}$$

Thus in the tree approximation, if we expand the left-handed fermion fields in renormalized fields for particles of definite mass, then the coefficients of the goldstino field in ψ_{Ln} and in λ_{AL} are proportional to $\sqrt{2}\mathscr{F}_{n0}$ and iD_{A0}, respectively, in agreement with Eq. (29.2.32).

If we normalize the spinor fields ψ_{Ln} and λ_A so that the matrix connecting these fields to the renormalized fields of particles of definite mass is unitary, then

$$\sum_n |N_n|^2 + \sum_A |N_A|^2 = 1\,, \tag{29.2.37}$$

so Eq. (29.2.32) gives the non-perturbative result

$$|F|^2 = 2\sum_n \left|\left\langle\mathscr{F}_n\right\rangle_{\rm VAC}\right|^2 + \sum_A \left|\left\langle D_A\right\rangle_{\rm VAC}\right|^2\,. \tag{29.2.38}$$

The result (29.2.38) for $|F|^2$ allows us to express the vacuum expectation density (29.2.8) in terms of the vacuum expectation values of auxiliary fields

$$\rho_{\rm VAC} = \sum_n \left|\left\langle\mathscr{F}_n\right\rangle_{\rm VAC}\right|^2 + \frac{1}{2}\sum_A \left|\left\langle D_A\right\rangle_{\rm VAC}\right|^2\,. \tag{29.2.39}$$

This is the non-perturbative generalization of the zeroth-order result (27.4.9). It confirms a result used in Section 27.6, that the conditions $\langle\mathscr{F}_n\rangle = \langle D_A\rangle = 0$ are sufficient as well as necessary for supersymmetry to be unbroken.

* * *

It is instructive to see how Eq. (29.2.8) can be derived without the device of a finite volume. For this purpose, consider the vacuum expectation value of the anticommutator of two supersymmetry currents. By using Lorentz invariance and the vanishing of anticommutators at spacelike separations in the same way as in the previous section, we find

$$\begin{aligned}\Big\langle {\rm VAC}\Big|\Big\{S^\mu(x), \bar{S}^\nu(y)\Big\}\Big|{\rm VAC}\Big\rangle &= -i\int dm^2\Big[H^{(1)}(m^2)\gamma^\mu\partial^\nu + H^{(2)}(m^2)\gamma^\nu\partial^\mu \\ &+H^{(3)}(m^2)\,\partial\!\!\!/\partial^\mu\partial^\nu + H^{(4)}(m^2)\,\partial\!\!\!/\eta^{\mu\nu} + H^{(5)}(m^2)\epsilon^{\mu\nu\lambda\rho}\partial_\lambda\gamma_\rho\Big]\Delta(x-y,m^2) \\ &+\cdots\end{aligned} \tag{29.2.40}$$

where

$$\int dX\, \delta^4(p - p_X)\left\langle \text{VAC}\middle|S^\mu_\alpha(0)\middle|X\right\rangle\left\langle X\middle|\bar{S}^\nu_\beta(0)\middle|\text{VAC}\right\rangle = H^{(1)}(-p^2)\gamma^\mu p^\nu$$
$$+ H^{(2)}(-p^2)\gamma^\nu p^\mu - H^{(3)}(-p^2)\, \not{p} p^\mu p^\nu + H^{(4)}(-p^2)\, \not{p}\eta^{\mu\nu}$$
$$+ H^{(5)}(-p^2)\epsilon^{\mu\nu\lambda\rho} p_\lambda \gamma_\rho + \cdots\,, \tag{29.2.41}$$

and the dots in Eqs. (29.2.40) and (29.2.41) denote a linear combination of the other independent Dirac covariant matrices 1, γ_5, $\gamma_5\gamma_\sigma$, and $[\gamma_\sigma, \gamma_\tau]$, which will not concern us here. The Majorana nature of the currents together with Eq. (26.A.20) tell us that the spectral functions $H^{(i)}$ are all real, and the conservation of the supersymmetry currents dictate that

$$H^{(1)}(m^2) = H^{(2)}(m^2) = -m^2\, H^{(3)}(m^2) - H^{(4)}(m^2)\,, \tag{29.2.42}$$

and

$$m^2 H^{(1)}(m^2) = 0\,. \tag{29.2.43}$$

Setting $\mu = \nu = 0$ and $x^0 = y^0$ in Eq. (29.2.40), integrating over $\mathbf{x}$, and using Eq. (29.2.42) then gives

$$\left\langle \text{VAC}\middle|\left\{Q,\, \bar{S}^0(0)\right\}\middle|\text{VAC}\right\rangle = (2\pi)^3\beta \int dm^2 \Big[H^{(1)}(m^2) + H^{(2)}(m^2)$$
$$+ m^2\, H^{(3)}(m^2) + H^{(4)}(m^2)\Big] + \cdots = (2\pi)^3\beta \int dm^2\, H^{(1)}(m^2) + \cdots\,. \tag{29.2.44}$$

To evaluate this anticommutator, we first write it as

$$\left\{Q,\, \bar{S}^\nu(x)\right\} = -2i\gamma_\mu T^{\mu\nu}(x) + \cdots\,, \tag{29.2.45}$$

where $T^{\mu\nu}(x)$ is some tensor operator and the dots again denote some linear combination of the other independent Dirac covariants. The Majorana character of Q and S^μ tells us that $T^{\mu\nu}(x)$ is Hermitian; the conservation of the supersymmetry current tells us that it is conserved, in the sense that

$$\partial_\nu T^{\mu\nu} = 0\,, \tag{29.2.46}$$

and the anticommutation relation (25.2.36) tells us that

$$\int d^3x\, T^{\mu 0}(x) = P^\mu\,. \tag{29.2.47}$$

These properties allow us to identify $T^{\mu\nu}$ as the energy-momentum tensor. It is not in general equal to the *symmetric* energy-momentum tensor $\Theta^{\mu\nu}$ discussed in Section 7.4, but Eq. (29.2.47) shows that the energy density T^{00} can differ from Θ^{00} only by spatial derivative terms that cannot contribute in states of zero three-momentum, so the vacuum energy density is given by

$$\rho_{\text{VAC}} = \left\langle \text{VAC}\middle|T^{00}\middle|\text{VAC}\right\rangle. \tag{29.2.48}$$

Thus Eqs. (29.2.44) and (29.2.45) yield

$$2\rho_{\rm VAC} = (2\pi)^3 \int dm^2\, H^{(1)}(m^2)\,. \tag{29.2.49}$$

But Eq. (29.2.43) tells us that $H^{(1)}(m^2)$ vanishes except perhaps at $m^2 = 0$, so

$$H^{(1)}(m^2) = 2(2\pi)^{-3}\delta(m^2)\rho_{\rm VAC}\,. \tag{29.2.50}$$

Thus we see again that a non-vanishing vacuum energy density entails the existence of a massless fermion, the goldstino. Using Eq. (29.2.7), a straightforward calculation of the one-goldstino contribution to the spectral functions yields

$$H^{(1)}(m^2) = (2\pi)^{-3}\delta(m^2)\,|F|^2\,. \tag{29.2.51}$$

Comparing this with Eq. (29.2.50) gives our previous result (29.2.8) for the vacuum energy density.

29.3 Non-Perturbative Corrections to the Superpotential

We saw in Section 27.6 that the superpotential in general supersymmetric theories of gauge and chiral superfields is not renormalized to any finite order of perturbation theory, so that if supersymmetry is not broken in the tree approximation then it can only be broken by non-perturbative corrections to the Wilsonian effective Lagrangian. We now take up a general analysis of these corrections. These were thoroughly studied in a series of papers by Affleck, Davis, Dine, and Seiberg in the early 1980s,[3] with special attention to the case of supersymmetric versions of quantum chromodynamics with arbitrary numbers of colors and flavors. Here we will present a somewhat simplified analysis of general supersymmetric gauge theories, based on the more recent holomorphy arguments of Seiberg[4] already used in Section 27.7.

To study non-perturbative effects, we shall here again consider a general renormalizable supersymmetric theory, but now including a possible θ-term in the Lagrangian density

$$\mathscr{L} = \Big[\Phi^\dagger e^{-V}\Phi\Big]_D + 2\,{\rm Re}\Big[f(\Phi)\Big]_{\mathscr{F}} + {\rm Re}\left[\frac{\tau}{8\pi i}\sum_{A\alpha\beta}\epsilon_{\alpha\beta}W_{A\alpha L}W_{A\beta L}\right]_{\mathscr{F}}\,, \tag{29.3.1}$$

where the superpotential $f(\Phi)$ is a gauge-invariant cubic polynomial in the left-chiral superfields and τ is the parameter (27.3.23):

$$\tau = \frac{4\pi i}{g^2} + \frac{\theta}{2\pi}\,. \tag{29.3.2}$$

As in Section 27.6, we introduce a pair of gauge-invariant left-chiral external superfields, now called Y and T, and replace the Lagrangian density with

$$\mathscr{L}^\sharp = \Big[\Phi^\dagger e^{-V}\Phi\Big]_D + 2\,\mathrm{Re}\Big[Y f(\Phi)\Big]_{\mathscr{F}} + \mathrm{Re}\left[\frac{T}{8\pi i}\sum_{A\alpha\beta}\epsilon_{\alpha\beta}W_{A\alpha L}W_{A\beta L}\right]_{\mathscr{F}} . \tag{29.3.3}$$

This becomes the same as (29.3.1) when we set the spinor and auxiliary components of Y and T equal to zero, and take their scalar components as $y = 1$ and $t = \tau$, respectively. Non-perturbative effects will in general invalidate both of the two symmetries on which the analysis of Section 27.6 was based. The translation operation, which in our present notation is $T \to T + \xi$ with real ξ, is not a symmetry because $\sum_A \epsilon_{\mu\nu\rho\sigma} f_A^{\mu\nu} f_A^{\rho\sigma}$ can have a non-vanishing integral over spacetime. The original R invariance (with T and Y having R values 0 and +2) is not a symmetry because the anomaly discussed in Chapter 22 gives the R-current a non-vanishing divergence. With θ_L and θ_R having $R = +1$ and $R = -1$, respectively, and V_A and Φ_n R-neutral, the fermion fields λ_{AL} and ψ_{nL} have $R = +1$ and $R = -1$, respectively, so Eq. (22.2.26) here gives

$$\partial_\mu J_R^\mu = -\frac{1}{32\pi^2}\Big(C_1 - C_2\Big)\sum_A \epsilon_{\mu\nu\rho\sigma} f_A^{\mu\nu} f_A^{\rho\sigma} , \tag{29.3.4}$$

where C_1 and C_2 are the constants defined in Eqs. (17.5.33) and (17.5.34):

$$\sum_{CD} C_{ACD}C_{BCD} = C_1\,\delta_{AB} , \qquad \mathrm{Tr}\{t_A t_B\} = C_2\,\delta_{AB} , \tag{29.3.5}$$

with the trace taken over all species of left-chiral superfield.* For instance, in the generalized supersymmetric version of quantum chromodynamics studied in Reference 3, with gauge group $SU(N_c)$ and N_f pairs of left-chiral quark superfields Q_n and $\overline{Q}_n$ in the defining representation and its complex conjugate, these constants have values given by Eq. (17.5.35) (with $n_f = 2N_f$) as

$$C_1 = N_c , \qquad C_2 = N_f .$$

Although T translation and R invariance are invalidated by non-perturbative effects, there is a remaining symmetry which is almost as

* A factor 32 instead of 16 appears in the denominator in Eq. (29.3.4) because gauginos do not have distinct antiparticles, and we are now counting antiparticles separately from particles in taking the trace in Eq. (29.3.5). Also, we are now adopting the convention, described at the end of Section 27.3, of including a gauge coupling factor in the gauge fields and not in structure constants and the matrix generators t_A. The gauge generators are thus normalized so that for t_A, t_B, and t_C in the standard $SU(2)$ subalgebra of the gauge algebra, the structure constant is $C_{ABC} = \epsilon_{ABC}$.

powerful. Consider a general R transformation

$$\theta_L \to e^{i\varphi}\theta_L\,, \quad \Phi \to \Phi\,, \quad V_A \to V_A\,, \quad Y \to e^{2i\varphi}Y\,, \tag{29.3.6}$$

with arbitrary real φ. This leaves the T-independent terms of the Lagrangian density (29.3.3) invariant, but according to Eq. (29.3.4), quantum effects violate this symmetry, just as if there were a term $\Delta\mathscr{L}$ in the Lagrangian density with a transformation

$$\Delta\mathscr{L} \to \Delta\mathscr{L} - \frac{1}{32\pi^2}\Big(C_1 - C_2\Big)\sum_A \epsilon_{\mu\nu\rho\sigma} f_A^{\mu\nu} f_A^{\rho\sigma}\,\varphi\,.$$

Recalling Eq. (27.3.18), this is cancelled if we give T a transformation

$$T \to T + (C_1 - C_2)\varphi/\pi\,. \tag{29.3.7}$$

Because $W_{A\alpha L}$ has $R = 2$, the whole theory including non-perturbative effects is invariant under the *combined* transformations (29.3.6) and (29.3.7). In particular, the superfield $\exp(2i\pi T)$, which for $T = \tau$ is periodic in θ, has $R = 2(C_1 - C_2)$.

We again introduce an ultraviolet cut-off, and consider the effective 'Wilsonian' Lagrangian

$$\mathscr{L}^\#_\lambda = \Big[\mathscr{A}_\lambda(\Phi, \Phi^\dagger, V, T, T^\dagger, Y, Y^\dagger, \mathscr{D}\cdots)\Big]_D$$
$$+2\,\mathrm{Re}\left[\frac{T}{8\pi i}\sum_{A\alpha\beta}\epsilon_{\alpha\beta}W_{AL\alpha}W_{AL\beta} + \mathscr{B}_\lambda(\Phi, W_L, T, Y)\right]_{\mathscr{F}}\,, \tag{29.3.8}$$

with $\mathscr{A}_\lambda$ and $\mathscr{B}_\lambda$ both gauge-invariant functions of the displayed arguments. The term proportional to T has been separated from the function $\mathscr{B}_\lambda$ in order that the translation (29.3.7) of T should continue to cancel the anomaly in the R transformation (29.3.6). Invariance under the combined transformation (29.3.6), (29.3.7) then tells us that terms in the function $\mathscr{B}_\lambda$ must be proportional to powers of $\exp(2i\pi T)$, which have definite R values.

Furthermore, it is only *positive* powers of $\exp(2i\pi T)$ that may appear in $\mathscr{B}_\lambda$. According to Eq. (27.3.24), it is only instantons with positive winding number $\nu \geq 0$ that can make contributions to the effective Lagrangian that are holomorphic in T rather than T^*, and these give rise to factors $\exp(2i\pi\nu T)$. More generally, for $T = \tau$ any power $\exp(2ia\pi T)$ will depend on the gauge coupling through a factor $\exp(-8\pi^2 a/g^2)$, so that a must be positive in order for non-perturbative effects to be suppressed for small g. In consequence, non-perturbative effects enter in $\mathscr{L}^\#_\lambda$ through operators $\exp(2i\pi aT)$ that have positive-definite, zero, or negative-definite values of R, depending on whether $C_1 > C_2$, $C_1 = C_2$, or $C_1 < C_2$. (In the generalized supersymmetric version of quantum chromodynamics described above, this corresponds to $N_c > N_f$, $N_c = N_f$, and $N_c < N_f$, respectively.) We shall now consider each of these cases.

$$C_1 > C_2$$

Here the powers $\exp(2ia\pi T)$ with $a > 0$ have positive-definite values of R, given by Eq. (29.3.7) as $R = 2(C_1 - C_2)a$. Lorentz invariance tells us that if any term in $\mathscr{B}_\lambda$ contains a factor $W_{A\alpha L}$, then it must contain at least two of them, so the only ways to construct terms in $\mathscr{B}_\lambda$ with $R = 2$ is to have two Ws and no dependence on Y or T, or one Y and no dependence on W or T, or one factor of $\exp(2i\pi T/(C_1 - C_2))$ and no dependence on W or Y:

$$\mathscr{B}_\lambda = Y f_\lambda(\Phi) + \sum_{\alpha\beta AB} \epsilon_{\alpha\beta} W_{A\alpha L} W_{B\beta B} \ell_{\lambda AB}(\Phi) + \exp\left(\frac{2i\pi T}{C_1 - C_2}\right) v_\lambda(\Phi) . \tag{29.3.9}$$

Because $f_\lambda(\Phi)$ does not depend on Y or T, it can only be the tree-approximation superpotential

$$f_\lambda(\Phi) = f(\Phi) , \tag{29.3.10}$$

just as in perturbation theory. Likewise, because $\ell_{\lambda AB}(\Phi)$ does not depend on Y or T it must have equal numbers of Φs and $\Phi^\dagger$s, so since it does not depend on $\Phi^\dagger$ it cannot depend on Φ either. Gauge invariance then requires (for a simple gauge group) that $\ell_{\lambda AB}(\Phi)$ is proportional to δ_{AB}, and since it does not depend on T or Y, the power-counting argument of Section 27.6 shows that the coefficient of δ_{AB} can only be the one-loop contribution to the running inverse-square Wilsonian gauge coupling.

To be more explicit about the running gauge coupling, recall that in non-supersymmetric gauge theories with fermions, Eq. (18.7.2) gives the one-loop renormalization group equation as

$$\lambda \frac{dg_\lambda}{d\lambda} = b g_\lambda^3 , \tag{29.3.11}$$

with

$$b = -\frac{1}{4\pi^2}\left(\frac{11}{12}C_1 - \frac{1}{6}C_2\right) , \tag{29.3.12}$$

with the coefficient of C_2 taken as $-1/6$ rather than $-1/3$ because we are now counting the left-chiral states of antifermions separately from the left-chiral states of particles. As we have seen in Section 28.2, the effect of gauginos is to multiply the C_1-term by a factor $9/11$, while the effect of the scalar components of the left-chiral superfields (such as squarks and sleptons) is to multiply the C_2-term by a factor $3/2$, so in supersymmetric theories Eq. (29.3.12) becomes instead

$$b = -\frac{1}{16\pi^2}(3C_1 - C_2) \ . \tag{29.3.13}$$

The solution of Eq. (29.3.11) for the running gauge coupling is then

$$g_\lambda^{-2} = g^{-2} + \frac{3C_1 - C_2}{8\pi^2}\ln\left(\frac{\lambda}{K}\right) , \tag{29.3.14}$$

where K is an ultraviolet cut-off, introduced to give meaning to the otherwise ultraviolet-divergent bare gauge coupling g.

To summarize the results so far, setting $T = \tau$ and $Y = 1$, the Wilsonian effective Lagrangian for $C_1 > C_2$ takes the form

$$\begin{aligned}\mathscr{L}^\sharp_\lambda = &\left[\mathscr{A}_\lambda(\Phi, \Phi^\dagger, V, \tau, \tau^*, \mathscr{D}\cdots)\right]_D + 2\,\mathrm{Re}\left[\frac{\tau_\lambda}{8\pi i}\sum_{A\alpha\beta}\epsilon_{\alpha\beta}W_{A\alpha L}W_{A\beta L}\right]_{\mathscr{F}} \\ &+2\,\mathrm{Re}\,[f(\Phi)]_{\mathscr{F}} + \exp\left(\frac{2i\pi\tau_\lambda}{C_1 - C_2}\right)[v_\lambda(\Phi)]_{\mathscr{F}} ,\end{aligned} \tag{29.3.15}$$

where

$$\tau_\lambda = \frac{4\pi i}{g_\lambda^2} + \frac{\theta}{2\pi} \ . \tag{29.3.16}$$

We have been able to replace τ with τ_λ in the exponential in the last term in Eq. (29.3.15), because the difference is a constant times $\ln\lambda$, which yields a power of λ that can be absorbed into the definition of v_λ.

Non-perturbative effects have now been isolated in the last term in Eq. (29.3.15). This term can be generated by instantons of winding number $\nu > 0$ if $C_1 - C_2 = 1/\nu$. (In general $C_1 - C_2$ is a rational number. For the generalized supersymmetric version of quantum chromodynamics, $C_1 - C_2 = N_c - N_f$ is an integer, so the condition $C_1 - C_2 = 1/\nu$ requires that $N_c = N_f - 1$, and then only $\nu = 1$ instantons contribute. Detailed calculations[5] in this model show that instantons actually do make such contributions.) Whether or not it is instantons that generate the non-perturbative contribution $v_\lambda(\Phi)$, we can determine its form by considering the non-anomalous symmetries of the theory. Since this function is independent of Y, it can be evaluated as if $Y = 0$, so it shares all the non-anomalous symmetries of the first term in Eq. (29.3.1). These include the gauge symmetry itself and a global symmetry under $\prod_d SU(n(d))$, where d labels the different irreducible representations of the gauge group furnished by the left-chiral superfields, and $n(d)$ is the number of times representation d occurs. (For instance, in generalized supersymmetric quantum chromodynamics d takes two values, labelling the N_c and $\bar{N}_c$ representations of $SU(N_c)$, and $n(N_c) = n(\bar{N}_c) = N_f$.) Let us label the Φs as $\Phi^{(d)}_{ai}$, where a is a gauge index, and i is a 'flavor' index, labelling the $n(d)$ different Φs that transform under the gauge group according to the representation d. The only way to construct a function of the Φs that is invariant under the global symmetry group $\prod_d SU(n(d))$ is as a product of Φs, with the $n(d)$ flavor indices contracted for each d with the antisymmetric $SU(n(d))$ tensor $\epsilon_{i_1 \dots i_{n(d)}}$, and with the gauge indices contracted with constant tensors of the gauge group. (For instance, for generalized supersymmetric quantum chromodynamics, v_λ must be a function of the sole invariant

$$D \equiv \mathrm{Det}_{ij} \sum_a Q_{ai} \overline{Q}_{aj} \,,$$

which is non-zero only for $N_c \geq N_f$.)

In addition to the anomaly-free $SU(n(d))$ flavor symmetries, there is also a $U_d(1)$ symmetry for each of the irreducible representations d furnished by the Φs, with all $\Phi^{(d)}_{ai}$ for a given d undergoing the transformation

$$\Phi^{(d)}_{ai} \to e^{i\varphi_d} \Phi^{(d)}_{ai} \,. \tag{29.3.17}$$

This symmetry is anomalous, with the effects of the anomaly the same as if the Lagrangian underwent the change

$$\mathscr{L} \to \mathscr{L} - \sum_d \frac{n(d) C_{2d}}{32\pi^2} \sum_A \epsilon_{\mu\nu\rho\sigma} f_A^{\mu\nu} f_A^{\rho\sigma} \varphi_d \,, \tag{29.3.18}$$

where C_{2d} is the contribution to C_2 of any one left-chiral scalar superfield belonging to the irreducible representation d of the gauge group. The symmetry is restored if we give T the transformation property

$$T \to T + n(d) C_{2d} \varphi_d / \pi \,. \tag{29.3.19}$$

Since $v_\lambda(\Phi)$ is accompanied in Eq. (29.3.9) by a factor $\exp(2i\pi T/(C_1 - C_2))$, which undergoes the transformation

$$\exp\left(\frac{2i\pi T}{C_1 - C_2}\right) \to \prod_d \exp\left(\frac{+2in(d) C_{2d} \varphi_d}{C_1 - C_2}\right) \exp\left(\frac{2i\pi T}{C_1 - C_2}\right) \,, \tag{29.3.20}$$

we conclude that *for each representation d of the gauge group furnished by the left-chiral scalars, $v_\lambda(\Phi)$ must be a homogeneous function of the $\Phi^{(d)}_{ai}$ of negative order $-2n(d)C_{2d}/(C_1 - C_2)$.* (For instance, in generalized supersymmetric quantum chromodynamics we have two irreducible representations of $SU(N_c)$, the defining and antidefining representations, each with $n(d) = N_f$ and $C_{2d} = 1/2$, so v_λ is a homogeneous function of order $-N_f/(N_c - N_f)$ in the Q belonging to the defining representation and of the same order in the $\bar{Q}$ belonging to the antidefining representation. Thus it must be proportional to $D^{-1/(N_c - N_f)}$, where D is the determinant introduced earlier.) In general $C_2 = \sum_d n(d) C_{2d}$, so $v_\lambda(\Phi)$ is a homogeneous function of all the Φs, of order $-2C_2/(C_1 - C_2)$.

This result satisfies an important consistency check. Recall from Section 27.4 that any superpotential has dimensionality +3 (counting powers of mass, with $\hbar = c = 1$), while the scalar superfields Φ like ordinary scalar fields have dimensionality +1, so the Φ-dependent part of v_λ must appear with a coefficient of dimensionality

$$3 + \frac{2C_2}{C_1 - C_2} = \frac{3C_1 - C_2}{C_1 - C_2} .$$

This coefficient does not depend on the gauge coupling or on any of the couplings or masses in the superpotential, and because we have replaced the bare coupling g with g_λ in the second term of Eq. (29.3.11) it cannot depend on the ultraviolet cut-off K used to define g either, so it can only depend on λ. Therefore

$$v_\lambda(\Phi) = \lambda^{(3C_1 - C_2)/(C_1 - C_2)} H(\Phi) , \tag{29.3.21}$$

where $H(\Phi)$ is a homogeneous function of order $-2n(d)C_{2d}/(C_1 - C_2)$ in the Φs that belong to each representation d of the gauge group and is independent of any parameters of the theory. We can rewrite Eq. (29.3.14) in the form

$$\tau_\lambda = i\frac{3C_1 - C_2}{2\pi} \ln\left(\frac{\lambda}{\Lambda}\right) + \frac{\theta}{2\pi} , \tag{29.3.22}$$

where Λ is an energy parameter that characterizes the running gauge coupling, like the $\Lambda \approx 200$ MeV of quantum chromodynamics. Thus the last term in Eq. (29.3.15) is

$$\exp\left(\frac{2i\pi\tau_\lambda}{C_1 - C_2}\right) v_\lambda(\Phi) = \exp\left(\frac{i\theta}{C_1 - C_2}\right) \Lambda^{(3C_1 - C_2)/(C_1 - C_2)} H(\Phi) . \tag{29.3.23}$$

The whole effective superpotential, including the non-perturbative contribution (29.3.23), is therefore independent of the floating cut-off λ.

The function $H(\Phi)$ is homogeneous and of negative order in Φ, so in the absence of a bare superpotential the potential is positive-definite at finite values of the scalar fields and vanishes only at infinite field values. In such a theory, there is no stable vacuum state, and the question of supersymmetry breaking is moot. The vacuum may be stabilized by adding a suitable bare superpotential. For instance, in generalized supersymmetric quantum chromodynamics with $N_f < N_c$ the only renormalizable bare superpotential is a sum of mass terms

$$f(Q, \overline{Q}) = \sum_{ija} m_{ij} \overline{Q}_{ai} Q_{aj} . \tag{29.3.24}$$

To seek a supersymmetric vacuum state, we need first of all to find what scalar components q_{ai} and $\overline{q}_{ai}$ satisfy Eq. (27.4.11), which here reads

$$\sum_{abi} q^*_{ai}(t_A)_{ab}q_{bi} - \sum_{abi} \overline{q}^*_{ai}(t_A)_{ab}\overline{q}_{bi} = 0\,, \tag{29.3.25}$$

for all generators t_A of $SU(N_c)$. The gauge interactions (but not the superpotential) are invariant under simultaneous $SU(N_c)$ transformations on the color indices a of both q_{ai} and $\overline{q}_{ai}$; under independent $SU(N_f)$ and $\overline{SU(N_f)}$ transformations on the flavor indices i of q_{ai} and $\overline{q}_{ai}$, respectively, and under a $U(1)$ transformation of both q_{ai} and $\overline{q}_{ai}$ by opposite phases. Using these symmetries, it is possible to put the general solution of these conditions in the form

$$q_{ai} = \overline{q}_{ai} = \begin{cases} u_i\delta_{ai} & a \leq N_f \\ 0 & a > N_f \end{cases}\,, \tag{29.3.26}$$

where the u_i are complex numbers with the same phase. (Here is the proof. The $SU(N_c)$ generators t_A span the space of all traceless Hermitian matrices, so Eq. (29.3.25) is equivalent to the requirement that

$$\sum_i q^*_{ai}q_{bi} - \sum_i \overline{q}^*_{ai}\overline{q}_{bi} = k\delta_{ab}\,, \tag{29.3.27}$$

for some constant k. By a combined color and flavor transformation $q \to UqV$ with U and V unitary and unimodular, we can put the matrix q in the diagonal form (29.3.26), and by a unimodular change of phase of the diagonal elements we can arrange that they all have the same phase. Then Eq. (29.3.27) becomes

$$\sum_i \overline{q}^*_{ai}\overline{q}_{bi} = \begin{cases} \left(u_a^2 - k\right)\delta_{ab} & a \leq N_f \\ -k\delta_{ab} & a > N_f \end{cases}\,.$$

The conditions for $a > N_f$ show that $k \leq 0$. If k were non-zero, then the $\overline{q}_{ai}$ would furnish N_c non-zero orthogonal vectors with N_f components, which is impossible for $N_f < N_c$, so $k = 0$. We can then put the $\overline{q}_{ai}$ in the diagonal form (29.3.26) by a series of unitary flavor transformations: first rotate $\overline{q}_{1i}$ into the 1-direction; then keeping the 1-direction fixed, rotate in the space perpendicular to this direction to put $\overline{q}_{2i}$ in the 2-direction; and so on; and then perform a unimodular phase transformation so that all diagonal elements have the same phase. Eq. (29.3.27) then shows that the absolute values of the diagonal elements of q_{ai} and $\overline{q}_{ai}$ are equal, and by a non-anomalous opposite phase change of q_{ai} and $\overline{q}_{ai}$ we can arrange that their common phases are equal, as was to be proved.)

The function H in Eq. (29.3.23) is here

$$H(q,\overline{q}) = \mathscr{I}\left[\text{Det}_{ij}\sum_a q_{ai}\overline{q}_{aj}\right]^{-1/(N_c-N_f)} = \mathscr{I}\left[\prod_i u_i\right]^{-2/(N_c-N_f)}\,, \tag{29.3.28}$$

where $\mathscr{I}$ is a purely numerical constant. (Detailed calculations show that $\mathscr{I} = 2$ for $N_f = 2$ and $N_c = 3$.) Adding the terms (29.3.23) and (29.3.24), the complete effective superpotential is now

$$f_{\rm total}(q,\overline{q}) = \mathscr{K}\left[\prod_i u_i\right]^{-2/(N_c-N_f)} + \sum_i m_i u_i^2\,, \tag{29.3.29}$$

where

$$\mathscr{K} \equiv \mathscr{I}\exp\left(\frac{i\theta}{N_c - N_f}\right)\Lambda^{(3N_c-N_f)/(N_c-N_f)} \tag{29.3.30}$$

and the m_i are the diagonal elements of the mass matrix that results when the original mass matrix is subjected to the $SU(N_f) \times \overline{SU(N_f)}$ transformation used to put the scalars in the form (29.3.26). The condition (27.4.10), that $f_{\rm total}(q,\overline{q})$ should be stationary, has the solution

$$u_i^2 = \frac{1}{m_i}\left(\frac{\mathscr{K}}{N_c - N_f}\right)^{1-N_f/N_c}\left(\prod_j m_j\right)^{-1/N_c}. \tag{29.3.31}$$

Because we have put the scalar fields in a basis in which the u_i have a common phase, the m_i must also have a common phase in this basis. But the common phase of the u_i^2 is not unique — the $1/N_c$ powers in Eq. (29.3.31) tell us that the solution is undetermined by a factor $\exp(2i\pi n/N_c)$, with n an integer ranging from 0 to $N_c - 1$. (The two signs of u_i for a given u_i^2 are physically equivalent, because the whole theory is invariant under a non-anomalous symmetry with $q_{ai} \to e^{i\pi} q_{ai}$ and $\overline{q}_{ai} \to e^{-i\pi}\overline{q}_{ai}$.) The fact that there are N_c physically inequivalent solutions will turn up again in our discussion of the Witten index in the next section.

$C_1 = C_2$

This case is of some interest because, as we saw in Eq. (27.9.3), the simplest $N = 2$ supersymmetric Yang–Mills theory, when written in terms of $N = 1$ superfields, contains a single left-chiral superfield in the adjoint representation, for which** of course $C_2 = C_1$.

For $C_2 = C_1$ the function $\exp(2i\pi T)$ has $R = 0$, so that its appearance in $\mathscr{L}_\lambda^\sharp$ is not restricted by R invariance. The general form of the F-term

** Note that here C_2 refers to the representation of the gauge group furnished by the chiral superfields, so it is the same as the quantity C_2^b in Section 27.9, which refers to the representation furnished by complex scalars, but half of C_2^f, which refers to the representation furnished by all spinor fields, including the gauginos.

in $\mathscr{L}^\sharp_\lambda$ is given here by Eq. (29.3.9), but with the last term absent:

$$\mathscr{B}_\lambda = Y f_\lambda(\Phi, \exp(2i\pi T)) + \sum_{\alpha\beta AB} \epsilon_{\alpha\beta} W_{A\alpha L} W_{B\beta B} \ell_{\lambda AB}(\Phi, \exp(2i\pi T)) . \tag{29.3.32}$$

Because f_λ may depend on T, we cannot now conclude that it is equal to the bare superpotential, but only that it depends linearly on whatever coupling coefficients and masses appear linearly in the bare superpotential. In particular, *if there is no superpotential to begin with, then none is generated by non-perturbative effects.*

To go further, we need to make use of the anomalous chiral symmetry under a $U(1)$ transformation of all the Φ_n. In order for the whole theory to be invariant under this symmetry, it is necessary to introduce a separate external left-chiral superfield Y_r for the terms in the bare superpotential of order r in the Φ_n. Then the theory is invariant under the combined transformations

$$\Phi_n \to e^{i\varphi}\Phi_n , \qquad T \to T + C_2\varphi/\pi , \qquad Y_r \to e^{-ir\varphi} Y_r . \tag{29.3.33}$$

This symmetry tells us that a term in $\mathscr{B}_\lambda$ that is of order $\mathscr{N}_r$ in the coefficients of the term in the superpotential of order r in Φ and proportional to $\exp(2ai\pi T)$ must be of an order $\mathscr{N}_\Phi$ in Φ, given by

$$\mathscr{N}_\Phi = \sum_r r\mathscr{N}_r - 2C_2 a . \tag{29.3.34}$$

The coefficients $\ell_{\lambda AB}$ of the terms in $\mathscr{B}_\lambda$ that are quadratic in W are shown in Eq. (29.3.32) to be independent of the parameters in the superpotential for $C_1 = C_2$, so in this case Eq. (29.3.34) becomes

$$\mathscr{N}_\Phi = -2C_2 a . \tag{29.3.35}$$

Thus there can be no terms in $\ell_{\lambda AB}$ of positive order in the Φ_r, and any term in $\ell_{\lambda AB}$ that is independent of the Φ_r must be independent of T. These Φ-independent terms in $\ell_{\lambda AB}$ are therefore again just the one-loop contribution to the running coupling parameter τ_λ.

The effective superpotential is shown by Eq. (29.3.32) to be linear in the parameters in the superpotential, so all of its terms have just one $\mathscr{N}_r = 1$ and the others zero. For such a term Eq. (29.3.34) gives

$$\mathscr{N}_\Phi = r - 2C_2 a . \tag{29.3.36}$$

A term in the effective superpotential with $\mathscr{N}_\Phi$ powers of Φ can therefore only arise from terms in the bare superpotential with $r \geq \mathscr{N}_\Phi$ powers of Φ. The terms with $r = \mathscr{N}_\Phi$ have $a = 0$, so they are given by the tree approximation as just the bare superpotential. The only other terms are non-perturbative corrections with $r > \mathscr{N}_\Phi$. Such a non-perturbative term, of a given order in the Φs, can only arise from terms in the bare superpotential of *higher* order in the Φs.

$$C_1 < C_2$$

Here the R value $2(C_1 - C_2)$ of $\exp(2i\pi T)$ is *negative*, so positive powers of $\exp(2i\pi T)$ can compensate for the positive R values of Y and W_α, and $\mathscr{B}_\lambda$ may therefore contain terms of arbitrary order in Y and W_α. Using the chiral symmetry condition (29.3.34) and the R invariance condition

$$2 = \mathscr{N}_W + 2\sum_r \mathscr{N}_r - 2a(C_2 - C_1)\,, \tag{29.3.37}$$

we can, however, set limits on the structure of terms of a given order $\mathscr{N}_\Phi$ in the Φs. Eqs. (29.3.34) and (29.3.37) have a trivial solution with $\mathscr{N}_r = 1$ for $r = \mathscr{N}_\Phi$, $\mathscr{N}_r = 0$ for other values of r, and $a = \mathscr{N}_W = 0$; this solution just represents the presence in $\mathscr{B}_\lambda$ of the original bare superpotential, with no radiative corrections. If there is no superpotential to begin with, then $\mathscr{N}_r = 0$ for all r, so Eq. (29.3.35) does not allow any terms in the Wilsonian Lagrangian with $\mathscr{N}_W = 0$, and so *no superpotential can be generated.* (In the supersymmetric version of quantum chromodynamics this conclusion is usually derived by noting that there is no possible term in the superpotential that would be consistent with all symmetries, but as we have now seen, the conclusion is much more general.) For renormalizable asymptotically free theories there is a useful limit on the structure of Φ-independent terms in $\mathscr{B}_\lambda$. The condition of renormalizability tells us that $\mathscr{N}_r = 0$ for $r > 3$, so by subtracting 2/3 of Eq. (29.3.34) from Eq. (29.3.37) we find

$$2 \geq \frac{2}{3}\mathscr{N}_\Phi + \mathscr{N}_W + 2a\left(C_1 - \frac{1}{3}C_2\right)\,. \tag{29.3.38}$$

Asymptotic freedom requires that $3C_1 > C_2$, so for $\mathscr{N}_\Phi = 0$ (or $\mathscr{N}_\Phi > 0$) each term on the right-hand side is positive. It follows that we can have no Φ-independent terms of higher than second order in W, and these terms are also independent of T, so they again represent the one-loop contribution to the running coupling parameter τ_λ. But for $C_1 < C_2$ there is no general prohibition against terms of second or higher order in W and negative order in Φ.

29.4 Supersymmetry Breaking in Gauge Theories

We now turn to a question of great physical interest: in what gauge theories is supersymmetry spontaneously broken?

Let us start with an Abelian gauge theory, the supersymmetric version of quantum electrodynamics described in Section 27.5. This is a $U(1)$ gauge theory with two chiral superfields $\Phi_\pm$ carrying $U(1)$ quantum numbers $\pm e$, and superpotential $f(\Phi) = m\Phi_+\Phi_-$. We saw in Section 27.5 that supersymmetry is broken in the tree approximation if we include a Fayet–Iliopoulos term $\xi[V]_D$ in the Lagrangian density, so the Witten index is zero for $\xi \neq 0$ and small e, and hence also for all values of e and ξ, including $\xi = 0$. Is supersymmetry unbroken for $\xi = 0$? It is unbroken in the tree approximation, but how can we tell if higher-order corrections or non-perturbative effects give the vacuum a finite energy in this case?

To answer this we shall use a symmetry principle of the theory for $\xi = 0$, in the way that was described in general terms in Section 29.1. The symmetry here is charge conjugation: the whole Lagrangian density is invariant under the charge conjugation transformation of the chiral and gauge superfields:

$$\mathsf{C}\Phi_\pm\mathsf{C}^{-1} = \Phi_\mp\,, \qquad \mathsf{C}V\mathsf{C}^{-1} = -V\,. \tag{29.4.1}$$

There is a massless fermion in the tree approximation, the photino, but if we take the vacuum to be even under C then the one-photino state is odd under C, so these states are not related to each other by multiplication with the supersymmetry generator. The chiral fermion here has mass m in the tree approximation, so for small e it is not paired with the vacuum either. With no fermionic state available to pair with, the vacuum state must have strictly zero energy, at least for e small enough so that perturbation theory gives a good qualitative picture of the spectrum. As we saw in Section 29.1, the zero energy of the vacuum implies that supersymmetry is not broken. Likewise, the photino must be strictly massless, since it has no bosonic state with which to pair.

Now, what about values of e that are so large that perturbation theory cannot be trusted at all? The Witten index itself is no help here, because it vanishes. Instead, let us consider the weighted Witten index, $\mathrm{Tr}\,\mathsf{C}(-1)^F$. We have seen that for small e and $\xi = 0$ the vacuum has zero energy, and there are also two zero-energy states containing a zero-momentum photino of spin up or down, and a bosonic zero-energy state containing two zero-momentum photinos of opposite spin. The vacuum makes a contribution $+1$ to $\mathrm{Tr}\,\mathsf{C}(-1)^F$; the two one-photino states make a contribution $+2$ (because both C and $(-1)^F$ are -1); and the two-photino state makes a contribution $+1$, giving a weighted Witten index $\mathrm{Tr}\,\mathsf{C}(-1)^F = 4$. This is independent of the value of e, so even for strong couplings the weighted Witten index is 4, and so supersymmetry is not broken.

There is a complication here.[2] In counting zero-momentum states in the tree approximation, we have not considered the zero-momentum components of the gauge field $V_\mu(x)$. A constant term in $V_0(x)$ is no problem, because it can be removed by a gauge transformation

$$V_\mu(x) \to V_\mu(x) + \partial_\mu \Lambda(x)\,, \tag{29.4.2}$$

with gauge parameter $\Lambda(x)$ proportional to x^0. On the other hand, we cannot simply remove a constant term in $V_i(x)$, because this would require a gauge transformation with $\Lambda(x)$ proportional to x^i, which would conflict with the assumed periodicity of the fields under translations by the box dimension $L \equiv V^{1/3}$. In the particular model under consideration here all fields have charges $\pm e$ or zero, so this periodicity is preserved if we limit ourselves to a lattice of gauge transformations with

$$\Lambda(x) = \frac{2\pi}{eL}\sum_i \ell_i x^i\,, \tag{29.4.3}$$

where ℓ_i are three positive or negative integers. Therefore although we cannot remove the zero-momentum components of $V_i(x)$ by a gauge transformation, we can freely shift them by amounts $2\pi\ell_i/eL$. The tree-approximation Lagrangian (*not* Lagrangian density) for the x^i-independent part of V_μ in a gauge with $V_0 = 0$ is simply $-\frac{1}{2}L^3\sum_i(\partial_0 V_i)^2$, so the Hamiltonian is $+\frac{1}{2}L^{-3}\sum_i(\pi_i)^2$, where π_i is the canonical conjugate to V_i: $\pi_i = L^3\partial_0 V_i$. The wave function $\Psi(\mathbf{V})$ for this field is then just like that of a free particle of unit mass in three dimensions, in a box of linear dimensions $2\pi/eL$ with periodic boundary conditions. The wave functions of definite energy are proportional to $\exp(i\mathbf{k}\cdot\mathbf{V})$, with energy $\mathbf{k}^2/2L^3$, and $k_i = eL\ell_i$, with ℓ_i integers. There is a unique zero-energy state of this field, with $k_i = 0$ and with a normalized wave function equal to the constant $(eL/2\pi)^{3/2}$. Because this state is unique, our counting of zero-energy states is unaffected by the gauge degree of freedom, and the weighted Witten index is indeed 4.

Now let us consider the theory of a simple non-Abelian gauge superfield, without chiral superfields. Witten's 1982 paper[2] on the Witten index presented an argument that for such theories this index is $r+1$ (or possibly $-r-1$), where r is the rank of the gauge group, the maximum number of commuting generators. In 1997 he found a correction to this calculation,[6] with the result that for the classical unitary, orthogonal, and symplectic groups the index is a Casimir invariant C_1, which for the unitary and symplectic groups is indeed equal to $r+1$, but for the orthogonal groups $O(N)$ with $N > 7$ and the exceptional groups takes a different value. In general, this Casimir invariant is defined by Eq. (17.5.33):

$$\sum_{CD} C_{ACD}C_{BCD} = g^2 C_1 \delta_{AB}\,, \tag{29.4.4}$$

where g is a coupling constant, whose definition can be made unambiguous by specifying that when the generators t_A, t_B, and t_C are restricted to the three generators of the 'standard' $SU(2)$ subalgebra that was used in the calculation of instanton effects in Section 23.5, the structure constant is simply $g\epsilon_{ABC}$. For the classical groups, we have

$$C_1 = \begin{cases} N & SU(N) \\ N-2 & SO(N) \text{ for } N > 3 \\ N+1 & USp(2N) \end{cases} . \tag{29.4.5}$$

For $SU(N)$ the rank is $r = N - 1$, while for $USp(2N)$ it is $r = N$, so in both cases $r + 1 = C_1$. But $SO(N)$ with $N > 6$ has rank $r = (N - 1)/2 = (C_1 + 1)/2$ for N odd and $r = N/2 = (C_1 + 2)/2$ for N even, and the index is C_1, not $r + 1$. Of course, this does not affect the main conclusion that, since the Witten index is not zero, supersymmetry is not spontaneously broken. Kac and Smilga showed in 1999 that the index is also equal to C_1 for exceptional groups.[7] We shall calculate the Witten index here only for $SU(N)$ and $USp(2N)$ supersymmetric gauge theories with no chiral superfields, but in the course of this calculation we will also see why the orthogonal and exceptional groups present special difficulties.

The general strategy of this calculation is the same as for the Abelian theories considered previously. We first examine the states of zero energy, to see if there are any that cannot be paired by action of the supersymmetry generator. If there are, then as long as the coupling is weak enough so that the tree approximation gives a good qualitative picture of the spectrum, we know that these unpaired states really have precisely zero energy. We can then find some non-zero weighted Witten index, which will be constant even for stronger couplings, and conclude that supersymmetry is not broken for any coupling strength.

There is another complication here, that did not enter in the earlier example of an Abelian theory with charged chiral superfields. As usually formulated, general renormalizable theories of gauge bosons and gauginos with no chiral superfields do not contain any dimensionless parameters at all, so that there is no coupling parameter that can be adjusted to make the couplings weak. Instead we have a running coupling constant, depending on the ratio of the energy to a characteristic energy scale Λ, like the scale $\Lambda \approx 200$ MeV for quantum chromodynamics discussed in Section 18.7. For non-Abelian theories the gauge coupling constant becomes strong at energies below Λ, however weak it may be at higher energies. But here we are working in a box of volume L^3, which provides an infrared cut-off at an energy $\approx 1/L$, that is normally lacking in theories with unbroken gauge symmetries. When we speak of a weakly coupled gauge theory, it should be understood that we mean one with a coupling that is small down to energies of order $1/L$. In this case, it is essential to be able to argue that supersymmetry is broken for strong as well as weak couplings, in order to be able to draw any conclusions at all about the realistic limit of infinite volume.

We will work in temporal gauge, where $V_A^0 = 0$. The Lagrangian density (27.3.1) is then

$$\mathscr{L} = -\tfrac{1}{4}\sum_{Aij} f_{Aij}^2 - \tfrac{1}{2}\sum_A (\partial_0 V_{Ai})^2 - \tfrac{1}{2}\sum_A \left(\bar{\lambda}_A (\not{D}\lambda)_A\right) + \tfrac{1}{2}\sum_A D_A^2 \,, \quad (29.4.6)$$

where, in temporal gauge,

$$f_{Aij} = \partial_i V_{Aj} - \partial_j V_{Ai} + \sum_{BC} C_{ABC} V_{Bi} V_{Cj} \,, \quad (29.4.7)$$

$$(D_i\lambda)_A = \partial_i \lambda_A + \sum_{BC} C_{ABC} V_{Bi} \lambda_C \,, \quad (29.4.8)$$

$$(D_0\lambda)_A = \partial_0 \lambda_A \,. \quad (29.4.9)$$

(As usual, the gauge coupling constant or constants are included as factors in the structure constants C_{ABC}.) Without chiral superfields, there is no other dependence on the auxiliary field D_A; since it enters quadratically, it may be put equal to the value at which the Lagrangian is stationary, $D_A = 0$, and ignored from now on. Retaining only the **x**-independent modes, the effective Lagrangian becomes

$$\int d^3x\, \mathscr{L} = L^3\Bigg[-\frac{1}{4}\sum_{Aij}\Big(\sum_{BC} C_{ABC} V_{Bi} V_{Cj}\Big)^2 - \frac{1}{2}\sum_{Ai} (\partial_0 V_{Ai})^2$$
$$- \frac{1}{2}\sum_{ABCi} C_{ABC}\left(\bar{\lambda}_A \gamma_i V_{Bi} \lambda_C\right) - \tfrac{1}{2}\sum_A \bar{\lambda}_A \gamma^0 \partial_0 \lambda_A \Bigg] \,. \quad (29.4.10)$$

The Hamiltonian is then

$$H = \frac{1}{2L^3}\sum_{Ai} \pi_{Ai}^2 + \frac{L^3}{4}\sum_{Aij}\Big(\sum_{BC} C_{ABC} V_{Bi} V_{Cj}\Big)^2 + \frac{L^3}{2}\sum_{ABCi} C_{ABC}\left(\bar{\lambda}_A \gamma_i V_{Bi} \lambda_C\right) , \quad (29.4.11)$$

where $\pi_{Ai} = L^3 \partial_0 V_{Ai}$ is the canonical conjugate to V_{Ai}.

The gauge field configurations with zero energy in the tree approximation are those for which $\sum_{BC} C_{ABC} V_{Bi} V_{Cj} = 0$ for all A, i, and j. This condition is always satisfied if V_{Bi} vanishes for all i except where t_B is in a Cartan subalgebra of the gauge Lie algebra.* For the unitary and

* A *Cartan subalgebra* is any subalgebra spanned by r-independent generators $t_{\mathscr{A}}$ that commute with one another, that is, for which $C_{A\mathscr{A}\mathscr{B}}$ vanishes for all t_A when $t_{\mathscr{A}}$ and $t_{\mathscr{B}}$ are in the Cartan subalgebra, where r is the rank, the maximum number of such generators. For instance, for the $SU(3)$ symmetry of strong interactions discussed in Section 19.7, the rank is $r = 2$, and the Cartan subalgebra can be taken to consist of the third component of isospin t_3 and the hypercharge t_8, which act on the light quarks with the commuting matrices denoted λ_3 and λ_8 in Eq. (19.7.2).

symplectic groups and their direct products this is the only way that it can be satisfied. The same is true for the orthogonal gauge algebras $O(N)$ with $N \leq 6$, which are all equivalent to symplectic and/or unitary Lie algebras (see the Appendix to Chapter 15), but not for orthogonal gauge algebras $O(N)$ with $N \geq 7$, which is why Witten's original calculation needed correction for this case.

In the rest of this calculation we will only consider gauge algebras like the symplectic and unitary algebras for which the condition of zero energy in the tree approximation does require the V_{Bi} to vanish for all i except where t_B is in a Cartan subalgebra of the gauge Lie algebra. With all $V_{Ai} = 0$ except for the $V_{\mathscr{A}i}$ for which $t_{\mathscr{A}}$ is in the Cartan subalgebra, the zero-energy modes of the fermion field are those also with $\lambda_A = 0$, except for the $\lambda_{\mathscr{A}}$ for which $t_{\mathscr{A}}$ is in the Cartan subalgebra.

Now we must count these states. The eigenvalues of the $t_{\mathscr{A}}$ in any representation of a semi-simple Lie algebra are quantized, so by gauge transformations all values of the non-zero gauge fields $V_{\mathscr{A}i}$ are equivalent to values in a finite box with periodic boundary conditions. The quantization of these modes is just like that carried out for the gauge field in the $U(1)$ model considered above, so the zero-energy state of these fields is again unique, with a constant wave function in the box.

The multiplicity of states with zero energy in the tree approximation comes entirely from the fermion degrees of freedom. It is convenient to use a two-component notation, in which instead of a four-component Majorana field $\lambda_{\mathscr{A}\alpha}$ for each generator of the Cartan subalgebra, we have two left-handed fields $\lambda_{\mathscr{A}La}$ with $a = \pm 1/2$, and their right-handed Hermitian adjoints $\lambda^*_{\mathscr{A}La}$. These gaugino fields satisfy the canonical anticommutation relations

$$\left\{\lambda_{\mathscr{A}La}\,,\,\lambda^*_{\mathscr{B}Lb}\right\} = \delta_{\mathscr{A}\mathscr{B}}\delta_{ab}$$

and

$$\left\{\lambda_{\mathscr{A}La}\,,\,\lambda_{\mathscr{B}Lb}\right\} = \left\{\lambda^*_{\mathscr{A}La}\,,\,\lambda^*_{\mathscr{B}Lb}\right\} = 0\,.$$

By operating on an arbitrary state vector with as many factors of $\lambda_{\mathscr{A}La}$ as necessary, we can construct a state vector $|0\rangle$ which is annihilated by all $\lambda_{\mathscr{A}La}$. The general zero-energy state vector is then a linear combination of products of the $\lambda^*_{\mathscr{A}La}$ acting on $|0\rangle$.

To see which of these states may be paired with each other by the action of the supersymmetry generator, we must take account of a symmetry of the theory. The zero-energy condition, that the gauge and gaugino fields lie only in directions corresponding to the Cartan subalgebra, is invariant under the subgroup of the original gauge group consisting of elements h that leaves this subalgebra invariant, that is, for which $h^{-1}t_{\mathscr{A}}h$ is a linear combination of the $t_{\mathscr{B}}$. These form a finite group, known as the *Weyl group*. For instance, in the defining representation of $SU(N)$ the Weyl group consists of permutations of the N coordinate axes, together with multiplication by a phase needed to make the transformations unimodular. These can be represented by products of the finite gauge transformations $W(i,j) = \exp(i\pi\sigma(ij)/2) = i\sigma(ij)$ with $i \neq j$ that permute the ith and jth coordinate axes, where $\sigma(ij)$ is the $U(N)$ generator with $[\sigma(ij)]^i{}_j = [\sigma(ij)]^j{}_i = 1$, and with all other elements zero. These transformations induce orthogonal transformations in the space spanned by the diagonal traceless Hermitian matrices $t_{\mathscr{A}}$ that generate the Cartan subalgebra in the adjoint representation. For instance, for the group $SU(2)$ we can take the Cartan subalgebra to consist of just t_3, and the Weyl group then consists of the unit element and a single non-trivial gauge transformation $W(1,2) = i\sigma(1,2) = i\sigma_1$, for which $W^{-1}t_3W = -t_3$. For $SU(3)$ the Cartan subalgebra has the two generators λ_3 and λ_8, and the Weyl group consists of the six gauge transformations 1, $W(1,2)$, $W(2,3)$, $W(1,3)$, $W(1,2)W(2,3)$, and $W(2,3)W(1,2)$, which generate rotations by multiples of 60° in the space spanned by t_3 and t_8.

Assuming the vacuum to be invariant under the Weyl group, it can be paired by action of the supersymmetry generators only with other states that are invariant under the Weyl group.** This may or may not include the previously constructed zero-energy state $|0\rangle$. The condition that the state $|0\rangle$ be annihilated by all $\lambda_{\mathscr{A}La}$ is obviously invariant under the Weyl group, so if it is unique this state must furnish a one-dimensional representation of the Weyl group. The Weyl group always acts on the generators of the Cartan subalgebra by orthogonal transformations, so there are two such representations: the invariant representation, in which each Weyl transformation is represented by unity, and the pseudoinvariant representation, in which each Weyl transformation is represented by the determinant of its action on the generators of the Cartan subalgebra.

Let us first consider the case where $|0\rangle$ is invariant under the Weyl group. Obviously no linear combination of the one-fermion states $\lambda^*_{\mathscr{A}La}|0\rangle$ can be invariant under the Weyl group. There is just one Weyl-invariant linear combination of the two-fermion states; it is of the form $U|0\rangle$, where

$$U \equiv \sum_{ab\mathscr{A}} e_{ab}\, \lambda^*_{\mathscr{A}La}\, \lambda^*_{\mathscr{A}Lb}\,. \tag{29.4.12}$$

(The spin indices a and b are contracted with the antisymmetric tensor e_{ab} defined by Eq. (25.2.9), because the anticommutation relations make the

** Witten remarked that physical states are necessarily Weyl-invariant, but we will not need to go into this here, because the Weyl invariance of the vacuum and of the supersymmetry generators means that only Weyl-invariant states are relevant to spontaneous supersymmetry breaking.

product $\lambda^*_{\mathscr{A}La}\lambda^*_{\mathscr{A}Lb}$ antisymmetric in a and b.) There are various Weyl-invariant linear combinations of products of three or more generators of the Cartan subalgebra, but the anticommutation of the $\lambda^*_{\mathscr{A}La}$ makes all of them vanish† except for powers of U. Also, in the product of more than r of the Us some of the $\lambda^*_{\mathscr{A}La}$ would have to appear twice, so $U^{r+1}=0$. We conclude then that the Weyl-invariant states are limited to the $r+1$ states

$$|0\rangle\,,\ U|0\rangle\,, U^2|0\rangle\,,\ \ldots\ , U^r|0\rangle\,. \tag{29.4.13}$$

These are all bosonic states, and there are no fermionic states with zero energy in the tree approximation with which they could be paired, so for sufficiently weak coupling these states must have precisely zero energy, and supersymmetry is unbroken. Also, the Weyl-invariant Witten index here is $r+1$, and this is independent of the coupling strength, so *supersymmetry is not spontaneously broken whatever the strength of the gauge coupling.*

In the case where the state $|0\rangle$ is pseudoinvariant, the only Weyl-invariant states are of the form

$$\sum_{\mathscr{A},\mathscr{B},\cdots} \epsilon_{\mathscr{A},\mathscr{B},\cdots}\lambda^*_{\mathscr{A}La}\lambda^*_{\mathscr{B}Lb}\cdots|0\rangle\,, \tag{29.4.14}$$

† To see this for $SU(N)$, note that each generator $t_{\mathscr{A}}$ of its Cartan subalgebra may be written as a linear combination $t_{\mathscr{A}}=\sum_i c_{\mathscr{A}i}T_i$ of generators T_i of the Cartan subalgebra of $U(N)$, with the only non-zero element of each T_i being $(T_i)^i{}_i=1$ (with indices not summed). In order for the $t_{\mathscr{A}}$ to be traceless, we must have $\sum_i c_{\mathscr{A}i}=0$. Since the fields $\lambda^*_{\mathscr{A}La}$ transform under the Weyl group like the $t_{\mathscr{A}}$, it follows that for a function $\sum_{\mathscr{A}\mathscr{B}\ldots} d_{\mathscr{A}\mathscr{B}\ldots}\lambda^*_{\mathscr{A}La}\lambda^*_{\mathscr{B}Lb}\cdots$ to be Weyl-invariant, the coefficients $d_{\mathscr{A}\mathscr{B}\ldots}$ must take the form

$$d_{\mathscr{A}\mathscr{B}\cdots}=\sum_{ij\cdots} c_{\mathscr{A}i}\,c_{\mathscr{B}j}\cdots D_{ij\cdots}\,,$$

where the $D_{ij\cdots}$ are invariant tensors, in the sense that for any vectors u_i, v_j, etc., the function $D(u,v,\ldots)\equiv\sum_{ij\ldots}D_{ij\ldots}u_iv_j\ldots$ is invariant under permutations of the coordinate axes. The most general such function is a linear combination of products of the function

$$S(x,y,z,\ldots)=\sum_i x_iy_iz_i\cdots\,,$$

with arguments $x, y, z, \ldots$ taken as various subsets of the u, v, etc. But because $\sum_i c_{\mathscr{A}i}=0$, in our case the sum of the components of each vector vanishes, so $D(u)=S(u)=0$; $D(u,v)$ is proportional to $S(u,v)$; $D(u,v,w)$ is proportional to $S(u,v,w)$; $D(u,v,w,x)$ is a linear combination of $S(u,v,w,x)$, $S(u,v)S(w,x)$, $S(u,w)S(v,x)$, and $S(u,x)S(v,w)$; and so on. The important point is that, even though the function $D(u,v,\ldots)$ may not be symmetric in its arguments (because different products of Ss may appear with different coefficients), the functions $S(u,v,\ldots)$ *are* symmetric. In our case, the vectors are the anticommuting quantities $u(a)_i=\sum_{\mathscr{A}}\lambda^*_{\mathscr{A}L,a}c_{\mathscr{A}i}$, for which the only non-vanishing S function is $S(u(1/2),u(-1/2))$. With a suitable normalization of the generators, this is the operator U.

where $\epsilon_{\mathscr{A},\mathscr{B},\ldots}$ is the totally antisymmetric tensor of rank r. This is completely symmetric among the a, b,..., so each state is characterized by the number of these indices that are $+1/2$ rather than $-1/2$, a number that can take any value from 0 to r, so the number of independent states is $r+1$. Depending on whether r is even or odd, these states are either all bosonic or all fermionic, so the Witten index here is $\pm(r+1)$, and again supersymmetry is not spontaneously broken.

There is an interesting relation between the value of the Witten index obtained here and ideas of how certain global symmetries become broken. The Lagrangian for a supersymmetric gauge theory is invariant under the transformations of a global $U(1)$ 'R-symmetry,' which change the left- and right-handed parts of the gaugino fields by opposite phases:

$$\lambda_{AL} \to e^{i\varphi}\lambda_{AL}\,, \qquad \lambda_{AR} \to e^{-i\varphi}\lambda_{AR}\,, \tag{29.4.15}$$

with φ an arbitrary real constant phase. The conservation of the current J_5^μ associated with this symmetry is violated by an anomaly

$$\partial_\mu J_5^\mu = -\frac{1}{32\pi^2}\sum_{ABCD} C_{ACD}\, C_{BCD}\, f_A^{\mu\nu}\, f_B^{\rho\sigma}\, \epsilon_{\mu\nu\rho\sigma}\,, \tag{29.4.16}$$

where as usual $\epsilon_{\mu\nu\rho\sigma}$ is the totally antisymmetric quantity with $\epsilon^{0123} \equiv 1$. (This is obtained from Eq. (22.2.24) by taking the gauge group generators t_A as $(t_A)_{BC} = -iC_{ABC}$, because the gauginos are in the adjoint representation of the gauge group, and by multiplying the anomaly with a factor 1/2, because the gauginos do not have distinct antiparticles.) With the definition of the gauge coupling specified above, Eq. (23.5.20) gives the integral of the product of field strengths in the anomaly as

$$\epsilon_{\mu\nu\rho\sigma}\int d^4x \sum_A f_A^{\mu\nu} f_A^{\rho\sigma} = 64\pi^2\nu/g^2\,, \tag{29.4.17}$$

where the 'winding number' ν is an integer characterizing the topological class to which the gauge field belongs. Putting together Eqs. (29.4.4), (29.4.16), and (29.4.17), we see that an instanton of winding number ν induces a change in $R \equiv \int d^3x\, J_5^0$ given by

$$\Delta R = \int d^4x\, \partial_\mu J_5^\mu = -2\nu C_1\,. \tag{29.4.18}$$

That is, the effective action contains terms $(\sum_{Aab}\lambda_{ALa}\lambda_{ALb}e_{ab})^{C_1}$ and its integer powers that, instead of being invariant under the R-symmetry transformations (29.4.15), are transformed by integer powers of the phase $\exp(2i\varphi C_1)$. Thus instantons invalidate invariance under the general $U(1)$ R-transformation (29.4.15), reducing it to the group Z_{2C_1} of transformations (29.4.15) with φ an integer multiple of π/C_1. We might expect that the growth of the gauge coupling at low energy would lead (as in quantum chromodynamics) to the appearance of vacuum expectation values for gaugino bilinears, which would mean that the discrete symmetry group Z_{2C_1} is spontaneously broken to its Z_2 subgroup, generated by a simple sign change of the gaugino multiplet. Then there would be C_1 zero-energy states $|n\rangle$, given by acting on any one vacuum state $|X\rangle$ with the elements $\exp(in\pi R/C_1)$ of Z_{2C_1}, with n running only over the values $0, 1, \ldots, C_1 - 1$ because we treat as equivalent any pair of states that differ only by action of the generator $\exp(i\pi R)$ of Z_2. Since $|X\rangle$ can only be a linear combination of states with even values of R, we can form C_1 states with all values $\mathscr{R} = 0, 2, \ldots, 2C_1 - 2$ of R by taking the linear combinations

$$\sum_n \exp\left(\frac{in\pi\mathscr{R}}{C_1}\right)|n\rangle\,.$$

In particular, for $SU(N)$ and $USp(2N)$ these states are the same as the Weyl-invariant states encountered in the calculation of the Witten index in the case where the state $|0\rangle$ is R-invariant, and Weyl-invariant rather than pseudoinvariant. The operator U defined by Eq. (29.4.12) has $R = 2$, so there are $r + 1 = C_1$ states $U^n|0\rangle$ with $R = 2n$ running from zero to $2r = 2C_1 - 2$. More generally, the presence of C_1 zero-energy states with the same statistics helps to explain why the Witten index is equal to $\pm C_1$ for the exceptional and orthogonal as well as the unitary and symplectic groups.

The fact that the Witten index is non-vanishing for all pure gauge supersymmetric theories means that to find examples of spontaneous supersymmetry breaking we must add chiral superfields to the theory. It does not help to add *massive* chiral superfields to the theory, since for weak coupling the introduction of massive fields does not change the menu of zero-energy states. We have already seen an example of this: the work of Affleck, Dine, and Seiberg reviewed in Section 29.3 showed that an $SU(N_c)$ gauge theory with $N_f < N_c$ left-chiral superfields Q_{ai} in the defining representation of $SU(N_c)$ and an equal number of left-chiral superfields $\overline{Q}_{ai}$ in the complex conjugate representation, with a mass term $\sum_{aij} m_{ij}\overline{Q}_{ai}Q_{aj}$, has N_c zero-energy bosonic states and no zero-energy fermionic states. (This is one case where the Witten index is not left unchanged when a mass term vanishes, because this mass term is the term in the superpotential of highest order in the superfields, so that a vanishing mass changes the behavior of the superpotential for large superfields. In fact, Eq. (29.3.31) shows that as the masses go to zero the scalar field values in the state of zero energy go to infinity.)

On the other hand, there is no difficulty in finding theories with left-chiral superfields subject to a symmetry that keeps them massless, in which supersymmetry *is* dynamically broken. For instance, consider an

$SU(N_c)$ gauge theory with N_f left-chiral superfields Q_{ai} and N_f left-chiral superfields $\overline{Q}_{ai}$ again in the N_c and $\overline{N}_c$ representations of the gauge group, but now also with N_f left-chiral superfields L_i that are neutral under the gauge group $SU(N_c)$. Assume a global (or weakly coupled local) $SU(N_f)$ symmetry, acting on the 'flavor' i indices of the Qs and the Ls but not the $\overline{Q}$s, which among other things forbids a mass term linking the Qs and $\overline{Q}$s. We take Q_{ai} and L_i in the representations N_f and $\overline{N}_f$ of $SU(N_f)$, respectively, while the $\overline{Q}$s are taken to be $SU(N_f)$ singlets. The only renormalizable superpotential is then of the form

$$f(Q,\overline{Q},L) = \sum_{ija} \mathscr{G}_j \overline{Q}_{aj} Q_{ai} L_i \,, \tag{29.4.19}$$

where the $\mathscr{G}_j$ are a set of coupling constants, which by an $SU(N_f)$ rotation may be chosen to have only one non-vanishing component, say the one with $j = N_f$, which can also be chosen to be positive. (This is a generalization of a model treated by Affleck, Dine, and Seiberg,[8] in which they took $N_c = 3$ and $N_f = 2$.) The gauge-neutral superfields L_i have no effect on the non-perturbative terms in the effective superpotential, so we can use the result of Section 29.3, that for $N_c > N_f$ the gauge interactions yield a total effective superpotential

$$f_{\rm total}(Q,\overline{Q},L) = \sum_{ija} \mathscr{G}_j \overline{Q}_{aj} Q_{ai} L_i + \mathscr{K}\left[{\rm Det}_{ij} \sum_a \overline{Q}_{aj} Q_{ai}\right]^{-1/(N_c-N_f)} , \tag{29.4.20}$$

with $\mathscr{K}$ a constant. In order for supersymmetry to be unbroken it is necessary (though not sufficient) that the scalar components q_{aj}, $\overline{q}_{aj}$, and ℓ_i of the chiral superfields satisfy the condition $\partial f_{\rm total}(q,\overline{q},\ell)/\partial \ell_i = 0$, so that, for all i,

$$\sum_{ja} \mathscr{G}_j \overline{q}_{aj}\, q_{ai} = 0 \,. \tag{29.4.21}$$

But this tells us that the matrix $\sum_a \overline{q}_{aj} q_{ai}$ has a zero eigenvalue, and therefore has zero determinant, so this is a singular point of the superpotential (29.4.20), at which it is impossible for $\partial f_{\rm total}/\partial q_{ai}$ or $\partial f_{\rm total}/\partial \overline{q}_{aj}$ to vanish. Supersymmetry is therefore necessarily broken in this class of models.

For instance, Affleck, Dine, and Seiberg[8] found the non-vanishing scalar components q_{aj}, $\overline{q}_{aj}$, and ℓ_i of the superfields Φ_{ai}, $\overline{\Phi}_{ai}$, L_i at the minimum of the potential in their model with $N_c = 3$ and $N_f = 2$ to be at the values

$$\begin{aligned} q_{11} &= \overline{q}_{11} = 1.286\,(\mathscr{K}/2\mathscr{G})^{1/7} \,, \\ q_{22} &= \overline{q}_{22} = 1.249\,(\mathscr{K}/2\mathscr{G})^{1/7} \,, \\ \ell_1 &= \sqrt{q_{11}^2 - q_{22}^2} \,, \end{aligned} \tag{29.4.22}$$

where the vacuum energy density is

$$\rho_{\text{VAC}} = 3.593\,(16\mathcal{G}^{10}/\mathcal{K}^4)^{1/7}\,. \tag{29.4.23}$$

Also, $\mathcal{K} = 2\Lambda^7$ for $\theta = 0$.

The fact that supersymmetry is spontaneously broken in this case encourages us to think that it will be broken by strong gauge forces in a wide range of asymptotically free gauge theories, and thus lends legitimacy to the speculations about supersymmetry breaking in Section 28.3.

29.5 The Seiberg–Witten Solution*

It often happens that the tree-approximation potential in a supersymmetric theory will take the value zero for a continuous range of scalar field values. (For one example, see Eq. (29.1.9).) In this case the theory has a number of scalar excitations with zero mass in the tree approximation, which since supersymmetry is unbroken must be accompanied with suitable fermionic superpartners. At low energies the theory will then be described by a family of supersymmetric effective Lagrangians, whose members are parameterized by one or more *moduli*, the scalar expectation values in the underlying theory. Quantum effects in the underlying theory can modify the dependence of the effective Lagrangian on these moduli, and even alter the topology of the space of moduli.[9]

In one of the most striking accomplishments of the 1990s in supersymmetry theory, Seiberg and Witten[10] were able to calculate the exact dependence of the low-energy effective Lagrangian on a modular parameter in gauge theories with $N = 2$ supersymmetry. The ideas behind this calculation can be made apparent by running through only the simplest special case, that of an $SU(2)$ gauge theory with $N = 2$ supersymmetry and no additional matter hypermultiplets.

We saw in Section 27.9 that the Lagrangian density for this theory is given after elimination of the auxiliary fields by

$$\mathcal{L} = \frac{1}{e^2}\Bigg[-\sum_A (D_\mu\phi)^*_A\,(D^\mu\phi)_A - \frac{1}{2}\sum_A \Big(\overline{\psi_A}\,(\not{D}\psi)_A\Big)$$
$$-2\sqrt{2}\,\mathrm{Re}\sum_{ABC}\epsilon_{ABC}\Big(\lambda^{\mathrm{T}}_{AL}\,\epsilon\,\psi_{CL}\Big)\,\phi^*_B - \frac{1}{4}\sum_A f_{A\mu\nu}f_A^{\mu\nu}$$

* This section lies somewhat out of the book's main line of development and may be omitted in a first reading.

$$-\frac{1}{2}\sum_A \left(\overline{\lambda_A}\,(\not{D}\lambda)_A\right)\Bigg] + \frac{\theta}{64\pi^2}\epsilon_{\mu\nu\rho\sigma}\sum_A f_A^{\mu\nu} f_A^{\rho\sigma} - V(\phi,\phi^*)\,, \tag{29.5.1}$$

where A, B, and C now run over the values 1, 2, and 3. We have now rescaled all the fields by multiplying each of them by a factor e, so that e does not appear in the covariant derivatives:

$$(D_\mu\psi)_A = \partial_\mu\psi_A + \sum_{BC}\epsilon_{ABC}V_{B\mu}\psi_C\,, \tag{29.5.2}$$

$$(D_\mu\lambda)_A = \partial_\mu\lambda_A + \sum_{BC}\epsilon_{ABC}V_{B\mu}\lambda_C\,, \tag{29.5.3}$$

$$(D_\mu\phi)_A = \partial_\mu\phi_A + \sum_{BC}\epsilon_{ABC}V_{B\mu}\phi_C\,, \tag{29.5.4}$$

$$f_{A\mu\nu} = \partial_\mu V_{A\nu} - \partial_\nu V_{A\mu} + \sum_{BC}\epsilon_{ABC}V_{B\mu}V_{C\nu}\,, \tag{29.5.5}$$

and the potential is

$$V(\phi,\phi^*) = 2\sum_A\left[\sum_{BC}\epsilon_{ABC}\,\mathrm{Re}\,\phi_B\,\mathrm{Im}\,\phi_C\right]^2\,. \tag{29.5.6}$$

This potential takes the value zero for a family of scalar field expectation values, which (up to a gauge transformation) may be parameterized as

$$\phi_1 = \phi_2 = 0\,, \qquad \phi_3 = a\,, \tag{29.5.7}$$

with a a complex parameter, known as the vacuum modulus. This vacuum expectation value gives masses $2|a|$ to the vector fields $V_{1\mu}$ and $V_{2\mu}$, the gauginos λ_1 and λ_2, the chiral fermions ψ_1 and ψ_2, and the scalars ϕ_1 and ϕ_2, leaving $V_{3\mu}$, λ_3, ψ_3, and $\phi_3 = a$ all massless.

Taking account only of these zero-mass modes (and dropping the subscript 3), the tree approximation gives the effective low-energy theory as simply the free-field theory with Lagrangian density

$$\mathscr{L}_{\text{eff}} = \frac{1}{e^2}\Bigg[-(\partial_\mu a)^*\,(\partial^\mu a) - \frac{1}{2}\left(\overline{\psi}\,(\not{\partial}\psi)\right)$$

$$-\frac{1}{4}f_{\mu\nu}f^{\mu\nu} - \frac{1}{2}\left(\overline{\lambda}\,(\not{D}\lambda)\right)\Bigg] + \frac{\theta}{64\pi^2}\epsilon_{\mu\nu\rho\sigma}f^{\mu\nu}f^{\rho\sigma}\,, \tag{29.5.8}$$

with $f_{\mu\nu} = \partial_\mu V_\nu - \partial_\nu V_\mu$. Indeed, any renormalizable theory of a single gauge boson and its gauge-neutral $N = 2$ superpartners must be a free-field theory, because $N = 2$ supersymmetry does not allow a superpotential for this theory.

But this is not the whole story. In integrating out the massive degrees of freedom in this theory, quantum corrections produce non-renormalizable interaction terms in the low-energy effective field theory. We can classify the interactions and Feynman diagrams containing them that predominate at low energy by the same sort of counting of powers of energy that we used in dealing with low-momentum pions and nucleons in Section 19.5. When we use the effective Lagrangian perturbatively to calculate low-energy scattering amplitudes, the number ν of powers of energy contributed by a connected graph with L loops, I_f external fermion lines, I_b internal boson (a or V_μ) lines, I_a internal auxiliary field lines, and V_i vertices of each type i, is

$$\nu = 4L + \sum_i V_i d_i - 2I_b - I_f \,, \tag{29.5.9}$$

where d_i is the number of derivatives in the interaction of type i. (Internal auxiliary field lines do not contribute in Eq. (29.5.9) because their propagators are independent of momentum.) These quantities are subject to the topological relations

$$L = I_b + I_f + I_a - \sum_i V_i + 1 \,, \tag{29.5.10}$$

and

$$2I_b + E_b = \sum_i V_i b_i \,, \quad 2I_f + E_f = \sum_i V_i f_i \,, \quad 2I_a + E_a = \sum_i V_i a_i \,, \tag{29.5.11}$$

where E_b, E_f, and E_a are the numbers of external boson, fermion, and auxiliary field lines, and b_i, f_i, and a_i are the numbers of boson, fermion, and auxiliary fields in interactions of type i. We can therefore write the number of powers of energy as

$$\nu = \sum_i V_i \left(d_i + \tfrac{1}{2} f_i + a_i - 2\right) + 2L - E_f - 2E_a + 2 \,. \tag{29.5.12}$$

According to Eqs. (26.8.4) and (27.4.42), both the D-term of a function of left-chiral scalar $N = 1$ superfields and their adjoints and the $\mathscr{F}$-term of a pair of $N = 1$ gauge superfields W_α times an arbitrary function of $N = 1$ left-chiral scalar superfields have $d_i + \frac{1}{2} f_i + a_i = 2$, while adding any additional W_α factors or superderivatives $\mathscr{D}_\alpha$ would give $d_i + \frac{1}{2} f_i + a_i > 2$, so in our case supersymmetry rules out any interactions with $d_i + \frac{1}{2} f_i + a_i$ less than 2. The dominant contribution to low-energy scattering amplitudes is thus given by the tree approximation ($L = 0$), calculated with an effective Lagrangian containing only terms with $d_i + \frac{1}{2} f_i + a_i = 2$, which takes the general form discussed in Section 27.4:

$$\mathscr{L}_{\text{eff}} = \frac{1}{2}\Big[K(\Phi, \Phi^*)\Big]_D - \frac{1}{2}\text{Re}\,\Big[T(\Phi)\left(W_L^{\rm T} \epsilon W_L\right)\Big]_{\mathscr{F}} \,. \tag{29.5.13}$$

Eq. (27.4.42) then gives the Lagrangian in terms of component fields as

$$\begin{aligned}\mathscr{L} = &\frac{\partial^2 K(a,a^*)}{\partial a\,\partial a^*}\Big[-\frac{1}{2}\big(\overline{\psi}\,\partial\!\!\!/\psi\big) + |\mathscr{F}|^2 - \partial_\mu a\,\partial^\mu a^*\Big]\\
&-\mathrm{Re}\left\{\frac{\partial^3 K(a,a^*)}{\partial^2 a\,\partial a^*}\big(\overline{\psi}\psi_L\big)\mathscr{F}^*\right\} - \frac{1}{2}\mathrm{Re}\left\{\frac{\partial^3 K(a,a^*)}{\partial^2 a\,\partial a^*}\big(\overline{\psi}\gamma^\mu\gamma_5\psi\big)\,\partial_\mu a\right\}\\
&+\frac{1}{4}\frac{\partial^4 K(a,a^*)}{\partial^2 a\,\partial^2 a^*}\big(\overline{\psi}\psi_L\big)\big(\overline{\psi}\psi_R\big)\\
&+\frac{1}{4}\mathrm{Re}\left\{\big(\overline{\lambda}\lambda_L\big)\big(\overline{\psi}\psi_L\big)\frac{d^2T(a)}{da^2}\right\} - \frac{1}{2}\mathrm{Re}\left\{\big(\overline{\lambda}\lambda_L\big)\mathscr{F}\frac{dT(a)}{da}\right\}\\
&+\mathrm{Re}\left\{T(a)\Big[-\frac{1}{2}\big(\overline{\lambda}\,\partial\!\!\!/(1-\gamma_5)\lambda\big) - \frac{1}{4}f_{\mu\nu}f^{\mu\nu}\right.\\
&\qquad\left.+\frac{1}{8}i\,\epsilon_{\mu\nu\rho\sigma}f^{\mu\nu}f^{\rho\sigma} + \frac{1}{2}D^2\Big]\right\}\\
&+\frac{\sqrt{2}}{4}\mathrm{Re}\left\{\frac{dT(a)}{da}\Big[-\big(\overline{\lambda}\gamma^\mu\gamma^\nu\psi_L\big)f_{\mu\nu} + 2i\big(\overline{\lambda}\psi_L\big)D\Big]\right\}\,. \end{aligned}\tag{29.5.14}$$

In order to implement $N=2$ supersymmetry, we now want to impose invariance under the discrete R-symmetry transformation (27.9.2):

$$\psi\to\lambda\,,\qquad \lambda\to-\psi\,,\tag{29.5.15}$$

with a and V_μ unchanged. The condition that the coefficients of $(\overline{\psi}\,\partial\!\!\!/\psi)$ and $(\overline{\lambda}\,\partial\!\!\!/\lambda)$ should be equal is

$$\frac{\partial^2 K(a,a^*)}{\partial a\,\partial a^*} = \mathrm{Re}\,T(a)\,.\tag{29.5.16}$$

The right-hand side is the sum of a function of a and a function of a^*, so $\partial^4 K/\partial^2 a\,\partial^2 a^* = 0$, and therefore the term quartic in ψ, which would have had no counterpart for λ, is absent. By an integration by parts, the term $\frac{1}{2}\mathrm{Re}\{T(a)(\overline{\lambda}\,\partial\!\!\!/\gamma_5\lambda)\}$ may be replaced with $-\frac{1}{4}\mathrm{Re}\{(\overline{\lambda}\gamma^\mu\gamma_5\lambda)\partial_\mu T(a)\}$. The condition that the coefficients of $(\overline{\psi}\gamma^\mu\gamma_5\psi)$ and $(\overline{\lambda}\gamma^\mu\gamma_5\lambda)$ should be equal is then satisfied if

$$\frac{1}{4}\partial_\mu T = \frac{1}{2}\frac{\partial^3 K}{\partial^2 a\,\partial a^*}\partial_\mu a\,,$$

which is also an automatic consequence of Eq. (29.5.16). According to Eq. (26.A.7), the terms proportional to $f_{\mu\nu}(\overline{\lambda}[\gamma^\mu,\gamma^\nu]\psi)$ and $f_{\mu\nu}(\overline{\lambda}[\gamma^\mu,\gamma^\nu]\gamma_5\psi)$ are automatically invariant under the transformation (29.5.15), as are also the terms proportional to $(\overline{\lambda}\lambda_L)(\overline{\psi}\psi_L)$ and its adjoint. On the other hand, the invariance of the terms proportional to $(\overline{\lambda}\psi)D$ and $(\overline{\lambda}\gamma_5\psi)D$ requires that we extend the transformation (29.5.15) so that

$$D \to -D\,, \tag{29.5.17}$$

which also leaves the term $\frac{1}{2}D^2$ invariant. Finally, Eq. (29.5.16) tells us that the coefficients of $(\overline{\psi}\psi_L)\mathscr{F}^*$ and $(\overline{\lambda}\lambda_L)\mathscr{F}$ are equal, so the transformation (29.5.15) must also be extended so that

$$\mathscr{F} \to \mathscr{F}^*\,, \tag{29.5.18}$$

which also leaves the term $|\mathscr{F}|^2$ invariant. We conclude then that the condition (29.5.16) ensures that the whole action derived from Eq. (29.5.14) is invariant under the combined transformation (29.5.15), (29.5.17), and (29.5.18). Since the Lagrangian is invariant under $N = 1$ supersymmetry with a left-chiral scalar supermultiplet $(a, \psi, \mathscr{F})$ and a gauge supermultiplet (V_μ, λ, D), it is also invariant under a second supersymmetry with a left-chiral scalar supermultiplet $(a, \lambda, \mathscr{F}^*)$ and a gauge supermultiplet $(V_\mu, -\psi, -D)$. Eq. (29.5.16) is therefore enough to ensure that the action obtained from Eq. (29.5.13) or (29.5.14) is invariant (without imposing the field equations) under $N = 2$ supersymmetry.

The general solution of Eq. (29.5.16) can be expressed as**

$$T(a) = \frac{1}{4\pi i}\frac{dh(a)}{da}\,, \qquad K(a,a^*) = \text{Im}\left\{\frac{a^* h(a)}{4\pi}\right\}\,, \tag{29.5.19}$$

with h a function of a alone. In terms of h, the Lagrangian density (29.5.14) now reads

$$\begin{aligned}
\mathscr{L} = \frac{1}{4\pi}\text{Im}\Bigg\{ &\left[\frac{dh}{da}\right]\Big[-\frac{1}{2}\big(\overline{\psi}\,\partial\!\!\!/(1-\gamma_5)\psi\big) - \frac{1}{2}\big(\overline{\lambda}\,\partial\!\!\!/(1-\gamma_5)\lambda\big) \\
&-\partial_\mu a\,\partial^\mu a^* + |\mathscr{F}|^2 + \frac{1}{2}D^2 - \frac{1}{4}f_{\mu\nu}f^{\mu\nu} + \frac{1}{8}i\,\epsilon_{\mu\nu\rho\sigma}f^{\mu\nu}f^{\rho\sigma}\Big]\Bigg\} \\
+\frac{1}{4\pi}\text{Im}\Bigg\{ &\left[\frac{d^2h}{da^2}\right]\Big[-\frac{1}{2}\big(\overline{\psi}\psi_L\big)\mathscr{F}^* - \frac{1}{2}\big(\overline{\lambda}\lambda_L\big)\mathscr{F} + \frac{\sqrt{2}}{2}i\big(\overline{\lambda}\psi_L\big)D \\
&-\frac{\sqrt{2}}{4}\big(\overline{\lambda}\gamma^\mu\gamma^\nu u\psi_L\big)f_{\mu\nu}\Big]\Bigg\} \\
+\frac{1}{16\pi}\text{Im}\Bigg\{ &\frac{d^3h(a)}{da^3}\big(\overline{\lambda}\lambda_L\big)\big(\overline{\psi}\psi_L\big)\Bigg\}\,.
\end{aligned} \tag{29.5.20}$$

This is invariant under an $SU(2)$ R-symmetry, under which (ψ, λ) transforms as a doublet and $(\text{Im}\,\mathscr{F}, \text{Re}\,\mathscr{F}, D/\sqrt{2})$ as a triplet. The discrete

** The factor $1/4\pi i$ is included in order to simplify the duality transformation introduced below. Seiberg and Witten introduced a function $\mathscr{F}(a)$ (not related to the auxiliary field $\mathscr{F}$), known as the *prepotential*, for which $h(a) = d\mathscr{F}(a)/da$.

transformation (29.5.15), (29.5.17), and (29.5.18) is a finite element of this $SU(2)$ group: a rotation of π radians around the 2-axis.

Comparing Eqs. (29.5.8) and (29.5.20), we see that in the tree approximation

$$h(a)_{\text{tree}} = \left[\frac{4\pi i}{e^2} + \frac{\theta}{2\pi}\right] a \,. \tag{29.5.21}$$

The achievement of Seiberg and Witten was to calculate $h(a)$ exactly.

The first step in this calculation is to recognize that there are various linear transformations on a and $dh(a)/da$ that give physically equivalent theories. This is because of a remarkable property of the low-energy effective theory, related to the duality property discussed briefly in Section 27.9. To demonstrate this property, let us return to the Lagrangian density (29.5.13) expressed in terms of $N = 1$ superfields, now using the relations (29.5.19) required by $N = 2$ supersymmetry

$$\mathscr{L}_{\text{eff}} = \frac{1}{8\pi}\text{Im}\left[\Phi^* h(\Phi)\right]_D - \frac{1}{8\pi}\text{Im}\left[h'(\Phi)\left(W_L^{\text{T}}\epsilon W_L\right)\right]_{\mathscr{F}} \,. \tag{29.5.22}$$

In path integrals the spinor field-strength superfield W_L is constrained by the supersymmetric extension (27.2.20) of the homogeneous Maxwell equations:

$$\text{Re}\left(\mathscr{D}_L^{\text{T}}\epsilon W_L\right) = 0 \,. \tag{29.5.23}$$

This is usually imposed by requiring that W_L take the form (27.2.15)

$$W_L = \frac{i}{4}\left(\mathscr{D}_R^{\text{T}}\epsilon\mathscr{D}_R\right)\mathscr{D}_L V \,, \tag{29.5.24}$$

where V is an unrestricted real superfield. Instead, we can implement the condition (29.5.23) by introducing a Lagrange multiplier term in the action

$$\Delta I_{\text{eff}} = \frac{1}{8\pi}\text{Re}\int d^4x\,\left[\tilde{V}\left(\mathscr{D}_L^{\text{T}}\epsilon W_L\right)\right]_D \,, \tag{29.5.25}$$

where $\tilde{V}$ is an unrestricted real superfield. (The numerical factor $1/8\pi$ serves to fix the normalization of $\tilde{V}$ in a way we shall find convenient later.) Then in path integrals we can integrate over both $\tilde{V}$ and W_L, with no restrictions on either except that $\tilde{V}$ is a real superfield and W_L is a left-handed spinor superfield satisfying the left-chirality condition $\mathscr{D}_{R\alpha}W_{L\beta} = 0$. Integrating by parts in superspace allows us to write the new term in the action as

$$\Delta I_{\text{eff}} = -\frac{1}{8\pi}\text{Re}\int d^4x\,\left[\left((\mathscr{D}_L\tilde{V})^{\text{T}}\epsilon W_L\right)\right]_D \,, \tag{29.5.26}$$

or, using Eq. (26.3.31) and the fact that W_L is left-chiral,

$$\Delta I_{\text{eff}} = \text{Re}\left[\frac{i}{4\pi}\int d^4x\,\left[\left(\tilde{W}_L^{\text{T}}\epsilon W_L\right)\right]_{\mathscr{F}}\right], \tag{29.5.27}$$

where $\tilde{W}_L$ is defined in terms of $\tilde{V}$ by the same relation that previously gave W_L in terms of V:

$$\tilde{W}_L = \frac{i}{4}\left(\mathscr{D}_R^{\text{T}}\epsilon\mathscr{D}_R\right)\mathscr{D}_L\tilde{V}\,. \tag{29.5.28}$$

But now we are integrating over W_L with no such constraint, with an action that is quadratic in W_L

$$I_{\text{eff}} + \Delta I_{\text{eff}} = \text{Im}\int d^4x\,\left[-\frac{1}{8\pi}\,h'(\Phi)\left(W_L^{\text{T}}\epsilon W_L\right) - \frac{1}{4\pi}\left(\tilde{W}_L^{\text{T}}\epsilon W_L\right)\right]_{\mathscr{F}}$$
$$+\frac{1}{8\pi}\text{Im}\int d^4x\,\left[\Phi^* h(\Phi)\right]_D. \tag{29.5.29}$$

This integration is done by setting W_L equal to the value where the action is stationary in W_L:

$$W_L = -\frac{\tilde{W}_L}{h'(\Phi)} \tag{29.5.30}$$

and the whole effective action becomes

$$\tilde{I}_{\text{eff}} = +\frac{1}{8\pi}\text{Im}\int d^4x\,\left[\frac{1}{h'(\Phi)}\left(\tilde{W}_L^{\text{T}}\epsilon\tilde{W}_L\right)\right]_{\mathscr{F}} + \frac{1}{8\pi}\text{Im}\int d^4x\,\left[\Phi^* h(\Phi)\right]_D. \tag{29.5.31}$$

Now, if we define a new left-chiral scalar superfield and a new h function

$$\tilde{\Phi} \equiv h(\Phi)\,, \qquad \tilde{h}(\tilde{\Phi}) \equiv -\Phi\,, \tag{29.5.32}$$

then

$$\frac{d\tilde{h}}{d\tilde{\Phi}}\frac{dh}{d\Phi} = -\frac{d\Phi}{d\tilde{\Phi}}\frac{d\tilde{\Phi}}{d\Phi} = -1\,, \tag{29.5.33}$$

so the action (29.5.31) may be written as

$$\tilde{I}_{\text{eff}} = -\frac{1}{8\pi}\text{Im}\int d^4x\,\left[\tilde{h}'(\tilde{\Phi})\left(\tilde{W}_L^{\text{T}}\epsilon\tilde{W}_L\right)\right]_{\mathscr{F}} + \frac{1}{8\pi}\text{Im}\int d^4x\,\left[\tilde{\Phi}^*\tilde{h}(\tilde{\Phi})\right]_D. \tag{29.5.34}$$

From the method by which we have derived it, we know that the theory based on this effective action is equivalent to the original effective field theory, so *the $N = 2$ effective field theory with scalar field value a and h function $h(a)$ is physically equivalent to one with scalar field value*[†] $a_D \equiv$

[†] The subscript D stands for 'dual,' and of course has nothing to do with the D-term of a superfield.

$h(a)$ and h function $\tilde{h}(a_D) \equiv -a$. This is the version of duality that applies in the present context.

There is another transformation of the function $h(\Phi)$ that also yields an equivalent Lagrangian (now with no change in Φ or W) and can be combined with the transformation (29.5.32) to give a larger group of duality transformations. Suppose we shift $h(\Phi)$ by a linear term with a real coefficient:

$$h(\Phi) \to h(\Phi) + b\,\Phi\,, \tag{29.5.35}$$

where b is a real constant. Then the first term in the effective Lagrangian density (29.5.22) is shifted by $b\,\mathrm{Im}\{[\Phi^*\Phi]_D\}/8\pi$, which vanishes since the D-term of $\Phi^*\Phi$ is real. The change in the effective Lagrangian density (29.5.22) is therefore given by the shift in the second term

$$\mathscr{L}_{\text{eff}} \to \mathscr{L}_{\text{eff}} - \frac{b}{8\pi}\mathrm{Im}\left[\left(W_L^{\mathrm{T}}\epsilon W_L\right)\right]_{\mathscr{F}}\,, \tag{29.5.36}$$

or, according to Eq. (27.2.13),

$$\mathscr{L}_{\text{eff}} \to \mathscr{L}_{\text{eff}} - \frac{b}{8\pi}\left[i\left(\bar{\lambda}\,\partial\!\!\!/\gamma_5\lambda\right) + \frac{1}{4}\epsilon^{\mu\nu\rho\sigma}f_{\mu\nu}f_{\rho\sigma}\right]\,. \tag{29.5.37}$$

The first term in the brackets on the right-hand side is a spacetime derivative, and therefore does not affect the effective action. The second term in the brackets can also be written as a spacetime derivative $(1/2)\partial_\mu(\epsilon^{\mu\nu\rho\sigma}A_\nu f_{\rho\sigma})$ wherever $f_{\mu\nu}$ can be written as $\partial_\mu A_\nu - \partial_\nu A_\mu$. However, as discussed in Section 23.3, the gauge transformation that we used to put ϕ_A in the 3-direction must be singular somewhere, so we cannot write $f_{\mu\nu}$ everywhere in terms of a single A_μ. In consequence, physical quantities *can* be affected by a term in the action of the form

$$-\frac{\theta}{64\pi^2}\int d^4x\;\epsilon^{\mu\nu\rho\sigma}f_{\mu\nu}f_{\rho\sigma}\,. \tag{29.5.38}$$

In particular, Witten[11] has shown that, in the theory of magnetic monopoles described in Section 23.3, the electric charge of a magnetic monopole with minimum monopole moment in the presence of such a term is $e\theta/2\pi$. But, as mentioned in Section 27.9, in this theory there are also dyons, particles with both magnetic monopole moments and charges given by any integer multiple of e, so the whole pattern of monopole and dyon charges is periodic in θ with period 2π. In fact, all physical quantities have this periodicity, because the θ of the low-energy effective theory is inherited from the θ appearing in the Lagrangian density (29.5.1) of the underlying theory (note Eq. (29.5.21)), and we have seen in Section 23.5 that all physical quantities are periodic in this θ. According to Eq. (29.5.37), the transformation (29.5.35) changes θ by $2\pi b$, and therefore with b equal to an arbitrary positive or negative integer it yields an equivalent effective action.

This sort of transformation is closely related to an exact invariance of the underlying theory. The Lagrangian (29.5.1) is invariant under a continuous R-symmetry, under which

$$\theta_L \to \exp(i\alpha)\theta_L\,, \quad W_{AL} \to \exp(i\alpha)W_{AL}\,, \quad \Phi_A \to \exp(2i\alpha)\Phi_A\,. \quad (29.5.39)$$

This symmetry is broken by an anomaly: both λ_{AL} and ψ_{AL} have R quantum numbers $+1$, and of course λ_{AR} and ψ_{AR} have R quantum numbers -1, so according to Eqs. (23.5.21) and (23.5.23), the measure in the integral over fermion fields changes under the transformation (29.5.39) by a factor $\exp(2i\alpha N\nu)$, where ν is an integer, the winding number of the vector field configuration, and N is defined by

$$\mathrm{Tr}\,(t_A t_B) = \frac{1}{2}N\delta_{AB}\,.$$

(The factor $\exp(2i\alpha N\nu)$ is the same as given in Section 23.5, even though here we have two fields ψ_A and λ_A in the same representation of the gauge group, because these are Majorana fields, in contrast with the Dirac fields used in Section 23.5.) The generators here are $(t_A)_{BC} = -i\epsilon_{ABC}$, so $N = 4$, and therefore the measure remains invariant when $\exp(8i\alpha) = 1$. In other words, the continuous R-symmetry is broken by instantons to a $\mathbb{Z}_8$ subgroup, generated by the transformation

$$\psi_{AL} \to \sqrt{i}\,\psi_{AL}\,, \quad \lambda_{AL} \to \sqrt{i}\,\lambda_{AL}\,, \quad \phi_A \to i\,\phi_A\,. \quad (29.5.40)$$

This symmetry then must carry over to the effective low-energy theory. However, it does *not* tell us that the effective Lagrangian (29.5.20) is invariant under the discrete transformation (29.5.40), which would require that $-ih(ia) = h(a)$ for all a. This condition is satisfied by the tree approximation (29.5.21), but (as we will see) it is violated even in one-loop order. The $\mathbb{Z}_8$ R-symmetry is realized in the effective theory by the condition that the effective theory with h function $-ih(ia)$ is *equivalent* to one with h function $h(a)$. That is, $-ih(ia)$ must be related to $h(a)$ by some combination of the transformations (29.5.32) and (29.5.35), with b an integer.

We have seen that the physical significance of the theory is left unchanged by two sorts of transformation: the transformation $\Phi \to \tilde{\Phi} = h(\Phi)$ and $h(\Phi) \to \tilde{h}(\tilde{\Phi}) = -\Phi$, which can be written as

$$\begin{pmatrix} \Phi \\ h(\Phi) \end{pmatrix} \to \begin{pmatrix} 0 & 1 \\ -1 & 0 \end{pmatrix} \begin{pmatrix} \Phi \\ h(\Phi) \end{pmatrix}, \quad (29.5.41)$$

and the transformations $h(\Phi) \to h(\Phi) + b\Phi$, $\Phi \to \Phi$, which can be written as

$$\begin{pmatrix} \Phi \\ h(\Phi) \end{pmatrix} \to \begin{pmatrix} 1 & 0 \\ b & 1 \end{pmatrix} \begin{pmatrix} \Phi \\ h(\Phi) \end{pmatrix} , \tag{29.5.42}$$

where b is an arbitrary integer. By combining these transformations, we can make generalized duality transformations

$$\begin{pmatrix} \Phi \\ h(\Phi) \end{pmatrix} \to \begin{pmatrix} n & m \\ k & l \end{pmatrix} \begin{pmatrix} \Phi \\ h(\Phi) \end{pmatrix} , \tag{29.5.43}$$

where n, m, k, and l are any integers satisfying the condition that the matrix in Eq. (29.5.43) should, like those in Eqs. (29.5.41) and (29.5.42), have unit determinant:

$$nl - mk = 1 . \tag{29.5.44}$$

These transformations therefore form the group $SL(2, \mathbb{Z})$.

The physical significance of the duality transformation can be brought forward by considering the central charge of the $N = 2$ supersymmetry algebra. As shown in Section 25.5, the eigenvalue Z_{12} of this central charge in any one-particle state sets a lower bound $M \geq |Z_{12}|/2$ on the mass of the state, a bound that is reached for particles that belong to 'short' supermultiplets. We saw in Section 27.9 that Z_{12} is given in the underlying $SU(2)$ gauge theory with $N = 2$ supersymmetry by Eq. (27.9.22):

$$Z_{12} = 2\sqrt{2}\, v\, [iq - \mathscr{M}] ,$$

where q and $\mathscr{M}$ are the charge and magnetic moment of the particle, and v is the vacuum expectation value of the conventionally normalized neutral scalar field. In the notation used in the present section, we have absorbed a factor e into the normalization of the field a, so here we have

$$Z_{12} = 2\sqrt{2}\, a\, [iq - \mathscr{M}]/e . \tag{29.5.45}$$

This theory contains particles like the massive elementary scalars, spinors, and vector bosons, which have charge e and zero magnetic monopole moment, so such particles are eigenstates of Z_{12} with eigenvalue

$$Z_{12}^{\text{charged elementary}} = 2\sqrt{2}\, i\, a . \tag{29.5.46}$$

As we have seen, the theory with scalar field a and h function $h(a)$ is equivalent to one with scalar field $n\,a + m\,h(a)$ and h function $k\,a + l\,h(a)$. Therefore for all integers n and m the theory with scalar field a and h function $h(a)$ must *also* contain a particle that looks like a massive particle of charge e and zero magnetic moment in the version of the theory with scalar field $na + mh(a)$ and h function $ka + lh(a)$, and that therefore has a central charge

$$Z_{12} = 2\sqrt{2}\, i \Big[n\,a + m\,h(a)\Big] . \tag{29.5.47}$$

Comparing this with Eq. (29.5.45), we see that, in the version of this theory with scalar field a and h function $h(a)$, this particle appears as one with a charge q and a magnetic monopole moment $\mathscr{M}$ given by

$$q/e = n + m\,\mathrm{Re}\,[h(a)/a]\,, \qquad \mathscr{M}/e = m\,\mathrm{Im}\,[h(a)/a]\,. \tag{29.5.48}$$

This is a dyon, with both charge and magnetic monopole moment. Note that the formula (29.5.48) for the charge of this particle bears out Witten's earlier result about the charge of magnetic monopoles: adding a term $b\,a$ to the function $h(a)$ changes the charge of a monopole with $m = 1$ by an amount $be = e\Delta\theta/2\pi$.

Using the tree-approximation result (29.5.21) for $h(a)$ in Eq. (29.5.48) shows that in this approximation the magnetic monopole moments in this theory are multiples of a value $4\pi/e$. This is the same as the magnetic monopole moment derived in Section 23.3 (recall that the magnetic monopole moment g used there is $\mathscr{M}/4\pi$), but this is a semi-classical result, subject to quantum corrections. Note that the duality transformation (29.5.41) with $n = 0$ and $m = 1$ takes an elementary particle with charge e and magnetic monopole moment zero into a non-elementary particle with magnetic monopole moment $\mathrm{Im}\,[h(a)/a]$, which in the tree approximation is $4\pi/e$.

The beta function for the electric charge in the underlying $N = 2$ supersymmetric $SU(2)$ gauge theory is given perturbatively by the one-loop result Eq. (27.9.50), with the first Casimir invariant taken now as $C_1 = 2e$ and the number of hypermultiplets set at $H = 0$:

$$\beta_{\mathrm{perturbative}}(e) = -\frac{e^3}{4\pi^2}\,. \tag{29.5.49}$$

We here take a as the renormalization scale, so the running charge $e(a)$ satisfies

$$a\frac{d}{da}e(a) = \beta\Big(e(a)\Big)\,. \tag{29.5.50}$$

Using the perturbation theory formula (29.5.49), this gives

$$[e^{-2}(a)]_{\mathrm{perturbative}} = \frac{1}{2\pi^2}\ln\left(\frac{a}{\Lambda}\right)\,, \tag{29.5.51}$$

where Λ is an integration constant. In the formalism we are adopting here, with a factor e absorbed into the definition of the gauge field, the quantity $e^{-2}(a)$ appears as the coefficient $h'(a)/4\pi i$ of the term $-\frac{1}{4}f_{\mu\nu}f^{\mu\nu}$ in the low-energy effective Lagrangian (29.5.20), so the function $h(a)$ is given in perturbation theory by[††]

$$[h(a)]_{\text{perturbative}} = \frac{2i}{\pi}\left[a\ln\left(\frac{a}{\Lambda}\right) - a\right] = 4\pi i a\left[[e^{-2}(a)]_{\text{perturbative}} - \frac{1}{2\pi^2}\right] . \tag{29.5.52}$$

This is a good approximation[‡] for sufficiently large values of $|a|$, where Eq. (29.5.51) gives a small value of $e(a)$. In this case, Eq. (29.5.48) gives the magnetic monopole moment of monopoles and dyons as

$$\mathscr{M}_{\text{perturbative}}/e = 4\pi m a\left[[e^{-2}(a)]_{\text{perturbative}} - \frac{1}{2\pi^2}\right] , \tag{29.5.53}$$

where m is an arbitrary integer. Note that this is *not* the same as would be found by simply replacing e with the running charge $e(a)$ on the right-hand side of the semi-classical formula $\mathscr{M}/e = 4\pi m/e^2$.

For sufficiently large values of $|a|$, Eq. (29.5.52) also satisfies a necessary consistency condition: the coefficient $\text{Im}\, h'(a)/4\pi$ of the kinematic terms $-\frac{1}{2}(\bar{\psi}\,\partial\!\!\!/\psi)$, $-\frac{1}{2}(\bar{\lambda}\,\partial\!\!\!/\lambda)$, $-\partial_\mu a^*\partial^\mu a$ and $-\frac{1}{4}f_{\mu\nu}f^{\mu\nu}$ must be *positive*. The same condition tells us also that Eq. (29.5.52) cannot be a good approximation for all values of a, because it gives a negative value for $\text{Im}\, h'(a)$ for $|a|$ sufficiently small.

Eq. (29.5.52) shows that if a is carried counterclockwise around a circle with a large value of $|a|$, where perturbation theory is valid, then $h(a)$ is shifted by $-4a$. This tells us that $h(a)$ must have one or more singularities at finite values of a.

We can easily rule out the possibility that $h(a)$ has just one singularity, because this would make it impossible to satisfy the condition that $h'(a)$ should have a positive imaginary part for all non-singular values of a, which as we have seen is necessary for the positivity of the coefficients of the kinematic terms in the effective Lagrangian. A single singularity would have to be at $a = 0$, because the $\mathbb{Z}_8$-symmetry tells us that if $h(a)$ is singular at a then it is also singular at ia. The function $h(a) - h_{\text{perturbative}}(a)$ would then be analytic except at infinity and perhaps at $a = 0$, but this function vanishes as $|a| \to \infty$, and in particular does not change when carried around a circle at large a, so it could at most be a polynomial in $1/a$, with no constant term. If the polynomial does not vanish, then $h'(a)$ goes as some negative power of a as $a \to 0$, which would not

[††] The additive constant needed in integrating $h'(a)$ has no effect on the low-energy effective action derived from Eq. (29.5.22), because $[\Phi^*]_D$ is a derivative. This constant can be fixed by reference to the $\mathbb{Z}_8$-symmetry discussed earlier. Eq. (29.5.52) satisfies $-ih(ia) = h(a) - 1$, which is a special case with $b = -1$ of the transformations (29.5.35) that leave the physical significance of the theory unchanged. This would not be the case if we had added a constant to Eq. (29.5.52).

[‡] There are non-perturbative contributions[12] to $\beta(e)$, arising from instantons, but these vanish rapidly for $e \to 0$.

have a positive imaginary part, while if the polynomial did vanish then $h'(a)$ would be equal to $h'_{\text{perturbative}}(a)$, which also does not have positive imaginary part for $a \to 0$.

In studying the singularity structure of $h(a)$, it is helpful to keep track of duality by treating a and $a_D = h(a)$ in the same way, expressing them both as functions of some complex variable u. Seiberg and Witten took u to be the expectation value of the gauge-invariant quantity $\frac{1}{2}\sum_A \phi_A\phi_A$. This changes sign under the $\mathbb{Z}_8$ transformations $\phi_A \to \pm i\phi_A$, and of course under this transformation $a \equiv \phi_3 \to \pm ia$, so

$$a(-u) = \pm i\, a(u)\,. \tag{29.5.54}$$

The sign here has no physical significance because a and $-a$ are related by a finite $SU(2)$ gauge transformation. (Where necessary it can be made definite by adopting the convention that the upper and lower signs apply if $\text{Re}\,u > 0$ or $\text{Re}\,u < 0$, respectively.) We can trust perturbation theory for large values of $|a|$, so for $|u| \to \infty$

$$a \to \sqrt{2u}\,, \qquad a_D \to \frac{i}{\pi}\left[\sqrt{2u}\,\ln\left(\frac{2u}{\Lambda^2}\right) - 2\sqrt{2u}\right]\,. \tag{29.5.55}$$

Note that when u is taken counterclockwise around a circle at a fixed large value of $|u|$, the logarithm $\ln(2u/\Lambda^2)$ is shifted by $2i\pi$, while $\sqrt{2u}$ changes sign, so the changes in a and a_D are given by a *monodromy matrix*:

$$\begin{pmatrix} a \\ a_D \end{pmatrix} \to \begin{pmatrix} -1 & 0 \\ 2 & -1 \end{pmatrix}\begin{pmatrix} a \\ a_D \end{pmatrix}. \tag{29.5.56}$$

The functions $a(u)$ and $a_D(u)$ must therefore have two or more singularities at finite values u_n of u, such that the combined effect of going counterclockwise around each singularity is the same as (29.5.56).

Let us consider the possibility that there are just two singularities. (This was shown to be the case in the second Seiberg–Witten paper of Reference 10.) Under the $\mathbb{Z}_8$-symmetry that takes $a_A \to ia_A$ we have $u \to -u$, so the singularities must be at a pair of u values u_0 and $-u_0$. Going from a non-singular base point P in the u plane counterclockwise around the singularity at $\pm u_0$ all the way back to P should yield an equivalent theory. Therefore it must take the form of a duality transformation, in general depending on P, in which the vector (a, a_D) is multiplied with an $SL(2,\mathbb{Z})$ *monodromy matrix* $M_\pm$, as in Eq. (29.5.43). The counterclockwise contour around the circle with fixed large u can be deformed into a contour that starts at P and goes counterclockwise back to P around $-u_0$ and then counterclockwise back to P around $+u_0$. (See Figure 29.1.) Since this deformation cannot change the integral, the product of the monodromy matrices *in this order* (reading from right to left) must be equal to the

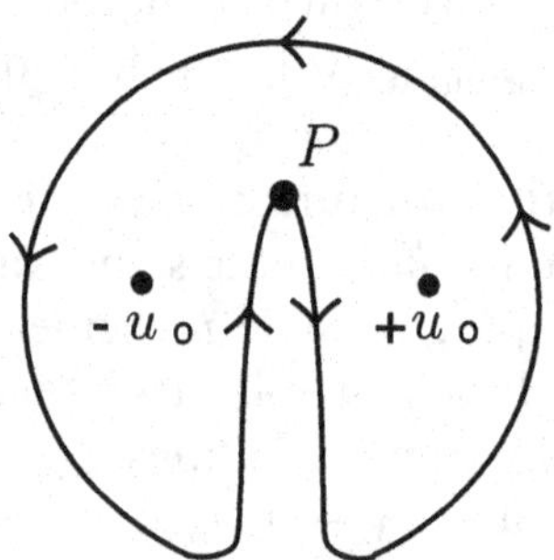

Figure 29.1. Deformation of a contour that circles the complex u plane counterclockwise at large $|u|$ into a contour that starts at a base point P, circles the singularity at $u = -u_0$ counterclockwise back to P, and then circles the singularity at $u = +u_0$ counterclockwise back to P.

matrix in Eq. (29.5.56):

$$M_+ M_- = M_\infty \equiv \begin{pmatrix} -1 & 0 \\ 2 & -1 \end{pmatrix} . \tag{29.5.57}$$

Singularities occur when a and a_D take values at which some particle has zero mass. For instance, the singularity at $a = 0$ in the perturbative formula (29.5.52) for $h(a)$ occurs because this is the value of a at which elementary charged particles become massless in perturbation theory. We have already ruled out the possibility of a single singularity at $a = 0$, so the singularities at $\pm u_0$ must arise from the vanishing of the mass of some other particles.

The most striking part of the Seiberg–Witten calculation was their realization that these particles are among the non-elementary magnetic monopoles or dyons found in the underlying $SU(2)$ supersymmetric theory. Semi-classical calculations of the sort done in Section 23.3 show that the stable monopoles and dyons have magnetic quantum number $m = \pm 1$ and any integer electric quantum number n, and belong to hypermultiplets, each consisting of a pair of Majorana spinors and a pair of complex scalars. These are 'short' multiplets, with a mass given by the BPS value, which according to Eqs. (27.9.24) and (29.5.47) is

$$M = |Z_{12}|/2 = \sqrt{2}\left| Na + h(a) \right| , \tag{29.5.58}$$

where $N \equiv \pm n$. The easiest way to calculate what happens when this mass goes to zero is to consider the more familiar problem of what happens when the mass of a hypermultiplet of ordinary charged particles goes to zero, and then use duality to switch to the case of a light monopole. The beta function for the $U(1)$ gauge coupling is given here by Eq. (27.9.45) with $C_1 = 0$, $C_2^f = C_2^s = 2$, so

$$\beta(e) = +\frac{e^3}{8\pi^2}\,.$$

The solution of the renormalization group equation is then

$$e^{-2}(a) = -\frac{1}{4\pi^2}\ln\left(\frac{a}{\text{constant}}\right)\,.$$

As we saw in deriving Eq. (29.5.52), this gives

$$h'(a) = 4\pi\, i\, e^{-2}(a) = -\frac{i}{\pi}\ln\left(\frac{a}{\text{constant}}\right)\,.$$

Because this gives a value of $e^2(a)$ that is positive and small for $a \to 0$, this *would* be a reliable result if the theory really contained a hypermultiplet of ordinary charged particles whose mass goes to zero in this limit. Instead, we are assuming that there is a hypermultiplet of monopoles or dyons whose mass goes to zero; to deal with this case, we can apply the duality transformation (29.5.43) that takes a into $Na + h(a)$

$$\begin{pmatrix} a \\ h(a) \end{pmatrix} \to \begin{pmatrix} \hat{a} \\ \hat{h}(\hat{a}) \end{pmatrix} = \begin{pmatrix} N & +1 \\ -1 & 0 \end{pmatrix}\begin{pmatrix} a \\ h(a) \end{pmatrix}\,. \tag{29.5.59}$$

We conclude that when u approaches a point u_0 where $\hat{a} = Na + h(a) \to 0$, we have

$$\frac{d\hat{h}(\hat{a})}{d\hat{a}} \to -\frac{i}{\pi}\ln\left(\frac{\hat{a}}{\text{constant}}\right)\,, \tag{29.5.60}$$

or, in other words,

$$\frac{da}{d\left(h(a) + Na\right)} \to +\frac{i}{\pi}\ln\left(\frac{h(a) + Na}{\text{constant}}\right)\,. \tag{29.5.61}$$

The solution is

$$a(u) = a_0 + \frac{i}{\pi}\Big(h(u) + Na(u)\Big)\ln\left(\frac{h(u) + Na(u)}{\Lambda_0}\right)\,, \tag{29.5.62}$$

where a_0 and Λ_0 are integration constants. We are also assuming that $h + Na \to 0$ for $u \to u_0$, so we can write the leading term as

$$h(u) + Na(u) \to c_0(u - u_0)\,. \tag{29.5.63}$$

Thus Eq. (29.5.62) has the leading term

$$a(u) \to a_0 + \frac{i\, c_0}{\pi}(u - u_0)\ln\left(\frac{c_0(u - u_0)}{\Lambda_0}\right)\,. \tag{29.5.64}$$

When we take u counterclockwise all the way around a circle surrounding u_0 there is no change in $h(a) + Na$, but a is shifted by $-2(h(a) + Na)$, so the monodromy matrix for this singularity is

$$M_+ = \begin{pmatrix} 1-2N & -2 \\ 2N^2 & 1+2N \end{pmatrix} . \tag{29.5.65}$$

Likewise, if the singularity at $-u_0$ is associated with the vanishing mass of a monopole or dyon with magnetic quantum number $\pm' 1$ (the prime distinguishing this sign from that for u_0) and electric quantum number n', then at this singularity $h(u) + N'a(u) \to 0$, and

$$a \to a'_0 + \frac{i}{\pi}\Big(h(u) + N'a(u)\Big) \ln\left(\frac{h(u) + N'a(u)}{\Lambda'_0}\right) , \tag{29.5.66}$$

where $N' \equiv \pm' n'$ and a'_0 and Λ'_0 are new integration constants. The leading terms are

$$h(u) + N'a(u) \to c'_0(u + u_0) , \tag{29.5.67}$$

$$a \to a'_0 + \frac{i\,c'_0}{\pi}(u + u_0) \ln\left(\frac{c'_0(u + u_0)}{\Lambda'_0}\right) . \tag{29.5.68}$$

The monodromy matrix for this singularity is

$$M_- = \begin{pmatrix} 1-2N' & -2 \\ 2N'^2 & 1+2N' \end{pmatrix} . \tag{29.5.69}$$

It is straightforward then to see that the condition (29.5.57) on these matrices is satisfied if and only if

$$N' = N - 1 . \tag{29.5.70}$$

It makes no difference what value we take for N, because we can shift it by an even integer $2M$ by going around the circle at infinity M times, and we can shift it by unity by reflecting $u \leftrightarrow -u$. Seiberg and Witten chose to take $N = 0$, so that $N' = -1$. Then for $u \to u_0$

$$h(u) \to c_0(u - u_0) , \tag{29.5.71}$$

$$a(u) \to a_0 + \frac{i\,c_0}{\pi}(u - u_0) \ln\left(\frac{u - u_0}{\Lambda_0}\right) , \tag{29.5.72}$$

and for $u \to -u_0$

$$h(u) - a(u) \to c'_0(u + u_0) , \tag{29.5.73}$$

$$a(u) \to a'_0 + \frac{i\,c'_0}{\pi}(u + u_0) \ln\left(\frac{u + u_0}{\Lambda'_0}\right) . \tag{29.5.74}$$

We now impose the unbroken $\mathbb{Z}_8$-symmetry condition (29.5.54). We can calculate $a(u)$ for $u \to -u_0$ by writing it as $-ia(-u)$ and using Eq. (29.5.72).

Then for $u \to -u_0$

$$a(u) \to -ia_0 + \frac{c_0}{\pi} \ln\left(\frac{-c_0(u+u_0)}{\Lambda_0}\right) . \tag{29.5.75}$$

Comparing this with Eq. (29.5.74), we see that $c'_0 = i\,c_0$, so Eq. (29.5.73) becomes, for $u \to -u_0$,

$$h(u) - a(u) \to i\,c_0\,(u+u_0) . \tag{29.5.76}$$

The field a is defined to include a factor of the gauge coupling e, which can be given any value by an appropriate choice of the renormalization point at which e is evaluated. Seiberg and Witten chose to define the scale of a and u (keeping $u = a^2/2$ at infinity) so that $u_0 = 1$; that is, the singularities are at $u = \pm 1$. With this convention, they obtained the solutions (defined in the complex plane cut from -1 to $+1$)

$$a_{\rm SW}(u) = \frac{\sqrt{2}}{\pi}\int_{-1}^{1} dx\,\sqrt{\frac{u-x}{1-x^2}}\,, \tag{29.5.77}$$

$$h_{\rm SW}(u) = \frac{i\sqrt{2}}{\pi}\int_{1}^{u} dx\,\sqrt{\frac{u-x}{x^2-1}}\,. \tag{29.5.78}$$

The mathematical methods originally used to obtain these results are beyond the scope of this book, but fortunately it is not difficult after the fact to check that they are correct.

First, we can check that $a_{\rm SW}(u)$ and $h_{\rm SW}(u)$ have a singular behavior near $u = \pm 1$ of the same form (29.5.71), (29.5.72), (29.5.75), and (29.5.76) as the true solution. For $u \to 1$, Eq. (29.5.77) gives

$$\begin{aligned} a_{\rm SW}(u) &\to \frac{\sqrt{2}}{\pi}\int_{-1}^{1}\frac{dx}{\sqrt{x+1}} + \frac{u-1}{2\pi}\int_{-1}^{1}\frac{dx}{\sqrt{(1-x)(u-x)}} \\ &= \frac{4}{\sqrt{\pi}} + \frac{u-1}{2\pi}\ln\left(\frac{u-1}{3+u-2\sqrt{2}\sqrt{1+u}}\right) \\ &\to \frac{4}{\sqrt{\pi}} - \frac{u-1}{2\pi}\ln\Big(4(u-1)\Big) . \end{aligned} \tag{29.5.79}$$

Also, Eq. (29.5.78) gives

$$h_{\rm SW}(u) \to \frac{i}{\pi}\int_{1}^{u} dx\,\sqrt{\frac{u-x}{x-1}} = \frac{i\,(u-1)}{2} . \tag{29.5.80}$$

Eqs. (29.5.79) and (29.5.80) agree with our previous results (29.5.71) and (29.5.72), with $c_0 = i$ and $a_0 = 4/\sqrt{\pi}$. Eq. (29.5.77) satisfies the $\mathbb{Z}_8$ reflection property (29.5.54), so $a_{\rm SW}(u)$ automatically has the behavior (29.5.75) for $u \to -1$, and (with due regard to the signs of square roots dictated by the cut from $u = -1$ to $u = +1$) it follows directly from

Eqs. (29.5.77) and (29.5.78) that $h_{SW}(u) - a_{SW}(u)$ is proportional to $u + 1$ for u near -1.

Now since $a_{SW}(u)$ and $h_{SW}(u)$ have the same singularity structure at $u \to \pm 1$ as $a(u)$ and $h(u)$, they have the same monodromy (Eqs. (29.5.59) and (29.5.69) with $N = 0$ and $N' = -1$): as u circles the point $+1$, we have

$$\begin{pmatrix} a_{SW} \\ h_{SW} \end{pmatrix} \to \begin{pmatrix} 0 & +1 \\ -1 & 0 \end{pmatrix} \begin{pmatrix} a_{SW} \\ h_{SW} \end{pmatrix} , \tag{29.5.81}$$

while as u circles the point -1,

$$\begin{pmatrix} a_{SW} \\ h_{SW} \end{pmatrix} \to \begin{pmatrix} 3 & -2 \\ 2 & -1 \end{pmatrix} \begin{pmatrix} a_{SW} \\ h_{SW} \end{pmatrix} . \tag{29.5.82}$$

Let us now consider the quantity

$$f(u) \equiv a(u)h_{SW}(u) - a_{SW}(u)h(u) . \tag{29.5.83}$$

This is an $SL(2, \mathbb{Z})$ invariant, so it has trivial monodromy: it returns to the same value when u circles either $+1$ or -1. The only finite singularities in $a(u)$, $h(u)$, $a_{SW}(u)$, or $h_{SW}(u)$ are the logarithmic singularities at ± 1, but since $f(u)$ has trivial monodromy it does not have these singularities, and it is therefore analytic at all finite points.

To evaluate the entire function $f(u)$, let's first check that the leading terms in the asymptotic behavior of the functions $a_{SW}(u)$ and $h_{SW}(u)$ are the same as those of $a(u)$ and $h(u)$, respectively. For $u \to \infty$, Eq. (29.5.77) gives

$$a_{SW}(u) \to \frac{\sqrt{2u}}{\pi} \int_{-1}^{1} \frac{dx}{\sqrt{1 - x^2}} = \sqrt{2u} , \tag{29.5.84}$$

while Eq. (29.5.78) gives

$$h_{SW}(u) \to \frac{i\sqrt{2u}}{\pi} \int_{1}^{u} \frac{dx}{\sqrt{x^2 - 1}} \to \frac{i\sqrt{2}}{\pi} \sqrt{u} \ln u . \tag{29.5.85}$$

This is the same as the leading behavior (29.5.55) of $a(u)$ and $h(u)$.

This in itself only shows that $f(u)/u \ln u \to 0$ for $u \to \infty$. But note that the reflection symmetry (29.5.54) for $a(u)$ and its counterpart for $a_{SW}(u)$ tell us that the next-to-leading terms in both functions are of order $\sqrt{u}/u^2$. Also, the next-to-leading terms in $h(u)$ and $h_{SW}(u)$ are of order $\sqrt{u}$. It follows that we can calculate the leading term in the asymptotic behavior of $f(u)$ by setting $a(u) = a_{SW}(u) = \sqrt{2u}$, so that $f(u) = O(u)$ for $u \to \infty$. Since $f(u)$ is an entire function, this means that $f(u)$ is linear in u. But $f(u)$ vanishes at $u = +1$, where $h(u)$ and $h_{SW}(u)$ are $O(u-1)$, and $f(u)$ also vanishes at $u = -1$, where $a(u) - h(u)$ and $a_{SW}(u) - h_{SW}(u)$ are $O(u - 1)$, so it must vanish everywhere. We conclude then that

$$\frac{a_{\rm SW}(u)}{a(u)} = \frac{h_{\rm SW}(u)}{h(u)} \equiv g(u) \,. \tag{29.5.86}$$

Now we must consider the properties of the function $g(u)$. Because $h_{\rm SW}(u)$ and $h(u)$ are analytic at all finite points except $u = -1$, and $a_{\rm SW}(u) - h_{\rm SW}(u)$ and $a(u) - h(u)$ are analytic at all finite points except $u = +1$, it follows that $g(u)$ (which can also be written as $[a_{\rm SW}(u) - h_{\rm SW}(u)]/[a(u) - h(u)]$) is analytic everywhere. (There are no zeroes of $a(u)$ or $h(u)$ at $u \neq \pm 1$, let alone zeroes of both, because a zero of $a(u)$ or $h(u)$ would be associated with a point where a charged particle or monopole mass vanishes, which has been assumed to be not the case for any $u \neq \pm 1$.) Also, the fact that the leading terms in the asymptotic behavior for $u \to \infty$ of the functions $a_{\rm SW}(u)$ and $h_{\rm SW}(u)$ are the same as those of $a(u)$ and $h(u)$, respectively, means that $g(u) \to 1$ for $u \to \infty$. It follows then that the entire function $g(u)$ must equal unity for all u, and therefore

$$a(u) = a_{\rm SW}(u) \,, \qquad h(u) = h_{\rm SW}(u) \,, \tag{29.5.87}$$

as was to be shown.

Problems

1. What is the Witten index of the model in Problem 4 of Chapter 26 for $a \neq 0$? For $a = 0$? Can supersymmetry be broken by higher-order effects in this model for $a = 0$? Explain.

2. Consider the renormalizable supersymmetric theory with $SO(N_c)$ gauge symmetry and N_f left-chiral scalar superfields Φ_n in the N-vector representation. What can you say about the structure of the non-perturbative Wilsonian Lagrangian density when the bare superpotential vanishes? What if there is a bare superpotential $\sum_n \Phi_n \Phi_n$?

3. In the theory with Lagrangian density (29.5.22), what is the relation between the component fields of the spinor field strength superfield W_α and its dual $\tilde{W}_\alpha$?

References

1. E. Witten, *Nucl. Phys.* **B185**, 513 (1981). This article is reprinted in *Supersymmetry*, S. Ferrara, ed. (North Holland/World Scientific, Amsterdam/Singapore, 1987).

1*a*. J. Hughes and J. Polchinski, *Nucl. Phys.* **B278**, 147 (1986); J. Hughes, J. Liu, and J. Polchinski, *Phys. Lett.* **B 180**, 370 (1986); J. Bagger and A. Galperin, *Phys. Lett.* **B336**, 25 (1994); *Phys. Rev.* **D55**, 1091 (1997); *Phys. Lett.* **B412**, 296 (1997); L. Antoniadis, H. Partouche, and T. R. Taylor, *Phys. Lett.* **B372**, 83 (1996); S. Ferrara, L. Girardello, and M. Porrati, *Phys. Lett.* **B376**, 275 (1996).

2. E. Witten, *Nucl. Phys.* **B202**, 253 (1982). This article is reprinted in *Supersymmetry*, Ref. 1.

3. A. C. Davis, M. Dine, and N. Seiberg, *Phys. Lett.* **125B**, 487 (1983); I. Affleck, M. Dine, and N. Seiberg, *Phys. Rev. Lett.* **51**, 1026 (1983); *Phys. Lett.* **137B**, 187 (1984), reprinted in *Supersymmetry*, Ref. 1; *Nucl. Phys.* **B241**, 493 (1984); *Nucl. Phys.* **B52**, 1677 (1984); *Phys. Lett.* **140B**, 59 (1984).

4. N. Seiberg, *Phys. Lett.* **B318**, 469 (1993).

5. V. A. Novikov, M. A. Shifman, A. I. Vainshtein, and V. I. Zakharov, *Nucl. Phys.* **B229**, 381 (1983); *Nucl. Phys.* **B260**, 157 (1985); G. C. Rossi and G. Veneziano, *Phys. Lett.* **138B**, 195 (1984), reprinted in *Supersymmetry*, Ref. 1; S. F. Cordes, *Nucl. Phys.* **B273**, 629 (1986). For a review, see M. A. Shifman and A. I. Vainshtein, hep-th/9902018, to be published.

6. E. Witten, *J. High Energy Phys.* **9802**, 006 (1998).

7. V. G. Kac and A. V. Smilga, hep-th/9902029, to be published.

8. I. Affleck, M. Dine, and N. Seiberg, *Nucl. Phys.* **B256**, 557 (1985).

9. N. Seiberg, *Phys. Rev.* **D49**, 6857 (1994).

10. N. Seiberg and E. Witten, *Nucl. Phys.* **B426**, 19 (1994); *erratum*, *Nucl. Phys.* **B430**, 485 (1994). The extension to an $N = 2$ theory with matter hypermultiplets is given by N. Seiberg and E. Witten, *Nucl. Phys.* **B431**, 484 (1994). For a review, see: K. Intrilligator and N. Seiberg, *Nucl. Phys. Proc. Suppl.* **45BC**, 1 (1996); also published in *Summer School in High Energy Physics and Cosmology, Trieste, 1995*, E. Gava, ed. (World Scientific, Singapore, 1997), and in *QCD and Beyond: Proceedings of the Theoretical Advanced Study Institute in Elementary Particle Physics, Boulder Colorado, 1995*, D. E. Soper, ed. (World Scientific, Singapore, 1996).

11. E. Witten, *Phys. Lett.* **86B**, 283 (1979).

12. N. Seiberg, *Phys. Lett.* **B206**, 75 (1988).

30
Supergraphs

The introduction of Feynman graphs in the late 1940s provided the enormous advantage of preserving manifest Lorentz invariance at every stage of perturbative calculations. For this purpose it was necessary that the exchange of all spin states of any virtual particle should be described by a single propagator. Fortunately, it has become possible to go a step further, and develop a supergraph formalism, in which supersymmetry as well as Lorentz invariance is kept manifest at every stage.[1] To do so, it is necessary to describe the exchange of all the particles described by a given superfield by a single superpropagator.

There is a problem here in that the left-chiral superfields Φ over which we integrate are subject to a differential constraint, $\mathscr{D}_R\Phi = 0$. This is analogous to the problem in electrodynamics that the field-strength tensor is subject to a differential constraint, the homogeneous Maxwell equation. In electrodynamics this problem is dealt with by expressing the field-strength tensor in terms of a vector potential, and doing path integrals over the vector potential instead of the field strength tensor. In much the same way, here we will impose the left-chiral constraint by expressing left-chiral superfields in terms of potential superfields, over which we integrate in path integrals. In this formalism we encounter problems like those produced by gauge invariance in electrodynamics, and deal with them in much the same way.

The most important result to come out of the supergraph formalism was the non-renormalization theorem for the superpotential.[2] We have already proved this theorem in Section 27.6, using a much easier indirect technique developed by Seiberg, which can be extended also to describe non-perturbative effects. Nevertheless, it is interesting to see how these surprising cancellations of renormalization effects occur in actual perturbative calculations.

30.1 Potential Superfields

Consider a theory of left-chiral superfields $\Phi_n(x,\theta)$ and their complex conjugates, but for simplicity without gauge superfields. All of the vacuum expectation values of time-ordered products of component fields may be calculated from vacuum expectation values of the time-ordered products of these superfields. We might try to calculate these from the path-integral formula

$$\left\langle T\left\{\Phi_{n_1}(x_1,\theta_1),\, \Phi_{n_2}(x_2,\theta_2),\, \cdots\right\}\right\rangle = \int \left[\prod_{n,x,\theta} d\Phi_n(x,\theta)\right] \exp\Big(iI[\Phi]\Big) \\ \times\, \Phi_{n_1}(x_1,\theta_1)\, \Phi_{n_2}(x_2,\theta_2)\cdots\,, \tag{30.1.1}$$

where $I[\Phi]$ is the action

$$I[\Phi] = \frac{1}{2}\int d^4x\Big[\sum_n \Phi_n^*(x,\theta)\Phi_n(x,\theta)\Big]_D + 2\,\mathrm{Re}\int d^4x\Big[f(\Phi)\Big]_{\mathscr{F}}\,. \tag{30.1.2}$$

(The factor 1/2 is introduced in the first term, as in Eq. (26.4.3), to give the component fields of Φ a conventional normalization.) But we *cannot* simply read off the supergraph Feynman rules from Eq. (30.1.1), because the functional integral over the superfields Φ_n must be constrained to satisfy the left-chiral condition $\mathscr{D}_R\Phi_n = 0$.

This is analogous to a similar problem in electrodynamics. As discussed in Section 12.3, at energies below the electron mass the interactions of soft photons with each other are described by an effective action of the form:

$$I[f] = -\frac{1}{4}\int d^4x\left[f_{\mu\nu}f^{\mu\nu} + c_1\Big(f_{\mu\nu}f^{\mu\nu}\Big)^2 + c_2\Big(\epsilon_{\mu\nu\rho\sigma}f^{\mu\nu}f^{\rho\sigma}\Big)^2\right].$$

But we cannot read off the Feynman rules from this formula without taking into account the fact that the path integral is constrained by the homogeneous Maxwell equations

$$\partial_\mu f_{\nu\rho} + \partial_\nu f_{\rho\mu} + \partial_\rho f_{\mu\nu} = 0\,.$$

As everyone knows, we deal with this constraint by introducing a four-vector potential A_μ, with $f_{\mu\nu} = \partial_\mu A_\nu - \partial_\nu A_\mu$, so that the constraint is automatically satisfied, and integrate over $A_\mu(x)$, not $f_{\mu\nu}(x)$.

In the same way, we may adopt a trick already used in Section 26.6 to derive the field equations for superfields, and introduce a non-chiral potential superfield $S_n(x,\theta)$, with

$$\Phi_n = \mathscr{D}_R^2 S_n\,, \tag{30.1.3}$$

where

$$\mathscr{D}_R^2 \equiv \sum_{\alpha\beta} \epsilon_{\alpha\beta} \mathscr{D}_{R\alpha} \mathscr{D}_{R\beta} \,, \tag{30.1.4}$$

so that Φ_n automatically satisfies the left-chiral constraint, $\mathscr{D}_{R\alpha}\Phi_n = 0$. In place of Eq. (30.1.1), we then have the path-integral formula

$$\left\langle T\Big\{\Phi_{n_1}(x_1,\theta_1),\, \Phi_{n_2}(x_2,\theta_2),\, \cdots\Big\}\right\rangle = \int \left[\prod_{n,x,\theta} dS_n(x,\theta)\right] \exp\Big(iI\,[\mathscr{D}_R^2 S]\Big)$$
$$\times\, \mathscr{D}_R^2 S_{n_1}(x_1,\theta_1)\, \mathscr{D}_R^2 S_{n_2}(x_2,\theta_2) \cdots . \tag{30.1.5}$$

In expressing the action (30.1.2) in terms of S_n, we may recall that the D-term of a superderivative does not contribute to the action, so we may shift the operator $(\mathscr{D}_R^2)^* = \mathscr{D}_L^2$ acting on S_n^* in the first term of Eq. (30.1.2) to act instead on S_n, which gives

$$I\,[\mathscr{D}_R^2 S] = \frac{1}{2}\int d^4x \left[\sum_n S_n^* \mathscr{D}_L^2 \mathscr{D}_R^2 S_n\right]_D + 2\,\mathrm{Re}\int d^4x \left[f(\mathscr{D}_R^2 S)\right]_{\mathscr{F}} . \tag{30.1.6}$$

The $\mathscr{D}_R^2$ operator in any one of the factors $\mathscr{D}_R^2 S$ in any term in $f(\mathscr{D}_R^2 S)$ may be taken to act on the whole term, since $\mathscr{D}_R$ gives zero when acting on any other factor of $\mathscr{D}_R^2 S$ in that term. In this way, we may write

$$f(\mathscr{D}_R^2 S) = \mathscr{D}_R^2\, \tilde{f}(S)\,, \tag{30.1.7}$$

where $\tilde{f}(S)$ is obtained from $f(\mathscr{D}_R^2 S)$ by omitting any *one* operator $\mathscr{D}_R^2$ in each term. For instance, for a single type of superfield, if $f(\Phi) = \sum_r c_r \Phi^r$, then

$$\tilde{f}(S) = \sum_r c_r S\, (\mathscr{D}_R^2 S)^{r-1} .$$

Using Eqs. (30.1.6), (30.1.7), and (26.3.31), we may put the whole action in the form of a D-term

$$I\,[\mathscr{D}_R^2 S] = \frac{1}{2}\int d^4x \sum_n \left[S_n^* \mathscr{D}_L^2 \mathscr{D}_R^2 S_n\right]_D + 2\,\mathrm{Re}\int d^4x \left[\tilde{f}(S)\right]_D . \tag{30.1.8}$$

Eq. (26.6.5) shows that this may also be written as a superspace integral:

$$I\,[\mathscr{D}_R^2 S] = -\frac{1}{4}\int d^4x \int d^4\theta \sum_n S_n^* \mathscr{D}_L^2 \mathscr{D}_R^2 S_n - \mathrm{Re}\int d^4x \int d^4\theta\, \tilde{f}(S) . \tag{30.1.9}$$

This is the action we will use in the path-integral derivation of the supergraph formalism.

30.2 Superpropagators

Usually the propagator can be obtained directly from the part of the action of second order in the fields. If we write this quadratic part for a complex field ϕ_i (where i is a compound index including spacetime coordinates as well as spin and species labels) in the form

$$I_{\text{quad}}[\phi] = -\sum_{ij} D_{ij}\phi_i^*\phi_j\,, \tag{30.2.1}$$

with D_{ij} Hermitian, then as explained in Section 9.4 the propagator is simply $\Delta = D^{-1}$. The problem with this arises when the quadratic part of the action is invariant under a (linearized) gauge transformation

$$\phi_i \to \phi_i + \xi_i\,, \tag{30.2.2}$$

for some class of 'vectors' ξ. In this case we have

$$\sum_i D_{ij}\xi_j = 0\,, \tag{30.2.3}$$

and we clearly cannot invert the 'matrix' D_{ij}. In electrodynamics the problem arises because the Lagrangian density is invariant under a gauge transformation $A_\mu \to A_\mu + \partial_\mu\Lambda$. We have this problem here too: since the action is actually a functional of $\mathscr{D}_R^2 S_n$ and not of S_n itself, it is invariant under the transformation

$$S_n \to S_n + \mathscr{D}_R X_n\,, \tag{30.2.4}$$

for any superfields X_n.

In the electrodynamics of charged particles the problems introduced by gauge invariance are typically handled by choosing a gauge, for instance by the Faddeev–Popov–de Witt method described in Section 15.5. But the problems raised here by invariance under the transformation (30.2.2) are much more like the problems of the effective field theory of photons at energies below the threshold for producing charged particles, where the theory is gauge-invariant simply because the action only involves gauge-invariant fields. In such theories there is a simpler option. In addition to the eigenvector ξ_i of D_{ij} with eigenvalue zero (taken for simplicity to be in a unique direction) we can find a set of orthonormal eigenvectors $u_{\nu i}$ with eigenvalues $d_\nu \neq 0$:

$$\sum_j D_{ij}\,u_{\nu j} = d_\nu\,u_{\nu i}\,, \qquad \sum_i u_{\nu i}^*\,u_{\nu' i} = \delta_{\nu\nu'}\,, \qquad \sum_i u_{\nu i}^*\,\xi_i = 0\,. \tag{30.2.5}$$

We can introduce a new set of integration variables ϕ' and ϕ'_ν:

$$\phi_i = \phi'\,\xi_i + \sum_\nu \phi'_\nu\,u_{\nu i}\,. \tag{30.2.6}$$

The sort of integral encountered in Feynman diagram calculations of quantum expectation values may then be written

$$\int \left[\prod_i d\phi_i\, d\phi_i^*\right] \exp\{iI_{\rm quad}[\phi]\}\phi_a \cdots \phi_b^* \cdots = \mathscr{J} \int d\phi'\, d\phi'^*$$
$$\times \int \left[\prod_\nu d\phi'_\nu\, d\phi'^*_\nu\right] \exp\left\{-i\sum_\nu d_\nu |\phi'_\nu|^2\right\} \left[\phi'\xi_a + \sum_\nu \phi'_\nu u_{\nu a}\right] \cdots$$
$$\times \left[\phi'\xi_b + \sum_\nu \phi'_\nu u_{\nu b}\right]^* \cdots , \tag{30.2.7}$$

where $\mathscr{J}$ is the Jacobian of the transformation (30.2.6). The integral over ϕ' and ϕ'^* is of course ill defined, because they do not appear in the argument of the exponential. But this doesn't matter if the action only involves gauge-invariant fields, because then ϕ_a, ϕ_b, etc. will be contracted with 'currents' J_a, J_b, etc. for which

$$\sum_a \xi_a J_a = 0\,. \tag{30.2.8}$$

We may thus write Eq. (30.2.7) as

$$\int \left[\prod_i d\phi_i\, d\phi_i^*\right] \exp\{iI_{\rm quad}[\phi]\}\phi_a \cdots \phi_b^* \cdots = \mathscr{C} \int \left[\prod_\nu d\phi'_\nu d\phi'^*_\nu\right]$$
$$\times \exp\left\{-i\sum_\nu d_\nu |\phi'_\nu|^2\right\} \left[\sum_\nu \phi'_\nu u_{\nu a}\right] \cdots \left[\sum_\nu \phi'_\nu u_{\nu b}\right]^* \cdots$$
$$+\,\xi\text{-terms}\,, \tag{30.2.9}$$

where 'ξ-terms' denotes terms proportional to one or more factors of ξ_a, ξ_b, etc., which vanish when contracted with Js satisfying Eq. (30.2.8), and $\mathscr{C}$ is the infinite constant $\mathscr{J} \int d\phi'$. The integral over the ϕ'_ν then gives

$$\int \left[\prod_i d\phi_i\, d\phi_i^*\right] \exp\{iI_{\rm quad}[\phi]\}\phi_a \cdots \phi_b^* \cdots \propto \sum_{\rm pairings} \left[-i\Delta_{ab}\right] \cdots$$
$$+\,\xi\text{-terms}\,, \tag{30.2.10}$$

where the sum over pairings is over all ways of pairing indices on the ϕs with indices on the ϕ^*s, and Δ_{ab} is the propagator

$$\Delta_{ab} = \sum_\nu \frac{u_{\nu a}\, u_{\nu b}^*}{d_\nu}\,. \tag{30.2.11}$$

Instead of actually evaluating the sum (30.2.11), we can use its defining property that

$$\sum_c D_{ac}\Delta_{cb} = \sum_\nu u_{\nu a}\, u_{\nu b}^* \equiv \Pi_{ab}\,, \tag{30.2.12}$$

with Π the projection operator on the space orthogonal to ξ:

$$\Pi^2 = \Pi\,, \qquad \Pi\xi = 0\,. \tag{30.2.13}$$

The solution of Eq. (30.2.12) is unique only up to ξ-terms, but these do not matter if the fields ϕ_i appear in the action only in gauge-invariant combinations.

For instance, in electrodynamics, we can write the kinematic part of the action as

$$I_{\rm quad}[A] = -\frac{1}{4}\int d^4x\, f_{\mu\nu}f^{\mu\nu} = +\frac{1}{2}\int d^4x\, A^\mu\left(\Box\delta^\nu_\mu - \partial_\mu\partial^\nu\right)A_\nu\,.$$

The differential operator $-\Box\delta^\nu_\mu + \partial^\nu\partial_\mu$ is not invertible, because it has a zero eigenvalue with eigenvectors of the form $\xi^\mu = \partial^\mu\Lambda$. The projection matrix onto the space orthogonal to these vectors is

$$\Pi_\mu{}^\nu(x,y) = [\delta_\mu{}^\nu - \partial_\mu\partial^\nu\Box^{-1}]\delta^4(x-y)\,,$$

where $\Box^{-1}\delta^4(x-y)$ is any solution of the equation $\Box[\Box^{-1}\delta^4(x-y)] = \delta^4(x-y)$. The defining equation for the propagator is

$$\left(-\Box\delta^\lambda_\mu + \partial_\mu\partial^\lambda\right)\Delta_\lambda{}^\nu(x,y) = \Pi_\mu{}^\nu(x,y)\,,$$

with the solution

$$\Delta_\lambda{}^\nu(x,y) = \delta^\nu_\mu\,\Delta_F(x-y) + \partial_\mu\partial^\nu\text{-terms}\,,$$

where $\Delta_F(x-y)$ is the usual Feynman propagator (6.2.16), satisfying $\Box\Delta_F(x-y) = -\delta^4(x-y)$. (The factor 1/2 in the action does not appear in the defining equation for the propagator here because A_μ is a real field. The origin of the $-i\epsilon$ in the denominator of the Fourier integral in Eq. (6.2.16) is explained in the path-integral formalism in Section 9.2.)

Inspection of the first term in Eq. (30.1.9) shows that the defining equation for the superpropagator of the potential superfields is

$$\frac{1}{4}\mathscr{D}_L^2\,\mathscr{D}_R^2\,\Delta^S_{nm}(x,\theta;x',\theta') = \mathscr{P}\delta^4(x-x')\delta^4(\theta-\theta')\delta_{nm}\,, \tag{30.2.14}$$

where $\mathscr{P}$ is a superspace differential operator satisfying the conditions for a projection operator

$$\mathscr{P}^2 = \mathscr{P}\,, \qquad \mathscr{P}\mathscr{D}_R = 0\,, \tag{30.2.15}$$

and $\delta^4(\theta-\theta')$ is the fermionic delta function introduced in Eq. (26.6.8). The solution is

$$\mathscr{P} = \frac{-1}{16\Box}\mathscr{D}_L^2\,\mathscr{D}_R^2\,. \tag{30.2.16}$$

(It is obvious that $\mathscr{P}\mathscr{D}_R = 0$. To check that $\mathscr{P}^2 = \mathscr{P}$, we need to use (26.6.12), which shows that $\mathscr{D}_R^2\mathscr{D}_L^2\mathscr{D}_R^2 = -16\Box\mathscr{D}_R^2$.) The solution of

Eq. (30.2.14) is

$$\Delta^S_{nm}(x,\theta;x',\theta') = -\frac{1}{4\Box}\delta^4(x-x')\delta^4(\theta-\theta')\delta_{nm}$$
$$= \frac{1}{4}\Delta_F(x-x')\delta^4(\theta-\theta')\delta_{nm} + \mathscr{D}_R\text{-terms}\,. \tag{30.2.17}$$

This is the superpropagator for a line created by a potential superfield $S^*_m(x',\theta')$ and destroyed by a potential superfield $S_n(x,\theta)$, which we must use in evaluating supergraphs for the action (30.1.9). To make contact with ordinary propagators, it is also of some interest to consider the superpropagator of the line created by a left-chiral superfield $\Phi^*_m(x',\theta')$ and destroyed by a left-chiral superfield $\Phi_n(x,\theta)$. These chiral superfields are obtained by operating on $S_n(x,\theta)$ with $\mathscr{D}^2_R$ and on $S^*_m(x',\theta')$ with $\mathscr{D}'^{2\,*}_R = \mathscr{D}'^2_L$, so the propagator for the left-chiral superfields is

$$\Delta^\Phi_{nm}(x,\theta;x',\theta') = \frac{1}{4}\mathscr{D}^2_R\mathscr{D}'^2_L\Delta_F(x-x')\,\delta^4(\theta-\theta')\,\delta_{nm}\,. \tag{30.2.18}$$

For instance, the term in the superpropagator of zeroth order in θ and θ' is

$$\Big[\Delta^\Phi_{nm}(x,\theta;x',\theta')\Big]_{\theta=\theta'=0} = \frac{1}{4}\left(\frac{\partial}{\partial\theta_R}\right)^2\left(\frac{\partial}{\partial\theta'_L}\right)^2\Delta_F(x-x')\,\delta^4(\theta-\theta')\,\delta_{nm}\,. \tag{30.2.19}$$

To evaluate this, we recall Eq. (26.6.8) for the fermionic delta function:

$$\delta^4(\theta-\theta') = \frac{1}{4}\Big((\theta_L-\theta'_L)^{\rm T}\epsilon(\theta_L-\theta'_L)\Big)\,\Big((\theta_R-\theta'_R)^{\rm T}\epsilon(\theta_R-\theta'_R)\Big)\,,$$

from which we find $(\partial/\partial\theta_R)^2(\partial/\partial\theta'_L)^2\delta^4(\theta-\theta') = 4$. Eq. (30.2.19) therefore yields

$$\Big[\Delta^\Phi_{nm}(x,\theta;x',\theta')\Big]_{\theta=\theta'=0} = \Delta_F(x-x')\,\delta_{nm}\,, \tag{30.2.20}$$

which is just the usual propagator for the scalar components of the superfield.

30.3 Calculations with Supergraphs

We now consider how to use the results of the previous sections to calculate the quantum effective action $\Gamma[S,S^*]$ for a set of classical potential superfields $S_n(x,\theta)$ and their adjoints. Following the prescription discussed in Section 16.1, this may be defined in terms of a sum over all connected one-particle-irreducible supergraphs, consisting of vertices to which are attached directed internal and external lines. For each external line of type n

ending or beginning in a vertex labelled x, θ, we include a c-number factor $S_n(x, \theta)$ or $S_n^*(x, \theta)$, respectively (but no propagator). A vertex labelled x, θ with N incoming or N outgoing lines labelled $n_1, n_2, \ldots, n_N$ yields a factor equal to i times the coefficient of the $S_1 S_2 \cdots S_N$ term in the superpotential $\tilde{f}(S)$ or times the complex conjugate of this coefficient, respectively. An internal line of any type coming out of a vertex labelled x, θ and going into a vertex labelled x', θ' yields a propagator factor, given by Eqs. (30.2.10) and (30.2.17) as

$$-\frac{i}{4}\delta^4(\theta - \theta')\,\Delta_F(x - x')\,. \tag{30.3.1}$$

In addition, as shown by Eq. (30.1.7), a superderivative $\mathscr{D}_R^2$ acts on the propagators or external line S-factors of all but one of the internal or external lines coming into any vertex, and a superderivative $\mathscr{D}_L^2$ acts on the propagators or external-line S^*-factors of all but one of the lines coming out of any vertex. The product of these factors must be integrated over all xs and θs; the quantum effective action is the sum of these integrals for all one-particle-irreducible graphs.

By integrating by parts in superspace, the $\mathscr{D}_L^2$ and/or $\mathscr{D}_R^2$ operators accompanying any one propagator (say, one that connects vertices labelled x, θ and x', θ') may be moved to other propagators or external line factors. This leaves the factor contributed by this internal line proportional to a factor $\delta^4(\theta - \theta')$. Integrating over θ' then eliminates this delta function, and lets us replace θ' with θ everywhere else. (In cases where several internal lines connect the same pair of vertices, we need to use the property of fermionic delta functions that $[\delta^4(\theta - \theta')]^2 = 0$.) Continuing in this way, we wind up with a single four-dimensional θ integral, with all $\mathscr{D}$s acting on the external line factors S_n and S_m^*. That is, though not usually local in spatial coordinates, $\Gamma[S, S^*]$ *is local in the fermionic coordinates.*

The structure of this functional is governed by its invariance under the 'gauge' transformations (30.2.4). This tells us that every one of the external line factors S_n or S_n^* must be acted on by an operator $\mathscr{D}_R^2$ or $\mathscr{D}_L^2$, respectively, with two possible exceptions. The exceptions are that a term with only incoming or only outgoing external lines, in which all but one of the external line factors S_n or S_n^* are acted on by $\mathscr{D}_R^2$ or $\mathscr{D}_L^2$, respectively, and by no other superderivatives, is still invariant under the transformation (30.2.4), even though it cannot be expressed as a four-dimensional θ-integral of a functional only of the Φ_n and/or Φ_n^* alone. This is because the change in such an amplitude under this transformation could only come from the change in the external line factor S_n or S_n^* that is *not* acted on by an operator $\mathscr{D}_R^2$ or $\mathscr{D}_L^2$, and this change is eliminated if we use integration by parts to move one of the other $\mathscr{D}_R^2$ or $\mathscr{D}_L^2$ operators to act on it. As we saw in Section 30.1, such a term in $\Gamma[S, S^*]$ is the $\mathscr{F}$-term of a functional of the Φ_n or of the Φ_n^*, and thus makes a correction to the superpotential or to its complex conjugate. Aside from these exceptions, every term in $\Gamma[S, S^*]$ may be written as a four-dimensional θ-integral of a functional only of the Φ_n and/or Φ_n^*, and therefore makes a correction to the D-term part of the effective action.

Also note that a term in the quantum effective action in which all but one of the external line factors S_n or S_n^* is acted on by $\mathscr{D}_R^2$ or $\mathscr{D}_L^2$, respectively, *and* also by additional $\mathscr{D}_R^2$ or $\mathscr{D}_L^2$ operators (such as in combinations like $\mathscr{D}_R^2\mathscr{D}_L^2\mathscr{D}_R^2 S_n$ or $\mathscr{D}_L^2\mathscr{D}_R^2\mathscr{D}_L^2 S_n^*$) may also be written by integration by parts as a term in which one of the extra $\mathscr{D}_R^2$ or $\mathscr{D}_L^2$ acts on the previously undifferentiated external line S_n or S_n^* factor. Such a term may therefore be expressed as a four-dimensional θ-integral of a functional only of the Φ_n and/or Φ_n^*, and therefore just makes another correction to the D-term part of the action. The only terms in $\Gamma[S, S^*]$ that cannot be written in this way are those that have E incoming external lines and *only* $E - 1$ operators $\mathscr{D}_R^2$ operating on the S_n external line factors, or E^* incoming external lines and *only* $E^* - 1$ operators $\mathscr{D}_L^2$ operating on the S_n^* external line factors. We can therefore tell whether a supergraph can make a contribution to the superpotential or its adjoint by simply counting the number of $\mathscr{D}_R^2$ or $\mathscr{D}_L^2$ operators contributed by the supergraph to the corresponding term in $\Gamma[S, S^*]$.

Let us count these superderivatives. Consider a connected graph with: V_n vertices with n lines coming in; V_n^* vertices with n lines going out; I internal lines; E external incoming lines; and E^* external outgoing lines. These numbers are related by

$$I + E = \sum_n nV_n \,, \qquad I + E^* = \sum_n nV_n^* \,. \tag{30.3.2}$$

The total numbers of $\mathscr{D}_R^2$ and $\mathscr{D}_L^2$ operators are then

$$N_R = \sum_n V_n(n-1) = I + E - V = L + V^* + E - 1 \,, \tag{30.3.3}$$

and

$$N_L = \sum_n V_n^*(n-1) = I + E^* - V^* = L + V + E^* - 1 \,, \tag{30.3.4}$$

where $V = \sum_n V_n$ is the total number of vertices with incoming lines; $V^* = \sum_n V_n^*$ is the total number of vertices with outgoing lines; and $L = I - V - V^* + 1$ is the number of loops. We see that a graph with any loops will have $N_R \geq E$ and $N_L \geq E^*$, so that there are at least enough $\mathscr{D}_R^2$ operators to convert all S_n into $\Phi_n = \mathscr{D}_R^2 S_n$ and its derivatives, and enough $\mathscr{D}_L^2$ operators to convert all S_n^* into $\Phi_n^* = \mathscr{D}_L^2 S_n^*$ and its derivatives. *Any graph with loops thus yields only a contribution to the four-dimensional*

θ-integral — in other words the D-term — of a functional of the left-chiral superfields Φ_n and their adjoints.

The only way to obtain a contribution to an $\mathscr{F}$-term or its adjoint is to have just $N_R = E - 1$ operators $\mathscr{D}_R^2$ or just $N_L = E^* - 1$ operators $\mathscr{D}_L^2$, respectively. According to Eqs. (30.3.3) and (30.3.4), such a graph would have $L = 0$ and $V^* = 0$ or $V = 0$, respectively. In other words, since we are considering only one-particle-irreducible graphs, we can only get an $\mathscr{F}$-term from a graph with a single vertex with incoming lines, and we can only get its adjoint from a graph with a single vertex with outgoing lines. This contribution is nothing but the integrated $\mathscr{F}$-term of the original superpotential, or its adjoint. Thus we see again that *there is no finite or infinite renormalization of $\mathscr{F}$-terms to any order of perturbation theory.*

Problems

1. Use Eq. (30.2.18) to calculate the propagators of the spinor and auxiliary components of a chiral superfield.

2. Consider a supersymmetric theory of a single left-chiral superfield Φ, with Lagrangian density
$$\mathscr{L} = \frac{1}{2}\Big[\Phi^*\Phi\Big]_D + 2\mathrm{Re}\left(g[\Phi^3]_{\mathscr{F}}\right),$$
with g an arbitrary complex constant. Use the supergraph formalism to calculate the one-loop contribution to the quantum effective action. Express the answer as an integral over coordinates and over a single Grassman coordinate θ.

3. What is the superpropagator for a gauge superfield $V(x,\theta)$ in a supersymmetric Abelian gauge theory with kinematic term (27.3.17)?

References

1. A. Salam and J. Strathdee, *Phys. Rev.* **D11**, 1521 (1975); *Nucl. Phys.* **B86**, 142 (1975); D. M. Capper, *Nuovo Cimento* **25A**, 259 (1975); R. Delbourgo, *Nuovo Cimento* **25A**, 646 (1975); D. M. Capper and G. Leibrandt, *Nucl. Phys.* **B85**, 492 (1975); F. Krause, M. Scheunert, J. Honerkamp, and M. Schlindwein, *Phys. Lett.* **53B**, 60 (1974); K. Fujikawa and W. Lang, *Nucl. Phys.* **B88**, 61 (1975); J. Honerkamp, M. Schlindwein, F. Krause, and M. Scheunert, *Nucl. Phys.* **B95**, 397 (1975); S. Ferrara and O. Piguet, *Nucl. Phys.* **B93**, 261 (1975); R.

Delbourgo, *J. Phys.* **G1**, 800 (1975). This formalism was extended to supergravity by W. Siegel, *Phys. Lett.* **84B**, 197 (1979).

2. M. T. Grisaru, W. Siegel, and M. Roček, *Nucl. Phys.* **B159**, 429 (1979).

31

Supergravity

Gravity exists, so if there is any truth to supersymmetry then any realistic supersymmetry theory must eventually be enlarged to a supersymmetric theory of matter and gravitation, known as *supergravity*. Supersymmetry without supergravity is not an option, though it may be a good approximation at energies far below the Planck scale.

There are two leading approaches to the construction of the theory of supergravity. First, supergravity can be presented as a theory of curved superspace.[1] This approach is analogous to the development of supersymmetric gauge theories in Sections 27.1–27.3; the gravitational field appears as a component of a superfield with unphysical as well as physical components, like the unphysical C, M, N, and ω components of the gauge superfield V. The task of deriving the full non-linear supergravity theory in this way is forbiddingly complicated, and so far has not been freed of steps that are apparently arbitrary. At one point or another in the derivation, it has been necessary simply to state that some set of constraints on the graviton superfield are the proper ones to adopt.

Here we will follow a second approach that is less elegant but more transparent.[2] In our discussion here, we begin in Sections 31.1–31.5 with the case where the gravitational field is weak,[3] analyzing supergravity by the same flat-space superfield methods that we used in Chapters 26 and 27 to study ordinary supersymmetry theories. In this way we can identify the physical components of the gravitational superfield (including auxiliary fields analogous to the D-component of the gauge superfield V.) The weak-field approximation will allow us to obtain some of the most important consequences of supergravity theory, including the general formula for the gravitino mass in Section 31.3 and the results for gaugino masses and the A and B parameters given by anomaly-mediated supersymmetry breaking in Section 31.4.

In Section 31.6 we add terms of higher order in G to the supersymmetry transformation rules for the physical fields and to the Lagrangian that describes their interactions, subject to the condition that the supersymmetry transformations and general coordinate transformations form a closed algebra, and that the Lagrangian is invariant under these transformations. This approach is in some respects analogous to the treatment of supersymmetric gauge theories in Section 27.8; we work only with the physical components of the gravitational superfield, and obtain transformation rules that involve covariant rather than ordinary derivatives. This is the approach that has been used in deriving the most important application of supergravity theory beyond the weak-field approximation: the derivation of the low-energy effective Lagrangian in theories of gravitationally mediated supersymmetry breaking. This application is the subject of Section 31.7.

31.1 The Metric Superfield

Supergravity necessarily involves spinor as well as tensor fields, so we will have to describe gravitational fields in terms of a vierbein (or tetrad) $e^a{}_\mu(x)$ rather than a metric, which is related to the vierbein by

$$g_{\mu\nu}(x) = \eta_{ab}\, e^a{}_\mu(x)\, e^b{}_\nu(x)\,. \tag{31.1.1}$$

The indices μ, ν, etc. label general coordinates, while indices a, b, etc. label coordinates in a locally inertial coordinate system, with η_{ab} the usual diagonal matrix with elements $+1, +1, +1, -1$. In the vierbein formalism, the action must be supposed to be invariant under two different sorts of symmetry transformations: general coordinate transformations $x^\mu \to x'^\mu(x)$, under which the vierbein $e^a{}_\mu(x)$ is transformed into $e'^a{}_\mu(x)$, where

$$e'^a{}_\mu(x') = \frac{\partial x^\nu}{\partial x'^\mu} e^a{}_\nu(x)\,, \tag{31.1.2}$$

and local Lorentz transformations, under which

$$e^a{}_\mu(x) \to \Lambda^a{}_b(x)\, e^b{}_\mu(x)\,, \tag{31.1.3}$$

with $\Lambda^a{}_b(x)$ an arbitrary real matrix subject to the constraints

$$\eta_{ab}\Lambda^a{}_c(x)\Lambda^b{}_d(x) = \eta_{cd}\,. \tag{31.1.4}$$

An elementary review of this formalism is given in the appendix to this chapter.

A weak gravitational field is one for which the vierbein is close to the unit matrix. In such a field, the vierbein may conveniently be written as

$$e^a{}_\mu(x) = \delta^a_\mu + 2\kappa\, \phi^a{}_\mu(x)\,, \tag{31.1.5}$$

where $\phi^a{}_\mu(x)$ is small. As we shall see in Section 31.2, if $\phi^a{}_\mu$ is to be a conventionally normalized field, then the constant κ should be expressed in terms of Newton's constant G by $\kappa = \sqrt{8\pi G}$. The closeness of the vierbein to the unit matrix is preserved by a small coordinate transformation

$$x^\mu \to x^\mu + \xi^\mu(x)\,, \tag{31.1.6}$$

and also by small local Lorentz transformations

$$\Lambda^a{}_b(x) = \delta^a_b + \omega^a{}_b(x)\,, \tag{31.1.7}$$

where $\xi^\mu(x)$ and $\omega^a{}_b(x)$ are of the same order as $\phi^a{}_\mu(x)$, and $\omega_{ab}(x) \equiv \eta_{ac}\omega^c{}_b(x)$ is constrained by Eq. (31.1.4) to satisfy

$$\omega_{ab}(x) = -\,\omega_{ba}(x)\,. \tag{31.1.8}$$

The combined transformations (31.1.2) and (31.1.3) then become

$$\phi_{\mu\nu}(x) \to \phi_{\mu\nu}(x) + \frac{1}{2\kappa}\left[-\frac{\partial\xi_\mu(x)}{\partial x_\nu} + \omega_{\mu\nu}(x)\right]\,. \tag{31.1.9}$$

We are now dropping the distinction between general coordinate indices μ, ν, etc. and local Lorentz coordinate indices a, b, etc., and raising and lowering all indices with $\eta^{\mu\nu}$ and $\eta_{\mu\nu}$. In terms of the metric (31.1.1), the weak field assumption (31.1.5) becomes

$$g_{\mu\nu}(x) = \eta_{\mu\nu} + 2\kappa\, h_{\mu\nu}(x)\,, \tag{31.1.10}$$

with

$$h_{\mu\nu}(x) \equiv \phi_{\mu\nu}(x) + \phi_{\nu\mu}(x)\,, \tag{31.1.11}$$

while the transformation law (31.1.9) reads

$$h_{\mu\nu}(x) \to h_{\mu\nu}(x) - \frac{1}{2\kappa}\left[\frac{\partial\xi_\mu(x)}{\partial x^\nu} + \frac{\partial\xi_\nu(x)}{\partial x^\mu}\right]\,. \tag{31.1.12}$$

By use of the supersymmetry algebra we showed in Section 25.4 that the graviton has a fermionic superpartner, the gravitino, with helicities $\pm 3/2$. We saw in Section 5.9 that a self-charge-conjugate massless particle with helicities ± 1 can only have low-energy interactions if described by a real field $A_\mu(x)$ whose interactions are invariant under a gauge transformation $A_\mu(x) \to A_\mu(x) + \partial_\mu\Lambda(x)$. In the same way, a self-charge-conjugate massless particle with helicities $\pm 3/2$ can only have low-energy interactions if it is described by a Majorana field $\psi_\mu(x)$ with an extra vector index μ, with invariance under a gauge transformation

$$\psi_\mu(x) \to \psi_\mu(x) + \partial_\mu\psi(x)\,, \tag{31.1.13}$$

where $\psi(x)$ is an arbitrary Majorana field.[3a] We need now to consider how fields $\phi_{\mu\nu}(x)$ and $\psi_\mu(x)$ with these transformation properties can be put into a superfield.

We found in Section 27.1 that a gauge field $V_\nu(x)$ can be regarded as the V_ν-component, defined in Eq. (26.2.10) as the coefficient of $i(\bar\theta\gamma_5\gamma_\nu\theta)/2$, of a real scalar superfield $V(x,\theta)$. Similarly, we would like to incorporate the vierbein field $\phi_{\mu\nu}(x)$ and the gravitino field $\psi_\mu(x)$ into a *vector* superfield $H_\mu(x,\theta)$, known as the *metric superfield.* The question is: how are $\phi_{\mu\nu}(x)$ and $\psi_\mu(x)$ related to the components of this superfield?

To address this question, note that supersymmetry requires the 'gauge' transformations (31.1.9) and (31.1.13) to be special cases of a transformation of the whole metric superfield

$$H_\mu(x,\theta) \to H_\mu(x,\theta) + \Delta_\mu(x,\theta)\,. \tag{31.1.14}$$

Furthermore, a *weak* gravitational field $h_{\mu\nu}$ interacts with an energy-momentum tensor $T^{\mu\nu}$ which as we found in Section 26.7 is a linear combination of components of a real vector superfield Θ^μ, so we expect the interaction of the whole superfield H_μ with matter to take the form

$$I_{\rm int} = 2\kappa \int d^4x \left[H_\mu\, \Theta^\mu\right]_D\,. \tag{31.1.15}$$

(Later in this section we will confirm the correctness of the coefficient 2κ, with Θ^μ normalized as in Section 26.7.) The supercurrent Θ^μ was shown in Section 26.7 to satisfy the conservation conditions

$$\gamma^\mu \mathscr{D}\Theta_\mu = \mathscr{D}X\,, \tag{31.1.16}$$

where X is a real chiral scalar superfield (the sum of a left-chiral scalar superfield and its complex conjugate) and $\mathscr{D}$ is the four-component superderivative (26.2.26). It follows that this interaction is invariant under a transformation of the form (31.1.14), with Δ_μ of the form

$$\Delta_\mu = \left(\bar{\mathscr{D}}\gamma_\mu\Xi\right)\,, \tag{31.1.17}$$

where $\Xi(x,\theta)$ is a superfield subject to the manifestly supersymmetric condition

$$\left(\bar{\mathscr{D}}\mathscr{D}\right)\left(\bar{\mathscr{D}}\Xi\right) = 0\,. \tag{31.1.18}$$

This can be seen by recalling from Section 26.7 that the chirality condition on X allows us to write it in the form

$$X = \left(\bar{\mathscr{D}}\mathscr{D}\right)\Omega\,, \tag{31.1.19}$$

with Ω in general a non-local superfield. From Eq. (31.1.16) we then find

$$\int \left[\Theta^\mu\left(\bar{\mathscr{D}}\gamma_\mu\Xi\right)\right]_D = -\int \left[\left((\bar{\mathscr{D}}\Theta^\mu)\gamma_\mu\Xi\right)\right]_D = -\int\left[\left((\bar{\mathscr{D}}X)\Xi\right)\right]_D$$
$$= +\int\left[\Omega\left(\bar{\mathscr{D}}\mathscr{D}\right)\left(\bar{\mathscr{D}}\Xi\right)\right]_D = 0\,,$$

so that the interaction (31.1.15) is invariant under the transformation $H_\mu \to H_\mu + (\bar{\mathscr{D}}\gamma_\mu\Xi)$. (We can also obtain the same result without using the representation (31.1.19) for X, by noting from Eq. (26.2.25) that the D-term of the product of a chiral superfield like X which satisfies Eqs. (26.3.1) and (26.3.2) and a linear superfield like $(\bar{\mathscr{D}}\Xi)$ which satisfies Eq. (26.3.45) is a spacetime derivative.)

At the end of this section we shall show that Eqs. (31.1.17) and (31.1.18) yield conditions on the components of the superfield Δ_μ:

$$V^\Delta_{\mu\nu}(x) + V^\Delta_{\nu\mu}(x) = \frac{\partial v_\mu(x)}{\partial x_\nu} + \frac{\partial v_\nu(x)}{\partial x_\mu} - 2\,\eta_{\mu\nu}\frac{\partial v^\lambda(x)}{\partial x^\lambda}\,, \tag{31.1.20}$$

$$\lambda^\Delta_\mu(x) - \tfrac{1}{3}\gamma_\mu\gamma^\rho\lambda^\Delta_\rho(x) - \tfrac{1}{3}\gamma_\mu\partial^\rho\omega^\Delta_\rho(x) = \partial_\mu\chi(x)\,, \tag{31.1.21}$$

$$-\tfrac{1}{2}\epsilon^{\nu\mu\kappa\sigma}\partial_\kappa V^\Delta_{\nu\mu}(x) = D^{\Delta\,\sigma}(x) + \partial^\sigma\partial^\rho C^\Delta_\rho(x)\,, \tag{31.1.22}$$

$$\partial^\mu M^\Delta_\mu(x) = \partial^\mu N^\Delta_\mu(x) = 0\,, \tag{31.1.23}$$

with $v_\mu(x)$ a real vector field, and $\chi(x)$ a Majorana spinor field. (Here we introduce a notation that will be used throughout this chapter; following Eq. (26.3.9), the components C^S, ω^S, M^S, N^S, V^S_ν, λ^S, and D^S of an arbitrary superfield $S(x,\theta)$ are defined by the expansion

$$\begin{aligned} S(x,\theta) = \;& C^S(x) - i\big(\bar\theta\,\gamma_5\,\omega^S(x)\big) - \frac{i}{2}\big(\bar\theta\,\gamma_5\,\theta\big)M^S(x) - \frac{1}{2}\big(\bar\theta\,\theta\big)N^S(x) \\ & + \frac{i}{2}\big(\bar\theta\,\gamma_5\,\gamma^\nu\,\theta\big)V^S_\nu(x) - i\big(\bar\theta\,\gamma_5\,\theta\big)\Big(\bar\theta\Big[\lambda^S(x) + \frac{1}{2}\,\partial\!\!\!/\,\omega^S(x)\Big]\Big) \\ & - \frac{1}{4}\big(\bar\theta\,\gamma_5\,\theta\big)^2\Big(D^S(x) + \frac{1}{2}\Box C^S(x)\Big)\,. \end{aligned} \tag{31.1.24}$$

Also, $V^\Delta_{\mu\nu}(x)$ is the V_ν-component of Δ_μ.) This leads us to define the fields

$$\phi_{\mu\nu}(x) \equiv V^H_{\mu\nu}(x) - \tfrac{1}{3}\eta_{\mu\nu}\,V^{H\,\lambda}{}_\lambda(x)\,, \tag{31.1.25}$$

$$\tfrac{1}{2}\psi_\mu(x) \equiv \lambda^H_\mu(x) - \tfrac{1}{3}\gamma_\mu\gamma^\rho\lambda^H_\rho(x) - \tfrac{1}{3}\gamma_\mu\partial^\rho\omega^H_\rho(x)\,, \tag{31.1.26}$$

$$b^\sigma(x) \equiv D^{H\,\sigma}(x) + \tfrac{1}{2}\epsilon^{\nu\mu\kappa\sigma}\partial_\kappa V^H_{\nu\mu}(x) + \partial^\sigma\partial^\rho C^H_\rho(x)\,. \tag{31.1.27}$$

(As discussed in Section 31.3, the factor 1/2 introduced on the left-hand side of Eq. (31.1.26) will give the field ψ_μ a conventional normalization.) From Eqs. (31.1.20) and (31.1.21) it follows that the transformation (31.1.14) induces on $\phi_{\mu\nu}(x)$ and $\psi_\mu(x)$ the gauge transformations (31.1.9) and (31.1.13), with

$$\xi_\mu = -2\kappa\,v_\mu\,,\quad \omega_{\mu\nu} = \kappa\Big[-\frac{\partial v_\mu}{\partial x^\nu} + \frac{\partial v_\nu}{\partial x^\mu} + V^\Delta_{\mu\nu} - V^\Delta_{\nu\mu}\Big]\,,\quad \psi = 2\chi\,, \tag{31.1.28}$$

while Eq. (31.1.22) shows that $b_\mu(x)$ is invariant. Also, Eq. (31.1.23) shows that the transformation (31.1.14) induces shifts of $M^H_\mu(x)$ and $N^H_\mu(x)$ that leave invariant the fields

$$s \equiv \partial^\mu M^H_\mu(x)\,, \qquad p \equiv \partial^\mu N^H_\mu(x)\,. \tag{31.1.29}$$

Finally, since $C^\Delta_\mu(x)$, $V^\Delta_{\mu\nu}(x) - V^\Delta_{\nu\mu}(x)$, and $\omega^\Delta_\mu(x)$ are unconstrained by supersymmetry, the transformation (31.1.14) allows us to make the components $C^H_\mu(x)$, $V^H_{\mu\nu}(x) - V^H_{\nu\mu}(x) = \phi_{\mu\nu} - \phi_{\nu\mu}$, and $\omega^H_\mu(x)$ anything we like. In particular, in analogy with the Wess–Zumino gauge for gauge superfields discussed in Section 27.1, we can take $C^H_\mu(x)$, $V^H_{\mu\nu}(x) - V^H_{\nu\mu}(x) = \phi_{\mu\nu} - \phi_{\nu\mu}$, and $\omega^H_\mu(x)$ all to vanish. The fields $h_{\mu\nu}(x)$ and $\psi_\mu(x)$ are identified by their transformation properties as the fields of the graviton and gravitino, respectively, while $b_\mu(x)$, $s(x)$, and $p(x)$ are auxiliary fields[4] that are important in understanding the coupling of the graviton superfield to matter.

Incidentally, note that the number of independent components of the symmetric tensor $h_{\mu\nu}$ modulo the gauge transformations (31.1.12) is $10 - 4 = 6$, which with the auxiliary fields s, p, and b_μ gives a total of $6+6 = 12$ independent physical bosonic fields, while the number of independent physical components of the Majorana spinor field ψ_μ modulo the gauge transformations (31.1.13) is $16 - 4 = 12$. This satisfies the condition, discussed at the end of Section 26.2, that in any supermultiplet of fields that furnishes a representation of the supersymmetry algebra, there must be equal numbers of independent bosonic and fermionic field components.

Let us now return to the interaction of matter and gravitation. In general, the integrated D-term of the product of two superfields Θ^μ and H_μ is given by Eq. (26.2.25) as

$$\begin{aligned}\int d^4x\,\Big[\Theta^\mu\,H_\mu\Big]_D = \int d^4x\,\Big[& -\partial_\mu C^{H\,\sigma}\,\partial^\mu C^\Theta_\sigma + C^{H\,\sigma}\,D^\Theta_\sigma + D^{H\,\sigma}\,C^\Theta_\sigma \\ & -\Big(\overline{\omega}^{H\,\sigma}\,[\lambda^\Theta_\sigma + \tfrac{1}{2}\,\partial\!\!\!/\,\omega^\Theta_\sigma]\Big) - \Big([\overline{\lambda}^{H\,\sigma} + \tfrac{1}{2}\overline{\omega}^{H\,\sigma}\,\partial\!\!\!/\,]\omega^\Theta_\sigma\Big) \\ & +M^{H\,\sigma}\,M^\Theta_\sigma + N^{H\,\sigma}\,N^\Theta_\sigma - V^{H\,\kappa\sigma}\,V^\Theta_{\kappa\sigma}\Big]\,. \end{aligned} \tag{31.1.30}$$

Using Eqs. (31.1.25)–(31.1.27), we may express $V^H_{\mu\nu}$, λ^H_μ, and D^H_μ respectively in terms of $\phi_{\mu\nu}$, ψ_μ, and b_μ, and find

$$\begin{aligned} I_{\rm int} &= 2\kappa \int d^4x\,\Big[\Theta^\mu\,H_\mu\Big]_D \\ &= 2\kappa \int d^4x\,\Big[C^{H\,\sigma}\Big[\Box C^\Theta_\sigma - \partial_\sigma\partial^\rho C^\Theta_\rho + D^\Theta_\sigma\Big] + b^\sigma\,C^\Theta_\sigma \end{aligned}$$

$$+\left(\overline{\omega}^{H\sigma}\left[-\lambda_\sigma^\Theta - \not{\partial}\omega_\sigma^\Theta + \partial_\sigma\gamma^\rho\omega_\rho^\Theta\right]\right) - \left(\overline{\psi}^\sigma\,\omega_\sigma^\Theta\right) + \left(\overline{\psi}^\sigma\,\gamma_\sigma\gamma^\rho\omega_\rho^\Theta\right)$$
$$+M^{H\sigma}M_\sigma^\Theta + N^{H\sigma}N_\sigma^\Theta$$
$$+\tfrac{1}{2}\epsilon^{\nu\mu\kappa\sigma}\phi_{\nu\mu}\partial_\kappa C_\sigma^\Theta - \phi^{\lambda\sigma}\left[V_{\lambda\sigma}^\Theta - \eta_{\lambda\sigma}V^{\Theta\,\rho}{}_\rho\right]\Bigg]\,. \qquad (31.1.31)$$

We saw in Section 26.7 that the conservation condition (31.1.16) yields conditions (26.7.44), (26.7.39), and (26.7.35):

$$D_\mu^\Theta = -\Box C_\mu^\Theta + \partial_\mu\partial^\nu C_\nu^\Theta\,,$$
$$\lambda_\nu^\Theta = -\not{\partial}\omega_\nu^\Theta + \partial_\nu\gamma^\mu\omega_\mu^\Theta\,,$$
$$0 = V_{\mu\nu}^\Theta - V_{\nu\mu}^\Theta + \epsilon_{\mu\nu\rho\sigma}\partial^\sigma C^{\Theta\,\rho}\,,$$

which respectively tell us that the coefficients in Eq. (31.1.31) of $C^{H\sigma}$, of $\overline{\omega}^{H\sigma}$, and of the antisymmetric part of $\phi^{\mu\nu}$ all vanish. Also, we may write the remaining terms of Eq. (31.1.31) in terms of the supersymmetry current (26.7.20), the energy-momentum tensor (26.7.42), the $\mathscr{R}$-current (26.7.51):

$$S^\mu = -2\omega^{\Theta\,\mu} + 2\gamma^\mu\gamma^\nu\,\omega_\nu^\Theta\,,$$
$$T_{\mu\nu} = -\tfrac{1}{2}V_{\mu\nu}^\Theta - \tfrac{1}{2}V_{\nu\mu}^\Theta + \eta_{\mu\nu}V^{\Theta\,\lambda}{}_\lambda\,,$$
$$\mathscr{R}^\mu = 2\,C^{\Theta\,\mu}\,,$$

and densities $\mathscr{M}$ and $\mathscr{N}$ defined by

$$M_\mu^\Theta = \partial_\mu\mathscr{M}\,, \qquad N_\mu^\Theta = \partial_\mu\mathscr{N}\,, \qquad (31.1.32)$$

and given by Eqs. (26.7.33) and (26.7.34) as:

$$\mathscr{N} = -A^X\,, \qquad \mathscr{M} = B^X\,, \qquad (31.1.33)$$

where X is the real chiral superfield appearing on the right-hand side of the conservation equation (31.1.16). (A label 'new' on the supersymmetry current will be understood in this chapter.) The first-order interaction between matter and the components of the metric superfield is then

$$2\kappa\int d^4x\,\left[\Theta^\mu\,H_\mu\right]_D = \kappa\int d^4x\,\left[\mathscr{R}_\sigma b^\sigma + \tfrac{1}{2}\bar{S}^\sigma\psi_\sigma - 2\mathscr{M}\,s - 2\mathscr{N}\,p + T^{\kappa\sigma}h_{\sigma\kappa}\right]. \qquad (31.1.34)$$

We note that the gravitino field interacts with the supersymmetry current, in much the same way as the gravitational field interacts with the energy-momentum tensor.

We can now check the constant factor appearing in this interaction. The usual definition of $T^{\mu\nu}$ is that the variation of the matter action under a change $\delta g_{\mu\nu}$ in the metric is[5]

$$\delta I_M = \frac{1}{2}\int d^4x\,\sqrt{\operatorname{Det} g}\,T^{\mu\nu}\,\delta g_{\mu\nu}\,.$$

The interaction between matter and a weak gravitational field given by Eq. (31.1.10) is then

$$\kappa \int d^4x \; T^{\mu\nu}(x)\, h_{\mu\nu}(x) \, ,$$

which agrees with the $h_{\mu\nu}$-dependent part of Eq. (31.1.34), thus confirming the normalization of the interaction (31.1.15).

* * *

We shall now check that Eqs. (31.1.17) and (31.1.18) yield the conditions (31.1.20)–(31.1.23). The superfield $\Upsilon \equiv \left(\bar{\mathscr{D}}\Xi\right)$ has the components

$$\begin{aligned}
C^\Upsilon &= -i\mathrm{Tr}\,(\epsilon\omega^\Xi)\,, \\
\omega^\Upsilon &= -i\gamma_5\, \partial\!\!\!/ C^\Xi + M^\Xi - i\gamma_5 N^\Xi + V\!\!\!/^\Xi\,, \\
M^\Upsilon &= -\mathrm{Tr}\,(\epsilon\gamma_5\lambda^\Xi)\,, \\
N^\Upsilon &= i\mathrm{Tr}\,(\epsilon\lambda^\Xi)\,, \\
V^\Upsilon_\nu &= -\mathrm{Tr}\,(\epsilon\gamma_5\lambda_\nu\lambda^\Xi) - \tfrac{1}{2}\mathrm{Tr}\,(\epsilon\gamma_5\,[\gamma_\nu\,,\, \partial\!\!\!/]\lambda^\Xi)\,, \\
\lambda^\Upsilon &= -\,\partial\!\!\!/ M^\Xi - i\gamma_5\, \partial\!\!\!/ N^\Xi - i\gamma_5(D^\Xi + \Box C^\Xi) - \partial_\nu V^{\Xi\nu}\,, \\
D^\Upsilon &= i\mathrm{Tr}\,(\epsilon\, \partial\!\!\!/ \lambda^\Xi) + i\mathrm{Tr}\,(\epsilon\Box\omega^\Xi)\,,
\end{aligned}$$

and the condition (31.1.18) yields

$$M^\Upsilon = N^\Upsilon = D^\Upsilon + \Box C^\Upsilon = \partial^\lambda V^\Upsilon_\lambda = \lambda^\Upsilon + \partial\!\!\!/ \omega^\Upsilon = 0\,.$$

The vanishing of M^Υ and N^Υ tells us that λ^Ξ is a linear combination of the form

$$\lambda^\Xi\epsilon = f_\mu\gamma^\mu + g_\mu\gamma_5\gamma^\mu + k_{\mu\nu}\,[\gamma^\mu\,,\,\gamma^\nu]\,. \tag{31.1.35}$$

The vanishing of $\partial^\lambda V^\Upsilon_\lambda$ and $D^\Upsilon + \Box C^\Upsilon$ then implies that

$$\partial_\mu f^\mu = \partial_\mu g^\mu = 0\,. \tag{31.1.36}$$

Also, the vanishing of $\lambda^\Upsilon + \partial\!\!\!/ \omega^\Upsilon$ yields

$$D^\Xi = \frac{i}{2}\gamma_5\,[\gamma^\nu\,,\, \partial\!\!\!/]\,V^\Xi_\nu\,. \tag{31.1.37}$$

The components of $\Delta_\mu(x,\theta)$ are

$$C^\Delta_\mu = i\,\mathrm{Tr}\,(\epsilon\gamma_5\gamma_\mu\omega^\Xi)\,, \tag{31.1.38}$$
$$\omega^\Delta_\mu = i\, \partial\!\!\!/ \gamma_\mu C^\Xi + \gamma_5\gamma_\mu M^\Xi - i\gamma_\mu N^\Xi + \gamma_5\gamma^\rho\gamma_\mu V^\Xi_\rho\,, \tag{31.1.39}$$
$$M^\Delta_\mu = -\,\mathrm{Tr}\,(\epsilon\gamma_\mu\lambda^\Xi)\,, \tag{31.1.40}$$
$$N^\Delta_\mu = i\,\mathrm{Tr}\,(\epsilon\gamma_5\gamma_\mu\lambda^\Xi)\,, \tag{31.1.41}$$
$$V^\Delta_{\mu\nu} = \mathrm{Tr}\,(\epsilon\gamma_\nu\gamma_\mu\lambda^\Xi) - \tfrac{1}{2}\mathrm{Tr}\,(\epsilon\,[\gamma_\nu\,,\, \partial\!\!\!/]\,\gamma_\mu\,\omega^\Xi)\,, \tag{31.1.42}$$
$$\lambda^\Delta_\mu = \gamma_5\, \partial\!\!\!/ \gamma_\mu M^\Xi + i\, \partial\!\!\!/ \gamma_\mu N^\Xi - i\gamma_\mu(D^\Xi + \Box C^\Xi) + \gamma_5\gamma_\mu\partial^\nu V^\Xi_\nu\,, \tag{31.1.43}$$

$$D^{\Delta}_{\mu} = i\,\mathrm{Tr}\,(\epsilon\, \partial\!\!\!/ \gamma_5\gamma_\mu\lambda^{\Xi}) + i\Box\,\mathrm{Tr}\,(\epsilon\gamma_5\gamma_\mu\omega^{\Xi})\,. \tag{31.1.44}$$

The symmetric part of Eq. (31.1.42) gives condition (31.1.20), with

$$v_\mu = -\,\mathrm{Tr}\,(\epsilon\gamma_\mu\omega^{\Xi}) + \text{constant}\,. \tag{31.1.45}$$

A linear combination of Eqs. (31.1.39) and (31.1.43) yields condition (31.1.21), with

$$\chi = 2\gamma_5 M^{\Xi} + 2i\,N^{\Xi} + \text{constant}\,. \tag{31.1.46}$$

We then use the antisymmetric part of Eq. (31.1.42), together with Eqs. (31.1.37), (31.1.38), and the identities

$$[\gamma_\nu\,,\gamma_\rho]\gamma_\mu - [\gamma_\mu\,,\gamma_\rho]\gamma_\nu = 2\eta_{\mu\rho}\gamma_\nu - 2\eta_{\nu\rho}\gamma_\mu + 2i\epsilon_{\nu\rho\mu\lambda}\,\gamma_5\gamma^\lambda\,,$$

$$\epsilon^{\nu\mu\kappa\sigma}[\gamma_\nu\,,\gamma_\mu] = 2i\gamma_5[\gamma^\kappa\,,\gamma^\sigma]\,,$$

and find the condition (31.1.22). Finally, Eqs. (31.1.40) and (31.1.41) along with Eqs. (31.1.35) and (31.1.36) yield the condition (31.1.23).

31.2 The Gravitational Action

To find a suitable gravitational action, we must construct a superfield that is invariant under the generalized gauge transformation $H_\mu \to H_\mu + \Delta_\mu$. As a starting point, we recall that the field b_μ defined by Eq. (31.1.27) is invariant under this transformation. By making successive supersymmetry transformations we can see that b_μ is the C-component of an 'Einstein' superfield E_μ, whose components are

$$C^E_\mu = b_\mu\,, \tag{31.2.1}$$

$$\omega^E_\mu = \frac{3}{2}L_\mu - \frac{1}{2}\gamma_\mu\gamma^\nu L_\nu\,, \tag{31.2.2}$$

$$M^E_\mu = \partial_\mu s\,, \qquad N^E_\mu = \partial_\mu p\,, \tag{31.2.3}$$

$$V^E_{\mu\nu} = -\frac{3}{2}E_{\mu\nu} + \frac{1}{2}\eta_{\mu\nu}E^\rho{}_\rho + \frac{1}{2}\epsilon_{\nu\mu\sigma\rho}\partial^\sigma b^\rho\,, \tag{31.2.4}$$

$$\lambda^E_\mu = \partial_\mu\gamma^\nu\omega^E_\nu - \partial\!\!\!/\omega^E_\mu\,, \tag{31.2.5}$$

$$D^E_\mu = \partial_\mu\partial^\nu b_\nu - \Box b_\mu\,, \tag{31.2.6}$$

where $E_{\mu\nu}$ is the linearized Einstein tensor

$$E_{\mu\nu} \equiv \frac{1}{2}\Big(\partial_\mu\partial_\nu h^\lambda{}_\lambda + \Box h_{\mu\nu} - \partial_\mu\partial^\lambda h_{\lambda\nu} - \partial_\nu\partial^\lambda h_{\lambda\mu} - \eta_{\mu\nu}\Box h^\lambda{}_\lambda + \eta_{\mu\nu}\partial^\lambda\partial^\rho h_{\lambda\rho}\Big) \tag{31.2.7}$$

and

$$L^{\nu} \equiv i\epsilon^{\nu\mu\kappa\rho}\gamma_5\gamma_\mu\partial_\kappa\psi_\rho\,, \tag{31.2.8}$$

which will be shown in Section 31.3 to be the left-hand side of the wave equation for a free massless field of spin 3/2. For instance, by applying the supersymmetry transformation rules (26.2.11), (26.2.15), and (26.2.17) for C^H, V^H_μ, and D^H to Eqs. (31.2.1) and (31.1.27) we find

$$\delta C^{E\,\sigma} = i\left(\bar{\alpha}\gamma_5\left[\partial\!\!\!/\lambda^{H\,\sigma} - \tfrac{1}{2}i\epsilon^{\nu\mu\kappa\sigma}\gamma_5\gamma_\mu\partial_\kappa\lambda^H_\nu + \partial^\sigma\partial^\kappa\omega^H_\kappa\right]\right)\,.$$

Comparison of this with the transformation rule (26.2.11) for C^E shows that

$$\omega^{E\,\sigma} = \partial\!\!\!/\lambda^{H\,\sigma} - \tfrac{1}{2}i\epsilon^{\nu\mu\kappa\sigma}\gamma_5\gamma_\mu\partial_\kappa\lambda^H_\nu + \partial^\sigma\partial^\kappa\omega^H_\kappa\,.$$

We can use Eq. (31.1.26) to express λ^H_μ in terms of ψ_μ and ω^H_μ:

$$\lambda^H_\mu = \psi_\mu - \gamma_\mu\gamma^\rho\psi_\rho - \gamma_\mu\partial^\rho\omega^H_\rho\,,$$

and find that ω^E_σ may be written in terms of ψ_σ alone

$$\omega^E_\sigma = \partial\!\!\!/\psi_\sigma - \partial_\sigma\gamma^\rho\psi_\rho - \tfrac{1}{2}i\epsilon_{\nu\mu\kappa\sigma}\gamma_5\gamma^\mu\partial^\kappa\psi^\nu\,.$$

Using the identity

$$\eta_{\mu\nu}\gamma_\lambda - \eta_{\mu\lambda}\gamma_\nu = i\epsilon_{\mu\lambda\nu\rho}\gamma_5\gamma^\rho - \tfrac{1}{2}i\gamma_\mu\gamma^\sigma\epsilon_{\sigma\lambda\nu\rho}\gamma_5\gamma^\rho\,,$$

we find Eq. (31.2.2) for ω^E_μ. Continuing in the same way gives the other formulas (31.2.3)–(31.2.6) for the components of E_μ and confirms that these components make up a real superfield.

We can now form a quadratic action for the metric superfield that is invariant under both supersymmetry and the extended gauge transformation $H_\mu \to H_\mu + \Delta_\mu$, by taking the Lagrangian density as

$$\mathscr{L}_E = \tfrac{4}{3}\left(E_\mu H^\mu\right)_D = E_{\mu\nu}h^{\mu\nu} - \tfrac{1}{2}\bar{\psi}_\mu L^\mu - \tfrac{4}{3}(s^2 + p^2 - b_\mu b^\mu)\,. \tag{31.2.9}$$

The factor 4/3 is chosen to give the kinematic Lagrangian for the graviton field a conventional sign and normalization: apart from terms involving $\partial^\mu h_{\mu\nu}$ or $h^\lambda{}_\lambda$, it is a sum of Klein–Gordon Lagrangians for the components of $h_{\mu\nu}$. The normalization of the gravitino field is discussed in the next section.

To see that the first two terms in the final expression in Eq. (31.2.9) are invariant under the gauge transformations (31.1.12) and (31.1.13), we note that $E_{\mu\nu}$ and L_μ are manifestly invariant under these transformations, and that the action is symmetric between the two factors of $h_{\mu\nu}$ or ψ_μ appearing in these terms. The absence in Eq. (31.2.9) of derivatives of s, p, and b_μ shows that these are auxiliary fields. The field equations make them vanish for pure gravitation, but not when gravitation is coupled to matter.

Before taking up the coupling of matter and gravitation, let us pause to consider what value we should give the normalization constant κ in Eqs. (31.1.5) and (31.1.10). The Einstein–Hilbert action for a pure gravitational field is

$$I_{GR} = -\frac{1}{16\pi G}\int d^4x\,\sqrt{g}\,R\,, \tag{31.2.10}$$

where G is the Newton constant of gravitation, $g(x)$ is the determinant of the metric tensor $g_{\mu\nu}(x)$, and $R(x)$ is the curvature scalar calculated from $g_{\mu\nu}(x)$. To calculate I_{GR} for a weak gravitational field with $g_{\mu\nu} = \eta_{\mu\nu} + 2\kappa h_{\mu\nu}$, we may recall that for arbitrary variations $\delta g_{\mu\nu}(x)$ in the gravitational field[6]

$$\delta I_{GR} = \frac{1}{16\pi G}\int d^4x\,\sqrt{g}\left[R^{\mu\nu} - \frac{1}{2}g^{\mu\nu}R\right]\delta g_{\mu\nu}\,, \tag{31.2.11}$$

where $R^{\mu\nu}(x)$ is the Ricci tensor calculated from $g_{\mu\nu}(x)$. For a weak field with $g_{\mu\nu} = \eta_{\mu\nu} + 2\kappa h_{\mu\nu}$, the Ricci tensor is[7]

$$R^{\mu\nu} = \kappa\Big(\Box h^{\mu\nu} - \partial_\lambda\partial^\mu h^{\lambda\nu} - \partial_\lambda\partial^\nu h^{\lambda\mu} + \partial^\mu\partial^\nu h^\lambda{}_\lambda\Big)\,, \tag{31.2.12}$$

so for weak fields

$$R^{\mu\nu} - \tfrac{1}{2}g^{\mu\nu}R = 2\kappa E^{\mu\nu}\,, \tag{31.2.13}$$

and therefore Eq. (31.2.10) gives

$$\delta I_{GR} = \frac{\kappa^2}{4\pi G}\int d^4x\,E^{\mu\nu}\delta h_{\mu\nu}\,.$$

On the other hand, taking account of the symmetry of $\int d^4x\,E_{\mu\nu}\,h^{\mu\nu}$ between the two factors of h it contains, we have

$$\delta\int d^4x\,E^{\mu\nu}\,h_{\mu\nu} = 2\int d^4x\,E^{\mu\nu}\,\delta h_{\mu\nu}\,.$$

In order for the term $E_{\mu\nu}\,h^{\mu\nu}$ in Eq. (31.2.9) to give the usual gravitational Lagrangian density, it is therefore necessary to take

$$\kappa = \sqrt{8\pi G}\,. \tag{31.2.14}$$

Let's now combine the Einstein Lagrangian density (31.2.9) with the interaction (31.1.34) between gravitation and matter and with the matter Lagrangian $\mathscr{L}_M$, to form a total Lagrangian density:

$$\begin{aligned}\mathscr{L} = {}&\mathscr{L}_M + E_{\mu\nu}h^{\mu\nu} - \tfrac{1}{2}\bar{\psi}_\mu L^\mu - \tfrac{4}{3}(s^2 + p^2 - b_\mu b^\mu)\\ &+2\kappa\Big[\tfrac{1}{2}\mathscr{R}_\sigma b^\sigma + \tfrac{1}{4}\bar{S}^\sigma\psi_\sigma - \mathscr{M}\,s - \mathscr{N}\,p + \tfrac{1}{2}T^{\kappa\sigma}h_{\sigma\kappa}\Big]\,.\end{aligned} \tag{31.2.15}$$

The field equations for the auxiliary fields give

$$s = -6\kappa\mathscr{M}/8\,, \qquad p = -6\kappa\mathscr{N}/8\,, \qquad b_\mu = -6\kappa\mathscr{R}_\mu/16\,. \tag{31.2.16}$$

Using this to eliminate the auxiliary fields, the Lagrangian density (31.2.15) now gives

$$\mathscr{L} = \mathscr{L}_M + E_{\mu\nu}h^{\mu\nu} - \tfrac{1}{2}\bar{\psi}_\mu L^\mu + \tfrac{3}{4}\kappa^2(\mathscr{M}^2 + \mathscr{N}^2 - \tfrac{1}{4}\mathscr{R}_\mu\mathscr{R}^\mu) + \tfrac{1}{2}\kappa\bar{S}^\sigma\psi_\sigma + \kappa T^{\kappa\sigma}h_{\sigma\kappa}\,. \tag{31.2.17}$$

The sources of the fields ψ_μ and $h_{\mu\nu}$ are of order κ, so we can regard these fields as being of this order, which makes all terms in Eq. (31.2.17) beyond $\mathscr{L}_M$ of order κ^2.

In the vacuum only the scalar fields s and p can have tree-approximation expectation values, giving the vacuum an energy density, to order G:

$$\rho_{VAC} = -\mathscr{L}_{VAC} = -\mathscr{L}_{M\,VAC} - \tfrac{3}{4}\kappa^2(\mathscr{M}^2 + \mathscr{N}^2)\,. \tag{31.2.18}$$

The negative sign of the term in the vacuum energy density of first order in G is a characteristic difference between theories of supergravity and ordinary supersymmetry.

For instance, in the theory of a single left-chiral superfield Φ with superpotential $f(\Phi)$, the zeroth-order vacuum energy is $-\mathscr{L}_{M\,VAC} = |df(\phi)/d\phi|^2$, while Eqs. (31.1.33) and (26.7.27) give

$$\mathscr{M} + i\mathscr{N} = -\frac{1}{3}\left[\phi\frac{df(\phi)}{d\phi} - 3\,f(\phi)\right]\,,$$

so to first order in G, the total vacuum energy is

$$\rho_{VAC} = \left|\frac{df(\phi)}{d\phi}\right|^2 - \frac{8\pi G}{3}\left|\phi\frac{df(\phi)}{d\phi} - 3\,f(\phi)\right|^2\,, \tag{31.2.19}$$

where ϕ is the scalar component of Φ. This is for the definition of the metric for which the energy-momentum tensor is given by Eq. (26.7.42), so that in particular $T^\lambda{}_\lambda$ vanishes for $f = 0$. For other definitions the vacuum energy would be changed by terms of order $8\pi G|\phi|^2|df(\phi)/d\phi|^2$. However, this ambiguity is unimportant in calculating the minimum value of the vacuum energy. Inspection of Eq. (31.2.19) shows that if $f(\phi)$ is stationary at some point ϕ_0, then ρ_{VAC} has a local minimum for $|\phi - \phi_0|$ and $|df/d\phi|$ of first order in G, and at any such point the vacuum energy to first order in G is

$$\rho_{VAC} = -24\pi G|f(\phi_0)|^2\,. \tag{31.2.20}$$

There is an algebraic reason why vacuum states with unbroken supersymmetry in supergravity theories cannot have positive energy density.

The solutions of the Einstein field equations for a uniform non-zero vacuum energy density ρ_V take the form of a de Sitter space for $\rho_V > 0$ and anti-de Sitter space for $\rho_V < 0$. These spaces may be described as the surfaces

$$x_5^2 \pm \eta_{\mu\nu} x^\mu x^\nu = R^2 \,, \tag{31.2.21}$$

in a quasi-Euclidean five-dimensional space with line element

$$ds^2 = \eta_{\mu\nu} x^\mu x^\nu \pm dx_5^2 \,, \tag{31.2.22}$$

with the upper sign for de Sitter space and the lower sign for anti-de Sitter space. The spacetime symmetry of these spaces is no longer the Poincaré group consisting of translations and Lorentz transformations, but instead $O(4,1)$ for de Sitter space and $O(3,2)$ for anti-de Sitter space, where $O(n,m)$ is the group of linear transformations that leave invariant a diagonal metric with n positive and m negative elements on the main diagonal. The supersymmetry that is unbroken in theories with a de Sitter or anti-de Sitter vacuum state therefore has the simple group $O(4,1)$ or $O(3,2)$, respectively, as its spacetime symmetry. Nahm[7a] has cataloged all supersymmetries with simple spacetime symmetries. There are simple $O(3,2)$ supersymmetries, as well as N-extended $O(3,2)$ supersymmetries, but for $O(4,1)$ there is only an $N = 2$ supersymmetry. We have been considering $N = 1$ supergravity theories here, so they can have vacuum states with unbroken supersymmetry and $\rho_V < 0$, which gives $O(3,2)$ spacetime symmetry, but not with unbroken supersymmetry and $\rho_V > 0$, which would give $O(4,1)$ spacetime symmetry.

The possibility of vacuum field configurations with negative energy density may seem at first sight to threaten the stability of our universe. It is common for $f(\phi)$ to have several stationary points, at which it takes different values. Even if we fine-tune the parameters in $f(\phi)$ so that $f(\phi)$ vanishes at one of these stationary points, accounting for the observed nearly flat space of our universe, any other stationary point with a non-zero value of $f(\phi)$ will yield a *lower* vacuum energy, raising the possibility of a collapse to a state of negative energy density, with a metric that would have to be of the 'anti-de Sitter' form rather than flat.

Fortunately, the value (31.2.20) is just barely insufficient to compensate for the positive energy that has to go into surface tension in making a bubble of anti-de Sitter space, so that the transition from ordinary flat space is not actually favored energetically. Coleman and de Luccia[8] have applied the equations of general relativity to a bubble of negative internal energy density $-\epsilon$ and positive surface tension S in a flat space of energy density zero, and have shown that any such bubble that does not entail gravitational singularities will have positive energy if

$$\epsilon \leq 6\pi G S^2 \,. \tag{31.2.23}$$

The surface tension is the energy per area in the bubble surface, given to zeroth order in G by the integral of the energy density through the bubble wall:

$$S_1 = \int_{r_-}^{r_+} dr \left[\left| \frac{d\phi}{dr} \right|^2 + \left| \frac{df(\phi)}{d\phi} \right|^2 \right] , \tag{31.2.24}$$

with r_- and r_+ taken just inside and just outside the bubble wall. This can be rewritten as

$$S_1 = \int_{r_-}^{r_+} dr \left[\left| \frac{d\phi}{dr} + \xi \left(\frac{df(\phi)}{d\phi} \right)^* \right|^2 - 2\text{Re} \left(\xi^* \frac{d\phi}{dr} \frac{df(\phi)}{d\phi} \right) \right] ,$$

where ξ is an arbitrary phase factor with $|\xi| = 1$. The integral in the second term is trivial: since we assume that $\phi(r_+)$ is at a value where $f(\phi)$ is stationary and vanishes, while $\phi(r_-)$ is at some value ϕ_0 where $f(\phi)$ is stationary but non-zero, the integral is

$$\int_{r_-}^{r_+} dr \, \frac{d\phi}{dr} \frac{df(\phi)}{d\phi} = -f(\phi_0) .$$

To maximize this term, we take $\xi = f(\phi_0)/|f(\phi_0)|$, and obtain the inequality[9]

$$S_1 \geq 2\,|f(\phi_0)| , \tag{31.2.25}$$

with equality attained only if (as is usual) there is a solution of the differential equation $d\phi/dr = -\xi(df/d\phi)^*$ with the appropriate boundary conditions. The inequality (31.2.23) is thus satisfied if the internal energy density is not less than $-24\pi G|f(\phi_0)|^2$, which is precisely the value (31.2.20). This calculation leaves open the possibility of an instability of flat space due to radiative corrections, but the reader need not worry: it has been proved that in supergravity theories in which there is a vacuum field configuration with zero vacuum energy, the energy of any disturbance in the fields which is limited to a finite region is positive.[10]

* * *

Eq. (26.7.48) shows that the energy-momentum tensor for a set of left-chiral scalar superfields Φ_n contains a term

$$\Delta T^{\mu\nu} = \frac{1}{3} \left(\eta^{\mu\nu} \Box - \partial^\mu \partial^\nu \right) \sum_n |\phi_n|^2 . \tag{31.2.26}$$

Integrating by parts, the corresponding interaction $\kappa h_{\mu\nu} \Delta T^{\mu\nu}$ makes a contribution to the action of the form

$$\frac{\kappa}{3} \int \delta^4 x \sum_n |\phi_n|^2 \left(\eta^{\mu\nu} \Box - \partial^\mu \partial^\nu \right) h_{\mu\nu} = \frac{1}{6} R^{(1)} \sum_n |\phi_n|^2 , \tag{31.2.27}$$

where $R^{(1)}$ is the curvature scalar in linear approximation. This is added to the usual Einstein–Hilbert action $-\sqrt{g}R/2\kappa^2$ (which appears in Eq. (31.2.17) as the term $E_{\mu\nu}h^{\mu\nu}$) and has the effect of replacing the coefficient of this term with

$$-\frac{1}{2\kappa^2}+\frac{1}{6}\sum_n|\phi_n|^2=-\frac{1}{2\kappa^2}\left(1-\frac{\kappa^2}{3}\sum_n|\phi_n|^2\right)\,. \tag{31.2.28}$$

In order to restore the usual constant of gravitation, we can subject the metric to a *Weyl transformation*, replacing the vierbein $e^a{}_\mu$ with

$$\tilde{e}^a{}_\mu=e^a{}_\mu\sqrt{1-\frac{\kappa^2}{3}\sum_n|\phi_n|^2}\,. \tag{31.2.29}$$

That is, we replace the metric $g_{\mu\nu}$ with

$$\tilde{g}_{\mu\nu}=\left(1-\frac{\kappa^2}{3}\sum_n|\phi_n|^2\right)g_{\mu\nu}\,, \tag{31.2.30}$$

or for weak fields

$$\tilde{h}_{\mu\nu}=h_{\mu\nu}-\frac{\kappa}{6}\sum_n|\phi_n|^2\eta_{\mu\nu}\,. \tag{31.2.31}$$

The weak field Einstein tensor (31.2.7) for the new metric is

$$\begin{aligned}\tilde{E}_{\mu\nu}&\equiv\frac{1}{2}\left(\partial_\mu\partial_\nu\tilde{h}^\lambda{}_\lambda+\Box\tilde{h}_{\mu\nu}-\partial_\mu\partial^\lambda\tilde{h}_{\lambda\nu}-\partial_\nu\partial^\lambda\tilde{h}_{\lambda\mu}-\eta_{\mu\nu}\Box\tilde{h}^\lambda{}_\lambda+\eta_{\mu\nu}\partial^\lambda\partial^\rho\tilde{h}_{\lambda\rho}\right)\\&=E_{\mu\nu}-\frac{\kappa}{6}\left(\partial_\mu\partial_\nu-\eta_{\mu\nu}\Box\right)\sum_n|\phi_n|^2\,.\end{aligned} \tag{31.2.32}$$

The sum of the Einstein term in the original action and the term (31.2.27) is then

$$\begin{aligned}\int d^4x\left[h_{\mu\nu}E^{\mu\nu}+\frac{\kappa}{3}\sum_n|\phi_n|^2\left(\eta^{\mu\nu}\Box-\partial^\mu\partial^\nu\right)h_{\mu\nu}\right]&=\int d^4x\left[\tilde{h}_{\mu\nu}\tilde{E}^{\mu\nu}\right.\\&\left.+\frac{\kappa^2}{12}\left(\partial_\mu\sum_n|\phi_n|^2\right)\left(\partial^\mu\sum_n|\phi_n|^2\right)\right]\,,\end{aligned} \tag{31.2.33}$$

so the effective gravitational constant is now actually constant. This redefinition of the metric also produces a change in the potential. The original Lagrangian density contains a term $-e\sum_n|\partial f(\phi)/\partial\phi_n|^2$, which in terms of the new vierbein reads

$$\begin{aligned}-e\sum_n|\partial f(\phi)/\partial\phi_n|^2&=-\tilde{e}\sum_n|\partial f(\phi)/\partial\phi_n|^2\\&-\frac{2\kappa^2}{3}\left(\sum_n|\phi_n|^2\right)\sum_n|\partial f(\phi)/\partial\phi_n|^2\,.\end{aligned}$$

With the new definition of the metric, the potential (31.2.19) is therefore replaced with

$$\rho_{VAC} = \sum_n \left| \frac{\partial f(\phi)}{\partial \phi_n} \right|^2 - \frac{\kappa^2}{3} \left| \sum_n \phi_n \frac{\partial f(\phi)}{\partial \phi_n} - 3 f(\phi) \right|^2 + \frac{2\kappa^2}{3} \left(\sum_n |\phi_n|^2 \right) \sum_n \left| \frac{\partial f(\phi)}{\partial \phi_n} \right|^2 . \quad (31.2.34)$$

The new term does not change the value of ρ_{VAC} at its stationary point to order κ^2, so no change is needed in our previous discussion of vacuum stability.

31.3 The Gravitino

In this section we will use the weak field formalism developed in Sections 31.1–31.3 to derive some properties of the gravitino. In particular, by using an argument of continuity as $G \to 0$ we shall obtain a formula for the gravitino mass when supersymmetry is spontaneously broken that is valid to first order in G but to all orders in all other interactions. (The original derivation of this formula will be given in Section 31.6.)

First we must verify that the term $-\frac{1}{2}\bar{\psi}_\mu L^\mu$ in Eq. (31.2.9) is the correct free-field Lagrangian for a massless self-charge-conjugate particle of spin 3/2. The time-honored approach to the construction of suitable free-field Lagrangians for particles with spin is to guess at the Lagrangian and then check that it gives a physically satisfactory field equation and propagator. This has led to some uncertainty for particles of spin 3/2 — for instance, what is usually called the Rarita–Schwinger Lagrangian in papers on supersymmetry is not the Lagrangian originally proposed by Rarita and Schwinger.[11] Here we shall follow an approach in the spirit of Section 6.2: we will first derive the propagator from the requirements of Lorentz invariance for a massive particle of spin 3/2 and then invert it to find the Lagrangian. We work here with massive gravitinos for the sake of simplicity and because in the real world we must take account of the breakdown of supersymmetry, but we will be able to apply the results obtained in this way to the case of massless gravitinos by noting that supersymmetry current conservation makes the zero-mass singularities in the propagator inconsequential.

A spinor field ψ^μ with an extra vector index belongs to the $[(\frac{1}{2},0)+(0,\frac{1}{2})] \times (\frac{1}{2},\frac{1}{2})$ representation of the homogeneous Lorentz group. To isolate the $(1,\frac{1}{2})+(\frac{1}{2},1)$ parts of the free field, we impose the irreducibility condition

$$\gamma_\mu \psi^\mu = 0\,. \tag{31.3.1}$$

Rotational invariance and Eq. (31.3.1) tell us that the matrix elements of this field between the vacuum and a massive spin 3/2 particle of momentum $\mathbf{q} = 0$ and spin z-component s will satisfy the conditions

$$\langle 0|\psi^0(0)|s\rangle = 0 \tag{31.3.2}$$

and

$$\sum_{s=-3/2}^{3/2} \langle 0|\psi^i(0)|s\rangle\, \langle 0|\psi^j(0)|s\rangle^* \propto \delta_{ij} - \tfrac{1}{3}\gamma_i\gamma_j\,, \tag{31.3.3}$$

with a coefficient that may depend on the rotationally invariant matrix $\beta \equiv i\gamma^0$. With the usual Dirac convention that $\beta\langle 0|\psi^i(0)|s\rangle = \langle 0|\psi^i(0)|s\rangle$ (chosen to simplify the space inversion property of the field) and a conventional choice of normalization of the $\beta = +1$ component analogous to Eq. (5.5.23), Eq. (31.3.3) may be written

$$\sum_{s=-3/2}^{3/2} \langle 0|\psi^i(0)|s\rangle\langle 0|\psi^j(0)|s\rangle^* = (2\pi)^{-3}\left(\frac{1+\beta}{2}\right)\left[\delta_{ij} - \tfrac{1}{3}\gamma_i\gamma_j\right]. \tag{31.3.4}$$

It follows that the momentum-space propagator for a spin 3/2 particle of mass m_g takes the form

$$\Delta^{\mu\nu}(q) = \frac{P^{\mu\nu}(q)}{q^2 + m_g^2 - i\epsilon}\,, \tag{31.3.5}$$

where $P^{\mu\nu}(q)$ is a Lorentz-covariant polynomial in the four-vector q, subject to the conditions that for $\mathbf{q} = 0$ and $q^0 = m_g$, we have

$$P^{ij} = \left(\frac{1+\beta}{2}\right)\left[\delta_{ij} - \tfrac{1}{3}\gamma_i\gamma_j\right], \qquad P^{i0} = P^{0i} = P^{00} = 0\,. \tag{31.3.6}$$

Apart from possible terms that vanish on the mass shell (and whose effect therefore would be the same as direct current–current interactions) the unique covariant function with this limit is

$$P^{\mu\nu}(q) = \left(\eta^{\mu\nu} + \frac{q^\mu q^\nu}{m_g^2}\right)\left(-i\not{q} + m_g\right) - \frac{1}{3}\left(\gamma^\mu - i\frac{q^\mu}{m_g}\right)\left(i\not{q} + m_g\right)\left(\gamma^\nu - i\frac{q^\nu}{m_g}\right). \tag{31.3.7}$$

(The difference between Eq. (31.3.7) and any other covariant function with the limit (31.3.6) would be a covariant function whose components all vanish for $\mathbf{q} = 0$ and $q^0 = m_g$ and hence vanish everywhere on the mass shell.) The free-field Lagrangian density is then of the form

$$\mathscr{L}_0 = -\tfrac{1}{2}\left(\bar{\psi}^\mu D_{\mu\nu}(-i\partial)\,\psi^\nu\right), \tag{31.3.8}$$

where

$$\Delta^{\mu\nu}(q)\,D_{\nu\lambda}(q) = \delta^{\mu}_{\lambda}\,. \qquad (31.3.9)$$

A tedious but straightforward calculation gives

$$D_{\nu\lambda}(q) = -\epsilon_{\nu\mu\kappa\lambda}\,\gamma_5\,\gamma^{\mu}\,q_{\kappa} - \tfrac{1}{2}m_g\,[\gamma_{\nu}\,,\,\gamma_{\lambda}]\,, \qquad (31.3.10)$$

so that the Lagrangian density (31.3.8) is

$$\mathscr{L}_0 = -\tfrac{1}{2}i\,\epsilon^{\nu\mu\kappa\lambda}\left(\bar{\psi}_{\nu}\,\gamma_5\,\gamma_{\mu}\,\partial_{\kappa}\,\psi_{\lambda}\right) + \tfrac{1}{4}m_g\left(\bar{\psi}_{\nu}\,[\gamma^{\nu}\,,\gamma^{\lambda}]\,\psi_{\lambda}\right). \qquad (31.3.11)$$

For $m_g = 0$ this result verifies that the term $-\frac{1}{2}\bar{\psi}_{\mu}L^{\mu}$ in Eq. (31.2.9) is the correct conventionally normalized free-field Lagrangian for a massless self-charge-conjugate particle of spin 3/2. In the limit $m_g \to 0$ the propagator given by Eqs. (31.3.5) and (31.3.7) is singular (which just reflects the impossibility shown in Section 5.9 of forming a $(1, \frac{1}{2}) + (\frac{1}{2}, 1)$ field from the creation and annihilation operators for a massless particle of helicity $\pm 3/2$), but the terms in Eq. (31.3.7) for $P^{\mu\nu}(q)$ that blow up for $m_g \to 0$ are all proportional to q^{μ} and/or q^{ν}, and hence give no contribution when the current with which ψ_{μ} interacts is conserved.

As a further check on the validity of the Lagrangian density (31.3.11), including the peculiar-looking mass term, we note that it yields a field equation

$$-i\,\epsilon^{\nu\mu\kappa\lambda}\,\gamma_5\,\gamma_{\mu}\,\partial_{\kappa}\,\psi_{\lambda} + \tfrac{1}{2}m_g\,[\gamma^{\nu}\,,\gamma^{\lambda}]\,\psi_{\lambda} = 0\,. \qquad (31.3.12)$$

Taking the divergence of this equation yields the result that

$$[\not{\partial}\,,\,\gamma^{\lambda}]\,\psi_{\lambda} = 0\,.$$

Also, contracting Eq. (31.3.12) with γ_{ν} shows that

$$\gamma_{\nu}\psi^{\nu} \propto \epsilon^{\nu\mu\kappa\lambda}\,\gamma_5\,\gamma_{\nu}\gamma_{\mu}\,\partial_{\kappa}\,\psi_{\lambda} \propto [\not{\partial}\,,\,\gamma^{\lambda}]\,\psi_{\lambda} = 0\,,$$

so that the irreducibility condition (31.3.1) is satisfied for free fields (though not necessarily when interactions are taken into account). From these two results there follows another irreducibility condition

$$\partial_{\lambda}\psi^{\lambda} = \tfrac{1}{2}\{\not{\partial}\,,\,\gamma_{\lambda}\}\,\psi^{\lambda} = \tfrac{1}{2}[\not{\partial}\,,\,\gamma_{\lambda}]\,\psi^{\lambda} = 0\,.$$

Using these irreducibility results allows the field equation (31.3.12) to be put in the form of a Dirac equation

$$\left(\not{\partial} + m_g\right)\psi^{\lambda} = 0\,, \qquad (31.3.13)$$

which among other things shows that this is the free field of a particle of mass m_g.

We will now consider the effects in supergravity theories of a spontaneous breakdown of supersymmetry. Broken global supersymmetry entails the existence of a massless particle of spin 1/2, the goldstino, but in supergravity theories the goldstino field χ can be eliminated by a gauge transformation $\psi_\mu \to \psi_\mu - \partial_\mu \chi$. Since the gauge is then fixed by the condition that the goldstino field is eliminated in this way, gauge invariance no longer keeps the gravitino massless, and it acquires a mass (to be denoted m_g from now on), in much the same way as we saw in Section 21.3 that the vector bosons $W^\pm$ and Z^0 get masses from the spontaneous breakdown of the $SU(2) \times U(1)$ gauge symmetry of the electroweak interactions.

As discussed at the beginning of Chapter 28, if supersymmetry is relevant at all to accessible phenomena then the characteristic energy scale at which it is broken must be much less than the Planck mass. In this case we may use a continuity argument to give a universal formula for the gravitino mass m_g. According to Eq. (31.1.34), the gravitino field ψ_μ couples to the supersymmetry current S^μ with a coupling constant $\frac{1}{2}\kappa = \frac{1}{2}\sqrt{8\pi G}$, so the exchange of a virtual gravitino of four-momentum q in a transition $A + B \to C + D$ contributes to the invariant amplitude a term

$$M(A + B \to C + D) = \tfrac{1}{4}(8\pi G)\,\langle C|\bar{S}_\mu|A\rangle_N\,\Delta^{\mu\nu}(q)\,\langle D|S_\nu|B\rangle_N\,, \tag{31.3.14}$$

where the subscript N indicates that the one-goldstino pole at $q^2 = 0$ has been removed from the matrix element of the supersymmetry current. For a supersymmetry-breaking scale sufficiently small compared with the Planck mass, there will be a range of momentum transfers that are much larger than the gravitino mass but much less than the Planck mass. For such momenta, the matrix element is dominated by the $1/m_g^2$ terms in the propagator numerator (31.3.7)

$$M(A + B \to C + D) \to \tfrac{1}{4}(8\pi G)\,\langle C|\bar{S}_\mu|A\rangle_N \left(\frac{-2i\,\not{q}\,q^\mu q^\nu}{3m_g^2 q^2}\right) \langle D|S_\nu|B\rangle_N\,. \tag{31.3.15}$$

But for sufficiently large Planck mass the coupling of the gravitino becomes negligible, and the matrix element must be the same as would have been produced by goldstino exchange in a theory without gravitinos. According to Eq. (29.2.10), this is

$$M(A+B \to C+D) \to \langle C|\bar{S}_\mu|A\rangle_N \left(\frac{-i\,\not{q}}{q^2}\right)\left(\frac{q^\mu q^\nu}{F^2}\right) \langle D|S_\nu|B\rangle_N\,, \tag{31.3.16}$$

where F is the parameter characterizing the strength of supersymmetry breaking (here taken real), defined so that the vacuum energy density is $F^2/2$. In order for Eqs. (31.3.15) and (31.3.16) to agree, the gravitino mass must have the value

$$m_g = \sqrt{\frac{4\pi\,G F^2}{3}}\,. \tag{31.3.17}$$

This formula is valid only to lowest order in GF^2, but to all orders (and even non-perturbatively) in the non-gravitational interactions responsible for the spontaneous breakdown of supersymmetry.

It is convenient for some purposes to express m_g in terms of the expectation values $\langle s\rangle$ and $\langle p\rangle$ of the spinless auxiliary gravitational fields. Note that for the vacuum state to have zero spacetime curvature, the vacuum energy density $F^2/2$ of matter fields must be balanced by the negative vacuum energy of gravitation and its interaction with the hidden sector fields, which is given in terms of $\langle s\rangle$ and $\langle p\rangle$ by Eqs. (31.2.18) and (31.2.16) as $-(4/3)(\langle s\rangle^2 + \langle p\rangle^2)$, so

$$F^2/2 = (4/3)(\langle s\rangle^2 + \langle p\rangle^2)\,. \tag{31.3.18}$$

We can therefore write Eq. (31.3.17) as

$$m_g = \frac{2\kappa}{3}\sqrt{\langle s\rangle^2 + \langle p\rangle^2}\,. \tag{31.3.19}$$

It is sometimes convenient to introduce a complex gravitino mass, defined as

$$\tilde{m}_g \equiv \frac{2\kappa}{3}\Big(\langle s\rangle + i\langle p\rangle\Big)\,, \tag{31.3.20}$$

whose absolute magnitude is the physical gravitino mass (31.3.19).

31.4 Anomaly-Mediated Supersymmetry Breaking

In Section 28.3 the possibility was raised that supersymmetry may be broken in some sort of hidden sector of superfields that do not carry the $SU(3)\times SU(2)\times U(1)$ quantum numbers of the standard model, and communicated to observable particles gravitationally. In this section we will deal with one class of supersymmetry-breaking effects in the minimum supersymmetric standard model, those of first order in $\kappa \equiv \sqrt{8\pi G}$. This includes the gaugino masses and the parameters A_{ij} and B in the Lagrangian density (28.4.1). Other supersymmetry-breaking effects such as squark and slepton squared masses are of second order in κ, and will be taken up in Section 31.7, when we consider gravity-mediated supersymmetry breaking using the general supergravity formalism described in Section 31.6.

We can find the effects of gravity-mediated supersymmetry breaking to first order in κ by simply replacing the component fields of the gravitational supermultiplet in the interaction (31.1.34) with their expectation values. The only ones of these component fields that can acquire non-vanishing vacuum expectation values from the spontaneous breakdown of supersymmetry in a hidden sector of matter superfields are the spinless auxiliary fields s and p, so with Eq. (31.1.33) this gives a first-order supersymmetry-breaking interaction

$$\mathscr{L}^{(1)} = 2\kappa\Big[-A^X\langle p\rangle + B^X\langle s\rangle\Big] = 3\,\mathrm{Im}\Big[\tilde{m}_g^*(A^X + iB^X)\Big]\,, \tag{31.4.1}$$

where $\tilde{m}_g$ is the complex gravitino mass (31.3.20), and A^X and B^X are the A- and B-components of the real chiral scale-non-invariance superfield X discussed in Section 26.7.

We showed in Section 26.7 that, for a renormalizable theory of left-chiral superfields Φ_n with superpotential $f(\Phi)$, the X superfield is given by

$$X = \frac{2}{3}\,\mathrm{Im}\left[\sum_n \Phi_n \frac{\partial f(\Phi)}{\partial \Phi_n} - 3\,f(\Phi)\right]\,. \tag{31.4.2}$$

This can be put in a form that allows an immediate extension to more general theories by writing the coupling parameters of the superpotential in terms of dimensionless parameters and a parameter $\mathscr{M}$ with the dimensions of mass. Since the superpotential has the dimensions of (mass)^3, we have

$$\mathscr{M}\frac{\partial f(\Phi)}{\partial \mathscr{M}} + \sum_n \Phi_n \frac{\partial f(\Phi)}{\partial \Phi_n} = 3\,f(\Phi)\,. \tag{31.4.3}$$

Therefore Eq. (31.4.2) may be written

$$X = \frac{2}{3}\,\mathrm{Im}\left[\mathscr{M}\frac{\partial f(\Phi)}{\partial \mathscr{M}}\right]\,. \tag{31.4.4}$$

This formula can be generalized to give the contribution to X of the scale dependence of any sort of $\mathscr{F}$-term in the Lagrangian. A term $2\mathrm{Re}\,[f(\Phi, W)]_{\mathscr{F}}$ in the Lagrangian makes a contribution to X given by the obvious generalization of Eq. (31.4.4):

$$X = \frac{2}{3}\mathrm{Im}\left[\mathscr{M}\frac{\partial f(\Phi, W)}{\partial \mathscr{M}}\right]\,. \tag{31.4.5}$$

(There is also a contribution to X from any mass scale dependence in the D-terms in the Lagrangian.) Comparing Eq. (31.4.5) with Eqs. (26.3.10) and (26.3.13), we see that

$$A^X + iB^X = \frac{2\mathscr{M}}{3i}\frac{\partial}{\partial \mathscr{M}}\Big[f(\Phi, W)\Big]_{\theta=0} = \frac{2\mathscr{M}}{3i}\frac{\partial f(\phi, \lambda_L)}{\partial \mathscr{M}}\,. \tag{31.4.6}$$

There will then be a supersymmetry-breaking term in the effective Lagrangian of first order in κ, given by Eqs. (31.4.1) and (31.4.6) as

$$\mathscr{L}_f^{(1)} = -2\,\mathrm{Re}\left[\tilde{m}_g^*\,\frac{\mathscr{M}\,\partial f(\phi, \lambda_L)}{\partial \mathscr{M}}\right]\,. \tag{31.4.7}$$

Eq. (31.4.7) applies not only for terms in the Lagrangian with explicit scale dependence, but also for the scale dependence of coupling constants described by the renormalization group.[11a] This scale dependence arises from the quantum mechanical anomaly that gives a non-zero value to the trace of the energy-momentum tensor and also to the divergence of the R-current, so the observable supersymmetry-breaking effects that arise in this way are said to be *anomaly-mediated.*

Consider in particular the kinematic term $\mathscr{L}_{\rm gauge}$ for renormalizable supersymmetric gauge theories, given by Eqs. (27.3.22) and (27.3.23) as

$$\mathscr{L}_{\rm gauge} = -\frac{1}{2g^2}{\rm Re}\sum_{A\alpha\beta}\left[\epsilon_{\alpha\beta}W_{A\alpha}W_{A\beta}\right]_{\mathscr{F}} . \qquad (31.4.8)$$

This does not depend explicitly on any mass scale, but the coupling constant g has the familiar dependence on a renormalization scale $\mathscr{M}$, given by the renormalization group equation

$$\mathscr{M}\frac{dg(\mathscr{M})}{d\mathscr{M}} = \beta\Big(g(\mathscr{M})\Big) . \qquad (31.4.9)$$

Then Eq. (31.4.7) shows that the gauge Lagrangian (31.4.8) yields a supersymmetry-breaking term in the Lagrangian

$$\mathscr{L}^{(1)}_{\rm gauge} = -\frac{\beta(g)}{g^3}{\rm Re}\left[\tilde{m}_g^*\sum_{A\alpha\beta}\epsilon_{\alpha\beta}\lambda_{AL\alpha}\lambda_{AL\beta}\right] , \qquad (31.4.10)$$

where $\lambda_{AL\alpha}$ is the left-handed part of the gaugino field, normalized like $W_{A\alpha}$ by multiplication with the gauge coupling g, so that g does not appear in the structure constants or the interaction of the gauge superfield with quark superfields. Taking account of this normalization convention, the gaugino mass equals g^2 times the absolute value of the coefficient of $\frac{1}{2}\sum_{A\alpha\beta}\epsilon_{\alpha\beta}\lambda_{AL\alpha}\lambda_{AL\beta}$, or

$$m_{\rm gaugino} = m_g\left|\frac{\beta(g)}{g}\right| . \qquad (31.4.11)$$

In this formula $m_{\rm gaugino}$ and g are cut-off-dependent bare parameters in the Lagrangian, governed by Wilsonian renormalization group equations. In Section 27.6 we saw that $\beta(g)$ arises purely from one-loop diagrams, so that $\beta(g) = bg^3$, with b a constant, and therefore

$$m_{\rm gaugino} = m_g\,|b|\,g^2 . \qquad (31.4.12)$$

The physical mass of the gaugino differs from Eq. (31.4.12) by corrections of higher order in g but, since we know that gauginos must be much heavier than the characteristic scale of quantum chromodynamics, these corrections are small for the gluino as well as for the wino and bino.

Supersymmetry does not allow any explicitly scale-dependent terms in the Lagrangian for the gluon and quark superfields, so, with electroweak interactions neglected, Eq. (31.4.12) gives the only contribution of order κ to the gluino mass. For three quark generations, Eq. (28.2.10) gives $b = -3g_s^3/16\pi^2$. Taking $g_s^2/4\pi = 0.118$, Eq. (31.4.12) gives the gluino mass as

$$m_{\text{gluino}} = \frac{3g_s^2 m_g}{16\pi^2} = 2.8 \times 10^{-2}\, m_g\,. \qquad (31.4.13)$$

On the other hand, there is a scale-dependent interaction in the Lagrangian for the Higgs superfields, produced by the μ-term $-\mu(H_2^{\rm T} e H_1)$ in Eq. (28.1.7), so there is a term in the Lagrangian given by Eq. (31.4.7) as $2{\rm Re}\,[\tilde{m}_g^* \mu(\mathscr{H}_2^{\rm T} e \mathscr{H}_1)]$. Comparing with the $B\mu$-term in Eq. (28.4.1), we see that this gives

$$B = -\tilde{m}_g^*\,. \qquad (31.4.14)$$

With three generations of quarks and lepton superfields and one pair of Higgs doublet superfields H_1 and H_2, Eqs. (28.2.8) and (28.2.9) give $b = 11/16\pi^2$ for the $U(1)$ gauge coupling g' and $b = 1/16\pi^2$ for the $SU(2)$ gauge coupling g. Taking $g'^2/4\pi = 0.0102$ and $g^2/4\pi = 0.0338$, Eq. (31.4.11) would give masses $8.9 \times 10^{-3} m_g$ and $2.7 \times 10^{-3} m_g$ for the bino and wino, respectively. However, the bino and wino masses also receive contributions of order $g'^2 m_g/16\pi^2$ and $g^2 m_g/16\pi^2$, respectively, from diagrams in which the bino or wino is attached to a higgs–higgsino loop, with supersymmetry breaking introduced by the term $2{\rm Re}\,[\tilde{m}_g^* \mu(\mathscr{H}_2^{\rm T} e \mathscr{H}_1)]$ in the Lagrangian density. This gives the bino and wino masses[11b]

$$m_{\text{bino}} = \frac{g'^2 m_g}{16\pi^2}\left|11 - f\left(\frac{\mu^2}{m_A^2}\right)\right|\,, \qquad (31.4.15)$$

$$m_{\text{wino}} = \frac{g^2 m_g}{16\pi^2}\left|1 - f\left(\frac{\mu^2}{m_A^2}\right)\right|\,, \qquad (31.4.16)$$

where m_A is the pseudoscalar particle mass defined by Eq. (28.5.21), and

$$f(x) \equiv \frac{2x \ln x}{x-1}\,. \qquad (31.4.17)$$

The implications of these results will be considered further in Section 31.7.

Finally, there is a scale-dependent field-renormalization factor Z_r multiplying the kinematic Lagrangian density $[\Phi_r^* e^{-V} \Phi_r]_D$ for any left-chiral superfield Φ_r. This can be moved from the kinematic D-terms to the superpotential $\mathscr{F}$-terms by absorbing a factor $Z_r^{1/2}$ in Φ_r. The Yukawa couplings h_{rst} (such as the h^E_{ij}, h^D_{ij} and h^U_{ij} in Eq. (28.1.7)) in the trilinear superpotential terms $\sum_{rst} h_{rst}\Phi_r\Phi_s\Phi_t$ are then multiplied with a factor $Z_r^{-1/2} Z_s^{-1/2} Z_t^{-1/2}$ that depends on the cut-off $\mathscr{M}$. According to

Eq. (31.4.7), the interaction $\mathscr{L}^{(1)}$ will then make a contribution to the Lagrangian density:

$$\mathscr{L}^{(1)}_{\text{Yukawa}} = -2\sum_{rst}\gamma_{rst}\text{Re}\left[\tilde{m}_g^* h_{rst}\phi_r\phi_s\phi_t\right] , \tag{31.4.18}$$

where

$$\begin{aligned}\gamma_{rst} &\equiv \mathscr{M}\frac{\partial \ln h_{rst}(\mathscr{M})}{\partial \mathscr{M}} \\ &= -\frac{1}{2}\mathscr{M}\frac{\partial \ln Z_r(\mathscr{M})}{\partial \mathscr{M}} - \frac{1}{2}\mathscr{M}\frac{\partial \ln Z_s(\mathscr{M})}{\partial \mathscr{M}} - \frac{1}{2}\mathscr{M}\frac{\partial \ln Z_t(\mathscr{M})}{\partial \mathscr{M}} .\end{aligned} \tag{31.4.19}$$

We see that the coefficients A^E_{ij}, A^D_{ij}, and A^U_{ij} in Eq. (28.1.7) are given by[11b]

$$A^N_{ij} = \tilde{m}_g^*\gamma^N_{ij} = \tilde{m}_g^*\mathscr{M}\frac{\partial \ln h^N_{ij}}{\partial \mathscr{M}} , \tag{31.4.20}$$

where $N = E$, D, or U. This gives A^D_{ij} and A^U_{ij} of order $g_s^2 m_g/8\pi^2$, while A^E_{ij} is of order $g^2 m_g/8\pi^2$ or $g'^2 m_g/8\pi^2$.

31.5 Local Supersymmetry Transformations

As a last step before considering the effects of higher order in G on supersymmetry transformation rules, we will now complete our discussion of the transformation rules of the physical components of the graviton superfield H_μ and other superfields to lowest order in G.

Let us first note the form that these transformation rules take when expressed in terms of the physical fields $h_{\mu\nu}$, ψ_μ, b_μ, s, and p in a 'Wess–Zumino' gauge in which

$$C^H_\mu = \omega^H_\mu = V^H_{\mu\nu} - V^H_{\nu\mu} = 0 . \tag{31.5.1}$$

Using the identification (31.1.25)–(31.1.27), (31.1.29), and (31.1.11) of the physical fields with components of the superfield H_μ, together with the general transformation rules (26.2.11)–(26.2.17) and the 'gauge' conditions (31.5.1), we find the transformation rules

$$\delta h_{\mu\nu} = \tfrac{1}{2}\Big(\bar{\alpha}\left[\gamma_\mu\psi_\nu + \gamma_\nu\psi_\mu\right]\Big) , \tag{31.5.2}$$

$$\begin{aligned}\delta\psi_\mu = \Big[&\tfrac{1}{2}[\gamma^\nu , \gamma^\lambda]\,\partial_\lambda h_{\mu\nu} + \partial_\mu h^\lambda{}_\lambda + 2i\gamma_5 b_\mu - \tfrac{2}{3}i\gamma_\mu\gamma_\rho\gamma_5 b^\rho \\ &+ \tfrac{2}{3}\gamma_\mu s - \tfrac{2}{3}i\gamma_\mu\gamma_5 p\Big]\alpha ,\end{aligned} \tag{31.5.3}$$

$$\delta s = \tfrac{1}{4}\Big(\bar{\alpha}\gamma_\lambda L^\lambda\Big) , \tag{31.5.4}$$

$$\delta p = -\tfrac{1}{4}i\Big(\bar{\alpha}\gamma_5\gamma_\lambda L^\lambda\Big) , \tag{31.5.5}$$

$$\delta b_\mu = \tfrac{3}{4}i\left(\bar{\alpha}\gamma_5 L_\mu\right) - \tfrac{1}{4}i\left(\bar{\alpha}\gamma_5\gamma_\mu\gamma^\nu L_\nu\right), \tag{31.5.6}$$

where L_μ is given by Eq. (31.2.8):

$$L^\mu \equiv i\epsilon^{\mu\nu\kappa\rho}\,\gamma_5\,\gamma_\nu\,\partial_\kappa\psi_\rho\,.$$

This transformation shifts the components C^H_μ, ω^H_μ, and $V^H_{\mu\nu} - V^H_{\nu\mu}$ away from zero, by the amounts

$$\delta C^H_\mu = 0\,, \quad \delta\omega^H_\mu = V^H_{\mu\nu}\gamma^\nu\alpha\,, \tag{31.5.7}$$

$$\delta[V^H_{\nu\mu} - V^H_{\mu\nu}] = \left(\bar{\alpha}\,[\gamma_\mu\lambda^H_\nu - \gamma_\nu\lambda^H_\mu]\right). \tag{31.5.8}$$

We can then return to the gauge satisfying Eq. (31.5.1) by making a suitable gauge transformation $H_\mu \to H_\mu + \Delta_\mu$, where Δ_μ is a superfield of the form (31.1.17), (31.1.18), with components

$$C^\Delta_\mu = 0\,, \quad \omega^\Delta_\mu = -\,V^H_{\mu\nu}\gamma^\nu\alpha\,, \tag{31.5.9}$$

$$V^\Delta_{\nu\mu} - V^\Delta_{\mu\nu} = \left(\bar{\alpha}\,[\gamma_\nu\lambda^H_\mu - \gamma_\mu\lambda^H_\nu]\right). \tag{31.5.10}$$

So far, this has been for global supersymmetry transformations, with α an infinitesimal constant Majorana spinor. At least in lowest order, this symmetry can be easily extended to *local* supersymmetry transformations, with $\alpha(x)$ given an arbitrary dependence on x^μ. According to Eq. (26.7.11), under such a transformation the matter action undergoes the change

$$\delta\int d^4x\,\mathscr{L}_M = -\int d^4x\,\left(\bar{S}^\mu(x)\,\partial_\mu\alpha(x)\right). \tag{31.5.11}$$

Inspection of Eq. (31.2.17) shows that this change in the action is cancelled if we add an inhomogeneous term $(2/\kappa)\partial_\mu\alpha(x)$ to the right-hand side of Eq. (31.5.3), so that the change in the gravitino field is now

$$\delta\psi_\mu(x) = (2/\kappa)\partial_\mu\alpha(x) + \Big[\tfrac{1}{2}[\gamma^\nu, \gamma^\lambda]\,\partial_\lambda h_{\mu\nu}(x) + \partial_\mu h^\lambda{}_\lambda(x) + 2i\gamma_5\,b_\mu(x)$$
$$- \tfrac{2}{3}i\gamma_\mu\gamma_\rho\gamma_5\,b^\rho(x) + \tfrac{2}{3}\gamma_\mu\,s(x) - \tfrac{2}{3}\,i\gamma_\mu\gamma_5\,p(x)\Big]\alpha(x)\,. \tag{31.5.12}$$

It is useful to rewrite Eq. (31.5.12) in a form that makes the generalization to general coordinates more transparent. First, we note that the term $\partial_\mu h^\lambda{}_\lambda\alpha$ on the right-hand side of Eq. (31.5.12) may be eliminated by replacing the parameter $\alpha(x)$ with

$$\tilde{\alpha} \equiv (\mathrm{Det}\,g)^{1/4}\alpha \simeq \alpha + \tfrac{1}{2}\kappa h^\lambda{}_\lambda\alpha\,. \tag{31.5.13}$$

Dropping the tilde, to zeroth order in κ Eq. (31.5.12) then becomes

$$\delta\psi_\mu(x) = (2/\kappa)\partial_\mu\alpha(x) + \Big[\tfrac{1}{2}[\gamma^\nu, \gamma^\lambda]\,\partial_\lambda h_{\mu\nu}(x) + 2i\gamma_5\,b_\mu(x)$$
$$- \tfrac{2}{3}i\gamma_\mu\gamma_\rho\gamma_5\,b^\rho(x) + \tfrac{2}{3}\gamma_\mu\,s(x) - \tfrac{2}{3}\,i\gamma_\mu\gamma_5\,p(x)\Big]\alpha(x)\,. \tag{31.5.14}$$

We can express this in terms of the covariant derivative of $\alpha(x)$, which in general coordinates takes the form

$$D_\mu \alpha(x) = \partial_\mu \alpha(x) + \tfrac{1}{2} i \mathscr{J}_{bc}\, \omega_\mu^{bc}(x)\, \alpha(x)\,, \qquad (31.5.15)$$

where $\mathscr{J}_{bc}$ is the matrix (5.4.6) representing the generator of Lorentz transformations in the Dirac representation

$$\mathscr{J}^{bc} \equiv -\frac{i}{4}\left[\gamma^b\,,\gamma^c\right], \qquad (31.5.16)$$

and $\omega_\mu^{bc}(x)$ is the spin connection,

$$\omega_\mu^{bc} = e^b{}_\lambda\, e^c{}_{\nu;\mu}\, g^{\lambda\nu} = e^b{}_\lambda \frac{\partial e^c{}_\nu}{\partial x^\mu} g^{\lambda\nu} - \Gamma^\rho_{\nu\mu}\, e^b{}_\lambda\, e^c{}_\rho\, g^{\lambda\nu}\,. \qquad (31.5.17)$$

Using the weak field approximations (31.1.5), (31.1.10), and (31.1.11), together with the gauge condition $\phi_{\mu\nu} = \phi_{\nu\mu}$, and in this approximation again ignoring the difference between local Lorentz indices a, b, etc. and spacetime indices μ, ν, etc., this gives

$$D_\mu \alpha(x) \simeq \partial_\mu \alpha(x) + \tfrac{1}{4}\kappa\,[\gamma^\nu\,,\gamma^\lambda]\,\partial_\lambda h_{\mu\nu}(x)\,\alpha(x)\,. \qquad (31.5.18)$$

Thus the local supersymmetry transformation rule (31.5.12) may be written

$$\delta\psi_\mu(x) = (2/\kappa) D_\mu \alpha(x) + \Big[2i\gamma_5\, b_\mu(x) - \tfrac{2}{3} i \gamma_\mu \gamma_\rho \gamma_5\, b^\rho(x) + \tfrac{2}{3}\gamma_\mu\, s(x) - \tfrac{2}{3} i \gamma_\mu \gamma_5\, p(x)\Big]\alpha(x)\,. \qquad (31.5.19)$$

We see that in Wess–Zumino gauge the derivatives in the supersymmetry transformation become covariant derivatives. In this sense, the approach outlined here is similar to the de Wit–Freedman approach to supersymmetric gauge theories, described in Section 27.8.

The transformation $\psi_\mu(x) \to \psi_\mu(x) + (2/\kappa)\partial_\mu\alpha(x)$ is a gauge transformation of the same type as Eq. (31.1.13), so it leaves the zeroth-order gravitino action $-\frac{1}{2}\int d^4x(\bar\psi_\mu\, L^\mu)$ invariant. The whole action obtained from the Lagrangian density (31.2.15) is therefore invariant to order zero in κ under the local supersymmetry transformations (31.5.2), (31.5.4)–(31.5.6), (31.5.12) (or (31.5.19)) and matter superfield transformations like (26.7.15). We conclude that *the combination of gravitation and supersymmetry automatically leads to local supersymmetry.*

31.6 Supergravity to All Orders

Although the action derived from the Lagrangian density (31.2.15) is invariant to zeroth order in $\kappa \equiv \sqrt{8\pi G}$ under the local supersymmetry transformations constructed in Sections 26.7 and 31.5, it is *not* invariant to first order in κ, because the interaction of matter with the gravitational supermultiplet introduces terms of order κ in $\partial_\mu S^\mu$. To make possible the invariance of the full action, we will have to add terms of higher order in κ to the Lagrangian and to the supersymmetry transformation rules for the components of both the matter and the gravitational supermultiplets. This can be done by first adding terms of higher order in κ to the transformation rules, chosen so that these transformations form a closed algebra along with local Lorentz transformations and general coordinate transformations, and then adding terms to the action to make it invariant under all these transformations.

This is a long and tedious process. Here we shall give only the results,[12] and then turn in the next section to what has been their most important application. The local supersymmetry transformation of the vierbein, gravitino field, and auxiliary fields takes the form:

$$\delta e^a{}_\mu = \kappa\left(\bar{\alpha}\gamma^a\psi_\mu\right), \tag{31.6.1}$$

$$\delta\psi_\mu = (2/\kappa)D_\mu\alpha + 2i\gamma_5(b_\mu - \tfrac{1}{3}\gamma_\mu\gamma_\rho b^\rho)\alpha + \tfrac{2}{3}\gamma_\mu(s - i\gamma_5 p)\alpha, \tag{31.6.2}$$

$$\delta s = \frac{1}{4e}\left(\bar{\alpha}\gamma_\mu L^\mu\right) + \frac{\kappa}{2}\left(\bar{\alpha}\,[i\gamma_5 b^\nu - s\gamma^\nu - ip\gamma_5\gamma^\nu]\psi_\nu\right), \tag{31.6.3}$$

$$\delta p = -\frac{i}{4e}\left(\bar{\alpha}\gamma_5\gamma_\mu L^\mu\right) + \frac{\kappa}{2}\left(\bar{\alpha}\,[b^\nu + is\gamma_5\gamma^\nu - p\gamma^\nu]\psi_\nu\right), \tag{31.6.4}$$

$$\begin{aligned}\delta b_\mu = {}& \frac{3i}{4e}\left(\bar{\alpha}\gamma_5\left(L_\mu - \tfrac{1}{3}\gamma_\mu\gamma_\rho L^\rho\right)\right) + \frac{\kappa}{2}b_\nu\left(\bar{\alpha}\gamma^\nu\psi_\mu\right) \\ & + i\frac{\kappa}{2}\left(\bar{\psi}_\mu\gamma_5(s - i\gamma_5 p)\alpha\right) - i\frac{\kappa}{4}\epsilon_{\mu\nu\kappa\sigma}b^\nu\left(\bar{\alpha}\gamma_5\gamma^\kappa\psi^\sigma\right).\end{aligned} \tag{31.6.5}$$

Here D_μ is again the covariant derivative given by (31.5.15) and (31.5.16):

$$D_\mu \equiv \partial_\mu + \tfrac{1}{8}[\gamma_a, \gamma_b]\,\omega_\mu^{ab}, \tag{31.6.6}$$

but now with the spin connection including terms bilinear in the gravitino field

$$\begin{aligned}\omega_\mu^{ab} = {}& e^a{}_\lambda e^b{}_{\nu;\mu} g^{\lambda\nu} \\ & + \frac{\kappa^2}{4}\left[e^b{}_\nu\left(\bar{\psi}_\mu\gamma^a\psi^\nu\right) + e^a{}_\nu e^b{}_\rho\left(\bar{\psi}^\nu\gamma_\mu\psi^\rho\right) - e^a{}_\nu\left(\bar{\psi}_\mu\gamma^b\psi^\nu\right)\right].\end{aligned} \tag{31.6.7}$$

Also, L^μ is the covariant version of the Rarita–Schwinger operator (31.2.8)

$$L^\mu = i\gamma_5\gamma_\nu D_\rho\psi_\sigma\epsilon^{\mu\nu\rho\sigma}, \tag{31.6.8}$$

γ_μ is defined in terms of the usual Dirac matrices γ_a as

$$\gamma_\mu = e^a{}_\mu\gamma_a, \tag{31.6.9}$$

and e is here the determinant of the vierbein

$$e \equiv \sqrt{\mathrm{Det}\,g}. \tag{31.6.10}$$

It is easy to see that these transformation rules reduce in the weak field limit to Eqs. (31.5.2), (31.5.19), and (31.5.4)–(31.5.6).

The action for pure supergravity that is invariant under these transformations, and that reduces to the form (31.2.9) in the weak field limit, is

$$I_{\text{SUGRA}} = \int d^4x \left[-\frac{e}{2\kappa^2} R - \frac{1}{2}\left(\bar{\psi}_\mu L^\mu\right) - \frac{4e}{3}\left(s^2 + p^2 - b_\mu b^\mu\right) \right] , \tag{31.6.11}$$

where R is the curvature scalar, calculated using the spin connection (31.6.7). In the absence of matter, the action would be stationary for $s = p = b_\mu = 0$, leaving us with the simpler action

$$I_{\text{SUGRA}} = \int d^4x \left[-\frac{e}{2\kappa^2} R - \frac{1}{2}\left(\bar{\psi}_\mu L^\mu\right) \right] .$$

The transformation of matter fields is also now more complicated. For the components of a general scalar supermultiplet, these are

$$\delta C = i\left(\bar{\alpha}\gamma_5\,\omega\right) , \tag{31.6.12}$$

$$\delta\omega = \left[-i\gamma_5\, \not{\mathscr{D}} C - M + i\gamma_5 N + \not{V}\right]\alpha , \tag{31.6.13}$$

$$\delta M = -\left(\bar{\alpha}\left[\lambda + \not{\mathscr{D}}\omega\right]\right) + \frac{2\kappa}{3}\left(\bar{\alpha}\left[s - i\gamma_5 p + i\gamma_5\, \not{b}\right]\omega\right) , \tag{31.6.14}$$

$$\delta N = i\left(\bar{\alpha}\gamma_5\left[\lambda + \not{\mathscr{D}}\omega\right]\right) + \frac{2i\kappa}{3}\left(\bar{\alpha}\left[s - i\gamma_5 p + i\gamma_5\, \not{b}\right]\gamma_5\,\omega\right) , \tag{31.6.15}$$

$$\delta V_a = \left(\bar{\alpha}\gamma_a\lambda\right) + \left(\bar{\alpha}\mathscr{D}_a\omega\right) + \frac{\kappa}{3}\left(\bar{\alpha}\left[s - i\gamma_5 p + i\gamma_5\, \not{b}\right]\gamma_a\,\omega\right) , \tag{31.6.16}$$

$$\delta\lambda = -\frac{1}{4}[\gamma^a , \gamma^b]\,\alpha\, F_{ab} + i D\,\gamma_5\,\alpha , \tag{31.6.17}$$

$$\delta D = i\left(\bar{\alpha}\gamma_5\, \not{\mathscr{D}}\lambda\right) , \tag{31.6.18}$$

where the covariant derivatives here are

$$\mathscr{D}_a C = e_a{}^\mu\left[\partial_\mu C - \frac{i\kappa}{2}\left(\bar{\psi}_\mu\gamma_5\,\omega\right)\right] , \tag{31.6.19}$$

$$\mathscr{D}_a\omega = e_a{}^\mu\left[\partial_\mu\omega + \frac{1}{8}\omega_\mu^{cb}\,[\gamma_c , \gamma_b]\,\omega - i\kappa\, b_\mu\gamma_5\,\omega \right.$$
$$\left. -\frac{\kappa}{2}\left(\not{V} - i\gamma_5\, \not{\partial} C - M + i\gamma_5 N\right)\psi_\mu\right] , \tag{31.6.20}$$

$$\mathscr{D}_a\lambda = e_a{}^\mu\left[\partial_\mu\lambda + \frac{1}{8}\omega_\mu^{cb}[\gamma_c , \gamma_b]\,\lambda \right.$$
$$\left. + i\kappa\, b_\mu\gamma_5\,\lambda + \frac{\kappa}{8}[\gamma_b , \gamma_c]\,\psi_\mu F_{bc} - \frac{i\kappa}{2}\gamma_5 D\,\psi_\mu\right] , \tag{31.6.21}$$

$$F_{ab} = e_a{}^\mu e_b{}^\nu\left[\partial_\mu V_\nu + \frac{\kappa}{2}\partial_\mu\left(\bar{\psi}_\mu\,\omega\right) - \frac{\kappa}{2}\left(\bar{\psi}_\mu\gamma_\nu\,\lambda\right)\right] - a \leftrightarrow b , \tag{31.6.22}$$

with $V_\mu \equiv e^a{}_\mu V_a$. The rules for multiplying general scalar multiplets are the same as Eqs. (26.2.19)–(26.2.25), except that ∂_μ is everywhere replaced with $\mathscr{D}_a$, so that the components of a supermultiplet $S = S_1 S_2$ are

$$C = C_1 C_2 \,, \tag{31.6.23}$$

$$\omega = C_1\omega_2 + C_2\omega_1 \,, \tag{31.6.24}$$

$$M = C_1 M_2 + C_2 M_1 + \tfrac{1}{2} i \left(\overline{\omega_1}\, \gamma_5\, \omega_2\right) , \tag{31.6.25}$$

$$N = C_1 N_2 + C_2 N_1 - \tfrac{1}{2}\left(\overline{\omega_1}\, \omega_2\right) , \tag{31.6.26}$$

$$V^a = C_1 V_2^a + C_2 V_1^a - \tfrac{1}{2} i \left(\overline{\omega_1}\, \gamma_5 \gamma^a \omega_2\right) , \tag{31.6.27}$$

$$\begin{aligned}\lambda = {} & C_1\lambda_2 + C_2\lambda_1 - \tfrac{1}{2}\gamma^a\omega_1 \mathscr{D}_a C_2 - \tfrac{1}{2}\gamma^a\omega_2 \mathscr{D}_a C_1 + \tfrac{1}{2} i \not{V}_1 \gamma_5\, \omega_2 \\ & + \tfrac{1}{2} i \not{V}_2 \gamma_5\, \omega_1 + \tfrac{1}{2}(N_1 - i\gamma_5 M_1)\,\omega_2 + \tfrac{1}{2}(N_2 - i\gamma_5 M_2)\,\omega_1 \,,\end{aligned} \tag{31.6.28}$$

$$\begin{aligned}D = {} & -\mathscr{D}_a C_1\, \mathscr{D}^a C_2 + C_1 D_2 + C_2 D_1 + M_1 M_2 + N_1 N_2 \\ & -\left(\overline{\omega_1}\,[\lambda_2 + \tfrac{1}{2}\, \not{\mathscr{D}}\omega_2]\right) - \left(\overline{\omega_2}\,[\lambda_1 + \tfrac{1}{2}\, \not{\mathscr{D}}\omega_1]\right) - V_{1a} V_2^a \,.\end{aligned} \tag{31.6.29}$$

Just as in flat space, there are supersymmetric constraints that can be imposed on these general supermultiplets. One such constraint is reality. Inspection of Eqs. (31.6.12)–(31.6.22) together with Eqs. (26.A.20)–(26.A.21) shows that if C, M, N, V_a, D, ω, and λ form a supermultiplet S, then C^*, M^*, N^*, V_a^*, D^*, $\beta\epsilon\gamma_5\omega^*$, and $\beta\epsilon\gamma_5\lambda^*$ are also the components of a supermultiplet, called S^*. In particular, a *real* supermultiplet is one with $S = S^*$, so that C, M, N, V_a, and D are real and ω and λ are Majorana spinors.

We can also impose a supersymmetric condition of chirality. Suppose we set

$$\lambda = 0\,, \qquad D = 0\,, \qquad V_\nu + \tfrac{1}{2}\kappa\left(\bar{\psi}_\nu\, \omega\right) = \partial_\nu Z\,, \tag{31.6.30}$$

for some field Z. Then Eq. (31.6.22) gives $F_{ab} = 0$, so Eq. (31.6.17) shows that $\delta\lambda = 0$, while Eq. (31.6.21) gives $\mathscr{D}_a\lambda = 0$, so Eq. (31.6.18) shows that $\delta D = 0$. The conditions $\lambda = D = 0$ are therefore preserved by local supersymmetry transformations. With a little more work one can show that

$$\delta\left[V_\nu + \tfrac{1}{2}\kappa\left(\bar{\psi}_\nu\, \omega\right)\right] = \partial_\nu\left(\bar{\alpha}\,\omega\right) , \tag{31.6.31}$$

so the remaining condition, that $V_\nu + \tfrac{1}{2}\kappa\left(\bar{\psi}_\nu\, \omega\right)$ is a spacetime gradient, is also supersymmetric. A supermultiplet of component fields satisfying Eq. (31.6.30) is said to be *chiral.* Just as in the case of global supersymmetry, the components of a chiral supermultiplet are conventionally renamed as A, B, ψ, F, and G, defined by

$$C = A\,, \qquad \omega = -i\gamma_5\psi\,, \qquad M = G\,, \qquad N = -F\,, \qquad Z = B\,. \tag{31.6.32}$$

A chiral supermultiplet is real if A, B, F, and G are real and ψ is a Majorana spinor. Such a real chiral supermultiplet can be written as the sum of a left-chiral supermultiplet Φ, with components conventionally defined by

$$\phi \equiv \frac{A+iB}{\sqrt{2}}\,, \qquad \psi_L \equiv \left(\frac{1+\gamma_5}{2}\right)\psi\,, \qquad \mathscr{F} \equiv \frac{F-iG}{\sqrt{2}}\,, \tag{31.6.33}$$

and its complex conjugate, a right-chiral supermultiplet. From these definitions and Eqs. (31.6.12)–(31.6.15) and (31.6.31) we see that the components of a left-chiral supermultiplet have the transformation properties

$$\delta\phi = \sqrt{2}\left(\bar{\alpha}\,\psi_L\right)\,, \tag{31.6.34}$$

$$\delta\psi_L = \sqrt{2}(\not{\partial}\phi)\alpha_R - \kappa\gamma^\mu\left(\bar{\psi}_\mu\psi_L\right)\alpha_R + \sqrt{2}\mathscr{F}\,\alpha_L\,, \tag{31.6.35}$$

$$\delta\mathscr{F} = \sqrt{2}\left(\bar{\alpha}\,\not{\mathscr{D}}\psi_L\right) - \frac{2\kappa}{3}\left(\bar{\alpha}\,[s - ip - i\not{b}]\psi_L\right)\,, \tag{31.6.36}$$

with $\mathscr{D}_a\psi$ given by Eqs. (31.6.20) and (31.6.32). The rules for multiplying left-chiral supermultiplets are the same as the corresponding rules (26.3.27)–(26.3.29) in the case of global supersymmetry: The product of left-chiral supermultiplets Φ_1 and Φ_2 is a left-chiral supermultiplet denoted $\Phi_1\Phi_2$, with components

$$\phi = \phi_1\phi_2\,, \tag{31.6.37}$$

$$\psi_L = \phi_1\psi_{2L} + \phi_2\psi_{1L}\,, \tag{31.6.38}$$

$$\mathscr{F} = \phi_1\mathscr{F}_2 + \phi_2\mathscr{F}_1 - \left(\psi_{1L}^{\mathrm{T}}\,\epsilon\,\psi_{2L}\right)\,. \tag{31.6.39}$$

Now we must consider how to construct actions that are invariant under local supersymmetry transformations as well as general coordinate transformations and local Lorentz transformations. Eqs. (31.6.18) and (31.6.21) show that the change under a supersymmetry transformation of the D-component of a general supermultiplet S is no longer simply a spacetime derivative, so the integral of such a D-component is not suitable as a term in the action. Instead, from an arbitrary superfield S we can form a density whose integral *is* supersymmetric

$$\begin{aligned}[S]_D = e\,\Bigg[\, & D^S - \frac{i\kappa}{2}\left(\bar{\psi}^\mu\gamma_\mu\gamma_5\lambda^S\right) + \frac{4\kappa}{3}[-sN^S + pM^S - b^\mu V^S_\mu] \\ & - \frac{i\kappa}{3}\left(\overline{\omega^S}\gamma_5\,\not{L}\right) - \frac{\kappa^2}{4}\epsilon^{\mu\rho\sigma\tau}V^S_\sigma\left(\bar{\psi}_\rho\gamma_\tau\psi_\mu\right) \\ & - \frac{\kappa^2}{8}\epsilon^{\mu\rho\sigma\tau}\left(\overline{\omega^S}\psi_\sigma\right)\left(\bar{\psi}_\rho\gamma_\tau\psi_\mu\right)\Bigg] - \frac{2\kappa^2}{3}C^S\mathscr{L}_{\mathrm{SUGRA}}\,, \end{aligned} \tag{31.6.40}$$

where $\mathscr{L}_{\rm SUGRA}$ is the supergravity Lagrangian density in Eq. (31.6.11):

$$\mathscr{L}_{\rm SUGRA} = -\frac{e}{2\kappa^2}R - \frac{1}{2}\left(\bar{\psi}_\mu L^\mu\right) - \frac{4e}{3}\left(s^2 + p^2 - b_\mu b^\mu\right) . \tag{31.6.41}$$

Likewise, Eqs. (31.6.36) and (31.6.20) show that the change under a supersymmetry transformation of the $\mathscr{F}$-component of a left-chiral superfield X is not a spacetime derivative, so terms must be added to this $\mathscr{F}$-term to form a density whose integral is supersymmetric:

$$[X]_{\mathscr{F}} = e\Bigg[\mathscr{F}^X + \frac{\kappa}{\sqrt{2}}\left(\bar{\psi}_{\mu R}\gamma^\mu \psi_L^X\right) + \frac{\kappa^2}{4}\left(\bar{\psi}_{\mu R}[\gamma^\mu , \gamma^\nu]\psi_{\nu R}\right)\phi^X$$
$$+2\kappa(s - ip)\phi^X\Bigg] . \tag{31.6.42}$$

There are some supermultiplets of special physical interest. One is the real non-chiral supermultiplet $\mathbf{1}$ whose only non-zero component is $C = 1$. According to Eq. (31.6.40), this supermultiplet has

$$[\mathbf{1}]_D = -\frac{2\kappa^2}{3}\mathscr{L}_{\rm SUGRA} , \tag{31.6.43}$$

so this gives nothing new.

A more interesting example is provided by a *left-chiral* supermultiplet $\mathbf{I}$ whose only non-zero component is $\phi = 1$. According to Eq. (31.6.42), this supermultiplet has

$${\rm Re}\,[\mathbf{I}]_{\mathscr{F}} = e\left[\frac{\kappa^2}{4}\left(\bar{\psi}_\mu[\gamma^\mu , \gamma^\nu]\psi_\nu\right) + 4\,\kappa\, s\right] . \tag{31.6.44}$$

If a term $c\int d^4x\,{\rm Re}\,[\mathbf{I}]_{\mathscr{F}}$ appears in the Lagrangian density with a real coefficient c, then the sum of this term and the supergravity action (31.6.11) is stationary with respect to variation of auxiliary fields at

$$s = 3\,c\,\kappa/2 , \qquad p = b^\mu = 0 , \tag{31.6.45}$$

so that after eliminating the auxiliary fields, the action will contain a cosmological constant term

$$3\,\kappa^2\,c^2 \int d^4x\, e , \tag{31.6.46}$$

corresponding to a vacuum energy density $-3\,\kappa^2\,c^2$. Such a term is needed to keep the vacuum state Lorentz-invariant when supersymmetry is spontaneously broken; to cancel the positive vacuum energy density $F^2/2$ associated with supersymmetry breaking, we must take

$$3\,\kappa^2\,c^2 = F^2/2 . \tag{31.6.47}$$

Looking back at Eq. (31.6.44) and comparing with Eq. (31.3.11), we see that this gives a gravitino mass[13]

$$m_g = c\kappa^2 = \frac{F\kappa}{\sqrt{6}} = \sqrt{\frac{4\pi G F^2}{3}}\,, \tag{31.6.48}$$

in agreement with our previous result (31.3.17).

As an illustration of the use of these formulas that will also provide results needed in the next section, let us calculate the bosonic part of the Lagrangian density*

$$\mathscr{L} = \mathscr{L}_{\rm SUGRA} + \tfrac{1}{2}[K(\Phi,\Phi^*)]_D + 2{\rm Re}\,[f(\Phi)]_{\mathscr{F}}\,, \tag{31.6.49}$$

where $\mathscr{L}_{\rm SUGRA}$ is here the bosonic part of the supergravity Lagrangian density (31.6.41), $K(\Phi,\Phi^*)$ is a real function of a set of left-chiral supermultiplets Φ_n and their complex conjugates, and $f(\Phi)$ is a function of the Φ_n alone. The purely bosonic terms in the multiplication rules (31.6.23), (31.6.25)–(31.6.27), and (31.6.29) are the same as in global supersymmetry, so we can use either these rules or the superspace formalism of Chapter 26 to calculate that the supermultiplet $K(\Phi,\Phi^*)$ has the bosonic components

$$C^K = K(\phi,\phi^*)\,, \tag{31.6.50}$$

$$M^K = -2\,{\rm Im}\sum_n\left(\frac{\partial K(\phi,\phi^*)}{\partial\phi_n}\mathscr{F}_n\right) + \cdots\,, \tag{31.6.51}$$

$$N^K = -2\,{\rm Re}\sum_n\left(\frac{\partial K(\phi,\phi^*)}{\partial\phi_n}\mathscr{F}_n\right) + \cdots\,, \tag{31.6.52}$$

$$V^K_\mu = 2\,{\rm Im}\sum_n\left(\frac{\partial K(\phi,\phi^*)}{\partial\phi_n}\partial_\mu\phi_n\right) + \cdots\,, \tag{31.6.53}$$

$$D^K = 2\sum_{nm}\frac{\partial^2 K(\phi,\phi^*)}{\partial\phi_n\partial\phi_m^*}\Big(-g^{\mu\nu}\,\partial_\mu\phi_n\,\partial_\nu\phi_m^* + \mathscr{F}_n\mathscr{F}_m^*\Big) + \cdots\,, \tag{31.6.54}$$

where the dots indicate terms involving fermion fields. Also, the rules (31.6.37)–(31.6.39) for multiplying left-chiral supermultiplets are the same as those in global supersymmetry, so we can use either these rules or the superfield formalism of Chapter 26 to calculate that the left-chiral supermultiplet $f(\Phi)$ has the bosonic components

$$\phi^f = f(\phi)\,, \tag{31.6.55}$$

$$\mathscr{F}^f = \sum_n\frac{\partial f(\phi)}{\partial\phi_n}\mathscr{F}_n + \cdots\,, \tag{31.6.56}$$

* Often the term $\mathscr{L}_{\rm SUGRA}$ is omitted, with the supergravity Lagrangian introduced instead by including a constant term $-3/\kappa^2$ in $K(\phi,\phi^*)$. We will not follow this practice; here, with a conventional normalization of the scalar fields, the term in $K(\phi,\phi^*)$ of leading order in κ is $\sum_n|\phi_n|^2$.

with dots again indicating terms involving fermions. Inserting these results into Eqs. (31.6.40) and (31.6.42) then gives the bosonic terms of the Lagrangian density (31.6.49) as

$$\begin{aligned}\mathscr{L}_{\text{bosonic}} = &\left[-\frac{e}{2\kappa^2}R - \frac{4e}{3}\left(s^2+p^2-b_\mu b^\mu\right)\right]\left[1-\frac{\kappa^2}{3}K(\phi,\phi^*)\right] \\ &-e\sum_{nm}\frac{\partial^2 K(\phi,\phi^*)}{\partial\phi_n\,\partial\phi_m^*}\left(g^{\mu\nu}\partial_\mu\phi_n\partial_\nu\phi_m^* - \mathscr{F}_n\mathscr{F}_m^*\right) \\ &+\frac{4\kappa e}{3}\text{Re}\sum_n\frac{\partial K(\phi,\phi^*)}{\partial\phi_n}\left(\mathscr{F}_n(s+ip)+ib^\mu\partial_\mu\phi_n\right) \\ &+2e\,\text{Re}\left(\sum_n\frac{\partial f(\phi)}{\partial\phi_n}\mathscr{F}_n + 2\kappa(s-ip)f(\phi)\right) . \end{aligned} \tag{31.6.57}$$

Now we must eliminate the auxiliary fields by setting them at values where the Lagrangian density is stationary.** This gives the auxiliary fields

$$\begin{aligned}\mathscr{F}_n = &\frac{\kappa^2}{3N}\sum_m\left(\mathscr{G}^{-1}\right)_{mn}\frac{\partial K}{\partial\phi_m^*}\left(-\sum_{k\ell}\left(\mathscr{G}^{-1}\right)_{k\ell}\frac{\partial K}{\partial\phi_\ell}\left(\frac{\partial f}{\partial\phi_k}\right)^* + 3f^*\right) \\ &-\sum_m\left(\mathscr{G}^{-1}\right)_{mn}\left(\frac{\partial f}{\partial\phi_m}\right)^* , \end{aligned} \tag{31.6.58}$$

$$s-ip = \frac{\kappa}{2N}\left(-\sum_{\ell k}\left(\mathscr{G}^{-1}\right)_{k\ell}\frac{\partial K}{\partial\phi_\ell}\left(\frac{\partial f}{\partial\phi_k}\right)^* + 3f^*\right) , \tag{31.6.59}$$

$$b_\mu = \frac{\kappa}{2(1-\kappa^2K/3)}\text{Im}\left(\sum_n\frac{\partial K}{\partial\phi_n}\partial_\mu\phi_n\right) , \tag{31.6.60}$$

where

$$N \equiv 1 - \frac{\kappa^2}{3}K + \frac{\kappa^2}{3}\sum_{k\ell}\left(\mathscr{G}^{-1}\right)_{k\ell}\frac{\partial K}{\partial\phi_\ell}\frac{\partial K}{\partial\phi_k^*}$$

and $\mathscr{G}(\phi,\phi^*)$ is the Kahler metric

$$\mathscr{G}_{nm}(\phi,\phi^*) \equiv \frac{\partial^2 K(\phi,\phi^*)}{\partial\phi_n\,\partial\phi_m^*} . \tag{31.6.61}$$

** Though this procedure is generally followed, it is not strictly correct, for even though the Lagrangian density is quadratic in the auxiliary fields, the coefficients of the terms of second order in the auxiliary fields are not field-independent. In consequence, in doing the path integration over auxiliary fields we encounter determinants of the coefficients of the quadratic terms, which are equivalent to adding terms to the Lagrangian proportional to $\delta^4(0) = (2\pi)^{-4}\int d^4k\,1$. Such terms can be eliminated by using dimensional regularization, for which $\int d^4k\,1 = 0$.

Using this in Eq. (31.6.57) gives the bosonic Lagrangian

$$\begin{aligned}\mathscr{L}_{\text{bosonic}} = &-\frac{e}{2\kappa^2}R\left[1-\frac{\kappa^2}{3}K\right] - e\sum_{nm}\mathscr{G}_{nm}\,g^{\mu\nu}\partial_\mu\phi_n\partial_\nu\phi_m^*\\ &+\frac{e\kappa^2}{3N}\left|\sum_{mn}\left(\mathscr{G}^{-1}\right)_{nm}\frac{\partial f}{\partial\phi_m}\frac{\partial K}{\partial\phi_n^*}-3f\right|^2\\ &-\frac{e\kappa^2}{3\left(1-\kappa^2K/3\right)}\,\text{Im}\left[\sum_n\frac{\partial K}{\partial\phi_n}\partial_\mu\phi_n\right]\,\text{Im}\left[\sum_n\frac{\partial K}{\partial\phi_n}\partial_\nu\phi_n\right]g^{\mu\nu}\\ &-e\sum_{mn}\left(\mathscr{G}^{-1}\right)_{nm}\frac{\partial f}{\partial\phi_m}\left(\frac{\partial f}{\partial\phi_n}\right)^*\,. \end{aligned} \qquad (31.6.62)$$

As already noted in the weak field case in Section 31.2, the Lagrangian density (31.6.62) has the uncomfortable feature that the Einstein–Hilbert term $-eR/2\kappa^2$ is multiplied with a factor $(1-\kappa^2K(\phi,\phi^*)/3)$, so that the effective gravitational constant varies from point to point in spacetime. To remedy this, we perform a Weyl transformation, defining a new metric by

$$\tilde{g}_{\mu\nu} = \left(1-\kappa^3K/3\right)\,g_{\mu\nu}\,. \qquad (31.6.63)$$

The Einstein Lagrangian density is given in terms of the new metric by

$$eg^{\mu\nu}R_{\mu\nu} = \left(1-\frac{\kappa^2}{3}K\right)^{-1}\tilde{e}\,\tilde{g}^{\mu\nu}\left(\tilde{R}_{\mu\nu}+\frac{3}{2}\partial_\mu\ln\left(1-\frac{\kappa^2}{3}\right)\partial_\nu\ln\left(1-\frac{\kappa^2}{3}\right)\right)\,,$$

where $\tilde{R}_{\mu\nu}$ is the curvature tensor calculated using a metric $\tilde{g}_{\mu\nu}$ in place of $g_{\mu\nu}$, and $\tilde{e}\equiv\sqrt{\text{Det}\,\tilde{g}}$. A straightforward calculation then gives the bosonic Lagrangian (31.6.62) as

$$\begin{aligned}\mathscr{L}_{\text{bosonic}} = &-\frac{\tilde{e}}{2\kappa^2}\tilde{g}^{\mu\nu}\,\tilde{R}_{\mu\nu}\\ &-\tilde{e}\,\tilde{g}^{\mu\nu}\sum_{nm}\partial_\mu\phi_n\partial_\nu\phi_m^*\left[\left(1-\frac{\kappa^2}{3}K\right)^{-1}\frac{\partial^2K}{\partial\phi_n\phi_m^*}\right.\\ &\left.\qquad+\frac{\kappa^2}{3}\left(1-\frac{\kappa^2}{3}K\right)^{-2}\frac{\partial K}{\partial\phi_n}\frac{\partial K}{\partial\phi_m^*}\right]\\ &+\frac{\tilde{e}\,\kappa^2}{3N}\left(1-\frac{\kappa^2}{3}K\right)^{-2}\left|\sum_{mn}\left(\mathscr{G}^{-1}\right)_{nm}\frac{\partial f}{\partial\phi_m}\frac{\partial K}{\partial\phi_n^*}-3f\right|^2\\ &-\tilde{e}\left(1-\frac{\kappa^2}{3}K\right)^{-2}\sum_{mn}\left(\mathscr{G}^{-1}\right)_{nm}\frac{\partial f}{\partial\phi_m}\left(\frac{\partial f}{\partial\phi_n}\right)^*\,. \end{aligned} \qquad (31.6.64)$$

The Weyl transformation has not only removed the factor $(1-\kappa^2K/3)$

from the Einstein–Hilbert term; it has also eliminated terms proportional to $\partial_\mu \phi_n \, \partial_\nu \phi_m$ and $\partial_\mu \phi^*_n \, \partial_\nu \phi^*_m$.

This result may be further simplified by introducing a *modified Kahler potential* $d(\phi, \phi^*)$ in place of $K(\phi, \phi^*)$, which we define by

$$1 - \frac{\kappa^2}{3} K \equiv \exp\left(-\frac{\kappa^2 d}{3}\right) . \tag{31.6.65}$$

We also introduce a new metric on the scalar field space

$$g_{nm} \equiv \frac{\partial^2 d}{\partial \phi_n \, \partial \phi^*_m} . \tag{31.6.66}$$

The reciprocals of the new and old metrics are related by

$$\mathscr{G}^{-1}_{\ell k} = \exp(\kappa^2 d/3) \left[g^{-1}_{\ell k} + \frac{\kappa^2}{3} \frac{\sum_{mn} g^{-1}_{\ell n} g^{-1}_{mk} (\partial d/\partial \phi_n)(\partial d/\partial \phi^*_m)}{1 - (\kappa^2/3) \sum_{mn} g^{-1}_{mn} (\partial d/\partial \phi_n)(\partial d/\partial \phi^*_m)} \right] .$$

The bosonic Lagrangian density (31.6.64) now takes the simpler form

$$\mathscr{L}_{\text{bosonic}} = -\frac{\tilde{e}}{2\kappa^2} \tilde{g}^{\mu\nu} \tilde{R}_{\mu\nu} - \tilde{e} \, \tilde{g}^{\mu\nu} \sum_{nm} g_{nm} \partial_\mu \phi_n \partial_\nu \phi^*_m - \tilde{e} V , \tag{31.6.67}$$

where $V(\phi, \phi^*)$ is the potential

$$V = \exp(\kappa^2 d) \left[\sum_{nm} g^{-1}_{nm} L_m \, L^*_n \; - 3 \, \kappa^2 \, |f|^2 \right] \tag{31.6.68}$$

and

$$L_m \equiv \frac{\partial f}{\partial \phi_m} + \kappa^2 f \frac{\partial d}{\partial \phi_m} . \tag{31.6.69}$$

The potential (31.6.68) has an obvious stationary point at field strengths satisfying the condition $L_m = 0$. However, as we found in the weak field case, at this point the vacuum energy in general takes the negative value $-3\kappa^2 |f|^2$. To have the stationary point $L_m = 0$ give a solution with flat space, it is necessary that both $f(\phi)$ and $\partial f(\phi)/\partial \phi_n$ should vanish at these field values. Inspection of Eqs. (31.6.58) and (31.6.59) shows that the scalar auxiliary fields $\mathscr{F}_n$, s, and p vanish for such field values, so that the vacuum expectation values of the variations (31.6.2) and (31.6.35) of the gravitino and chiral spinor fields under global supersymmetry transformations with constant α vanish. Thus a vacuum field value for which $f(\phi)$ and $\partial f(\phi)/\partial \phi_n$ all vanish is one for which global supersymmetry is unbroken in the classical limit. In the next section we shall consider vacuum configurations in which supersymmetry *is* broken.

We will not show it here, but the additional terms in the bosonic Lagrangian required by the inclusion of gauge superfields are unaffected by gravitation, aside from an over-all determinantal factor $\tilde{e}$ and metric factors needed to raise and lower indices. After elimination of auxiliary fields and a Weyl transformation, the complete bosonic Lagrangian for a theory with gauge as well as chiral and gravitational superfields is

$$\mathscr{L}_{\text{bosonic}}/\tilde{e} = -\frac{1}{2\kappa^2}\tilde{R}^\mu{}_\mu - \sum_{nm} g_{nm} D_\mu \phi_n D^\mu \phi_m^* - \tfrac{1}{4}\sum_{AB} \text{Re}\, f_{AB}\, F^A_{\mu\nu} F^{B\mu\nu}$$
$$-\tfrac{1}{8}\sum_{AB} \text{Im}\, f_{AB}\, F^A_{\mu\nu} F^B_{\rho\sigma}\epsilon^{\mu\nu\rho\sigma} - V\,. \tag{31.6.70}$$

Here D_μ, $F^A_{\mu\nu}$, and t_A denote the gauge-covariant derivatives, the field-strength tensors, and the representatives of the gauge generators on the chiral scalar superfields, using the notation described in Section 15.1; f_{AB} is an independent holomorphic function of the ϕ_n; all spacetime indices are raised and lowered with $\tilde{g}_{\mu\nu}$; and the potential V now takes the form

$$V = \exp(\kappa^2 d)\left[\sum_{nm} g^{-1}_{nm} L_m L^*_n - 3\,\kappa^2 |f|^2\right]$$
$$+\frac{1}{2}\text{Re}\sum_{AB} f^{-1}_{AB}\left(\sum_{nm}\frac{\partial d}{\partial \phi_n}(t_A)_{nm}\phi_m\right)\left(\sum_{kl}\frac{\partial d}{\partial \phi_k}(t_B)_{kl}\phi_l\right)^*. \tag{31.6.71}$$

The form (31.6.71) of the bosonic potential is simple enough to make it apparent that terms in d that depend only on ϕ_n or only on ϕ^*_n may be traded for corrections to the superpotential. Specifically, if we write

$$d(\phi,\phi^*) = \tilde{d}(\phi,\phi^*) + a(\phi) + a(\phi)^*\,, \qquad f(\phi) = \tilde{f}(\phi)\exp\left(-\kappa^2 a(\phi)\right), \tag{31.6.72}$$

with $a(\phi)$ an arbitrary holomorphic function satisfying the gauge invariance condition

$$\sum_{nm}\frac{\partial a(\phi)}{\partial \phi_n}(t_A)_{nm}\phi_m = 0\,,$$

then the potential (31.6.71) takes the same form in terms of $\tilde{d}$ and $\tilde{f}$ as it did in terms of d and f. With a suitable redefinition of the superpotential, we can then eliminate any terms in the power series expansion of $d(\phi,\phi^*)$ that depend only on ϕ_n or only on ϕ^*_n. With this understanding, the leading term in the power series expansion of $d(\phi,\phi^*)$ (now dropping the tilde) is of the form $\sum_{nm} d_{nm}\phi_n\phi^*_m$. By a suitable linear transformation of the superfields, we can then make the matrix d_{nm} equal to δ_{nm}, so that the power series expansion of $d(\phi,\phi^*)$ begins

$$d(\phi,\phi^*) = \sum_n |\phi_n|^2 + \cdots\,, \tag{31.6.73}$$

and the power series expansion of the metric (31.6.66) begins

$$g_{nm} = \delta_{nm} + \cdots\,. \tag{31.6.74}$$

Inspection of the second term on the right-hand side of Eq. (31.6.67) shows that the scalar fields defined in this way are canonically normalized.

The fermion terms are much more complicated. Here we will quote only the terms quadratic in the gaugino fields

$$\mathscr{L}^{(2)}_{\rm gaugino}/\tilde{e} = -\frac{1}{2}\,{\rm Re}\sum_{AB} f_{AB}\left(\bar{\lambda}_A \not{D} \lambda_B\right)$$
$$+\frac{1}{2}\exp(\kappa^2 d/2)\,{\rm Re}\sum_{mn}\sum_{AB} g^{-1}_{nm} L_m \left(\frac{\partial f_{AB}}{\partial \phi_n}\right)^* \left(\bar{\lambda}_A \lambda_B\right)\,, \quad (31.6.75)$$

with L_m given by Eq. (31.6.69). We see that if the gauge fields are canonically normalized, then the constant term in the expansion of the function f_{AB} in powers of the scalar fields is δ_{AB}, and then the gaugino fields λ_A are also canonically normalized.

* * *

Instead of moving all the holomorphic terms in $d(\phi,\phi^*)$ and their complex conjugates into the superpotential, we can use the transformation (31.6.72) to make the new superpotential $\tilde{f}(\phi)$ equal to a constant, which can be chosen to be equal to unity, by taking $a(\phi) = -\kappa^{-2}\ln f(\phi)$. The potential then depends only on the function

$$\mathscr{D}(\phi,\phi^*) \equiv d(\phi,\phi^*) + 2\kappa^{-2}{\rm Re}\,\ln f(\phi)\,, \quad (31.6.76)$$

and takes the form

$$V = \exp(\kappa^2 \mathscr{D})\left[\kappa^4 \sum_{nm} g^{-1}_{nm}\left(\frac{\partial \mathscr{D}}{\partial \phi_m}\right)\left(\frac{\partial \mathscr{D}}{\partial \phi_n}\right)^* - 3\kappa^2\right]$$
$$+\frac{1}{2}{\rm Re}\sum_{AB} f^{-1}_{AB}\left(\sum_{nm}\frac{\partial \mathscr{D}}{\partial \phi_n}(t_A)_{nm}\phi_m\right)\left(\sum_{kl}\frac{\partial \mathscr{D}}{\partial \phi_k}(t_B)_{kl}\phi_l\right)^*\,. \quad (31.6.77)$$

Also, the metric (31.6.66) for the scalar fields may be written

$$g_{nm} = \frac{\partial^2 \mathscr{D}}{\partial \phi_n \partial \phi^*_m}\,. \quad (31.6.78)$$

Although we have not shown it here, the symmetry that allows us to replace the Kahler potential and superpotential with the single function $\mathscr{D}(\phi,\phi^*)$ also allows us to make this replacement in the whole Lagrangian, including all terms involving fermions and gauge fields.

There is an interesting class of 'no-scale' theories,[13a] in which the potential V vanishes for all values of ϕ_m. For instance, this is the case for a single gauge-neutral chiral scalar superfield, with

$$\mathscr{D} = -3\kappa^{-2}\ln\left(h(\phi) + h(\phi)^*\right)\,, \quad (31.6.79)$$

where $h(\phi)$ is an arbitrary function of ϕ. But there is no known principle that would require $\mathscr{D}$ to take this form.

31.7 Gravity-Mediated Supersymmetry Breaking

We now take up again the problem of supersymmetry breaking. As discussed at the beginning of Chapter 28, if supersymmetry is to be of use in solving the hierarchy problem — that is, in understanding the large ratio of the Planck mass $m_{\rm Pl} \equiv 1/\sqrt{8\pi G}$ to the mass scale of observed particles — then supersymmetry must be unbroken at the Planck scale, and broken spontaneously only at some much lower mass scale. The only plausible mechanism known that would naturally produce a very large ratio of mass scales is the non-perturbative effect of asymptotically free gauge interactions. If these interactions are moderately weak at the Planck scale then their slow growth with decreasing energy will make them become strong at a much lower scale $\Lambda \ll m_{\rm Pl}$. The known elementary particles do not feel such strong forces, so whatever supersymmetry breakdown is produced directly or indirectly by these strong gauge interactions must be communicated to the observed particles by some interaction in which they do participate.

In Section 28.3 we noted two possible mechanisms for communicating the breakdown of supersymmetry to observable particles. One mechanism, gauge-mediated supersymmetry breaking, was discussed in detail in Section 28.6. We are now ready to consider the other mechanism, mediation of supersymmetry breaking by effects of gravitational strength.

In the early 1980s, when gravitation was first considered as the mediator of supersymmetry breaking,[14] it was generally assumed that the superpotential consisted of two terms: a function $f(\Phi)$ of various left-chiral superfields Φ_r of an *observable sector*, including all the superfields of observable particles, plus a function $\tilde{f}(Z)$ of various left-chiral superfields Z_k of a *hidden sector*,[15] all of which are neutral under the $SU(3)\times SU(2)\times U(1)$ gauge group of the standard model. Further, the superpotential of the hidden sector was assumed to take the form

$$\tilde{f}(Z) = \epsilon^3 F(\kappa Z)\,, \tag{31.7.1}$$

where ϵ is some mass that is much less than the Planck mass, and $F(\kappa Z)$ is a power series in κZ with coefficients of order unity. The assumption that the total superpotential should be a sum $f(\Phi) + \tilde{f}(Z)$ is somewhat arbitrary but, as we shall see, it is not difficult to think of reasons why this should be at least approximately true. A more serious criticism of this approach is that it did not offer any hope of solving the hierarchy problem; the energy ϵ was simply assumed to be much less than the Planck mass.

After the development of these first models of gravity-mediated supersymmetry breaking there appeared other models in which the hierarchy of energy scales is explained naturally, in terms of the slow growth with decreasing energy of a gauge coupling which becomes strong at an energy $\Lambda \ll m_{\rm Pl}$. These models come in two versions that are distinguished by different assumptions regarding the source of supersymmetry breaking. As we shall see, in both versions the squarks and sleptons get supersymmetry-breaking masses of the order of the gravitino mass m_g, but they differ in the formula for m_g; in the first version $m_g \approx \kappa\Lambda^2$, while in the second version $m_g \approx \kappa^2\Lambda^3$, giving $\Lambda \approx 10^{11}$ GeV and $\Lambda \approx 10^{13}$ GeV, respectively. The two versions will turn out also to differ in the formulas for other soft supersymmetry-breaking parameters, including $B\mu$, the A-parameters, and the gaugino masses.

First Version[16]

In this version of gravity-mediated supersymmetry breaking it is assumed that the superfields of the theory fall into two sectors:

Observable Sector: These are the superfields of the minimum supersymmetric standard model: the $SU(3) \times SU(2) \times U(1)$ gauge superfields together with the quark, antiquark, lepton, antilepton, and Higgs left-chiral superfields which we will generically call Φ_r.

Hidden Sector: These are gauge superfields of an asymptotically free gauge interaction that becomes strong at an intermediate energy scale Λ with $m_W \ll \Lambda \ll m_{\rm Pl}$, together with left-chiral superfields Z_k that feel this gauge interaction.

The Z_k must be assumed to be neutral under the $SU(3) \times SU(2) \times U(1)$ gauge group, since otherwise we would be back in the case of gauge-mediated supersymmetry breaking. Also, we know enough about the observable sector to be sure that its chiral superfields do not feel the gauge interactions of the hidden sector.

In order naturally to have the renormalizable part of the total superpotential of the form $f(\Phi) + \tilde{f}(Z)$, we can assume that the symmetries that survive below the Planck scale include a group G_H (which may be part of the gauge group of the hidden sector), under which all of the fields of the observable sector and none of the fields of the hidden sector are invariant, and a group G_O (which may be part of the $SU(3) \times SU(2) \times U(1)$ gauge group of the observable sector), under which all of the fields of the hidden sector and none of the fields of the observable sector are invariant. In this case if any observable sector fields appear in a term of the superpotential then there must be at least two of them, and also if any hidden sector fields appear in a term of the superpotential then there must be at least two of them, so there can be no term in a cubic polynomial superpotential that involves both hidden and observable sector fields. This argument leaves open the possibility of non-renormalizable terms in the superpotential that involve two or more factors of both hidden and observable sector superfields, a possibility to which we will return later. Of course, we are assuming that the strong interactions of the hidden sector produce additional non-perturbative terms in the total hidden sector superpotential $\tilde{f}(Z)$, but these too depend only on the hidden sector superfields.

Assuming then that the superpotential takes the form $f(\Phi) + \tilde{f}(Z)$, the potential of the scalar components of these superfields is given by Eq. (31.6.71) as

$$\begin{aligned} V = e^{\kappa^2 d}\Bigg[& \sum_{rs} g^{-1}_{rs}\left(\frac{\partial f}{\partial \phi_r} + \kappa^2 (f+\tilde{f})\frac{\partial d}{\partial \phi_r}\right)\left(\frac{\partial f}{\partial \phi_s} + \kappa^2 (f+\tilde{f})\frac{\partial d}{\partial \phi_s}\right)^* \\ & + 2\mathrm{Re}\sum_{rk} g^{-1}_{\bar{r}k}\left(\frac{\partial f}{\partial \phi_r} + \kappa^2 (f+\tilde{f})\frac{\partial d}{\partial \phi_r}\right)\left(\frac{\partial \tilde{f}}{\partial z_k} + \kappa^2 (f+\tilde{f})\frac{\partial d}{\partial z_k}\right)^* \\ & + \sum_{kl} g^{-1}_{\bar{k}l}\left(\frac{\partial \tilde{f}}{\partial z_k} + \kappa^2 (f+\tilde{f})\frac{\partial d}{\partial z_k}\right)\left(\frac{\partial \tilde{f}}{\partial z_l} + \kappa^2 (f+\tilde{f})\frac{\partial d}{\partial z_l}\right)^* \\ & - 3\kappa^2 \left|f+\tilde{f}\right|^2 \Bigg] \\ & + \frac{1}{2}\mathrm{Re}\sum_{AB} f^{-1}_{AB}\left(\sum_{kl.}\frac{\partial d}{\partial z_k}(t_A)_{kl} z_l\right)\left(\sum_{mn}\frac{\partial d}{\partial z_m}(t_B)_{mn} z_n\right)^* \\ & + \frac{1}{2}\mathrm{Re}\sum_{AB} f^{-1}_{AB}\left(\sum_{rs}\frac{\partial d}{\partial \phi_r}(t_A)_{rs}\phi_s\right)\left(\sum_{tu}\frac{\partial d}{\partial \phi_t}(t_B)_{tu}\phi_u\right)^* . \end{aligned} \tag{31.7.2}$$

In writing the terms arising from gauge interactions, we are assuming here that there is no mixing of the gauge bosons of the hidden and observable sectors — that is, f^{-1}_{AB} vanishes for any pair of gauge generators t_A and t_B, for which t_A acts non-trivially on the ϕ_r and t_B acts non-trivially on the z_k, or vice-versa.

We are interested in exploring a region of field space where the scalar fields z_k of the hidden sector are of order Λ, the variable part of the superpotential $\tilde{f}(z)$ of the hidden sector is of order Λ^3, and, on dimensional grounds, $\partial\tilde{f}/\partial z_k$ is of order Λ^2. We will leave the magnitude of the constant part of $\tilde{f}$ open for the moment; as we shall see, we must include a constant term in $\tilde{f}$ that is much larger than Λ^3 in order to cancel the cosmological constant.

We further define the region of field space to be explored as one in which the fields ϕ_r of the observable sector are of order $\kappa\Lambda^2$ because, as we shall see, this is the characteristic mass scale arising in the observable sector from gravitational effects of supersymmetry breaking in the hidden sector. The observable sector superpotential $f(\phi)$ for fields of order $\kappa\Lambda^2$ is assumed to be of order $\kappa^3\Lambda^6$, and its derivatives $\partial f(\phi)/\partial\phi_r$ are taken to be of order $\kappa^2\Lambda^4$.

In consequence of the definition of the superpotential and the scalar fields discussed at the end of the previous section, and the symmetry $G_H \times G_O$ assumed above, the modified Kahler potential takes the form*

$$
\begin{aligned}
d(\phi,\phi^*,z,z^*) = &\sum_r |\phi_r|^2 + \sum_k |z_k|^2 + O(\kappa^2 z^{*2} z^2) \\
&+O(\kappa^2 z^* z^3) + O(\kappa^2 z^{*3} z) + O(\kappa^2 \phi^{*2} z^2) + O(\kappa^2 \phi^{*2} z^* z) \\
&+O(\kappa^2 z^{*2}\phi^2) + O(\kappa^2 z^* z \phi^2) + O(\kappa^2 \phi^* \phi z^2) \\
&+O(\kappa^2 z^{*2}\phi^*\phi) + O(\kappa^2 \phi^* z^* \phi z) + O(\kappa^2 \phi^{*2}\phi^2) \\
&+O(\kappa^2 \phi^* \phi^3) + O(\kappa^2 \phi^{*3}\phi) + \cdots \,,
\end{aligned}
\tag{31.7.3}
$$

with dots indicating terms of higher order. The metric (31.6.66) then has components

$$
\begin{aligned}
g_{rs} = \delta_{rs} &+ O(\kappa^2 z^2) + O(\kappa^2 z^{*2}) + O(\kappa^2 z^* z) \\
&+O(\kappa^2\phi^2) + O(\kappa^2\phi^{*2}) + O(\kappa^2\phi^*\phi) + \cdots \,,
\end{aligned}
\tag{31.7.4}
$$

$$
\begin{aligned}
g_{kl} = \delta_{kl} &+ O(\kappa^2 z^2) + O(\kappa^2 z^{*2}) + O(\kappa^2 z^* z) \\
&+O(\kappa^2\phi^2) + O(\kappa^2\phi^{*2}) + O(\kappa^2\phi^*\phi) + \cdots \,,
\end{aligned}
\tag{31.7.5}
$$

$$
g_{rk} = g^*_{kr} = O(\kappa^2\phi z^*) + O(\kappa^2\phi z) + O(\kappa^2\phi^* z^*) + O(\kappa^2\phi^* z) + \cdots \,. \tag{31.7.6}
$$

(The characteristic energy scale in d is assumed to be $1/\kappa$, because d is the modified Kahler potential produced by unknown dynamical effects at the Planck scale, in contrast with $\tilde{f}$, which gets its structure from dynamical effects of the strong gauge couplings at scale Λ.) The g_{rs} and g_{kl} are generically of order unity, while the mixed components g_{rk} and g_{kr} are of order of $\kappa^2(\kappa\Lambda^2)\Lambda = \kappa^3\Lambda^3 \ll 1$. It follows that the same is also true of the components of g^{-1}: $(\mathrm{g}^{-1})_{rs}$ and $(\mathrm{g}^{-1})_{kl}$ are generically of order unity, while the $(\mathrm{g}^{-1})_{rk}$ and $(\mathrm{g}^{-1})_{kr}$ are of order of $\kappa^3\Lambda^3 \ll 1$.

It follows from these estimates that unless cancellations intervene the dominant terms in the potential (31.7.2) will be at least of order Λ^4, and

* By $O(\kappa^2 z^{*2} z^2)$ is meant a term of the form $\kappa^2 \sum_{klmn} C_{klmn} z^*_k z^*_l z_m z_n$ with constant coefficients C_{klmn} of order unity, and likewise for the other terms in Eqs. (31.7.3)–(31.7.6).

take the form

$$[V]_{\Lambda^4} = \sum_k \left|\frac{\partial \tilde{f}}{\partial z_k}\right|^2 - 3\kappa^2 \left|\tilde{f}^0\right|^2 , \tag{31.7.7}$$

with $\tilde{f}^0$ the constant term in $\tilde{f}$, which will be needed to cancel the vacuum energy. We assume that supersymmetry is spontaneously broken in the hidden sector, which requires that there is a point z_k^0 at which $\sum_k |\partial\tilde{f}/\partial z_k|^2$ is at least a local minimum, but not zero. Then in order to cancel the vacuum energy to this order, we must take

$$3\kappa^2 \left|\tilde{f}^0\right|^2 = \sum_k \left|\left(\frac{\partial \tilde{f}}{\partial z_k}\right)^0\right|^2 , \tag{31.7.8}$$

the superscript 0 on the right-hand side indicating that the quantity is to be evaluated at $z = z^0$. Hence $\tilde{f}^0$ must be given an anomalously large value, of order Λ^2/κ. This is a more extreme fine tuning than will turn out to be necessary in the second version of gravitationally mediated supersymmetry breaking but, in the absence of a real understanding of the cosmological constant, some fine tuning will be necessary in any theory of supersymmetry breaking.

We can calculate the supersymmetry-breaking parameter F in our formula (31.3.17) for the gravitino mass by setting the vacuum energy density $F^2/2$ equal to the flat-space value $\sum_k |(\partial\tilde{f}/\partial z_k)^0|^2$. The gravitino mass is then given by Eqs. (31.3.17) and (31.7.8) as

$$m_g = \kappa\sqrt{\frac{1}{3}\sum_k \left|\left(\frac{\partial \tilde{f}(z)}{\partial z_k}\right)^0\right|^2} - \kappa^2 \left|\tilde{f}^0\right| . \tag{31.7.9}$$

This is of the same order $\approx \kappa\Lambda^2$ as the scalar fields of the observable sector.

Now let us turn to the terms in Eq. (31.7.2) that *do* depend on the observable sector scalars ϕ_r. We are considering field values for which the usual supersymmetric term $\sum_r |\partial f/\partial\phi_r|^2$ is of order $m_g^4 \approx \kappa^4\Lambda^8$, so we have to collect all ϕ-dependent terms in Eq. (31.7.2) of this order or greater. Let us look in turn at each of the six lines on the right-hand side of Eq. (31.7.2).

The leading term in $\kappa^2(f+\tilde{f})\partial d/\partial\phi_r$ is $\kappa^2\tilde{f}^0\phi_r^*$, which like $\partial f/\partial\phi_r$ is of order $\kappa^2(\Lambda^2/\kappa)(\kappa\Lambda^2) = \kappa^2\Lambda^4$, while other terms in $\kappa^2(f+\tilde{f})\partial d/\partial\phi_r$ are much smaller. To leading order we can approximate $\exp(\kappa^2 d)$ by unity and g_{rs}^{-1} by δ_{rs}, so in this order the first line of Eq. (31.7.2) gives

$$\sum_r \left|\frac{\partial f}{\partial\phi_r} + \kappa^2\tilde{f}^0\phi_r^*\right|^2 .$$

This is of the desired order $\kappa^4\Lambda^8$, so it is not necessary to consider higher-order corrections.

The leading term in $g^{-1}_{\bar{r}k}$ is of order $\kappa^3\Lambda^3$; the leading term in $\partial\tilde{f}/\partial z_k$ is of order Λ^2, while $\kappa^2(f+\tilde{f})\partial d/\partial z_k$ is smaller, of order $\kappa\Lambda^3$; and we have seen that the leading term in $\partial f/\partial\phi_r + \kappa^2(f+\tilde{f})\partial d/\partial\phi_r$ is of order $\kappa^2\Lambda^4$, so the term on the second line of Eq. (31.7.2) is of order $(\kappa^3\Lambda^3)(\Lambda^2)(\kappa^2\Lambda^4) = \kappa^5\Lambda^9$, which may be neglected in comparison with the terms in Eq. (31.7.2) of order $\kappa^4\Lambda^8$.

The leading terms in the third and fourth lines of Eq. (31.7.2) are of order Λ^4, but these are independent of ϕ_r. There is a ϕ dependence coming from the terms on the third line involving f; the leading terms of this sort are $2\kappa^2\text{Re}\,[f\,\sum_k z_k^*(\partial\tilde{f}/\partial z_k)^*]$, which are of order $\kappa^5\Lambda^9 \ll \kappa^4\Lambda^8$, and hence may be neglected. Also, there is a ϕ dependence arising from the terms on the fourth line of Eq. (31.7.2) involving f. The leading terms of this sort are

$$-6\kappa^2\text{Re}\,[f\tilde{f}^{0*}]$$

which are of order $\kappa^2(\kappa\Lambda^2)^3\Lambda^2/\kappa = \kappa^4\Lambda^8$. The factor $\exp(\kappa^2 d)$ contains ϕ-dependent terms of order $\kappa^2(\kappa\Lambda^2)^2$, but these multiply a potential whose leading terms (31.7.7) have been adjusted to cancel to order Λ^4, so the ϕ-dependent terms arising from this source are much less than of order $\kappa^4\Lambda^8$. There is one other type of ϕ-dependent term, arising from the ϕ-dependent terms in $g^{-1}_{\bar{k}l}$. According to Eqs. (31.7.4)–(31.7.6), these terms may be written as $\kappa^2 u_{kl}(\phi,\phi^*)$, with u_{kl} a homogeneous quadratic polynomial in ϕ_r and ϕ_r^* with coefficients of order unity. These give rise to a ϕ-dependent term in the third line of Eq. (31.7.2), of the form

$$\kappa^2\sum_{kl} u_{kl}(\phi,\phi^*)\left(\frac{\partial\tilde{f}}{\partial z_k}\right)^0\left(\frac{\partial\tilde{f}}{\partial z_l}\right)^{0*},$$

which is of order $\kappa^2(\kappa\Lambda^2)^2\Lambda^4 = \kappa^4\Lambda^8$.

The leading ϕ-dependent term in the fifth line of Eq. (31.7.2) arises either from $\kappa^2\phi^2$ terms in f^{-1}_{AB}, with both ds given by their leading-order terms, of order z^*z, or from terms in one of the ds with two factors of κ, two factors of ϕ and/or ϕ^*, and two factors of z and/or z^*, with f_{AB} and the other d given by their leading-order terms, of order 1 and z^*z respectively. Both types of ϕ-dependent term make a contribution of order $\kappa^2(\kappa\Lambda^2)^2\Lambda^4 = \kappa^4\Lambda^8$, so higher-order terms may be neglected.

The leading term in the sixth line of Eq. (31.7.2) arises from the leading term in f^{-1}_{AB}, which is of order unity, and the leading ϕ-dependent terms in d, which are of order $\phi^*\phi$. This makes a contribution to the potential of order $(\kappa\Lambda^2)^4$, so here too higher-order terms may be neglected.

The potential (31.7.2) also contains ϕ-independent terms of order $\kappa^2\Lambda^6$, $\kappa^4\Lambda^8$, and so on. The terms of order $\kappa^2\Lambda^6$ may be cancelled and the terms of order $\kappa^4\Lambda^8$ can be given an arbitrary value $\mathscr{C}$ by making a small shift of the constant term in $\tilde{f}(z)$ away from the value given by Eq. (31.7.8).

Putting these results together, to order $\kappa^4\Lambda^8 \approx m_g^4$ the potential of the observable sector is now

$$V_O(\phi,\phi^*) = \sum_r \left|\frac{\partial f(\phi)}{\partial \phi_r} + \kappa^2 \tilde{f}^0 \phi_r^*\right|^2 - 6\kappa^2 \text{Re}\left[f(\phi)\tilde{f}^{0*}\right]$$
$$+\frac{1}{2}\sum_A\left|\sum_{rs}\phi_r^*(t_A)_{rs}\phi_s\right|^2$$
$$+Q(\phi,\phi^*)+\mathscr{C}\,, \tag{31.7.10}$$

where $Q(\phi,\phi^*)$ is a quadratic polynomial in ϕ and/or ϕ^* with coefficients of order $\kappa^2\Lambda^4 \approx m_g^2$, which arises from the ϕ-dependent terms in g_{kl}^{-1} on the third line of Eq. (31.7.2) and from the ϕ-dependent terms in f_{AB}^{-1} and d on the fifth line of Eq. (31.7.2). We have normalized the gauge superfields so that $f_{AB} = \delta_{AB}$ when all scalar fields vanish. (There is also a ϕ dependence arising from a ϕ-dependent shift in the equilibrium value of the hidden sector scalars z_k, but this shift is at most of order $(\kappa\Lambda^2)^4/\Lambda^3$, and since (31.7.7) is supposed to be stationary at $z = z^0$ this shift enters quadratically in the effective potential of the observable sector fields, and may therefore be neglected.) The constant $\mathscr{C}$ may be chosen so that the value of this potential at its minimum is zero.

Finally, let us return to the non-renormalizable terms in the superpotential. As already mentioned, the leading Φ-dependent terms of this sort may be expected to be of order $\kappa\Phi^2 Z^2$. When the hidden sector superfields Z_k are set equal to their equilibrium values z_k^0 these terms become a second-order polynomial in the Φ_r, with coefficients of order $\kappa\Lambda^2$. Thus to leading order, we can take account of these non-renormalizable terms by simply including a quadratic polynomial function of the Φ_r in the superpotential, with coefficients of order $\kappa\Lambda^2 \approx m_g$.

In this way theories of gravity-mediated supersymmetry breaking avoid the problem of the μ-term, discussed in Sections 28.1 and 28.5. Recall that the $SU(3) \times SU(2) \times U(1)$ symmetry of the standard model allows a single superrenormalizable term in the superpotential of the minimal supersymmetric standard model, a term $\mu(H_1^{\rm T} e H_2)$. In order to explain in a natural way why the coefficient μ is not of the order of the Planck mass, it is necessary to impose some sort of symmetry, like the 'Peccei–Quinn' symmetry discussed in connection with strong CP violation in Section 23.6, under which the product $(H_1^{\rm T} e H_2)$ is not neutral. But a μ-term with μ of the same order m_g as other supersymmetry-breaking masses was found phenomenologically necessary in Section 28.5. Such a term can arise naturally from the breaking of the Peccei–Quinn symmetry by the vacuum expectation value of the hidden sector fields,** if the superpotential contains a non-renormalizable term in which $(H_1^{\rm T} eH_2)$ appears multiplied with two powers of the hidden sector fields z_k, with a coefficient of order κ.

We assume then that the effective superpotential $f(\phi)$ consists of a homogeneous polynomial $f^{(3)}(\phi)$ of third order in the fields ϕ_r, with coefficients very roughly of order unity, plus a μ-term that takes the form of a homogeneous polynomial $f^{(2)}(\phi)$ of second order in the fields ϕ_r, with coefficients of order $m_g \approx \kappa\Lambda^2$. The potential (31.7.10) then becomes

$$\begin{aligned} V_O(\phi,\phi^*) = & \sum_r \left|\frac{\partial f(\phi)}{\partial \phi_r}\right|^2 + \frac{1}{2}\sum_{AB}\left|\sum_{rs}\phi_r^*(t_A)_{rs}\phi_s\right|^2 \\ & -2\kappa^2 {\rm Re}\left[f^{(2)}(\phi)\tilde{f}^{0*}\right] + \kappa^4|\tilde{f}^0|^2\sum_r|\phi_r|^2 \\ & +Q(\phi,\phi^*) + \mathscr{C}\,. \end{aligned} \tag{31.7.11}$$

The terms on the first line of the right-hand side give a supersymmetric potential, while the terms on the second and third lines represent a soft breaking of supersymmetry. With $\tilde{f}^0 \approx \Lambda^2/\kappa$ and $(\partial\tilde{f}/\partial z)^0 \approx \Lambda^2$, the dimensional constants in the soft supersymmetry-breaking terms in Eq. (31.7.10) are all powers of $\kappa\Lambda^2 \approx m_g$, so that this is where we expect to find the expectation values of the observable sector scalar fields, justifying our choice of this as the region of field space to explore.

Setting $\kappa\Lambda^2$ equal to a typical mass $\approx$ 1 TeV in the effective Lagrangian of the supersymmetric standard model, we now find $\Lambda \approx 10^{11}$ GeV. It is mildly encouraging that, as discussed in Section 23.6, the spontaneous breakdown of the Peccei–Quinn symmetry at a scale $\Lambda \approx 10^{11}$ GeV is just what is needed to resolve the strong CP problem with a symmetry-breaking scale in the window from 10^{10} GeV to 10^{12} GeV allowed by astronomical observations.

Comparing the potential with the scalar field terms in the Lagrangian density (28.4.1) of the minimum supersymmetric standard model, we see that this version of gravity-mediated supersymmetry breaking predicts that supersymmetry is broken only by soft scalar mass terms with coefficients

** This is known as the Giudice–Masiero mechanism[17]. It is often described in terms of non-renormalizable holomorphic and antiholomorphic terms in the modified Kahler potential d but, as discussed at the end of the previous section, any such terms can be traded for holomorphic factors in the superpotential. Here we have *defined* d to not contain holomorphic and antiholomorphic terms, and with this definition the μ-term can only arise from non-renormalizable terms in the superpotential.

(including $B\mu$) of order m_g^2. To leading order in $\kappa\Lambda$, the coefficients A and C of trilinear supersymmetry-breaking terms vanish.

One serious problem with these results is that there is no reason why the quadratic polynomial $Q(\phi, \phi^*)$ in Eq. (31.7.11) should respect the degeneracy among squark masses and among slepton masses discussed in Section 28.4, that would avoid unobserved flavor-changing processes. However, the fourth term in Eq. (31.7.11) for the potential makes an additional contribution $\kappa^4|\tilde{f}^0|^2 \approx \kappa^2\Lambda^4$ to scalar squared masses that is the same for all scalars, so the constraints imposed by experimental upper bounds on flavor-changing processes might be satisfied if the coefficients in $Q(\phi, \phi^*)$ (which arise from terms on the third and fifth lines of Eq. (31.7.2)) happen to be small compared with $\kappa^2\Lambda^4$.

We would have an interesting relation among the parameters of the minimum supersymmetric standard model if it were really true that $Q(\phi, \phi^*)$ could be neglected. Taking the quadratic part $f^{(2)}$ of the superpotential as $\mu(\phi_1^{\rm T} e\phi_2)$, the coefficient $B\mu$ in Eq. (28.4.1) would be given by the second term of Eq. (31.7.11) as $B\mu = -\kappa^2\mu\tilde{f}^{0*}$, so

$$|B| = \kappa^2|\tilde{f}^0| = m_g \,,$$

in agreement with Eq. (31.4.13). Also, all the squark and slepton masses M_s would be given by the third term in Eq. (31.7.11) as $\kappa^2|\tilde{f}^0|$, and so we would have the new relation

$$|B| = M_s \,. \tag{31.7.12}$$

With all squark and slepton masses equal, there would be no inconsistency with limits on flavor-changing processes. Furthermore, with Q neglected there would be only one complex parameter $\tilde{f}^0$ in the supersymmetry-breaking part of the potential (31.7.11), which can be chosen to be real by a redefinition of the over-all phase of the superpotential, so now the supersymmetry-breaking part of the potential would introduce no new violation of CP invariance. But there is no known reason why Q should be small.

Another serious problem with this version of gravitationally-mediated supersymmetry breaking is that it does not yield sufficiently large gaugino masses.[18] According to Eq. (31.6.75), the mass matrix of the $SU(3) \times SU(2) \times U(1)$ gauginos is given in the tree approximation by

$$m_{AB} = \exp(\kappa^2 d/2) \sum_{NM} [g^{-1}]_{NM} L_N \left(\frac{\partial f_{AB}}{\partial \varphi_M}\right)^* \,, \tag{31.7.13}$$

where here φ_N runs over all the scalar fields ϕ_r and z_k on which f_{AB} may depend, with g_{NM} and L_M given by Eqs. (31.6.66) and (31.6.69). According to the estimates we have made here, $\kappa^2 d = O(\kappa^2\Lambda^2) \ll 1$; $L_k = O(\Lambda^2)$ while L_r is much smaller; and g^{-1}_{kl} is of order unity. Also, we assume that

f_{AB} is a term of order unity plus a term of order κ^2 times a bilinear in scalar fields and their complex conjugates, so $\partial f_{AB}/\partial z_k$ is of order $\kappa^2\Lambda$. This gives gaugino masses of order $\Lambda^2 \times \kappa^2\Lambda$. This is much smaller than the gravitino mass $m_g \approx \kappa\Lambda^2$ (which sets the scale of the supersymmetry-breaking terms in the potential (31.7.11) of the observable sector scalar fields) by a factor of order $\kappa\Lambda \approx 10^{-7}$, so if the scalar masses produced by supersymmetry breaking are of order 1 TeV, then the gaugino masses will be of order 100 keV, which is far too low to be consistent with the fact that gauginos have not yet been observed.

There are several ways that this problem might be avoided. One is to include gauge-singlet scalar fields among the z_k of the hidden sector, which can appear linearly in f_{AB}.[19] In this case $\partial f_{AB}/\partial z_k$ will be of order κ rather than $\kappa^2\Lambda$, yielding gaugino masses of order $\kappa\Lambda^2$, which is comparable to the squark and slepton masses. One trouble with this approach is that the inclusion of scalars that are neutral with respect to all gauge groups would make it no longer natural for the renormalizable part of the superpotential to take the form $f(\phi) + \tilde{f}(z)$.

Even without gauge singlets, there are one-loop contributions to the gluino, wino, and bino masses (and to the A-parameters), calculated in Section 31.4. For instance, if we take m_g at the largest value $\approx$ 10 TeV allowed by the 'naturalness' bound discussed in Section 28.1, then with $g_s^2/4\pi = 0.118$, Eq. (31.4.13) would give a gluino mass $3g_s^2 m_g/16\pi^2 = 280$ GeV, which is certainly high enough to allow the gluino to have escaped detection. The bino and wino masses depend on the unknown ratio of the μ parameter and the pseudoscalar Higgs mass m_A. Taking this ratio equal to unity and $m_g < 10$ TeV, Eqs. (31.4.15) and (31.4.16) would give $m_{\rm bino} = 9g'^2 m_g/16\pi^2 < 73$ GeV and $m_{\rm wino} = g^2 m_g/16\pi^2 < 27$ GeV.[19a] This bound on the wino mass is in conflict with the fact that wino pairs have not been seen at LEP in e^+–e^- collisions at an energy sufficiently high to produce W boson pairs, so that $m_{\rm wino} > m_W$. In order to avoid this contradiction it would be necessary[11b] either to take $m_g > 30$ TeV, which is awkward from the point of view of naturalness, or $\mu^2/m_A^2 > 8$. In any case, this model has the general consequence that the gauginos are much lighter than the squarks and sleptons.

If the polynomial $Q(\phi, \phi^*)$ can be neglected then, as shown in Section 31.4, the A-parameters are also given by one-loop corrections. These are of order $g_s^2 m_g/16\pi^2$ for squarks and $g^2 m_g/16\pi^2$ or $g'^2 m_g/16\pi^2$ for sleptons.

Second Version[20]

This version of gravity-mediated supersymmetry breaking has an observable sector with chiral superfields Φ_r and a hidden sector with chiral superfields Z_k, just as in the first version. The difference is that supersymmetry is now assumed to be *not* spontaneously broken in the hidden sector. Instead, the gauge couplings of the hidden sector, which become strong at an energy $\Lambda \ll m_{\rm Pl}$, produce a non-perturbative superpotential for the scalar fields of a third sector of superfields, the *modular superfields.* In various theories such as modern superstring theories there are extra dimensions that are not observed because they have been 'rolled up' into a tiny compact manifold, with size roughly of the order of κ. Typically some of the parameters that are needed to describe this compact manifold are not fixed in any order of perturbation theory. The values of these parameters may vary from point to point in four-dimensional spacetime and appear at energies far below the Planck scale κ^{-1} as gauge-invariant scalar fields y_a, known as *modular fields.* (The indices a, b, etc. here of course have nothing to do with the local Lorentz frame indices used in Section 31.6.) Assuming that supersymmetry is not broken in the compactification of the extra dimensions, these fields must be accompanied by fermionic superpartners and auxiliary fields, which together form gauge-invariant left-chiral modular superfields Y_a and their adjoints.†

Just below the compactification scale we have a supersymmetric theory with a superpotential that can depend on all the superfields, but in which κ is the only dimensional parameter. The circumstance that the y_a are not fixed in perturbation theory typically arises because the compactification does not result in any superpotential for the modular superfields Y_a alone. As we saw in Section 28.1, the $SU(3) \times SU(2) \times U(1)$ gauge symmetries rule out any terms in the superpotential with just one or two factors of observable sector superfields, except for a possible term bilinear in the Higgs superfields H_1 and H_2. We will again assume that this bilinear term in the bare superpotential is either accidentally absent (in which case it does not appear in any order of perturbation theory), or is ruled out by some symmetry, such as the 'Peccei–Quinn' symmetry discussed in connection with strong CP violation in Section 23.6. The gauge symmetry of the hidden sector rules out any terms in the superpotential with just one factor of the Z_k, and we shall assume that either accident or some symmetry (perhaps the same Peccei–Quinn symmetry) also rules out terms with two factors of the Z_k.

The bare superpotential therefore takes the form

$$f_{\rm bare}(\Phi, Y, Z) = \sum_{rst} f_{rst}(\kappa Y)\, \Phi_r \Phi_s \Phi_t + \sum f_{klm}(\kappa Y) Z_k Z_l Z_m + \cdots \,, \tag{31.7.14}$$

† Of course, modular fields may exist even under the assumptions of the first version of gravity-mediated supersymmetry breaking, but because of the smaller value of Λ in that case the couplings of the modular fields are too weak to be of interest there.

where f_{rst} and f_{klm} are power series in their arguments with coefficients roughly of order unity, and the dots indicate terms involving $n > 3$ factors of the Φs and Zs, as well as any number of factors of κY_a, suppressed by factors proportional to κ^{n-3}.

We assume that non-perturbative effects in the hidden sector such as 'gaugino condensation' (the appearance of expectation values of bilinear functions of gaugino fields) without themselves breaking supersymmetry produce a superpotential for the modular superfields. Since Λ is the only scale in the problem (aside from gravitational effects suppressed by factors of $\kappa\Lambda$), this superpotential would have to be of the form

$$\hat{f}(Y) = \Lambda^3 F(\kappa Y) . \tag{31.7.15}$$

Such terms can also be produced by expectation values of the scalar components of the Z superfields in the Z^3 term in Eq. (31.7.14). At the same time, the replacement of the Z_k by the expectation values of their scalar components in the non-renormalizable terms indicated by dots in Eq. (31.7.14) will produce additional Φ-dependent terms in the effective superpotential, about which more later. The superpotential (31.7.15) is of the form (31.7.1) originally assumed in theories of gravitationally mediated supersymmetry breaking, but with ϵ now identified as the intermediate scale Λ at which the gauge interactions of the hidden sector become strong.

It is plausible that the superpotential (31.7.15) can lead to supersymmetry breaking by the appearance of $\mathscr{F}$-terms for the modular chiral superfields. For the moment, let's ignore the other superfields, leaving the justification for later. The potential for the modular scalars is given by Eqs. (31.6.68) and (31.6.69) as

$$\hat{V}(y, y^*) = \exp\left(\kappa^2 \hat{d}(y, y^*)\right) \left[\sum_{ab} [\hat{g}^{-1}(y, y^*)]_{ab} \hat{L}_a(y) \hat{L}_b(y)^* - 3\kappa^2 |\hat{f}(y)|^2\right] , \tag{31.7.16}$$

where $\hat{d}(y, y^*)$ is the Kahler d function with the scalar components of the Z_k and Φ_r neglected, and

$$\hat{g}_{ab} = \frac{\partial^2 \hat{d}}{\partial y_a \partial y_b^*} , \tag{31.7.17}$$

$$\hat{L}_a = \frac{\partial \hat{f}}{\partial y_a} + \kappa^2 \hat{f} \frac{\partial \hat{d}}{\partial y_a} . \tag{31.7.18}$$

We are assuming here that $\hat{V}$ has a stationary point, labelled with a superscript 0, where $\hat{L}_a^0 \neq 0$, so that supersymmetry is broken, but that $\hat{V}^0$ is very small, so that it can be cancelled by terms arising from the observable sector, leaving us with a flat spacetime. Since $\hat{f}$ is of the form (31.7.15) and $\hat{d}$ equals κ^{-2} times a power series in κy_a and κy_a^* with coefficients of order unity, the whole potential is of the form $\kappa^2\Lambda^6$ times a power series in κy_a and κy_a^*, again with coefficients of order unity. The orders of magnitude of the various ingredients in the potential are therefore

$$\begin{aligned} &y_a^0 = O(\kappa^{-1})\,, \qquad \hat{f}^0 = O(\Lambda^3)\,, \qquad \hat{d}^0 = O(\kappa^{-2})\,, \\ &\hat{L}_a^0 = O(\kappa\Lambda^3)\,, \qquad \hat{g}_{ab}^0 = O(1)\,. \end{aligned} \tag{31.7.19}$$

With the modular and hidden sector fields fixed at their expectation values, the superpotential of the observable sector is now of the form

$$f(\Phi) = \sum_{rs} \mu_{rs}\Phi_r\Phi_s + \sum_{rst} g_{rst}\Phi_r\Phi_s\Phi_t + \cdots\,, \tag{31.7.20}$$

where g_{rst} is $f_{rst}(\kappa y_0)$, which is assumed to be roughly of order unity, plus terms suppressed by powers of $\kappa\Lambda$. The dots here denote terms with more than three factors of Φ that are suppressed by additional factors of $\kappa\Phi$. The coefficient μ_{rs} arises from the non-renormalizable terms denoted by dots in Eq. (31.7.14); if it comes from a term with $n > 1$ factors of Zs as well as two factors of Φs, then it has an order of magnitude

$$\mu_{rs} = O(\kappa^{n-1}\Lambda^n)\,. \tag{31.7.21}$$

We shall see that the desired order of magnitude of μ is $m_g = O(\kappa^2\Lambda^3)$, which would come from a term with $n = 3$ factors of Zs.

The supersymmetry breaking in the modular sector will be transmitted to the observable sector by effects of the gravitational field and its superpartners. Eq. (31.6.68) gives the potential of the observable sector as

$$\begin{aligned} V_O = e^{\kappa^2 d^0}\Bigg[&\sum_{rs}[g^{0\,-1}]_{rs}\left(\frac{\partial f}{\partial\phi_r} + \kappa^2(f+\hat{f}^0)\frac{\partial d^0}{\partial\phi_r}\right)\left(\frac{\partial f}{\partial\phi_s} + \kappa^2(f+\hat{f}^0)\frac{\partial d^0}{\partial\phi_s}\right)^* \\ &+2\,\mathrm{Re}\sum_{ra}[g^{0\,-1}]_{ra}\left(\frac{\partial f}{\partial\phi_r} + \kappa^2(f+\hat{f}^0)\frac{\partial d^0}{\partial\phi_r}\right)L_a^{0*} \\ &+\sum_{ab}[g^{0\,-1}]_{ab}\hat{L}_a^0\hat{L}_b^{0*} - 3\kappa^2\left|f+\hat{f}^0\right|^2\Bigg] \\ &+\frac{1}{2}\mathrm{Re}\sum_{AB}[f_{AB}^{-1}]^0\left(\sum_{rs}\frac{\partial d^0}{\partial\phi_r}(t_A)_{rs}\phi_s\right)\left(\sum_{tu}\frac{\partial d^0}{\partial\phi_t}(t_A)_{tu}\phi_u\right)^*\,, \end{aligned} \tag{31.7.22}$$

with the superscript zero again indicating that the modular and hidden sector scalar fields are fixed at their equilibrium values. (This will be reconsidered later.) Note that, although Eq. (31.7.22) involves terms like $\hat{L}_a^0$ that arise in the modular sector, there are no terms here that refer explicitly to the hidden sector. This is because in this version of gravity-mediated supersymmetry breaking, supersymmetry is assumed not to be broken in this sector, so that $L_k^0 = 0$, and $D_A^0 = 0$ for any gauge field that interacts with the hidden sector fields.

We are interested in exploring a region of field space in which the observable sector fields are of order $\kappa^2\Lambda^3$, because, as we shall see, this is the characteristic mass arising in the observable sector from gravitational effects of supersymmetry breaking in the hidden sector. According to Eq. (31.6.48) and our estimate $\hat{f}^0 \approx \Lambda^3$, this is also the order of magnitude of the gravitino mass m_g:

$$m_g \approx \kappa^2\Lambda^3 \, .$$

In order to calculate the potential for fields of this order, we note that the Kahler d function for the observable and modular scalar fields takes the form

$$d(\phi,\phi^*,y,y^*) = \kappa^{-2}\hat{d}(\kappa y,\kappa y^*) + \sum_{rs} \phi_r\phi_s^* A_{rs}(\kappa y,\kappa y^*)$$
$$+ \sum_{rs} \phi_r\phi_s B_{rs}(\kappa y,\kappa y^*) + \sum_{rs} \phi_r^*\phi_s^* B_{rs}^*(\kappa y,\kappa y^*) + \cdots \, , \qquad (31.7.23)$$

where $\hat{d}$, A_{rs}, and B_{rs} are power series in their arguments with coefficients of order unity, and the dots indicate terms with $n > 2$ factors of ϕ and/or ϕ^*, suppressed by factors κ^{n-2}. According to Eq. (31.6.72), we can remove any holomorphic term in d along with its complex conjugate by multiplying the total superpotential by a suitable holomorphic factor. In particular, by multiplying the total superpotential with a factor $\exp[\kappa^2\hat{d}^0 + \kappa^2\sum_{rs} B_{rs}^0\phi_r\phi_s]$, we can arrange that the transformed d function has

$$\hat{d}^0 = 0 \, , \qquad B_{rs}^0 = 0 \, . \qquad (31.7.24)$$

We shall assume that this has been done. Note that, because the total superpotential contains a constant term $\hat{f}^0$ of order Λ^3, while B_{rs}^0 was of order unity, this transformation generates a term in the superpotential that is quadratic in ϕ_r, making a contribution to its coefficient μ_{rs} of order $\kappa^2\Lambda^3$, of the same order as ϕ and m_g.

There can be another contribution to this coefficient of the same order of magnitude, arising from terms in the superpotential with $n > 1$ factors of Ys as well as two factors of Φs. Eq. (31.7.21) tells us that, in order for the contribution to the μ_{rs} coefficients from this also to be of order $\kappa^2\Lambda^3$, we must have $n = 3$. We can arrange that this is allowed, while terms with two factors of Φs and $n = 2$ factors of Ys are forbidden, by giving H_1 and H_2 Peccei–Quinn quantum numbers +1 and the Ys quantum numbers −2/3. The breaking of this Peccei–Quinn symmetry by the vacuum expectation values of the Y_k then produces an axion.

With μ_{rs} of order $\kappa^2\Lambda^3$, both the bilinear and trilinear terms in the superpotential of the observable sector are of order $\kappa^6\Lambda^9$. The usual supersymmetric potential term $\sum_r |\partial f/\partial\phi_r|^2$ is then of order $\kappa^8\Lambda^{12}$, so we have to collect all terms in Eq. (31.7.22) of this order or greater.

With ϕ_r of order $\kappa^2\Lambda^3$ and y_a fixed at its equilibrium value $y_a^0 \approx \kappa^{-1}$, Eq. (31.7.23) gives

$$g^0_{rs} = A^0_{rs} + O(\kappa^3\Lambda^3)\,, \tag{31.7.25}$$

$$g^0_{ab} = \left(\frac{\partial^2\hat{d}}{\partial y_a \partial y_b^*}\right)^0 + \sum_{rs}\phi_r\phi_s^*\left(\frac{\partial^2 A_{rs}}{\partial y_a\partial y_b^*}\right)^0 + \sum_{rs}\phi_r\phi_s\left(\frac{\partial^2 B_{rs}}{\partial y_a\partial y_b^*}\right)^0$$
$$+\sum_{rs}\phi_r^*\phi_s^*\left(\frac{\partial^2 B^*_{rs}}{\partial y_a\partial y_b^*}\right)^0 + O(\kappa^6\Lambda^6)\,, \tag{31.7.26}$$

$$g^0_{ra} = \sum_s \phi_s\left(\frac{\partial A_{rs}}{\partial y_a^*}\right)^0 + O(\kappa^6\Lambda^6) = g^{0\,*}_{ar}\,, \tag{31.7.27}$$

where the zero superscript again indicates that the y_a are fixed at their equilibrium values y_a^0. We will subject the observable-sector superfields Φ_r and the modular superfields Z_a to separate linear transformations, designed so that

$$A^0_{rs} = \delta_{rs}\,, \qquad \left(\frac{\partial^2\hat{d}}{\partial y_a\partial y_b^*}\right)^0 = \delta_{ab}\,. \tag{31.7.28}$$

With the metric (31.7.25)–(31.7.27) given by the unit matrix plus terms that are much less than unity, it is easy to calculate the inverse:

$$g^{0\,-1}_{rs} = \delta_{rs} + O(\kappa^3\Lambda^3)\,, \tag{31.7.29}$$

$$g^{0\,-1}_{ab} = \delta_{ab} - \sum_{rs}\phi_r\phi_s^*\left(\frac{\partial^2 A_{rs}}{\partial y_a\partial y_b^*}\right)^0 - \sum_{rs}\phi_r\phi_s\left(\frac{\partial^2 B_{rs}}{\partial y_a\partial y_b^*}\right)^0$$
$$-\sum_{rs}\phi_r^*\phi_s^*\left(\frac{\partial^2 B^*_{rs}}{\partial y_a\partial y_b^*}\right)^0 + O(\kappa^6\Lambda^6)\,, \tag{31.7.30}$$

$$g^{0\,-1}_{ra} = -\sum_s \phi_s\left(\frac{\partial A_{rs}}{\partial y_a^*}\right)^0 + O(\kappa^6\Lambda^6) = g^{0\,-1\,*}_{ar}\,. \tag{31.7.31}$$

In particular, our assumption about the form of the functions A_{rs} and B_{rs} gives the order-of-magnitude estimates

$$\left(\frac{\partial^2 A_{rs}}{\partial y_a\partial y_b^*}\right)^0 = O(\kappa^2)\,, \qquad \left(\frac{\partial^2 B_{rs}}{\partial y_a\partial y_b^*}\right)^0 = O(\kappa^2)\,, \qquad \left(\frac{\partial A_{rs}}{\partial y_a^*}\right)^0 = O(\kappa)\,, \tag{31.7.32}$$

so that, for $\phi_r = O(\kappa^2\Lambda^3)$,

$$[g^{0\,-1}]_{rs} = O(1)\,, \quad [g^{0\,-1}]_{ab} = O(1)\,, \quad [g^{0\,-1}]_{ra} = O(\kappa^3\Lambda^3)\,. \tag{31.7.33}$$

Further,

$$f = O(\phi^3) = O(\kappa^6\Lambda^9)\,, \qquad \frac{\partial f}{\partial \phi_r} = O(\phi^2) = O(\kappa^4\Lambda^6)\,, \tag{31.7.34}$$

and

$$\frac{\partial d^0}{\partial \phi_r} = O(\phi) = O(\kappa^2\Lambda^3)\,. \tag{31.7.35}$$

With $\hat{f}^0$ of order Λ^3, the quantities $\kappa^2(f+\hat{f}^0)\partial d/\partial\phi_r$ in Eq. (31.7.22) are dominated by $\kappa^2\hat{f}^0\phi_r^*$, which is of order $\kappa^2 \times \Lambda^3 \times \kappa^2\Lambda^3 = \kappa^4\Lambda^6$. This is of the same order of magnitude as $\partial f/\partial\phi_r$, so in leading order we must keep both terms:

$$\frac{\partial f}{\partial \phi_r} + \kappa^2(f+\hat{f}^0)\frac{\partial d}{\partial \phi_r} \simeq \frac{\partial f}{\partial \phi_r} + \kappa^2\hat{f}^0\,\phi_r^* = O(\kappa^4\Lambda^6)\,. \tag{31.7.36}$$

In this approximation, and with $g^{0\,-1}_{rs}$ replaced by its dominant term δ_{rs}, the first term in the square brackets in Eq. (31.7.22) is already of the desired order $\kappa^8\Lambda^{12}$, so we may use these approximations to write

$$\begin{aligned}\sum_{rs}[g^{0\,-1}]_{rs}&\left(\frac{\partial f}{\partial \phi_r} + \kappa^2(f+\hat{f}^0)\frac{\partial d^0}{\partial \phi_r}\right)\left(\frac{\partial f}{\partial \phi_s} + \kappa^2(f+\hat{f}^0)\frac{\partial d^0}{\partial \phi_s}\right)^*\\ &\simeq \sum_r\left|\frac{\partial f}{\partial \phi_r} + \kappa^2\hat{f}^0\,\phi_r^*\right|^2\,.\end{aligned} \tag{31.7.37}$$

The second term in the square brackets in Eq. (31.7.22) is of order $\kappa^3\Lambda^3 \times \kappa^4\Lambda^6 \times \kappa\Lambda^3 = \kappa^8\Lambda^{12}$, so we may evaluate it using only leading terms and find

$$\begin{aligned}2\mathrm{Re}\sum_{ra}[g^{0\,-1}]_{ra}&\left(\frac{\partial f}{\partial \phi_r} + \kappa^2(f+\hat{f}^0)\frac{\partial d^0}{\partial \phi_r}\right)\hat{L}_a^{0*}\\ &\simeq -2\mathrm{Re}\sum_{ras}\phi_s\left(\frac{\partial A_{rs}}{\partial y_a^*}\right)^0\left[\frac{\partial f}{\partial \phi_r} + \kappa^2\hat{f}^0\,\phi_r^*\right]\hat{L}_a^{0*}\,.\end{aligned} \tag{31.7.38}$$

The third and fourth terms in the square brackets in Eq. (31.7.22) are individually of order $\kappa^2\Lambda^6$, but are assumed nearly to cancel, so non-leading terms must be included in evaluating their contribution to the quantity in square brackets. One contribution comes from the terms in Eq. (31.7.30) for $g^{0\,-1}_{ab}$ that are of second order in ϕ and/or ϕ^*; this contribution is of order $\kappa^2\phi^2 \times (\kappa\Lambda^3)^2$, which is of order $\kappa^8\Lambda^{12}$. Another non-cancelling contribution to the potential from the last term in square brackets in Eq. (31.7.22) comes from the interference between f and $\hat{f}^0$, which makes a contribution of order $\kappa^2\phi^3\Lambda^3$, which is also of order $\kappa^8\Lambda^{12}$. There is also a constant contribution $\mathscr{C}$ from a possible failure of the last two terms in the square brackets in Eq. (31.7.22) to cancel when $\phi_r = 0$;

to avoid a large cosmological constant, we shall have to assume that $\mathscr{C}$ is also of order $\kappa^8\Lambda^{12}$. (This is an unnatural fine-tuning, which so far has been necessary in any theory that avoids a huge cosmological constant.) Putting these estimates together, to order $\kappa^8\Lambda^{12}$ the last two terms in the square brackets in Eq. (31.7.22) are

$$\sum_{ab}[g^{-1}_{ab}]^0\hat{L}^0_a\hat{L}^{0*}_b - 3\kappa^2\left|f+\hat{f}^0\right|^2 \simeq -\sum_{abrs}\left[\phi_r\phi^*_s\left(\frac{\partial^2 A_{rs}}{\partial y_a\partial y^*_b}\right)^0 + \phi_r\phi_s\left(\frac{\partial^2 B_{rs}}{\partial y_a\partial y^*_b}\right)^0 + \phi^*_r\phi^*_s\left(\frac{\partial^2 B^*_{rs}}{\partial y_a\partial y^*_b}\right)^0\right]\hat{L}^0_a\hat{L}^{0*}_b - 6\kappa^2\mathrm{Re}(f\,\hat{f}^{0*}) + \mathscr{C}\,. \tag{31.7.39}$$

All these terms are of the desired order $\kappa^8\Lambda^{12}$, while Eq. (31.7.24) gives $\kappa^2 d^0 = O(\kappa^2\phi^2) = O(\kappa^6\Lambda^6)$, so we can ignore the factor $\exp(\kappa^2 d^0)$ in Eq. (31.7.22). Finally, the gauge terms on the last line of Eq. (31.7.22) are of order $\phi^4 = O(\kappa^8\Lambda^{12})$, so these can be evaluated using the leading term ϕ^*_r in $\partial d^0/\partial\phi_r$.

Putting this all together, the complete scalar potential of the observable sector to order $\kappa^8\Lambda^{12}$ is

$$\begin{aligned} V_O \simeq{}& \sum_r\left|\frac{\partial f}{\partial\phi_r} + \kappa^2\hat{f}^0\,\phi^*_r\right|^2 \\ &-2\mathrm{Re}\sum_{ras}\phi_s\left(\frac{\partial A_{rs}}{\partial y^*_a}\right)^0\left[\frac{\partial f}{\partial\phi_r} + \kappa^2\hat{f}^0\,\phi^*_r\right]\hat{L}^{0*}_a \\ &-\sum_{abrs}\left[\phi_r\phi^*_s\left(\frac{\partial A_{rs}}{\partial y_a\partial y^*_b}\right)^0 + \phi_r\phi_s\left(\frac{\partial B_{rs}}{\partial y_a\partial y^*_b}\right)^0 \right. \\ &\left. \quad +\phi^*_r\phi^*_s\left(\frac{\partial B^*_{rs}}{\partial y_a\partial y^*_b}\right)^0\right]\hat{L}^0_a\hat{L}^{0*}_b - 6\kappa^2\mathrm{Re}(f\,\hat{f}^{0*}) \\ &+\frac{1}{2}\mathrm{Re}\sum_{AB}[f^{-1}_{AB}]^0\left(\sum_{rs}\phi^*_r(t_A)_{rs}\phi_s\right)\left(\sum_{tu}\phi^*_t(t_B)_{tu}\phi_u\right)^* + \mathscr{C}\,. \end{aligned} \tag{31.7.40}$$

The potential (31.7.40) of the observable sector scalar fields takes the form assumed in the minimal supersymmetric standard models discussed in Section 28.4: it is the sum of a supersymmetric term $V_{\rm susy}$ and a soft supersymmetry-breaking term $V_{\rm soft}$

$$V_O = V_{\rm susy} + V_{\rm soft}\,. \tag{31.7.41}$$

The supersymmetric term is, as usual,

$$V_{\rm susy} = \sum_r \left|\frac{\partial f}{\partial \phi_r}\right|^2 + \frac{1}{2}{\rm Re}\sum_{AB}[f_{AB}^{-1}]^0 \left(\sum_{rs}\phi_r^*(t_A)_{rs}\phi_s\right)\left(\sum_{tu}\phi_t^*(t_B)_{tu}\phi_u\right)^* \tag{31.7.42}$$

and the soft supersymmetry breaking term here is

$$\begin{aligned} V_{\rm soft} \simeq\, & 2\kappa^2{\rm Re}\sum_r\left(\phi_r\frac{\partial f}{\partial \phi_r}\hat{f}^{0*}\right) + \kappa^4\left|\hat{f}^0\right|^2\sum_r|\phi_r|^2 \\ & -2{\rm Re}\sum_{ras}\phi_s\left(\frac{\partial A_{rs}}{\partial y_a^*}\right)^0\left[\frac{\partial f}{\partial \phi_r} + \kappa^2\hat{f}^0\,\phi_r^*\right]\hat{L}_a^{0*} \\ & -\sum_{abrs}\left[\phi_r\phi_s^*\left(\frac{\partial A_{rs}}{\partial y_a \partial y_b^*}\right)^0 + \phi_r\phi_s\left(\frac{\partial B_{rs}}{\partial y_a\partial y_b^*}\right)^0\right. \\ & \left. +\phi_r^*\phi_s^*\left(\frac{\partial B_{rs}^*}{\partial y_a\partial y_b^*}\right)^0\right]\hat{L}_a^0\hat{L}_b^{0*} - 6\kappa^2{\rm Re}(f\,\hat{f}^{0*}) + \mathscr{C}\,. \end{aligned} \tag{31.7.43}$$

With $f(\phi)$ given by Eq. (31.7.20), this takes the form

$$V_{\rm soft} = \sum_{rs} M_{rs}^2\phi_r\phi_s^* + 2{\rm Re}\sum_{rs}N_{rs}^2\phi_r\phi_s + 2{\rm Re}\sum_{rst}A_{rst}\phi_r\phi_s\phi_t + \mathscr{C}\,, \tag{31.7.44}$$

where

$$M_{rs}^2 = \kappa^4|\hat{f}^0|^2\delta_{rs} - 2\kappa^2{\rm Re}\left[\hat{f}^0\sum_a\left(\frac{\partial A_{sr}}{\partial y_a^*}\right)^0\hat{L}_a^{0*}\right] - \sum_{ab}\left(\frac{\partial^2 A_{rs}}{\partial y_a\partial y_b^*}\right)^0\hat{L}_a^0\hat{L}_b^{0*}\,, \tag{31.7.45}$$

$$\begin{aligned} N_{rs}^2 =\, & 2\kappa^2\mu_{rs}\hat{f}^{0*} - \frac{1}{2}\sum_{at}\mu_{tr}\left(\frac{\partial A_{ts}}{\partial y_a^*}\right)^0\hat{L}_a^{0*} - \frac{1}{2}\sum_{at}\mu_{ts}\left(\frac{\partial A_{tr}}{\partial y_a^*}\right)^0\hat{L}_a^{0*} \\ & -\sum_{ab}\left(\frac{\partial^2 B_{rs}}{\partial y_a\partial y_b^*}\right)^0\hat{L}_a^0\hat{L}_b^{0*}\,, \end{aligned} \tag{31.7.46}$$

and

$$A_{rst} = -\sum_{au}\left[\left(\frac{\partial A_{ur}}{\partial y_a^*}\right)^0 g_{ust} + \left(\frac{\partial A_{us}}{\partial y_a^*}\right)^0 g_{urt} + \left(\frac{\partial A_{ut}}{\partial y_a^*}\right)^0 g_{urs}\right]\hat{L}_a^{0*}\,. \tag{31.7.47}$$

Our previous order-of-magnitude estimates give

$$M_{rs}^2 = O(\kappa^4\Lambda^6)\,, \qquad N_{rs}^2 = O(\kappa^4\Lambda^6)\,, \qquad A_{rst} = O(\kappa^2\Lambda^3)\,. \tag{31.7.48}$$

Together with our estimates $g_{rst} = O(1)$ and $\mu_{rs} = O(\kappa^2\Lambda^3)$ for the constants in $V_{\rm susy}$, this shows that if the potential has a stationary point at some $\phi_a^0 \neq 0$, then $\phi_a^0 = O(\kappa^2\Lambda^3)$, justifying our decision to explore fields of this order. For an equilibrium value of the ϕ_r of this order, the equilibrium value of the various terms in the potential is $O(\phi^4) = O(\kappa^8\Lambda^{12})$, so this is the order of magnitude of the constant $\mathscr{C}$ that is needed to cancel the vacuum energy.

In order for the characteristic mass $\kappa^2\Lambda^3$ in the minimal supersymmetric standard model to be of order 1 TeV we need $\Lambda \approx 10^{13}$ GeV. If it is a Peccei–Quinn symmetry that forbids a term in the bare superpotential with two superfield factors each from the observable and hidden sectors, as suggested above, then vacuum expectation values of the hidden sector scalars will break this symmetry, with a symmetry-breaking scale (called M in Section 23.6) of order 10^{13} GeV, and an axion mass which is then given by Eq. (23.6.26) as of order 10^{-6} eV. This value for the symmetry-breaking scale is somewhat above the upper bound of 10^{12} GeV quoted in Section 23.6, but given the uncertainty of cosmological arguments, this contradiction is not decisive.

Before considering further physical implications of these results, let's pause to reconsider a short-cut that we took in deriving them. In calculating the potential for the observable sector scalars ϕ_r, we fixed the modular fields at the equilibrium values y_a^0 that they would have in the absence of the observable sector fields ϕ_r. Instead, we ought to set the modular fields at their equilibrium values $y_a(\phi)$ for the actual values of the ϕ_r, by finding the stationary point of the potential

$$V_{\text{total}}(\phi,\phi^*,y,y^*) = \hat{V}(y,y^*) + V_O(\phi,\phi^*,y,y^*)\,, \tag{31.7.49}$$

and only then seek an equilibrium value for the ϕ_r. Since V_O is much smaller than $\hat{V}$ for fields of interest, the equilibrium value of y_a can be written as

$$y_a(\phi,\phi^*) = y_a^0 + \delta y_a(\phi,\phi^*)\,, \tag{31.7.50}$$

where y_a^0 is at the minimum of $\hat{V}(y,y^*)$ and

$$\sum_b \frac{\partial^2\hat{V}}{\partial y_a \partial y_b}\delta y_b + \sum_b \frac{\partial^2\hat{V}}{\partial y_a \partial y_b^*}\delta y_b^* = -\frac{\partial V_O}{\partial y_a}\,. \tag{31.7.51}$$

The second derivatives of $\hat{V}$ are of order $\kappa^2 \times (\kappa\Lambda^3)^2 = \kappa^4\Lambda^6$, while the first derivatives of V_O are of order $\kappa \times \kappa^8\Lambda^{12} = \kappa^9\Lambda^{12}$, so the δy_a are of order $\kappa^5\Lambda^6$. The change in the potential due to this ϕ-dependent shift in the equilibrium values of the y_a is quadratic in the δy_a and δy_a^*, with coefficients given by second derivatives of $\hat{V}$ with respect to y_a and/or y_a^*, and is therefore of order

$$(\kappa^5\Lambda^6)^2 \times \kappa^2 \times (\kappa\Lambda^3)^2 = \kappa^{14}\Lambda^{18}\,,$$

which is less than the potential we have calculated by a factor $(\kappa\Lambda)^6 \ll 1$.

Without further assumptions regarding the functions $A_{rs}(y, y^*)$ and $B_{rs}(y, y^*)$, we can learn nothing whatever from Eqs. (31.7.45)–(31.7.47) about the precise values of the coefficients M^2_{rs}, N^2_{rs}, and A_{rst} in the soft supersymmetry-breaking potential (31.7.44). The only definite prediction that emerges from these results is that the coefficients C_{ij} in the soft supersymmetry-breaking Lagrangian (28.4.1) are all negligible, as has generally been assumed anyway.

The greatest problem presented by the results (31.7.44)–(31.7.48) is that without further assumptions they do not insure the degeneracy of squark masses and of slepton masses that would avoid the quark and lepton flavor-changing processes discussed in Section 28.4. The $\phi_r\phi_s$-terms in the soft supersymmetry-breaking potential (31.7.44) and in the superpotential (31.7.20) are not allowed by $SU(3) \times SU(2) \times U(1)$ to depend on anything but the Higgs scalars, so they cannot lead to flavor-changing processes. The problem therefore arises only from the fact that the coefficients M^2_{rs} and A_{rst} in Eq. (31.7.44) may not conserve flavor in the same basis as the Yukawa couplings g_{rst}. One way to avoid this problem is to assume that for some reason the functions $A_{rs}(y, y^*)$ happen to depend only weakly on the y_a and y^*_a, so that Eq. (31.7.45) gives $M^2_{rs} \propto \delta_{rs}$ and Eq. (31.7.47) makes A_{rst} (and hence the A_{ij} coefficients in Eq. (28.4.1)) anomalously small. Another possibility is that, although not slowly varying, for some reason the whole function $A_{rs}(y, y^*)$ (or at least its first and second derivatives at $y_a = y^0_a$) is proportional to δ_{rs}. In this case, Eq. (31.7.45) again gives $M^2_{rs} \propto \delta_{rs}$, and now Eq. (31.7.47) gives trilinear couplings $A_{rst} \propto g_{rst}$, so that the A_{ij} coefficients in Eq. (28.4.1) would be all equal.[21]

We must also check the gaugino masses produced by this version of gravitationally mediated supersymmetry breaking. According to Eq. (31.6.75), the mass matrix of the $SU(3) \times SU(2) \times U(1)$ gauginos is given in general by

$$m_{AB} = \exp(\kappa^2 d/2) \sum_{NM} [g^{-1}]_{NM} L_N \left(\frac{\partial f_{AB}}{\partial \varphi_M}\right)^* , \tag{31.7.52}$$

where φ_N here runs over all the fields on which f_{AB} may depend, with g_{NM} and L_M given by Eqs. (31.6.66) and (31.6.69). According to the estimates we have made here, $\kappa^2 d \ll 1$; $L_a = O(\kappa\Lambda^3)$ for the modular fields y_a and is much smaller for the other fields; and $g^{-1}_{ab} \simeq \delta_{ab}$. Also, we assume that f_{AB} is a power series in κy with coefficients of order unity, so $\partial f_{AB}/\partial y_a$ is of order κ. The gaugino masses (31.7.52) are therefore of order $\kappa^2\Lambda^3$, which is the same as the order of magnitude of the scalar masses and expectation values, and hence is likely to be large enough to avoid conflict with observation. The one-loop corrections considered in Section 31.4 are much smaller here, and do not need to be taken into account.

In summary, the first version of gravity-mediated supersymmetry breaking has the advantage of giving an axion mass that is within cosmological bounds, while the second version has the advantage of giving the gauginos masses that are comparable to the masses of the squarks and sleptons. Both versions of gravity-mediated supersymmetry breaking have an advantage over theories of gauge-mediated supersymmetry breaking: they naturally give μ-terms of the experimentally necessary order of magnitude. On the other hand, theories of gauge-mediated supersymmetry breaking have the advantage of naturally yielding generation-independent squark and slepton masses.

Theories with either version of gravity-mediated supersymmetry breaking may naturally entail the existence of slowly decaying superheavy particles, which could have interesting astrophysical effects.[22] It is plausible that the gauge interactions of the hidden sector that become strong at energy Λ can bind composite particles with masses of order Λ. These superheavy particles may be long-lived if their decay is forbidden by accidental symmetries of the renormalizable part of the hidden sector Lagrangian and only occurs through non-renormalizable terms in the Lagrangian, which are suppressed by factors of $\kappa\Lambda$.

Appendix The Vierbein Formalism

The familiar formulation of gravity in terms of a metric tensor $g_{\mu\nu}$ is adequate for theories with matter fields restricted to scalars, vectors, and tensors, but not for supergravity, where spinors are an indispensable ingredient. Unlike vectors and tensors, spinors have a Lorentz transformation rule that has no natural generalization to arbitrary coordinate systems. Instead, to deal with spinors, we have to introduce systems of coordinates $\xi^a_X(x)$ with $a = 0, 1, 2, 3$ that are locally inertial at any given point X in an arbitrary coordinate system. The Principle of Equivalence tells us that gravitation has no effect in these locally inertial coordinates, so the action may then be expressed in terms of matter fields like spinors, vectors, etc. that are defined in these locally inertial frames, as well as the vierbein, which arises from the transformation between the locally inertial and general coordinates

$$e^a{}_\mu(X) \equiv \left.\frac{\partial \xi^a_X(x)}{\partial x^\mu}\right|_{x=X} . \tag{31.A.1}$$

The action will be invariant under general coordinate transformations $x^\mu \to x'^\mu$ *and* local Lorentz transformations $\xi^a \to \xi'^a = \Lambda^a{}_b(x)\xi^b$ with $\Lambda^a{}_c(x)\Lambda^b{}_d(x)\eta_{ab} = \eta_{cd}$. The definition (31.A.1) of the vierbein shows that, under general coordinate transformations $x \to x'$, it transforms as

$$e^a{}_\mu(x) \to e'^a{}_\mu(x') = \frac{\partial x^\nu}{\partial x'^\mu} e^a{}_\nu(x)\,, \tag{31.A.2}$$

while, under a local Lorentz transformation $\xi^a(x) \to \Lambda^a{}_b(x)\xi^b(x)$, it transforms as

$$e^a{}_\mu(x) \to \Lambda^a{}_b(x) e^b{}_\mu(x)\,. \tag{31.A.3}$$

For instance, theories of pure gravitation may be expressed in terms of a field that is invariant under local Lorentz transformations and that transforms as a tensor under general coordinate transformations, the metric

$$g_{\mu\nu} \equiv e^a{}_\mu e^b{}_\nu\, \eta_{ab}\,. \tag{31.A.4}$$

Vectors may be regarded either as quantities V^a that transform as vectors under local Lorentz transformations

$$V^a(x) \to \Lambda^a{}_b(x) V^b(x) \tag{31.A.5}$$

but as scalars under general coordinate transformations, or as quantities v^μ that transform as scalars under local Lorentz transformations but as vectors under general coordinate transformations, the two being related by

$$V^a = e^a{}_\mu v^\mu\,.$$

But the supergravity action also involves spinor fields, which necessarily transform like scalars under general coordinate transformations but as spinors under local Lorentz transformations

$$\psi_\alpha(x) \to D_{\alpha\beta}(\Lambda(x))\psi_\beta(x)\,, \tag{31.A.6}$$

where $D_{\alpha\beta}(\Lambda)$ is the spinor representation of the homogeneous Lorentz group.

Because the Lorentz transformations in Eqs. (31.A.5) and (31.A.6) depend on the coordinate x^μ, the spacetime derivative of a quantity like $V^a(x)$ or $\psi_\alpha(x)$ is not just another quantity that transforms in the same way under local Lorentz transformations and as a covariant vector under general coordinate transformations. For instance, the derivative of Eq. (31.A.6) gives the local Lorentz transformation rule

$$\partial_\mu \psi_\alpha \to D_{\alpha\beta}(\Lambda)\left\{\partial_\mu \psi_\beta + \left[D^{-1}(\Lambda)\partial_\mu D(\Lambda)\right]_{\beta\gamma}\psi_\gamma\right\}\,.$$

To cancel the second term in the brackets on the right-hand side we introduce a connection matrix Ω_μ with the local Lorentz transformation property

$$\Omega_\mu \to D(\Lambda)\Omega_\mu D^{-1}(\Lambda) - \Big(\partial_\mu D(\Lambda)\Big)D^{-1}(\Lambda) \tag{31.A.7}$$

and define a covariant derivative

$$\mathscr{D}_\mu\psi \equiv \partial_\mu\psi + \Omega_\mu\psi\,, \tag{31.A.8}$$

which transforms under local Lorentz transformations like ψ itself:

$$\mathscr{D}_\mu\psi \to D(\Lambda)\mathscr{D}_\mu\psi\,. \tag{31.A.9}$$

Also, Ω_μ must transform like a covariant vector under general coordinate transformations, so that $\mathscr{D}_\mu$ will give a covariant vector when acting on a coordinate scalar. In order that $\mathscr{D}_\mu$ should give a tensor with one extra lower index when acting on a tensor, it must be supplemented with the usual affine connection term. For instance, when acting on the gravitino field ψ_μ, the covariant derivative is defined as

$$\mathscr{D}_\mu\psi_\nu \equiv \psi_{\nu;\mu} + \Omega_\mu\psi_\nu \equiv \partial_\mu\psi_\nu - \Gamma^\lambda_{\mu\nu}\psi_\lambda + \Omega_\mu\psi_\nu\,. \tag{31.A.10}$$

Eqs. (31.A.8)–(31.A.10) apply not only for spinors, but for fields that transform under local Lorentz transformations according to arbitrary representations $D(\Lambda)$ of the Lorentz group. The matrix Ω_μ depends on this representation, but in any representation it can be written in the form

$$[\Omega_\mu]_{\alpha\beta}(x) = \frac{1}{2}i[\mathscr{J}_{ab}]_{\alpha\beta}\,\omega_\mu^{ab}(x)\,, \tag{31.A.11}$$

where $\mathscr{J}_{ab}$ are the matrices representing the generators of the homogeneous Lorentz group in the representation furnished by the fields in question:

$$i\,[\mathscr{J}_{ab}\,,\,\mathscr{J}_{cd}] = \eta_{bc}\mathscr{J}_{ad} - \eta_{ac}\mathscr{J}_{bd} + \eta_{bd}\mathscr{J}_{ca} - \eta_{ad}\mathscr{J}_{cb}\,, \tag{31.A.12}$$

and ω_μ^{ab} is a representation-independent field known as the *spin connection* that transforms as a covariant vector under general coordinate transformations. To satisfy the inhomogeneous local Lorentz transformation rule (31.A.7) we can take

$$\omega_\mu^{ab} = g^{\nu\lambda}e^a{}_\nu e^b{}_{\lambda;\mu}\,, \tag{31.A.13}$$

with the semi-colon again denoting an ordinary covariant derivative, constructed using the affine connection $\Gamma^\lambda_{\mu\nu}$. (This is antisymmetric in a and b because Eq. (31.A.4) gives $g^{\nu\lambda}e^a{}_\nu e^b{}_\lambda = \eta^{ab}$, a quantity with vanishing covariant derivative.) This is not the unique spin-connection for which Eq. (31.A.7) is satisfied; to it, we can add any field that is a covariant vector under general coordinate transformations and a tensor under local Lorentz transformations, a freedom of some importance in supergravity theories.

For any choice of spin connection there is a corresponding curvature tensor. From Eq. (31.A.7) it is straightforward to show that the quantity $\partial_\nu\Omega_\mu - \partial_\mu\Omega_\nu + [\Omega_\nu, \Omega_\mu]$ transforms homogeneously under local Lorentz transformations

$$\partial_\nu \Omega_\mu - \partial_\mu \Omega_\nu + [\Omega_\nu, \Omega_\mu] \to D(\Lambda)\left(\partial_\nu \Omega_\mu - \partial_\mu \Omega_\nu + [\Omega_\nu, \Omega_\mu]\right) D^{-1}(\Lambda) . \tag{31.A.14}$$

Using Eqs. (31.A.11) and (31.A.12), this matrix can be expressed as

$$\partial_\nu \Omega_\mu - \partial_\mu \Omega_\nu + [\Omega_\nu, \Omega_\mu] = \tfrac{1}{2} i \mathscr{J}_{ab} R_{\mu\nu}{}^{ab} , \tag{31.A.15}$$

where

$$R_{\mu\nu}{}^{ab} \equiv \partial_\nu \omega_\mu^{ab} - \partial_\mu \omega_\nu^{ab} + \omega_\nu^{ac} \omega_{\mu\, c}{}^{b} - \omega_\mu^{ac} \omega_{\nu\, c}{}^{b} . \tag{31.A.16}$$

From Eq. (31.A.14) it follows that $R_{\mu\nu}{}^{ab}$ transforms as a tensor under local Lorentz transformations

$$R_{\mu\nu}{}^{ab} \to D(\Lambda)^a{}_c D(\Lambda)^b{}_d R_{\mu\nu}{}^{cd} . \tag{31.A.17}$$

It also obviously transforms as a tensor under general coordinate transformations

$$R_{\mu\nu}{}^{ab} \to \frac{\partial x^\rho}{\partial x'^\mu} \frac{\partial x^\sigma}{\partial x'^\nu} R_{\rho\sigma}{}^{ab} . \tag{31.A.18}$$

We can therefore form a coordinate tensor of fourth rank by writing

$$R_{\mu\nu}{}^{ab} = e^a{}_\kappa \, e^b{}_\lambda \, R_{\mu\nu}{}^{\kappa\lambda} . \tag{31.A.19}$$

The tensor $R_{\mu\nu}{}^{\kappa\lambda}$ constructed in this way is the Riemann–Christoffel curvature tensor corresponding to the particular spin connection ω_ν^{ab}.

Problems

1. Derive formulas (31.2.3)–(31.2.6) for components of the Einstein superfield.

2. Suppose that supersymmetry is unbroken. Show how to calculate the amplitude for emission of a gravitino of very low energy in a general scattering process in terms of the amplitude for this process without the gravitino.

3. Check that the supergravity action (31.6.11) is invariant under the local supersymmetry transformations (31.6.1)–(31.6.6) to all orders in G.

4. Calculate the change of the generalized D-component (31.6.40) under a general local supersymmetry transformation.

5. Calculate the fermionic part of the Lagrangian density (31.6.49).

6. Consider the theory of a single chiral scalar superfield Φ interacting with supergravity, with modified Kahler potential $d(\Phi, \Phi^*) = \Phi^*\Phi$ and superpotential $f(\Phi) = M^2(\Phi + \beta)$, where M and β are constants. Find a value of β for which the classical field equations have a solution with flat spacetime. What is the value of ϕ for this solution?

References

1. P. Nath and R. Arnowitt, *Phys. Lett.* **56B**, 177 (1975); B. Zumino, in *Proceedings of the Conference on Gauge Theories and Modern Field Theories at Northeastern University, 1975*, R. Arnowitt and P. Nath, eds. (MIT Press, Cambridge, MA, 1976). This approach is described in detail by J. Wess and J. Bagger, *Supersymmetry and Supergravity*, 2nd edition (Princeton University Press, Princeton, NJ, 1992).

2. D. Z. Freedman, P. van Nieuwenhuizen, and S. Ferrara, *Phys. Rev.* **D13**, 3214 (1976); S. Deser and B. Zumino, *Phys. Lett.* **62B**, 335 (1976); S. Ferrara, J. Scherk, and P. van Nieuwenhuizen, *Phys. Rev. Lett.* **37**, 1035 (1976); S. Ferrara, F. Gliozzi, J. Scherk, and P. van Nieuwenhuizen, *Nucl. Phys.* **B117**, 333 (1976). These articles are reprinted in *Supersymmetry*, S. Ferrara, ed. (North Holland/World Scientific, Amsterdam/Singapore, 1987). Some of these results had been obtained by D. V. Volkov and V. A. Soroka, *JETP Lett.* **18**, 312 (1973); *Theor. Math. Phys.* **20**, 829 (1974). For a clear description of this approach, see P. West, *Introduction to Supersymmetry and Supergravity*, 2nd edition (World Scientific, Singapore, 1990).

3. S. Ferrara and B. Zumino, *Nucl. Phys.* **B134**, 301 (1978).

3*a*. M. T. Grisaru and H. N. Pendleton, *Phys. Lett.* **67B**, 323 (1977).

4. K. Stelle and P. C. West, *Phys. Lett.* **74B**, 330 (1978); S. Ferrara and P. van Nieuwenhuizen, *Phys. Lett.* **74B**, 333 (1978). These articles are reprinted in *Supersymmetry*, Ref. 2.

5. See, for example, S. Weinberg, *Gravitation and Cosmology* (Wiley, New York, 1972), Sec. 12.5.

6. See, for example, *Gravitation and Cosmology*, Ref. 5: Eq. (12.4.3).

7. See, for example, *Gravitation and Cosmology*, Ref. 5, Section 10.1.

7*a*. W. Nahm, *Nucl. Phys.* **B135**, 149 (1978).

8. S. Coleman and F. de Luccia, *Phys. Rev.* **D21**, 3305 (1980).

9. S. Weinberg, *Phys. Rev. Lett.* **48**, 1776 (1982).

10. S. Deser and C. Teitelboim, *Phys. Rev. Lett.* **39**, 249 (1977); M. Grisaru, *Phys. Lett.* **73B**, 207 (1978); E. Witten, *Commun. Math. Phys.* **80**, 381 (1981).

11. W. Rarita and J. Schwinger, *Phys. Rev.* **60**, 61 (1941).

11*a*. L. Randall and R. Sundrum, hep-th/9810155, to be published; G. F. Giudice, M. Luty, R. Rattazzi, and H. Murayama, *JHEP* **12**, 027 (1998); A. Pomerol and R. Rattazzi, hep-ph/9903448, to be published; E. Katz, Y. Shadmi, and Y. Shirman, hep-ph/9906296, to be published.

11*b*. G. F. Giudice, M. Luty, R. Rattazzi, and H. Murayama, Ref. 11a.

12. K. Stelle and P. C. West, Ref. 4; S. Ferrara and P. van Nieuwenhuizen, Ref. 4; E. Cremmer, B. Julia, J. Scherk, S. Ferrara, L. Girardello, and P. van Nieuwenhuizen, *Phys. Lett.* **79B**, 231 (1978); *Nucl. Phys.* **B147**, 105 (1979) (reprinted in *Supersymmetry*, Ref. 2.); D. G. Boulware, S. Deser, and J. H. Kay, *Physica* **280**, 141 (1979); E. Cremmer, S. Ferrara, L. Girardello, and A. Van Proeyen, *Phys. Lett.* **116B**, 231 (1982); *Nucl. Phys.* **B212**, 413 (1983) (reprinted in *Supersymmetry*, Ref. 2). For this construction in two spacetime dimensions, see S. Deser and B. Zumino, *Phys. Lett.* **65B**, 369 (1976).

13. S. Deser and B. Zumino, *Phys. Rev. Lett.* **38**, 1433 (1977). This article is reprinted in *Supersymmetry*, Ref. 2. Also see D. Z. Freedman and A. Das, *Nucl. Phys.* **B120**, 221 (1977); P. K. Townsend, *Phys. Rev.* **D 15**, 2802 (1977).

13*a*. E. Cremmer, S. Ferrara, C. Kounnas, and D. V. Nanopoulos, *Phys. Lett.* **133B**, 61 (1983). For a review of models based on this idea, see A. B. Lahanas and D. V. Nanopoulos, *Phys. Rept.* **145**, 1 (1987).

14. H. P. Nilles, *Phys. Lett.* **115B**, 193 (1982); A. Chamseddine, R. Arnowitt, and P. Nath, *Phys. Rev. Lett.* **49**, 970 (1982); R. Barbieri, S. Ferrara, and C. A. Savoy, *Phys. Lett.* **119B**, 343 (1982); E. Cremmer, P. Fayet, and L. Girardello, *Phys. Lett.* **122B**, 41 (1983); L. Ibañez, *Phys. Lett.* **118B**, 73 (1982); H. P. Nilles, M. Srednicki, and D. Wyler, *Phys. Lett.* **120B**, 346 (1983); L. Hall, J. Lykken, and S. Weinberg, *Phys. Rev.* **D27**, 2359 (1983); L. Alvarez-Gaumé, J. Polchinski, and M. B. Wise, *Nucl. Phys.* **B221**, 495 (1983). The above articles are reprinted in *Supersymmetry*, Ref. 2. Also see S. Ferrara, D. V. Nanopoulos, and C. A. Savoy, *Phys. Lett.* **12B**, 214 (1983); J. M. Leon, M. Quiros, and M. Ramon Medrano, *Phys. Lett.* **127B**, 85 (1983); *Phys. Lett.* **129B**, 61 (1983); N. Ohta, *Prog. Theor. Phys.* **70**, 542 (1983); P. Nath, R. Arnowitt, and A. Chamseddine, *Phys. Lett.* **121B**, 33 (1983); J. Ellis, D. V. Nanopoulos, and K. Tamvakis, *Phys.*

Lett. **121B**, 123 (1983). For a review, see H. P. Nilles, *Phys. Rept.* **110**, 1 (1984).

15. The breakdown of supersymmetry in a supergravity theory with an unobserved sector of chiral scalars seems to have been first proposed in an unpublished University of Budapest preprint by J. Polonyi (1977).
16. I. Affleck, M. Dine, and N. Seiberg, *Nucl. Phys.* **B256**, 557 (1985).
17. G. F. Giudice and A. Masiero, *Phys. Lett.* **B206**, 480 (1988). Also see J. A. Casas and C. Muñoz, *Phys. Lett.* **B306**, 288 (1993); J. E. Kim, hep-ph/9901204, to be published.
18. M. Dine and D. A. MacIntire, *Phys. Rev.* **D46**, 2594 (1992).
19. T. Banks, D. B. Kaplan, and A. Nelson, *Phys. Rev.* **D49**, 779 (1994); K. I. Izawa and T. Yanagida, *Prog. Theor. Phys.* **94**, 1105 (1995); A. Nelson, *Phys. Lett.* **B369**, 277 (1996).

19*a*. The cosmological implications of a wino that is lighter than the bino are discussed by T. Moroi and L. Randall, hep-ph/9906527, to be published.

20. V. Kaplunovsky and J. Louis, *Phys. Lett.* **B306**, 269 (1993); *Nucl. Phys.* **B422**, 57 (1994). Specific models of this sort have been proposed by P. Binétruy, M. K. Gaillard, and Y-Y. Wu, *Nucl. Phys.* **B493**, 27 (1997); *Phys. Lett.* **B412**, 288 (1997).
21. Theories with extra dimensions in which flavor-changing effects are suppressed have been suggested by L. Randall and R. Sundrum, Ref. 11a. Problems with such theories were pointed out by Z. Chacko, M. A. Luty, I. Maksymyk, and E. Pontón, hep-ph/9905390, to be published.
22. K. Hamaguchi, K.-I. Izawa, Y. Nomura, and T. Yanagida, hep-ph/9903207, to be published.

32
Supersymmetry Algebras in Higher Dimensions

Ever since the ground-breaking work of Kaluza[1] and Klein,[2] theorists have from time to time tried to formulate a more nearly fundamental physical theory in spacetimes of higher than four dimensions. This approach was revived in superstring theories, which take their simplest form in 10 spacetime dimensions.[3] More recently, it has been suggested that the various versions of string theory may be unified in a theory known as *M theory*, which in one limit is approximately described by supergravity in 11 spacetime dimensions.[4] In this chapter we shall catalog the different types of supersymmetry algebra possible in higher dimensions, and use them to classify supermultiplets of particles.

32.1 General Supersymmetry Algebras

Our analysis of the general supersymmetry algebra in higher dimensions will follow the same logical outline as the work of Haag, Lopuszanski, and Sohnius[5] on supersymmetry algebras in four spacetime dimensions, described in Section 25.2. The proof of the Coleman–Mandula theorem in the appendix of Chapter 24 makes it clear that the list of possible bosonic symmetry generators is essentially the same in $d > 2$ spacetime dimensions as in four spacetime dimensions: in an S-matrix theory of particles, there are only the momentum d-vector P^μ, a Lorentz generator $J^{\mu\nu} = -J^{\nu\mu}$ (with μ and ν here running over the values $1, 2, \ldots, d-1, 0$), and various Lorentz scalar 'charges.' (In some theories there are topologically stable extended objects such as closed strings, membranes, etc., in addition to particles, which make possible other conserved quantities, to which we will return in Section 32.3.) The anticommutators of the fermionic symmetry generators with each other are bosonic symmetry generators, and therefore must be a linear combination of P^μ, $J^{\mu\nu}$, and various conserved scalars. This puts severe limits on the Lorentz transformation properties of the fermionic generators, and on the superalgebra to which they belong.

We will first prove that the general fermionic symmetry generator must transform according to the fundamental spinor representations of the Lorentz group, which are reviewed in the appendix to this chapter, and not in higher spinor representations, such those obtained by adding vector indices to a spinor. As we saw in Section 25.2, the proof for $d = 4$ by Haag, Lopuszanski, and Sohnius made use of the isomorphism of $SO(4)$ to $SU(2) \times SU(2)$, which has no analog in higher dimensions. Here we will use an argument of Nahm,[6] which is actually somewhat simpler and applies in any number of dimensions.

Since the Lorentz transform of any fermionic symmetry generator is another fermionic symmetry generator, the fermionic symmetry generators furnish a representation of the homogeneous Lorentz group $O(d-1, d)$ (or, strictly speaking, of its covering group $Spin(d-1,1)$). Assuming that there are at most a finite number of fermionic symmetry generators, they must transform according to a finite-dimensional representation of the homogeneous Lorentz group. All of these representations can be obtained from the finite-dimensional *unitary* representations of the corresponding orthogonal group $O(d)$ (actually $Spin(d)$) by setting $x^d = ix^0$. So let us first consider the transformation of the fermionic generators under $O(d)$. For d even or odd, we can find $d/2$ or $(d-1)/2$ Lorentz generators J_{d1}, J_{23}, $J_{45}, \ldots$, which all commute with each other, and classify fermionic generators Q according to the values σ_{d1}, $\sigma_{23}, \ldots$ that they destroy:

$$[J_{d1}\,, Q] = -\sigma_{d1} Q\,, \qquad [J_{23}\,, Q] = -\sigma_{23} Q\,, \qquad [J_{45}\,, Q] = -\sigma_{45} Q\,, \ldots\,. \tag{32.1.1}$$

Since the finite-dimensional representations of $O(d)$ are all unitary, the σs are all real.

Let us focus on one of these quantum numbers, $\sigma_{d1} \equiv w$ and refer to any fermionic or bosonic operator O as having *weight* w if

$$[J_{d1}\,, O] = -w\, O\,, \tag{32.1.2}$$

or, in terms of the Minkowski component $J_{01} = iJ_{d1}$,

$$[J_{01}\,, O] = -iw\, O\,. \tag{32.1.3}$$

The reason for concentrating on this particular quantum number is that it has the special property of being the same for an operator and its Hermitian adjoint. This is because J_{01} must be represented on Hilbert space (though not on field variables or symmetry generators) by a Hermitian operator, so that (remembering that w is real) the Hermitian adjoint of Eq. (32.1.3) is

$$-\,[J_{01}\,, O^*] = +iwO^*\,, \tag{32.1.4}$$

so O^* has the same weight as O.

Now consider the anticommutator $\{Q, Q^*\}$ of any fermionic symmetry generator Q with its Hermitian adjoint. According to the Coleman–Mandula theorem, it is at most a linear combination of P_μ, $J_{\mu\nu}$, and scalars. To calculate the weights of the components of P_μ, we recall the commutation relation (2.4.13)

$$i[P_\mu, J_{\rho\sigma}] = \eta_{\mu\rho}P_\sigma - \eta_{\mu\sigma}P_\rho \,,$$

which shows that $P_0 \pm P_1$ has weight $w = \pm 1$, while the other components $P_2, P_3, \ldots, P_{d-1}$ all have weight zero. In the same way, the commutation relation (2.4.12) of the $J_{\mu\nu}$ with each other show that $J_{0i} \pm J_{1i}$ with $i = 2, 3, \ldots d-1$ have weight $w = \pm 1$, the J_{ij} with both i and j between 2 and $d-1$ have weight zero, J_{10} has weight zero, and of course all scalars have weight zero. We conclude then that all bosonic symmetry generators have weight ± 1 or 0 and the anticommutator $\{Q, Q^*\}$ must be a linear combination of operators with such weights. If Q has weight w then $\{Q, Q^*\}$ has weight $2w$, and it is manifestly non-zero for any non-zero Q, so each fermionic generator can only have weight $\pm 1/2$. (Weight zero is excluded by the connection between spin and statistics — fermionic operators can only be constructed from odd numbers of operators with half-integer weights.) Going back to the Euclidean formalism, since the commutators of the particular $O(d)$ generator J_{01} with all generators Q in a representation of $O(d)$ are given by Eq. (32.1.2) with $w = \pm 1/2$, and there is nothing special about the 01 plane, $O(d)$ invariance requires that the same is true for all $O(d)$ generators J_{ij}, so that all the σs in Eq. (32.1.1) are $\pm 1/2$. The only irreducible representations of the homogeneous Lorentz group with all σs equal to $\pm 1/2$ are the fundamental spinor representations, so Q must belong to some direct sum of these representations.

We can also use this approach to show that the fermionic generators Q all commute with the d-momentum P_μ. For this purpose, note that the double commutator of a momentum operator $P_0 \pm P_1$ of weight ± 1 with any fermionic generator Q would have weight either $\pm 5/2$ if Q has weight $\pm 1/2$ or weight $\pm 3/2$ if Q has weight $\mp 1/2$, and since we have found that there are no fermionic symmetry generators of weight $\pm 3/2$ or $\pm 5/2$, these double commutators must all vanish:

$$[P_0 \pm P_1, [P_0 \pm P_1, Q]] = 0 \,.$$

It follows then that

$$[P_0 \pm P_1, [P_0 \pm P_1, \{Q, Q^*\}]] = -2\,\{Q_\pm, Q_\pm^*\} \,,$$

where

$$Q_\pm \equiv [P_0 \pm P_1, Q] \,.$$

Now, $\{Q, Q^*\}$ is at most a linear combination of Js, Ps, and scalar symmetry generators. The commutators of $P_0 \pm P_1$ with the Ps and scalar symmetry generators vanish, while the commutators of $P_0 \pm P_1$ with the Js are linear combinations of Ps, which commute with the other $P_0 \pm P_1$, so the double commutator $[P_0 \pm P_1, [P_0 \pm P_1, \{Q, Q^*\}]]$ must vanish and therefore $\{Q_\pm, Q_\pm^*\} = 0$, which implies that $Q_\pm = 0$. Since *all* members of the representation of the Lorentz group provided by the Qs thus commute with P_0 and P_1, Lorentz invariance implies that all Qs commute with all Ps, as was to be shown.

There is an important corollary that since the Lorentz generators $J_{\mu\nu}$ do not commute with the momentum operators, they cannot appear on the right-hand side of the anticommutation relations. For the moment let us label the Qs as Q_n, where n runs over the labels for the different (not necessarily inequivalent) irreducible spinor representations among the Qs, now *including* their adjoints Q^*, and also over the index labelling members of these representations. The general anticommutation relation is then of the form

$$\{Q_n\,, Q_m\} = \Gamma^\mu_{nm} P_\mu + Z_{nm}\,, \tag{32.1.5}$$

where the Γ^μ_{nm} are c-number coefficients and the Z_{nm} are conserved scalar symmetry generators, which commute with the P_μ and $J_{\mu\nu}$. We now want to show that the Z_{nm} are *central charges* of the supersymmetry algebra — that is, that they commute with the Q_ℓ and each other as well as with the P_μ and $J_{\mu\nu}$ and all other symmetry generators.

To prove this for $d \geq 4$, note that for a given Z_{nm} to be non-zero, since it is a scalar all of the σs in Eq. (32.1.1) must be opposite for Q_n and Q_m. Consider another fermionic symmetry generator Q_ℓ, for which the σs of Eq. (32.1.1) are not all the same as those of either Q_n or Q_m. (For $d \geq 4$ there is always such a Q_ℓ in each set of Qs forming an irreducible spinor representation of $O(d)$.) We apply the super-Jacobi identity

$$[Q_\ell\,, \{Q_m\,, Q_n\}] + [Q_m\,, \{Q_n\,, Q_\ell\}] + [Q_n\,, \{Q_\ell\,, Q_m\}] = 0\,. \tag{32.1.6}$$

The anticommutators $\{Q_n\,, Q_\ell\}$ and $\{Q_\ell\,, Q_m\}$ are operators that have some σs non-zero, so they can only be linear combinations of Ps rather than Zs, and so must commute with all Qs. This leaves just

$$0 = [Q_\ell\,, \{Q_m\,, Q_n\}] = [Q_\ell\,, Z_{mn}]\,. \tag{32.1.7}$$

Thus in each set of Qs forming an irreducible spinor representation of $O(d)$ there is at least one that commutes with the given Z_{mn}. But Z_{mn} is a Lorentz scalar, so it must then commute with all Qs. It follows then immediately from Eq. (32.1.5) that they also commute with each other.

The fermionic generators must form a representation (perhaps trivial) of the algebra $\mathscr{A}$ consisting of *all* scalar bosonic symmetry generators. It follows then by precisely the same argument used in Section 25.2 that the central charges Z_{mn} furnish an invariant Abelian subalgebra of $\mathscr{A}$. The Coleman–Mandula theorem tells us that $\mathscr{A}$ must be a direct sum of a compact semi-simple Lie algebra, which by definition contains no invariant Abelian subalgebras, together with $U(1)$ generators, so the Z_{mn} must be $U(1)$ generators, which commute with all other bosonic symmetry generators, not just with each other.

To obtain more detailed information about the structure of the anti-commutation relations (32.1.5), we must be more specific about the Lorentz transformation and reality properties of the fermionic symmetry generators Q_n. These are very different for spacetimes of even and odd dimensionality.

Odd Dimensionality

The appendix to this chapter shows that for odd spacetime dimensions d there is just one fundamental spinor representation of the Lorentz algebra, by matrices $\mathscr{J}_{\mu\nu}$ given in terms of Dirac matrices by Eq. (32.A.2), so we must label the fermionic generators as $Q_{\alpha r}$, where α is a $2^{(d-1)/2}$-valued Dirac index, and $r = 1, 2, \ldots, N$ labels different spinors in the case of N-extended supersymmetry. With this notation, the Lorentz transformation properties of the Qs imply that

$$[J_{\mu\nu}, Q_{\alpha r}] = -\sum_{\beta}(\mathscr{J}_{\mu\nu})_{\alpha\beta}\, Q_{\beta r}\,, \tag{32.1.8}$$

so that the anticommutators of these generators have the transformation rule

$$[J_{\mu\nu}, \{Q_{\alpha r}, Q_{\beta s}\}] = -\sum_{\bar{\alpha}}(\mathscr{J}_{\mu\nu})_{\alpha\bar{\alpha}}\{Q_{\bar{\alpha} r}, Q_{\beta s}\} - \sum_{\bar{\beta}}(\mathscr{J}_{\mu\nu})_{\beta\bar{\beta}}\{Q_{\alpha r}, Q_{\bar{\beta} s}\}\,.$$

Recalling the Lorentz transformation rule (2.4.13) for the momentum operator P_λ, we see that the matrix Γ^λ_{rs} and the operator Z_{rs} in Eq. (32.1.5) (with Dirac indices now suppressed) must satisfy the conditions

$$\mathscr{J}_{\mu\nu}(\Gamma_\lambda)_{rs} + (\Gamma_\lambda)_{rs}\mathscr{J}^{\mathrm{T}}_{\mu\nu} = -i\,(\Gamma_\mu)_{rs}\eta_{\nu\lambda} + i\,(\Gamma_\nu)_{rs}\eta_{\mu\lambda}\,, \tag{32.1.9}$$

$$\mathscr{J}_{\mu\nu}Z_{rs} + Z_{rs}\mathscr{J}^{\mathrm{T}}_{\mu\nu} = 0\,. \tag{32.1.10}$$

But Eq. (32.A.38) gives $\mathscr{J}^{\mathrm{T}}_{\mu\nu} = -\mathscr{C}^{-1}\mathscr{J}_{\mu\nu}\mathscr{C}$, so Eqs. (32.1.9) and (32.1.10) may be expressed as the requirement that $(\Gamma_\mu)_{rs}\mathscr{C}^{-1}$ satisfies the same commutation relation (32.A.32) with $\mathscr{J}_{\mu\nu}$ as γ_μ, while $Z_{rs}\mathscr{C}^{-1}$ commutes with $\mathscr{J}_{\mu\nu}$. For odd d the matrices satisfying these conditions are unique up to multiplication with constants, so we can conclude that

$$\Gamma^\lambda_{\alpha r\,\beta s} = i\,g_{rs}\,(\gamma^\lambda\mathscr{C})_{\alpha\beta} \tag{32.1.11}$$

and

$$Z_{\alpha r\,\beta s} = \mathscr{C}_{\alpha\beta}\, z_{rs}\,, \qquad (32.1.12)$$

with the factor i inserted in Eq. (32.1.11) for later convenience. With Dirac indices suppressed, the anticommutation relations (32.1.5) now read

$$\{Q_r\,, Q_s^{\mathrm{T}}\} = i\, g_{rs}\gamma^\lambda \mathscr{C}\, P_\lambda + z_{rs}\mathscr{C}\,. \qquad (32.1.13)$$

Both $\Gamma^\lambda_{\alpha r\,\beta s}$ and $Z_{\alpha r\,\beta s}$ are symmetric under interchange of α and r with β and s, while Eqs. (32.A.30) and (32.A.31) (with $d = 2n+1$) show that: $\gamma^\lambda\mathscr{C}$ is symmetric and $\mathscr{C}$ is symmetric for $d = 1$ (mod 8); $\gamma^\lambda\mathscr{C}$ is symmetric and $\mathscr{C}$ is antisymmetric for $d = 3$ (mod 8); $\gamma^\lambda\mathscr{C}$ is antisymmetric and $\mathscr{C}$ is antisymmetric for $d = 5$ (mod 8); and $\gamma^\lambda\mathscr{C}$ is antisymmetric and $\mathscr{C}$ is symmetric for $d = 7$ (mod 8). It follows that: g_{rs} is symmetric and z_{rs} is symmetric for $d = 1$ (mod 8); g_{rs} is symmetric and z_{rs} is antisymmetric for $d = 3$ (mod 8); g_{rs} is antisymmetric and z_{rs} is antisymmetric for $d = 1$ (mod 8); and g_{rs} is antisymmetric and z_{rs} is symmetric for $d = 1$ (mod 8).

The complex conjugate of the matrices $\mathscr{J}_{\mu\nu}$ is given by Eq. (32.A.37). By taking the Hermitian adjoint of Eq. (32.1.8), we see that $\sum_\beta(\mathscr{C}\beta)_{\alpha\beta}Q^*_{\beta r}$ has the same Lorentz transformation properties as any $Q_{\alpha s}$, and therefore must be a linear combination of them

$$\sum_\beta(\mathscr{C}\beta)_{\alpha\beta}Q^*_{\beta r} = \sum_s \mathscr{S}_{rs}Q_{\alpha s}\,. \qquad (32.1.14)$$

Taking the Hermitian adjoint of this equation and using Eqs. (32.A.28) and (32.A.29) with $d = 2n+1$ yields

$$\mathscr{S}^*\mathscr{S} = (-1)^a \cdot 1\,, \qquad a = (d-1)(d-3)/8\,. \qquad (32.1.15)$$

For $d = 1$ (mod 8) and $d = 3$ (mod 8) the spinor representation of the Lorentz algebra is real, and we can choose a basis for the fermionic generators such that $\mathscr{S} = 1$. In contrast, for $d = 5$ (mod 8) and $d = 7$ (mod 8) the spinor representation of the Lorentz algebra is pseudoreal, and it is evidently impossible to choose a basis with $\mathscr{S} \propto 1$. By taking the determinant of Eq. (32.1.15), we see that in this case $\mathrm{Det}\,(-1) > 0$, so for $d = 5$ (mod 8) and $d = 7$ (mod 8) there must be an even number N of fermionic generators. In this case we can choose a basis in which $\mathscr{S} = \Omega$, where Ω is the real antisymmetric block-diagonal matrix

$$\Omega = \begin{pmatrix} e & 0 & 0 & \cdots \\ 0 & e & 0 & \cdots \\ 0 & 0 & e & \cdots \\ & \cdots & & \end{pmatrix}, \qquad e = \begin{pmatrix} 0 & 1 \\ -1 & 0 \end{pmatrix}. \qquad (32.1.16)$$

We can deduce the reality and positivity properties of g_{rs} and z_{rs} by using Eq. (32.1.14) to rewrite the anticommutation relation (32.1.13) as

$$\{Q_r\,,\,Q_s^\dagger\} = i(g\mathscr{S}^{\rm T})_{rs}\gamma_\lambda\mathscr{C}(\mathscr{C}\beta)^{{\rm T}-1}P^\lambda + (z\mathscr{S}^{\rm T})_{rs}\mathscr{C}(\mathscr{C}\beta)^{{\rm T}-1}\,.$$

Eqs. (32.A.12), (32.A.16), and (32.A.30) with $d = 2n+1$ show that $\beta^{\rm T} = -\beta$ and $\mathscr{C}\mathscr{C}^{{\rm T}-1} = (-1)^{(d-1)(d+1)/8}\cdot 1$, so

$$\{Q_r\,,\,Q_s^\dagger\} = -(-1)^{(d-1)(d+1)/8}\Big[i(g\mathscr{S}^{\rm T})_{rs}\gamma_\lambda\beta P^\lambda + (z\mathscr{S}^{\rm T})_{rs}\beta)\Big]\,. \qquad (32.1.17)$$

Recalling that $\gamma_0 = i\beta$, we note that the operator matrix $-i\gamma_\lambda\beta\,P^\lambda$ is positive, and positive-definite aside from the vacuum state. By considering a state of sufficiently large momentum so that the central charge term in Eq. (32.1.17) may be neglected, we conclude that the matrix $(-1)^{(d-1)(d+1)/8}g\mathscr{S}^{\rm T}$ is positive and Hermitian. Then considering arbitrary momenta, we also find that the array of operators $(z\mathscr{S}^{\rm T})_{rs}$ is Hermitian. (For non-zero central charges there is a lower bound on the mass, analogous to Eq. (25.5.22), which will not be given here.) From the Hermiticity of $g\mathscr{S}^{\rm T}$ we have

$$g^\dagger = \mathscr{S}^{{\rm T}\dagger-1}g\mathscr{S}^{\rm T} = (-1)^a\mathscr{S}g\mathscr{S}^{\rm T}. \qquad (32.1.18)$$

We are now in a position by an appropriate choice of basis to put the anticommutation relations in a convenient canonical form.

For $d = 1$ (mod 8) we have g and z symmetric and $(-1)^a = +1$, so if we choose a basis in which $\mathscr{S} = 1$ then g is real and the individual z_{rs} are Hermitian operators. We may introduce new Qs without changing $\mathscr{S} = 1$ by multiplying the old Qs by any real matrix $\mathscr{A}$, with the result that g is changed to $\mathscr{A}g\mathscr{A}^{\rm T}$. Since g is a positive matrix for $d = 1$ (mod 8), by a well-known theorem[6a] we may choose $\mathscr{A}$ to make $g = 1$.

For $d = 3$ (mod 8) we have g symmetric, z antisymmetric, and $(-1)^a = +1$, so if we choose a basis in which $\mathscr{S} = 1$ then g is real and the individual z_{rs} are anti-Hermitian operators. As in the case of $d = 1$ (mod 8), we may further adapt the basis so that $g = 1$.

For $d = 5$ (mod 8) we have g antisymmetric, so, with the choice $\mathscr{S} = \Omega$, Eq. (32.1.18) reads $g^* = -\Omega g\Omega$, where Ω is the standard antisymmetric matrix (32.1.16). Here we may introduce new Qs while keeping $\mathscr{S} = \Omega$ by multiplying the old ones by any matrix $\mathscr{B}$ with $\mathscr{B}^* = -\Omega\mathscr{B}\Omega$, with the effect that g is changed to $\mathscr{B}g\mathscr{B}^{\rm T}$. Since $(-1)^a = -1$, $g\Omega$ is positive, and so in this way we may arrange that $g = -\Omega$. Also, z is antisymmetric and $z\Omega$ is Hermitian, so $z^* = -\Omega\,z\,\Omega$.

For $d = 7$ (mod 8) we again have g antisymmetric, but now $(-1)^a = +1$, so by the same method as for $d = 5$ (mod 8), we may choose a basis in which $g = +\Omega$. Also, z is now symmetric and $z\Omega$ is again Hermitian, so now $z^* = +\Omega\,z\,\Omega$.

Even Dimensionality

The appendix to this chapter shows that for even spacetime dimensions d there are two inequivalent fundamental spinor representations of the Lorentz algebra by matrices $\mathscr{J}^{\pm}_{\mu\nu}$ given in terms of Dirac matrices by Eqs. (32.A.22), (32.A.2), and (32.A.17). Therefore here we must label the fermionic generators as $Q^{\pm}_{\alpha r}$, where α is a $2^{d/2}$-valued Dirac index, r labels different Qs belonging to equivalent representations of the Lorentz algebra in the case of extended supersymmetry, and

$$\sum_{\beta}(\gamma_{d+1})_{\alpha\beta}\, Q^{\pm}_{\beta r} = \pm\, Q^{\pm}_{\alpha r}\,, \tag{32.1.19}$$

where $\gamma_{d+1} \equiv i^{d/2-1}\gamma_1 \cdots \gamma_{d-1}\gamma_0$. With this notation, the Lorentz transformation properties of the Qs imply that

$$[J_{\mu\nu}\,,\, Q^{\pm}_{\alpha r}] = -\sum_{\beta}(\mathscr{J}^{\pm}_{\mu\nu})_{\alpha\beta}\, Q^{\pm}_{\beta r}\,, \tag{32.1.20}$$

where $\mathscr{J}^{\pm}_{\mu\nu}$ are the matrices (32.A.22). Taking account of Eq. (32.1.19) and the relation $\mathscr{C}^{-1}\gamma_{d+1}\mathscr{C} = (-1)^{d/2}\gamma_{d+1}$, the same arguments that gave the anticommutation relations (32.1.13) for odd d now give

$$\{Q^{\pm}_{r}\,,\, Q^{\mp(-1)^{d/2}\,\mathrm{T}}_{s}\} = i\, g^{\pm}_{rs}\left(\frac{1 \pm \gamma_{d+1}}{2}\right)\gamma^{\lambda}\mathscr{C}\, P_{\lambda}\,, \tag{32.1.21}$$

$$\{Q^{\pm}_{r}\,,\, Q^{\pm(-1)^{d/2}\,\mathrm{T}}_{s}\} = z^{\pm}_{rs}\left(\frac{1 \pm \gamma_{d+1}}{2}\right)\mathscr{C}\,. \tag{32.1.22}$$

Eqs. (32.A.30) and (32.A.31) show that $\mathscr{C}\gamma^{\lambda}$ is symmetric for $d = 0$ (mod 8) and $d = 2$ (mod 8) and antisymmetric for $d = 4$ (mod 8) and $d = 6$ (mod 8), while $\mathscr{C}$ is symmetric for $d = 0$ (mod 8) and $d = 6$ (mod 8) and antisymmetric for $d = 2$ (mod 8) and $d = 4$ (mod 8). Hence Eq. (32.1.21) requires the symmetry properties

$$g^{\pm}_{rs} = \begin{cases} g^{\mp(-1)^{d/2}}_{sr} & d = 0, 2 \pmod 8 \\ -g^{\mp(-1)^{d/2}}_{sr} & d = 4, 6 \pmod 8 \end{cases}\,, \tag{32.1.23}$$

while Eq. (32.1.22) requires that

$$z^{\pm}_{rs} = \begin{cases} z^{\pm(-1)^{d/2}}_{sr} & d = 0, 6 \pmod 8 \\ -z^{\pm(-1)^{d/2}}_{sr} & d = 2, 4 \pmod 8 \end{cases}\,. \tag{32.1.24}$$

In particular, $z^{\pm}$ is symmetric for $d = 0$ (mod 8), $g^{\pm}$ is symmetric for $d = 2$ (mod 8), $z^{\pm}$ is antisymmetric for $d = 4$ (mod 8), and $g^{\pm}$ is antisymmetric for $d = 6$ (mod 8).

Taking the Hermitian adjoint of Eq. (32.1.20) and using Eq. (32.A.25) shows that $\mathscr{C}\beta Q^{\pm *}_{r}$ has the same Lorentz transformation properties as the operators $Q_s^{\mp(-1)^{d/2}}$ and is therefore a linear combination of them:

$$\mathscr{C}\beta\, Q_r^{\pm *} = \sum_s \mathscr{S}_{rs}^{\pm}\, Q_s^{\mp(-1)^{d/2}} . \tag{32.1.25}$$

Taking the Hermitian adjoint of this equation and using Eqs. (32.A.28) and (32.A.29) with $d = 2n$ gives

$$\mathscr{S}^{\pm *}\mathscr{S}^{\mp(-)d/2} = (-1)^a \cdot 1 , \qquad a = d(d-2)/8 . \tag{32.1.26}$$

For $d = 0$ (mod 8) and $d = 4$ (mod 8) Eq. (32.1.25) relates one irreducible representation to another, and we may choose bases in which $\mathscr{S}^{\pm} = 1$ for $d = 0$ (mod 8) and $\mathscr{S}^{\pm} = \pm 1$ for $d = 4$ (mod 8). For $d = 2$ (mod 8) Eq. (32.1.25) relates real representations to themselves, and we may choose bases with $\mathscr{S}^{\pm} = 1$. For $d = 6$ (mod 8) Eq. (32.1.25) relates pseudoreal representations to themselves; the determinant of Eq. (32.1.26) shows that there must be an even number of Q^+ and an even number (not necessarily the same!) of Q^-, and we may choose bases with $\mathscr{S}^{\pm} = \Omega^{\pm}$, where $\Omega^{\pm}$ are standard real antisymmetric matrices of the form (32.1.16).

We can deduce the reality and positivity properties of $g_{rs}^{\pm}$ and the reality properties of $z_{rs}^{\pm}$ by using Eq. (32.1.25) to rewrite the anticommutation relations (32.1.21) and (32.1.22) in the form

$$\{Q_r^{\pm}, Q_s^{\pm\dagger}\} = \left(\frac{1 \pm \gamma_{d+1}}{2}\right)\gamma_\lambda \beta\, P^\lambda \mathscr{C}(\mathscr{C}\beta)^{\mathrm{T}-1}\left(g^{\pm}\mathscr{S}^{\mp\mathrm{T}}\right)_{rs} , \tag{32.1.27}$$

$$\{Q_r^{\pm}, Q_s^{\mp\dagger}\} = \left(\frac{1 \pm \gamma_{d+1}}{2}\right)\mathscr{C}(\mathscr{C}\beta)^{\mathrm{T}-1}\left(z^{\pm}\mathscr{S}^{\mp\mathrm{T}}\right)_{rs} . \tag{32.1.28}$$

We again use the relations $\mathscr{C}\mathscr{C}^{\mathrm{T}-1} = (-1)^{d(d+2)/8}$, $\beta^{\mathrm{T}} = -\beta$, and $\gamma_0 = i\beta$, and conclude that $(-1)^{d(d+2)/8} g^{\pm}\mathscr{S}^{\pm\mathrm{T}}$ is Hermitian and positive, while $(z^+\mathscr{S}^{-\mathrm{T}})^\dagger = z^-\mathscr{S}^{+\mathrm{T}}$. For $d = 0$ (mod 8) we can adopt a basis with $\mathscr{S}^{\pm} = 1$, $g^{\pm} = 1$, and $z^{+\dagger} = z^-$; for $d = 2$ (mod 8) we can adopt a basis with $\mathscr{S}^{\pm} = 1$, $g^{\pm} = -1$, and $z^{+\dagger} = z^-$; for $d = 4$ (mod 8) we can adopt a basis with $\mathscr{S}^{\pm} = \pm 1$, $g^{\pm} = \mp 1$, and $z^{+\dagger} = -z^-$; and for $d = 6$ (mod 8) we can adopt a basis with $\mathscr{S}^{\pm} = \Omega^{\pm}$, $g^{\pm} = \Omega^{\pm}$, and $(z^+\Omega^-)^\dagger = z^-\Omega^+$.

To summarize, in appropriate bases the anticommutation relations and reality and symmetry conditions are as follows:[6b]

$d = 0$ (mod 8)

$$\{Q_r^{\pm}, Q_s^{\mp\,\mathrm{T}}\} = i\,\delta_{rs}\left(\frac{1 \pm \gamma_{d+1}}{2}\right)\gamma^\lambda \mathscr{C} P_\lambda , \tag{32.1.29}$$

$$\{Q_r^{\pm}, Q_s^{\pm\,\mathrm{T}}\} = z_{rs}^{\pm}\left(\frac{1 \pm \gamma_{d+1}}{2}\right)\mathscr{C} , \tag{32.1.30}$$

$$\mathscr{C}\beta\, Q_r^{\pm *} = Q_r^{\mp} , \qquad z_{rs}^{\pm} = z_{sr}^{\pm} = (z_{rs}^{\mp})^* . \tag{32.1.31}$$

$d = 1$ (mod 8)

$$\{Q_r\,, Q_s^{\rm T}\} = i\,\delta_{rs}\gamma^\lambda \mathscr{C} P_\lambda + z_{rs}\mathscr{C}\,, \tag{32.1.32}$$

$$\mathscr{C}\beta Q_r^* = Q_r\,, \qquad z_{rs} = z_{sr} = z_{rs}^*\,. \tag{32.1.33}$$

$d = 2$ (mod 8)

$$\{Q_r^\pm\,, Q_s^{\pm\,\rm T}\} = -i\,\delta_{rs}\left(\frac{1 \pm \gamma_{d+1}}{2}\right)\gamma^\lambda \mathscr{C} P_\lambda\,, \tag{32.1.34}$$

$$\{Q_r^\pm\,, Q_s^{\mp\,\rm T}\} = z_{rs}^\pm\left(\frac{1 \pm \gamma_{d+1}}{2}\right)\mathscr{C}\,, \tag{32.1.35}$$

$$\mathscr{C}\beta\, Q_r^{\pm *} = Q_r^\pm\,, \qquad z_{rs}^\pm = -z_{rs}^{\pm *} = -z_{sr}^\mp\,. \tag{32.1.36}$$

$d = 3$ (mod 8)

$$\{Q_r\,, Q_s^{\rm T}\} = i\,\delta_{rs}\gamma^\lambda \mathscr{C} P_\lambda + z_{rs}\mathscr{C}\,, \tag{32.1.37}$$

$$\mathscr{C}\beta Q_r^* = Q_r\,, \qquad z_{rs} = -z_{sr} = -z_{rs}^*\,. \tag{32.1.38}$$

$d = 4$ (mod 8)

$$\{Q_r^\pm\,, Q_s^{\mp\,\rm T}\} = \mp i\,\delta_{rs}\left(\frac{1 \pm \gamma_{d+1}}{2}\right)\gamma^\lambda \mathscr{C} P_\lambda\,, \tag{32.1.39}$$

$$\{Q_r^\pm\,, Q_s^{\pm\,\rm T}\} = z_{rs}^\pm\left(\frac{1 \pm \gamma_{d+1}}{2}\right)\mathscr{C}\,, \tag{32.1.40}$$

$$\mathscr{C}\beta\, Q_r^{\pm *} = \pm Q_r^\mp\,, \qquad z_{rs}^\pm = z_{rs}^{\mp *} = -z_{sr}^\pm\,. \tag{32.1.41}$$

$d = 5$ (mod 8)

$$\{Q_r\,, Q_s^{\rm T}\} = -i\,\Omega_{rs}\gamma^\lambda \mathscr{C} P_\lambda + z_{rs}\mathscr{C}\,, \tag{32.1.42}$$

$$\mathscr{C}\beta Q_r^* = \sum_s \Omega_{rs} Q_s\,, \qquad z_{rs} = -z_{sr}\,, \qquad z^* = -\Omega z \Omega\,. \tag{32.1.43}$$

$d = 6$ (mod 8)

$$\{Q_r^\pm\,, Q_s^{\pm\,\rm T}\} = i\,\Omega_{rs}^\pm\left(\frac{1 \pm \gamma_{d+1}}{2}\right)\gamma^\lambda \mathscr{C} P_\lambda\,, \tag{32.1.44}$$

$$\{Q_r^\pm\,, Q_s^{\mp\,\rm T}\} = z_{rs}^\pm\left(\frac{1 \pm \gamma_{d+1}}{2}\right)\mathscr{C}\,, \tag{32.1.45}$$

$$\mathscr{C}\beta\, Q_r^{\pm *} = \sum_s \Omega_{rs}^{\pm} Q_s^{\pm}\,, \qquad z_{rs}^{\pm *} = \Omega^{\pm} z_{rs}^{\pm} \Omega^{\mp} = z_{sr}^{\mp}\,. \tag{32.1.46}$$

$d = 7 \pmod 8$

$$\{Q_r\,, Q_s^{\mathrm{T}}\} = i\,\Omega_{rs}\gamma^{\lambda}\mathscr{C}\,P_{\lambda} + z_{rs}\mathscr{C}\,, \tag{32.1.47}$$

$$\mathscr{C}\beta Q_r^{*} = \sum_s \Omega_{rs} Q_s\,, \qquad z_{rs} = z_{sr}\,, \qquad z^{*} = +\Omega z \Omega\,. \tag{32.1.48}$$

Inspection of these anticommutation relations reveals that in the absence of central charges they are invariant under groups of linear transformations on the fermionic generators, of the form $Q_r \to \sum_s V_{rs} Q_s$ for d odd and $Q_r^{\pm} \to \sum_s V_{rs}^{\pm} Q_s^{\pm}$ for d even. In order to preserve the relations (32.1.29)–(32.1.48) it is necessary that the Vs should satisfy the conditions:

$d = 0$ and $d = 4 \pmod 8$

$$V^{\pm} V^{\mp\mathrm{T}} = 1\,, \qquad V^{\pm *} = V^{\mp}\,. \tag{32.1.49}$$

$d = 1$ and $d = 3 \pmod 8$

$$V V^{\mathrm{T}} = 1\,, \qquad V^{*} = V\,. \tag{32.1.50}$$

$d = 2 \pmod 8$

$$V^{\pm} V^{\pm\mathrm{T}} = 1\,, \qquad V^{\pm} = V^{\pm *}\,. \tag{32.1.51}$$

$d = 5 \pmod 8$

$$V \Omega V^{\mathrm{T}} = \Omega\,, \qquad V^{*} = -\Omega V \Omega\,. \tag{32.1.52}$$

$d = 6 \pmod 8$

$$V^{\pm} V^{\pm\mathrm{T}} = 1\,, \qquad V^{\pm *} = -\Omega^{\pm} V^{\pm} \Omega^{\pm}\,. \tag{32.1.53}$$

$d = 7 \pmod 8$

$$V \Omega V^{\mathrm{T}} = \Omega\,, \qquad V^{*} = -\Omega V \Omega\,. \tag{32.1.54}$$

These matrices form the groups:

$d = 0$ and $d = 4 \pmod 8$ $\quad U(N)$.

$d = 1$ and $d = 3 \pmod 8$ $\quad O(N)$.

$d = 2 \pmod 8$ $\quad O(N_+) \times O(N_-)$.

$d = 5$ and $d = 7 \pmod 8$ $\quad USp(N)$ $\quad N$ even .

$d = 6 \pmod 8$ $\quad USp(N_+) \times USp(N_-)$ $\quad N_{\pm}$ even .

Here N is the number of fundamental spinor representations among the Qs for d odd and the number of fundamental spinor representations of each chirality among the Qs for $d = 0$ (mod 8) and $d = 4$ (mod 8). For $d = 2$ (mod 8) and $d = 6$ (mod 8) the numbers of fundamental spinor representations among the Qs need not be the same for each chirality, and so these are denoted N_+ and N_-.

32.2 Massless Multiplets

We will now consider how the supersymmetry algebras constructed in the previous section may be used to construct supermultiplets of massless particle states in $d \geq 4$ spacetime dimensions. The momentum operator P^μ commutes with all fermionic symmetry generators, so we can work in the one-particle subspace of Hilbert space where P^μ has a definite lightlike eigenvector p^μ, which may be taken in a direction with $p^1 = p^0$ and all other spatial components of p^μ zero. Just as in the case of four spacetime dimensions, discussed in Section 2.5, these one-particle states are classified according to the finite-dimensional representations they provide of the little group, the subgroup of the homogeneous Lorentz group that leaves p^μ invariant. The little group contains combined boosts along directions perpendicular to $\mathbf{p}$ and rotations in planes in which $\mathbf{p}$ lies, such as the transformations (2.5.6) for four spacetime dimensions, but these form an invariant Abelian subgroup and therefore in a finite-dimensional representation must be represented by the unit operator. With this subgroup omitted, the reduced little group in d dimensions is $O(d-2)$, consisting of rotations in planes orthogonal to $\mathbf{p}$. We therefore classify massless particle states according to the representations they provide of $O(d-2)$ and of the automorphism groups described at the end of the previous section.

These representations are more complicated than those we have dealt with in four spacetime dimensions, where the reduced little group is $O(2)$ and the representations are one-dimensional, characterized by a single number, the helicity. Nevertheless it is useful to label the representations of the reduced little group $O(d-2)$ with a 'spin,' defined as the *maximum* absolute value of the eigenvalue of any generator J_{ij} in the representation.

It is widely believed that there are no consistent quantum field theories involving massless particles with spin greater than 2. It is known[7] that *soft* massless particles with spin $j > 1/2$ can only interact with conserved currents carrying spin j. For $j = 1$ these are the currents of ordinary conserved scalars, like electric charge; for $j = 3/2$ they are the one or more supercurrents associated with supersymmetry; for $j = 2$ there is a single current, the energy-momentum tensor; but for $j > 5/2$ there is no conserved current with which a soft massless particle could interact. We may derive stringent limits on the dimensions in which supersymmetry is possible by adopting a prohibition against there being more than one type of massless particle of spin 2, or any massless particles whatever of spin greater than 2.

Let us return for a moment to the classification of operators used in Section 32.1, according to a weight equal to the value of the $O(d)$ generator J_{d1} that they destroy. (Recall that $J_{01} = iJ_{d1}$.) The fermionic supersymmetry generators have weights $1/2$ and $-1/2$, so the anticommutator of any of these generators with its Hermitian adjoint can only have weight $+1$ or -1, respectively, and therefore must be respectively proportional to the operator $P^0 + P^1$ or $P^0 - P^1$. But we are working in a subspace of Hilbert space in which the operator $P^0 - P^1$ vanishes, so in this subspace the fermionic supersymmetry generators of weight $-1/2$ all vanish. To classify the one-particle states we therefore have available just half of the supersymmetry generators, the 2^{n-1} generators with weight $\sigma_{d1} = +1/2$.

We can further divide the remaining supersymmetry generators into two classes, those in which $\sigma_{23} = 1/2$ or $\sigma_{23} = -1/2$ as well as $\sigma_{d1} = +1/2$. Since the operator $P^0 + P^1$ has $\sigma_{23} = 0$, the fermionic supersymmetry generators of each class anticommute with each other, though not necessarily with their adjoints or with generators of the other class.

Now consider a representation of the little group $O(d-2)$ with spin j, and consider any state $|\lambda\rangle$ that is an eigenstate of J_{23} with eigenvalue $\lambda > 0$ and is annihilated by all supersymmetry generators with $\sigma_{23} = -1/2$. (Any state that has the maximum eigenvalue j for J_{23} is of this type, but in general there may be other such states.) We may form states with $J_{23} = \lambda - k/2$ by acting on $|\lambda\rangle$ with k fermionic generators having $\sigma_{23} = +1/2$ as well as $\sigma_{d1} = +1/2$. (It can be shown that none of these states vanishes, because acting on them with the adjoints of k of these fermionic generators gives back the state $|\lambda\rangle$.) If there is a total of $\mathcal{N}$ fermionic supersymmetry generators of all types, then there are $\mathcal{N}/4$ of them with $\sigma_{23} = +1/2$ and $\sigma_{d1} = +1/2$, and since these operators all anticommute the number of states formed in this way with $J_{23} = \lambda - k/2$ will be given by the binomial coefficient

$$\binom{\mathcal{N}/4}{k}, \tag{32.2.1}$$

which when summed from $k = 0$ to the maximum value $k = \mathcal{N}/4$ gives a total of $2^{\mathcal{N}/4}$ components. The minimum eigenvalue of J_{23} obtained in this way is $\lambda - \mathcal{N}/8$, reached by multiplying the state $|\lambda\rangle$ by $k = \mathcal{N}/4$ supersymmetry generators. Taking $\lambda = j$, we see that to avoid having eigenvalues of J_{23} greater than $+2$ or less than -2, we must have $j \leq 2$ and $j - \mathcal{N}/8 \geq -2$, which requires a total number of fermionic generators $\mathcal{N}$ no greater than 32.

Further, for $\mathcal{N} = 32$ supersymmetry generators there can be at most a single supermultiplet of massless particles formed in this way by acting on the state $|2\rangle$ with products of supersymmetry generators having $\sigma_{23} = +1/2$ and $\sigma_{d1} = +1/2$. These states have eigenvalues for any generator of the little group running in steps of $1/2$ from -2 to $+2$. It is only for $\mathcal{N} < 32$ that there can be 'matter' supermultiplets, supermultiplets that do not contain the graviton.

A single fundamental spinor representation in $2n$ or $2n+1$ dimensions has 2^n components, so in order to have no more than 32 fermionic generators we must have $n \leq 5$. The spacetime dimensionality can thus be no larger than $d = 11$ and in this case must have $N = 1$. Supergravity in 11 dimensions is of special interest, because it may be the 'low-energy' limit of a fundamental theory known as *M theory*,[4] which is also believed to yield various string theories in other limits. We will now work out the spin content of $N = 1$ supersymmetry in $d = 11$ dimensions in detail, as an example of how this can be done by enlightened counting.

We can construct all the states of the massless multiplet for $d = 11$ by acting on an eigenstate $|2\rangle$ of J_{23} having eigenvalue 2 with products of $k = 0, 1, \ldots, 8$ supersymmetry generators having $\sigma_{23} = +1/2$ and $\sigma_{2n-1\;2n} = +1/2$. According to Eq. (32.2.1), we obtain one state each with $J_{23} = \pm 2$, eight states each with $J_{23} = \pm 3/2$, twenty-eight states each with $J_{23} = \pm 1$, fifty-six states each with $J_{23} = \pm 1/2$, and seventy states with $J_{23} = 0$.

For $d = 11$ the spin 2 graviton representation of the little group $O(9)$ is a symmetric traceless tensor with $9 \times 10/2 - 1 = 44$ independent components: there is one $2 \pm i3\,,\, 2 \pm i3$ component with $J_{23} = \pm 2$; seven $2 \pm i3\,,\, k$ components with $J_{23} = \pm 1$; and twenty-eight $k\,,\,\ell$ components with $J_{23} = 0$. (Here k and ℓ run over the seven values 4, 5, ..., 10. We do not count the $2 + i3\,,\, 2 - i3$ component because it is related to the $k\,,\,\ell$ components by the tracelessness condition in this representation.)

There is also a single spin 3/2 gravitino representation. This consists of a spinor ψ_i with an extra nine-vector index i, subject to an irreducibility condition $\sum_i \gamma_i \psi_i = 0$ which excludes spin 1/2 components, and so has $9 \times 16 - 16 = 128$ independent components.

By subtracting the number of components with each value of J_{23} contained in the graviton and gravitino states from those formed acting on $|2\rangle$ with supersymmetry generators, we see that we need one or more additional states having a total of $28 - 7 = 21$ components with $J_{23} = \pm 1$ and $70 - 28 = 42$ components with $J_{23} = 0$. The only representations of orthogonal groups that have no eigenvalues of J_{ij}s other than ± 1 and 0 are the antisymmetric tensors. An antisymmetric tensor $T_{i_1\cdots i_p}$ of rank p in nine dimensions has

$$\binom{7}{p} \text{ components } T_{k_1\cdots k_p} \text{ with } J_{23}=0\,,$$

$$\binom{7}{p-1} \text{ components } T_{2\pm i3\,k_2\cdots k_p} \text{ with } J_{23}=\pm 1\,,$$

$$\binom{7}{p-2} \text{ components } T_{2+i3\,2-i3\,k_3\cdots k_p} \text{ with } J_{23}=0\,,$$

where $k_1,\ldots,k_p$ run over the seven values $4,5,\ldots,10$. For $O(9)$ the only independent antisymmetric tensors are of rank $p=0$, 1, 2, 3, and 4. The antisymmetric tensor of rank 4 has 35 components with $J_{23}=\pm1$, which is more than needed, so it must be excluded. Any combination of $p=0$, $p=1$, and $p=2$ tensors with the twenty-one needed components with $J_{23}=\pm1$ would have too many components with $J_{23}=0$ (147 for 21 one-forms and no two-forms; 120 for 14 one-forms and 1 two-forms; 53 for 7 one-forms and 2 two-forms; and 66 for no one-forms and 3 two-forms), so we must include at least one three-form. The antisymmetric tensor with rank $p=3$ has just 21 components with $J_{23}=\pm1$ and 42 components with $J_{23}=0$, which is just what is needed. We conclude that *the unique massless particle multiplet for* $N=1$ *supersymmetry in* $d=11$ *contains a graviton, a gravitino, and a particle whose states transform under the little group as a single antisymmetric tensor of rank 3.*

There is a richer variety of possibilities for $d=10$. Here there are two ways to have $\mathcal{N}=32$ generators: the fermionic generators can comprise two 16-component Weyl spinors of the same chirality, with an automorphism group $O(2)$, or two of opposite chirality, with no automorphism group. For $d=10$ it is also possible to have a single Weyl fermionic generator, with just $\mathcal{N}=16$ independent components. These three possibilities play an important role in modern superstring theories — they represent the massless particle spectrum of three kinds of superstring: type IIA for 16 generators of each chirality; type IIB for 32 generators of the same chirality; and the heterotic superstring for 16 generators of just one chirality.

The type IIA case of $d=10$ and opposite chirality is just like the case of $d=11$, except that the irreducible representations of the little group $O(9)$ for $d=11$ break up into separate irreducible representations of the little group $O(8)$ for $d=10$. Thus the $O(9)$ graviton multiplet is decomposed into an $O(8)$ graviton with $8\times 9/2-1=35$ components, an $O(8)$ vector with 8 components, and a scalar with 1 component; the $O(9)$ gravitino multiplet is decomposed into $O(8)$ gravitinos of each chirality with $(16\times 8-16)/2=56$ components each and $O(8)$ spinors of each chirality with 8 components each; and the $O(9)$ three-form is decomposed into a $O(8)$ three-form with 56 components and an $O(8)$ two-form with 28 components.

In the type IIB case of $d = 10$ with $N_+ = 2$ and $N_- = 0$ we must classify states according to the representation of the little group $O(8)$ and a quantum number q that labels the representations of the automorphism group $O(2)$, under which the supersymmetry generators transform as a 2-vector. Since there is only one graviton, it must have $q = 0$. Acting on these states with a supersymmetry generator gives *two* gravitinos with $q = \pm 1$ and 56 components each; acting with another supersymmetry generator gives 2 two-form tensors with $q = \pm 2$ and 28 components each; acting with another supersymmetry generator gives 2 Weyl spinors with $q = \pm 3$ and 8 components each; and acting with another supersymmetry generator gives 2 scalars with $q = \pm 4$ together with a self-dual four-form with $q = 0$ and 35 components.

In the heterotic case of $d = 10$ with a single Weyl fermionic generator there are just $\mathcal{N} = 16$ independent components. In this case there is a graviton supermultiplet consisting of a graviton transforming under $O(8)$ as a symmetric traceless tensor with 35 independent components; a gravitino with 56 independent components; an $O(8)$ two-form with 28 independent components; a Weyl spinor with 8 components; and a scalar. (This graviton supermultiplet is constructed by acting with supersymmetry generators on one state $|2\rangle$, *six* states $|1\rangle$, and one state $|0\rangle$, giving altogether $8 \times 2^4 = 128 = 35+56+28+8+1$ components.) Here we also have the possibility of gauge supermultiplets that contain no particles having values greater than 1 or less than -1 for the eigenvalue of any J_{ij}. These gauge supermultiplets are formed by acting with supersymmetry generators on a state $|1\rangle$, and contain one gauge particle belonging to the vector representation of $O(8)$, with 8 components, and one particle transforming as a fundamental Weyl spinor of $O(8)$, also with 8 components.

32.3 *p*-Branes

In some theories in addition to particles there are stable extended objects, either of infinite extent or stabilized by 'wrapping' around a topologically non-trivial spacetime. The study of supersymmetry and supergravity in higher-dimensional theories of this sort has opened up remarkable opportunities for the construction of string theories and supersymmetric field theories in lower-dimensional spacetime and for the proof of equivalencies among these theories,[4,8] which are beyond the scope of this book. The feature of these extended objects that concerns us here is that they can carry conserved bosonic quantities other than those allowed by the Coleman–

Mandula theorem. These new conserved quantities may appear on the right-hand side of the anticommutation relations of supersymmetry, along with the momentum operator and ordinary conserved quantities.[9]

In the cases that have been studied so far, the new conserved bosonic quantities are all *forms* – antisymmetric tensors. For instance, an object of spatial dimensionality p (known as a 'p-brane') in a spacetime of dimensionality d is described by specifying the d spacetime coordinates $x^\mu(\sigma, t)$ (generally in overlapping patches covering the object) as functions of the time t and of a set of p coordinates σ^r that parameterize positions on this object. If the manifold $x^\mu = x^\mu(\sigma, t)$ at a given time is topologically non-trivial, in the sense that it cannot be continuously deformed to a single point, then it may have a non-vanishing value for the topologically invariant integral*

$$I^{\mu_1\mu_2\cdots\mu_p} = \int d\sigma^1\, d\sigma^2 \cdots d\sigma^p \sum_{r_1=1}^{p}\sum_{r_2=1}^{p}\cdots\sum_{r_p=1}^{p} \epsilon^{r_1 r_2 \cdots r_p} \times \frac{\partial x^{\mu_1}(\sigma,t)}{\partial\sigma^{r_1}}\frac{\partial x^{\mu_2}(\sigma,t)}{\partial\sigma^{r_2}}\cdots\frac{\partial x^{\mu_p}(\sigma,t)}{\partial\sigma^{r_p}}\,. \tag{32.3.1}$$

The invariance of such integrals under small changes in the functions $x^\mu(\sigma, t)$ shows in particular that they are invariant under spacetime translations, and hence may appear along with P^μ and central charges on the right-hand side of the anticommutation relations for the supersymmetry generators.[10] The calculation of the coefficients of such tensors on the right-hand side of the anticommutation relations is analogous to the Olive–Witten calculation of the scalar central charge Z_{rs} in $N = 2$ supersymmetry theories in four spacetime dimensions, discussed in Section 27.9. We will make no attempt in this section to evaluate these coefficients or to survey the other non-topological p-forms that may appear in the anticommutation relations,[11] but will simply consider the effects on the supersymmetry algebra of including conserved antisymmetric tensors that commute with the momentum operators.

---

* To see that this integral is topologically invariant, note that the effect of an infinitesimal change $\delta x^\mu(\sigma, t)$ in the function $x^\mu(\sigma, t)$ is to change $I^{\mu_1\mu_2\cdots\mu_p}$ by the amount

$$\delta I^{\mu_1\mu_2\cdots\mu_p} = \sum_{n=1}^{p}\sum_{r_1=1}^{p}\sum_{r_2=1}^{p}\cdots\sum_{r_p=1}^{p}\int d\sigma^1\, d\sigma^2\cdots d\sigma^p\,\frac{\partial}{\partial\sigma^{r_n}}\Bigg[\epsilon^{r_1 r_2\cdots r_p} \times \frac{\partial x^{\mu_1}}{\partial\sigma^{r_1}}\frac{\partial x^{\mu_2}}{\partial\sigma^{r_2}}\cdots\frac{\partial x^{\mu_{n-1}}}{\partial\sigma^{r_{n-1}}}\delta x^{\mu_n}\frac{\partial x^{\mu_{n+1}}}{\partial\sigma^{r_{n+1}}}\cdots\frac{\partial x^{\mu_p}}{\partial\sigma^{r_p}}\Bigg]\,,$$

which vanishes when the integral is taken over a compact manifold. It also vanishes if the integral is over all σ, provided $\delta x^\mu(\sigma, t)$ is constrained to vanish rapidly when $\sigma^r \to \infty$.

It is important that this possibility does not affect the key result that the supersymmetry generators always belong to the fundamental spinor representations of the Lorentz group. This is because a totally antisymmetric tensor in Euclidean coordinates can have at most one spacetime index equal to 1 and at most one spacetime index equal to d, and therefore its 'weight' defined by Eq. (32.1.2) can only be ± 1 or 0. Just as before, this means that the weight of a supersymmetry generator can only be $\pm 1/2$; Lorentz invariance then implies that all the σs defined by Eq. (32.1.1) are $\pm 1/2$, which is only possible if the supersymmetry generators belong to a fundamental spinor representation of $O(d-1,1)$. Also, because the new terms in the anticommutators of supersymmetry generators commute with momentum, the same argument that was given in Section 32.1 shows again that the supersymmetry generators also commute with momentum.

Lorentz invariance tells us that for non-zero values of the p-form 'charges,' the anticommutation relations (32.1.13) and (32.1.21)–(32.1.22) can only take the forms (in the same notation as in Section 32.1):

***d* odd**

$$\{Q_r\,,\,Q_s^{\mathrm{T}}\} = g_{rs}\gamma^\lambda \mathscr{C} P_\lambda + \sum_p z_{rs}^{\mu_1\mu_2\cdots\mu_p}\gamma_{\mu_1}\gamma_{\mu_2}\cdots\gamma_{\mu_p}\mathscr{C}\,. \tag{32.3.2}$$

***d* even**

$$\{Q_r^\pm\,,\,Q_s^{\mp(-1)^{d/2}\,\mathrm{T}}\} = \left(\frac{1\pm\gamma_{d+1}}{2}\right) \times\left[g_{rs}^\pm\gamma^\lambda\mathscr{C}P_\lambda + \sum_{\text{odd } p} z_{rs}^{\mu_1\mu_2\cdots\mu_p\,\pm}\gamma_{\mu_1}\gamma_{\mu_2}\cdots\gamma_{\mu_p}\mathscr{C}\right]\,, \tag{32.3.3}$$

$$\{Q_r^\pm\,,\,Q_s^{\pm(-1)^{d/2}\,\mathrm{T}}\} = \left(\frac{1\pm\gamma_{d+1}}{2}\right) \times\sum_{\text{even } p} z_{rs}^{\mu_1\mu_2\cdots\mu_p\,\pm}\gamma_{\mu_1}\gamma_{\mu_2}\cdots\gamma_{\mu_p}\mathscr{C}\,. \tag{32.3.4}$$

(Recall that $\mathscr{C}$ appears in the anticommutation relations because $\mathscr{J}^{\mathrm{T}}_{\mu\nu} = -\mathscr{C}^{-1}\mathscr{J}_{\mu\nu}\mathscr{C}$; the $Q_r^\pm$ are supersymmetry generators in the case of even d for which $\gamma_{d+1}Q_r^\pm = \pm Q_r^\pm$; and $\mathscr{C}^{-1}\gamma_{d+1}\mathscr{C} = (-1)^{d/2}\gamma_{d+1}$.) For even d we have

$$\epsilon^{\mu_1\mu_2\cdots\mu_d}\gamma_{\mu_1}\gamma_{\mu_2}\cdots\gamma_{\mu_p} \propto \gamma_{d+1}\gamma_{\mu_{p+1}}\gamma_{\mu_{p+2}}\cdots\gamma_{\mu_d}\,,$$

while for d odd

$$\epsilon^{\mu_1\mu_2\cdots\mu_d}\gamma_{\mu_1}\gamma_{\mu_2}\cdots\gamma_{\mu_p} \propto \gamma_{\mu_{p+1}}\gamma_{\mu_{p+2}}\cdots\gamma_{\mu_d}\,.$$

Thus p-branes and $d - p$ branes make the same contributions in Eqs. (32.3.1)–(32.3.3) for any d, so that we can restrict p to run only over values from 0 to $d/2$ for d even and from 0 to $(d-1)/2$ for d odd.

The symmetry of anticommutators is reflected in symmetry conditions on the p-brane central charges z^p_{rs} in Eqs. (32.3.2)–(32.3.4). Eqs. (32.A.15) and (32.A.30) give

$$\gamma^{\rm T}_\mu = (-1)^n \mathscr{C}^{-1}\gamma_\mu \mathscr{C}\,, \qquad \mathscr{C}^{\rm T} = (-1)^{n(n+1)/2}\mathscr{C}\,, \tag{32.3.5}$$

for both $d = 2n$ and $d = 2n+1$. These give the antisymmetrized products $\gamma_{[\mu_1}\gamma_{\mu_2}\cdots\gamma_{\mu_p]}$ the symmetry property

$$\begin{aligned}\gamma_{[\mu_1}\gamma_{\mu_2}\cdots\gamma_{\mu_p]}\mathscr{C} &= (-1)^{pn}(-1)^{n(n+1)/2}\Big[\gamma_{[\mu_p}\gamma_{\mu_{p-1}}\cdots\gamma_{\mu_1]}\mathscr{C}\Big]^{\rm T} \\ &= (-1)^{pn}(-1)^{n(n+1)/2}(-1)^{p(p-1)/2}\Big[\gamma_{[\mu_1}\gamma_{\mu_2}\cdots\gamma_{\mu_p]}\mathscr{C}\Big]^{\rm T}\,.\end{aligned} \tag{32.3.6}$$

It follows immediately that for odd d,

$$z^{\mu_1\mu_2\cdots\mu_p}_{rs} = (-1)^{pn}(-1)^{n(n+1)/2}(-1)^{p(p-1)/2} z^{\mu_1\mu_2\cdots\mu_p}_{sr}\,, \tag{32.3.7}$$

while for even d,

$$z^{\mu_1\mu_2\cdots\mu_p\,\pm}_{rs} = (-1)^{pn}(-1)^{n(n+1)/2}(-1)^{p(p-1)/2} z^{\mu_1\mu_2\cdots\mu_p\,(-1)^n(-1)^p\mp}_{sr}\,. \tag{32.3.8}$$

Consider for instance the important case of $N = 1$ supersymmetry in $d = 11$ spacetime dimensions, which is one version of the popular M-theory generalization of string theories. Eq. (32.3.8) shows that the single p-form central charge $z^{\mu_1\mu_2\cdots\mu_p}$ vanishes unless

$$-(-1)^p(-1)^{p(p-1)/2} = +1\,, \tag{32.3.9}$$

which is satisfied only for p equal to 1, 2, and 5. The value $p = 1$ is realized by the momentum operator itself, which arises from particles as well as extended objects. The other possibilities, $p = 2$ and $p = 5$, arise in theories with 2-branes and 5-branes, respectively. Note that there can be no other independent tensor central charges, such as a 1-form arising from 1-branes, because the number of independent components in P^μ and in a 2-form and a 5-form is

$$11 + \binom{11}{2} + \binom{11}{5} = 528\,,$$

while the number of independent components in an anticommutator of two 32-component fundamental spinors is $32 \times 33/2 = 528$.

Just as the 0-form electric charge is the source of a 1-form gauge field $A_\mu(x)$, so also a p-form conserved quantity $z^{\mu_1\mu_2\cdots\mu_p}$ may serve as the source of a $p+1$-form gauge field $A_{\mu_1\mu_2\cdots\mu_{p+1}}$ of the sort discussed in Section 8.8. In fact, such gauge fields do appear in supergravity theories. For instance, as remarked in the previous section, the $N = 1$ supergravity theory in $d = 11$ spacetime dimensions includes a massless particle whose states are in the 3-form representation of the $O(9)$ little group, and therefore must be described by a 3-form gauge field $A_{\mu\nu\rho}(x)$. The study of solutions of this supergravity theory shows that there are two-branes[12] that indeed do provide sources for $A_{\mu\nu\rho}(x)$. Also, as noted in Section 8.8, this gauge theory is equivalent to one with a $(d-p-2=6)$-form gauge field, which can have the 5-form $z^{\mu_1\cdots\mu_5}$ as a source, and there are indeed 5-brane solutions which provide such sources for this 6-form gauge field.[13] The $N = 1$ eleven-dimensional supersymmetry algebra does in fact receive contributions from these 2-branes and 5-branes.[11]

Appendix Spinors in Higher Dimensions

This appendix describes the fundamental spinor representations of the Lie algebra of the Lorentz group $O(d-1,1)$ in any number d of spacetime dimensions. These are obtained from the corresponding Clifford algebra, consisting of an irreducible set of finite matrices γ_μ that satisfy the anticommutation relations

$$\{\gamma_\mu , \gamma_\nu\} = 2\eta_{\mu\nu} \,, \tag{32.A.1}$$

where $\eta_{\mu\nu}$ is diagonal, with elements $+1$ on the diagonal except for $\eta_{00} = -1$, where x^0 is the time-component. From these we can construct matrices

$$\mathscr{J}_{\mu\nu} \equiv \frac{1}{4i}[\gamma_\mu , \gamma_\nu] = -\mathscr{J}_{\nu\mu} \,, \tag{32.A.2}$$

that satisfy the commutation relations (2.4.12) of the Lorentz group generators

$$i\left[\mathscr{J}_{\mu\nu}, \mathscr{J}_{\rho\sigma}\right] = \eta_{\nu\rho}\mathscr{J}_{\mu\sigma} - \eta_{\mu\rho}\mathscr{J}_{\nu\sigma} - \eta_{\sigma\mu}\mathscr{J}_{\rho\nu} + \eta_{\sigma\nu}\mathscr{J}_{\rho\mu} \,. \tag{32.A.3}$$

As we will see, although Eq. (32.A.2) always gives a representation of the Lorentz algebra, it is not always an irreducible representation.

We must now distinguish between the cases of even and odd dimensionality.

Even Dimensions: $d = 2n$

To construct a convenient specific representation for the gamma matrices in $d = 2n$ dimensions, we introduce n matrices

$$a_u \equiv \frac{1}{2}\left(\gamma_{2u-1} + i\,\gamma_{2u}\right) \qquad u = 1, 2, \ldots, n \,, \tag{32.A.4}$$

and take $\gamma_1, \ldots, \gamma_{2n}$ as Hermitian, it being understood that as usual

$$\gamma_{2n} \equiv -i\gamma_0 \,. \tag{32.A.5}$$

These have the anticommutation relations

$$\{a_u\,,\, a_v^\dagger\} = \delta_{uv}\,, \qquad \{a_u\,,\, a_v\} = \{a_u^\dagger\,,\, a_v^\dagger\} = 0\,. \tag{32.A.6}$$

We introduce a vector $|0\rangle$ in the representation space of the γs, defined by the condition

$$a_u^\dagger\,|0\rangle = 0\,, \tag{32.A.7}$$

and define the basis vectors

$$|s_1\, s_2\, \cdots s_n\rangle = a_1^{s_1} a_2^{s_2} \cdots a_n^{s_n}|0\rangle\,. \tag{32.A.8}$$

Because $a_u^2 = 0$, the operator a_u raises the value of s_u to $+1$ if $s_u = 0$ and annihilates the vector if $s_u = +1$ (as well as yielding a sign $(-1)^S$ where $S \equiv \sum_{v<u} s_v$), so all s_u take only the values 0 and $+1$, and the vectors therefore span a space of dimensionality 2^n. In this basis, the matrices a_u take the form

$$a_u = \begin{pmatrix} -1 & 0 \\ 0 & 1 \end{pmatrix} \otimes \cdots \otimes \begin{pmatrix} -1 & 0 \\ 0 & 1 \end{pmatrix} \otimes \begin{pmatrix} 0 & 1 \\ 0 & 0 \end{pmatrix} \otimes 1 \cdots \otimes 1\,, \tag{32.A.9}$$

with the last 2×2 matrix in the uth place. Taking the Hermitian and anti-Hermitian parts then gives the gamma matrices

$$\gamma_{2u-1} = \begin{pmatrix} -1 & 0 \\ 0 & 1 \end{pmatrix} \otimes \cdots \otimes \begin{pmatrix} -1 & 0 \\ 0 & 1 \end{pmatrix} \otimes \begin{pmatrix} 0 & 1 \\ 1 & 0 \end{pmatrix} \otimes 1 \cdots \otimes 1\,, \tag{32.A.10}$$

$$\gamma_{2u} = \begin{pmatrix} -1 & 0 \\ 0 & 1 \end{pmatrix} \otimes \cdots \otimes \begin{pmatrix} -1 & 0 \\ 0 & 1 \end{pmatrix} \otimes \begin{pmatrix} 0 & -i \\ i & 0 \end{pmatrix} \otimes 1 \cdots \otimes 1\,. \tag{32.A.11}$$

(Note that this does not give the same representation of the gamma matrices in four spacetime dimensions as was introduced in Section 5.4, and has been used throughout this book.)

The representation (32.A.10)–(32.A.11) gives the Euclidean γs the simple reality and symmetry properties

$$\gamma_i^* = \gamma_i^{\mathrm{T}} = \begin{cases} \gamma_i & \text{for } i \text{ odd} \\ -\gamma_i & \text{for } i \text{ even} \end{cases}\,, \tag{32.A.12}$$

where $i = 1, 2, \ldots, 2n$. This can be expressed as a similarity relation

$$\mathscr{C}^{-1}\gamma_i\mathscr{C} = (-1)^n\,\gamma_i^{\mathrm{T}} = (-1)^n\,\gamma_i^*\,, \tag{32.A.13}$$

where $\mathscr{C}$ is the matrix

$$\mathscr{C} \equiv \gamma_2\gamma_4 \cdots \gamma_{2n}\,. \tag{32.A.14}$$

Taking account of the factor $-i$ in Eq. (32.A.5), we can write this in terms of Minkowskian components as

$$\gamma_\mu^* = -\beta\gamma_\mu^{\mathrm{T}}\beta = -(-1)^n(\mathscr{C}\beta)^{-1}\gamma_\mu(\mathscr{C}\beta)\,, \tag{32.A.15}$$

where

$$\beta \equiv \gamma_{2n} = -i\gamma_0\,. \tag{32.A.16}$$

In any even dimension we may define a matrix γ_{2n+1} that plays a role similar to that of γ_5 in four dimensions. We take

$$\gamma_{2n+1} \equiv i^n\gamma_1\gamma_2\cdots\gamma_{2n}\,. \tag{32.A.17}$$

The phase here is chosen so that

$$\gamma_{2n+1}^2 = 1\,. \tag{32.A.18}$$

From the anticommutation relations (32.A.1), it follows immediately that γ_{2n+1} anticommutes with the other gamma matrices

$$\{\gamma_{2n+1}\,,\,\gamma_\mu\} = 0 \quad \text{for } \mu = 1, 2, \ldots, 2n-1, 0\,. \tag{32.A.19}$$

It is straightforward to check that γ_{2n+1} is real and symmetric

$$\gamma_{2n+1}^\dagger = \gamma_{2n+1}^* = \gamma_{2n+1}^{\mathrm{T}} = \gamma_{2n+1}\,. \tag{32.A.20}$$

From Eq. (32.A.19) we see that γ_{2n+1} commutes with the generators (32.A.2) of the $O(2n-1,1)$ algebra:

$$[\gamma_{2n+1}\,,\,\mathscr{J}_{\mu\nu}] = 0\,, \tag{32.A.21}$$

so that the $\mathscr{J}_{\mu\nu}$ cannot furnish an *irreducible* representation of the algebra of $O(2n-1,1)$. Instead, we may define a pair of 'Weyl' irreducible representations by projecting out the subspaces with $\gamma_{2n+1} = \pm 1$:

$$\mathscr{J}_{\mu\nu}^{\pm} \equiv \mathscr{J}_{\mu\nu}\left(\frac{1\pm\gamma_{2n+1}}{2}\right)\,. \tag{32.A.22}$$

From Eq. (32.A.15) and the relation $(\mathscr{C}\beta)^{-1}\gamma_{2n+1}\mathscr{C}\beta = -(-1)^n\gamma_{2n+1}$ we find that the complex conjugate and transpose of the Weyl Lorentz generators are

$$(\mathscr{J}_{\mu\nu}^{\pm})^* = -(\mathscr{C}\beta)^{-1}\mathscr{J}_{\mu\nu}^{\mp(-1)^n}(\mathscr{C}\beta)\,, \tag{32.A.23}$$

$$(\mathscr{J}_{\mu\nu}^{\pm})^T = -\mathscr{C}^{-1}\mathscr{J}_{\mu\nu}^{\pm(-1)^n}\mathscr{C}\,. \tag{32.A.24}$$

Hence, for n even, the Weyl irreducible representations are equivalent to complex conjugates of each other, while for n odd each is equivalent to its own complex conjugate.* For n odd, we still have to decide whether the Weyl representations are *real*, which would mean that there is a matrix $\mathcal{S}$ such that

$$-(\mathcal{S}\mathcal{J}^{\pm}_{\mu\nu}\mathcal{S}^{-1})^* = \mathcal{S}\mathcal{J}^{\pm}_{\mu\nu}\mathcal{S}^{-1}\,, \tag{32.A.25}$$

or *pseudoreal*, in which case there is no such $\mathcal{S}$. Using Eq. (32.A.23), the condition (32.A.25) may be written as the requirement that $\mathcal{S}^{-1}\mathcal{S}^*(\mathcal{C}\beta)^{-1}$ commutes with all $\mathcal{J}^{\pm}_{\mu\nu}$. Since the matrices $\mathcal{J}^{\pm}_{\mu\nu}$ form an irreducible set, this would require that $\mathcal{S}^{-1}\mathcal{S}^*(\mathcal{C}\beta)^{-1}$ be proportional to the unit matrix

$$\mathcal{C}\beta = \alpha\mathcal{S}^{-1}\mathcal{S}^*\,, \tag{32.A.26}$$

with α some constant. For this to be possible, we must have

$$\mathcal{C}\beta\,(\mathcal{C}\beta)^* = |\alpha|^2\cdot 1\,. \tag{32.A.27}$$

But $\mathcal{C}\beta = \gamma_2\gamma_4\cdots\gamma_{2n-2}$, and since all γ_i with i even are imaginary, we have

$$\mathcal{C}\beta(\mathcal{C}\beta)^* = (-1)^{n-1}(\gamma_2\gamma_4\cdots\gamma_{2n-2})^2 = (-1)^a\cdot 1\,, \tag{32.A.28}$$

where

$$a = (n-1)+(n-2)+\cdots+1 = n(n-1)/2\,. \tag{32.A.29}$$

Hence the Weyl representations can only be real for $n = 1$ (mod 4) and must be pseudoreal for $n = 3$ (mod 4).

We also note for use in Section 32.1 that

$$\mathcal{C}^* = (-1)^n\mathcal{C}\,,\qquad \mathcal{C}^{\mathrm{T}} = (-1)^{n(n+1)/2}\mathcal{C}\,,\qquad \mathcal{C}^{-1} = (-1)^{n(n-1)/2}\mathcal{C}\,, \tag{32.A.30}$$

and therefore Eq. (32.A.13) gives

$$(\mathcal{C}\gamma_\mu)^{\mathrm{T}} = (-1)^{n(n-1)/2}\mathcal{C}\gamma_\mu\,. \tag{32.A.31}$$

The γ_μ form a vector, in the sense that

$$[\mathcal{J}_{\mu\nu},\gamma_\rho] = -i\gamma_\mu\eta_{\nu\rho} + i\gamma_\nu\eta_{\mu\rho}\,, \tag{32.A.32}$$

and they have normal parity, in the sense that

$$\beta\,\gamma_0\,\beta = +\gamma_0\,,\qquad \beta\,\gamma_i\,\beta = -\gamma_i\ \text{ for }\ i = 1,\ldots,2n-1\,. \tag{32.A.33}$$

The anticommutation relation (32.A.1) prevents us from constructing new tensors by taking symmetric products of γs, but it allows us to construct

* We say that one representation of the Lorentz algebra by matrices $\mathcal{L}_{\mu\nu}$ (such as $\mathcal{J}_{\mu\nu}$ or $\mathcal{J}^+_{\mu\nu}$ or $\mathcal{J}^-_{\mu\nu}$) is the complex conjugate of another representation by matrices $\mathcal{L}'_{\mu\nu}$ if $\mathcal{L}'_{\mu\nu} = -\mathcal{L}^*_{\mu\nu}$. The minus sign is included because the matrices that represent elements of the Lorentz group in the neighborhood of the identity are of the form $1+\frac{1}{2}i\,\omega^{\mu\nu}\mathcal{L}_{\mu\nu}$, with $\omega^{\mu\nu}$ real infinitesimals.

antisymmetric tensors of rank up to $2n$

$$\gamma_{[\mu_1}\gamma_{\mu_2}\cdots\gamma_{\mu_p]}\,, \tag{32.A.34}$$

where the square brackets indicate antisymmetrization, and $p \leq 2n$. Each has a number of independent spacetime components given by the binomial coefficient $\binom{2n}{p}$, so the total number of matrices of this type is

$$\sum_{p=0}^{2n}\binom{2n}{p} = 2^{2n}\,. \tag{32.A.35}$$

None of these vanishes (as can be seen by calculating their squares) and they all have different Lorentz and/or parity transformation rules and are therefore linearly independent, so any $2^n \times 2^n$ matrix may be written as a linear combination of the 2^{2n} antisymmetric tensors (32.A.34).

Odd Dimensions: $d = 2n + 1$

Now let us consider an odd spacetime dimensionality $d = 2n + 1$. We can easily find a set of $2n + 1$ Dirac $n \times n$ matrices satisfying the anticommutation relations (32.A.1): we simply use the same γ_μ with $\mu = 1, 2, \ldots 2n - 1, 0$ as for $d = 2n$, and add the matrix γ_{2n+1} defined by Eq. (32.A.17). According to Eqs. (32.A.18) and (32.A.19) these gamma matrices satisfy the anticommutation relations (32.A.1), with μ and ν running over the values $1, 2, \ldots, 2n - 1, 0, 2n + 1$, and again $\gamma_0 = i\gamma_{2n}$.

Unlike the case of even dimensionality, we cannot here find any nontrivial matrix that commutes with all the Lorentz generators, because Eqs. (32.A.17) and (32.A.18) show that the product of the $2n + 1$ gamma matrices is trivial:

$$\gamma_1\gamma_2\cdots\gamma_{2n}\gamma_{2n+1} = i^{-n}\cdot 1\,. \tag{32.A.36}$$

The Lorentz generators (32.A.2) with μ and ν running over the values $1, 2, \ldots, 2n - 1, 0, 2n + 1$ therefore furnish an irreducible representation of the Lorentz group by themselves. To test their reality properties, we note that γ_{2n+1} is real and symmetric and satisfies $(\mathscr{C}\beta)^{-1}\gamma_{2n+1}\mathscr{C}\beta = -(-1)^n\gamma_{2n+1}$, so Eq. (32.A.15) applies for $\mu = 2n + 1$ as well as for $\mu = 1, 2, \ldots, 2n - 1, 0$. The Lorentz generators therefore satisfy

$$\mathscr{J}^*_{\mu\nu} = -(\mathscr{C}\beta)^{-1}\mathscr{J}_{\mu\nu}\mathscr{C}\beta\,, \tag{32.A.37}$$

$$\mathscr{J}^{\rm T}_{\mu\nu} = -\mathscr{C}^{-1}\mathscr{J}_{\mu\nu}\mathscr{C}\,, \tag{32.A.38}$$

so in each odd dimension the fundamental spinor representation is either real or pseudoreal. Exactly the same argument as in the case where $d = 2n$

tells us that the spinor representations for $d = 2n + 1$ are again real or pseudoreal according to whether the sign $(-1)^a$ in Eq. (32.A.28) is positive or negative, and therefore according to Eq. (32.A.29) they are real for $n = 0$ (mod 4) and $n = 1$ (mod 4) and pseudoreal for $n = 2$ (mod 4) and $n = 3$ (mod 4).

We can again construct antisymmetric tensors (32.A.34), now of rank p up to $2n + 1$, but only half of these are independent, because they obey relations

$$\epsilon^{\mu_1\mu_2\cdots\mu_{2n+1}}\gamma_{[\mu_1}\gamma_{\mu_2}\cdots\gamma_{\mu_p]} \propto \gamma^{[\mu_{p+1}}\gamma^{\mu_2}\cdots\gamma^{\mu_{2n+1}]}\,, \tag{32.A.39}$$

with $\epsilon^{\mu_1\mu_2\cdots\mu_{2n+1}}$ as usual totally antisymmetric. (For $d = 2n$ no such relations are possible, because the left- and right-hand sides of Eq. (32.A.39) have opposite parity, but this argument does not apply for $d = 2n + 1$, where $\epsilon^{\mu_1\mu_2\cdots\mu_{2n+1}}$ has even spatial parity.) The total number of independent matrices of the form (32.A.34) is now given by

$$\sum_{p=0}^{n}\binom{2n+1}{p} = 2^{2n}\,, \tag{32.A.40}$$

so any $2n \times 2n$ matrix may be written as a linear combination of the $n+1$ independent antisymmetric tensors (32.A.34) with $0 \leq p \leq n + 1$.

Finally, we note that for either $d = 2n$ or $d = 2n + 1$, the $O(d - 1, 1)$ Dirac and Lorentz algebras are related to the corresponding $O(d)$ algebras by setting

$$\gamma_{2n} \equiv -i\gamma_0\,, \qquad \mathscr{J}_{i\,2n} \equiv -i\mathscr{J}_{i0}\,, \tag{32.A.41}$$

so that

$$\{\gamma_i, \gamma_j\} = 2\delta_{ij}\,, \tag{32.A.42}$$

and

$$\mathscr{J}_{ij} = \frac{1}{4i}[\gamma_i, \gamma_j] = -\mathscr{J}_{ji}\,, \tag{32.A.43}$$

with i and j running from 1 to d. It follows from Eq. (32.A.42) that for $i \neq j$, $\mathscr{J}_{ij}^2 = 1/4$, so the eigenvalues of each $\mathscr{J}_{ij}$ are limited to $\pm 1/2$. To be more specific, in the fundamental spinor representation the generators of the Cartan subalgebra are represented by

$$\mathscr{J}_{2u-1\,2u} = \frac{1}{2}[a_u, a_u^*] = a_u a_u^* - \frac{1}{2}\,, \tag{32.A.44}$$

for which the basis vectors (32.A.8) are eigenvectors, with

$$\mathscr{J}_{2u-1\,2u}\Big|s_1\, s_2\,\cdots s_n\Big\rangle = \left(s_u - \frac{1}{2}\right)\Big|s_1\, s_2\,\cdots s_n\Big\rangle\,. \tag{32.A.45}$$

The difference between dimensionalities $d = 2n$ and $d = 2n + 1$ is that for $d = 2n$ we have two fundamental spinor representations in which the eigenvalue $(-2\sigma_1)(-2\sigma_2)\cdots(-2\sigma_n)$ of γ_{2n+1} is constrained to be $+1$ or -1, while for $d = 2n + 1$ there is one fundamental spinor representation with no such limitation on the σ_u.

It was the limitation of the eigenvalues of each $\mathscr{J}_{ij}$ to $\pm 1/2$ that in Section 32.1 identified the fundamental spinor representations as the only possible representations of the Lorentz algebra that could be furnished by fermionic symmetry generators. Indeed, from this condition we could have inferred that the $O(d)$ generators may be represented in a basis of the form (32.A.8), with $s_u = \sigma_u + 1/2$, and carried out the derivations of this appendix in reverse order, using Eqs. (32.A.4)–(32.A.7) (along with Eq. (32.A.17) for d odd) to express the Lorentz generators in terms of a set of γ_μ satisfying the anticommutation relations (32.A.1).

Problems

1. Classify the massless particle multiplets for each allowed kind of supersymmetry in six spacetime dimensions, when all central charges vanish.

2. Suppose that it were possible to have massless particles for all spins up to $j = 3$, but no higher. Taking into account the fact that massless particles of spin 2 do exist, what is the maximum spacetime dimensionality in which supersymmetry is possible? What is the maximum number of supersymmetry generators for each allowed value of the spacetime dimensionality?

3. Consider types IIA and IIB supersymmetry in ten spacetime dimensions. Assume that only scalar central charges appear in the extended supersymmetry anticommutation relations. Find a lower bound on particle masses in terms of these central charges. Describe the 'BPS' massive particle multiplets allowed for particles whose masses are at this lower bound.

4. List the possible independent scalar and/or tensor central charges for $N = 1$ supersymmetry in nine spacetime dimensions.

References

1. T. Kaluza, *Sitz. Preuss. Akad. Wiss.* **K1**, 966 (1921).
2. O. Klein, *Z. Phys.* **37**, 895 (1926); *Nature* **118**, 516 (1926).

3. J. H. Schwarz, *Nucl. Phys.* **B46**, 61 (1972); R. C. Brower and K. A. Friedman, *Phys. Rev.* **D7**, 535 (1972).

4. The supergravity field theory in 11 spacetime dimensions was formulated by E. Cremmer, B. Julia, and J. Scherk, *Phys. Lett.* **76B**, 409 (1978). The idea that weakly coupled Type IIA string theories in 10 spacetime dimensions have an 11-dimensional origin is due to M. J. Duff, P. S. Howe, T. Inami, and K. Stelle, *Phys. Lett.* **B191**, 70 (1987). This was shown for strongly coupled Type IIA string theories in ten dimensions by P. K. Townsend, *Phys. Lett.* **B350**, 184 (1995). Connections between these theories and all the other ten-dimensional string theories were then pointed out by E. Witten, *Nucl. Phys.* **B445**, 85 (1995).

5. R. Haag, J. Lopuszanski, and M. Sohnius, *Nucl. Phys.* **B88**, 257 (1975).

6. W. Nahm, *Nucl. Phys.* **B135**, 149 (1978).

6*a*. See, for example, H. W. Turnbull and A. C. Aitkens, *An Introduction to the Theory of Canonical Matrices* (Dover Publications, New York, 1961).

6*b*. A useful summary is given by J. Strathdee, *Int. J. Mod. Phys.* **A2**, 273 (1987).

7. For these arguments in four spacetime dimensions, see S. Weinberg, *Phys. Rev.* **135**, B1049 (1964); *Phys. Rev.* **138**, B988 (1965); R. P. Feynman, unpublished; M. T. Grisaru and H. N. Pendleton, *Phys. Lett.* **67B**, 323 (1977). The arguments in higher spacetime dimensions are similiar.

8. See, for example, J. Hughes, J. Liu, and J. Polchinski, *Phys. Lett.* **B180**, 370 (1986); E. Bergshoeff, E. Sezgin, P. K. Townsend, *Phys. Lett.* **B189**, 75 (1987); A. Achúcarro, J. M. Evans, P. K. Townsend, and D. L. Wiltshire, *Phys. Lett.* **B198**, 441 (1987); P. K. Townsend, *Phys. Lett.* **B202**, 53 (1988); P. K. Townsend, in *Particles, Strings, and Cosmology: Proceedings of Workshop on Current Problems in Particle Theory 19 at Johns Hopkins University, March 1995* (World Scientific, Singapore, 1996); J. Maldacena, *Adv. Theor. Math. Phys.* **2**, 231 (1998). For reviews, see M. J. Duff, R. R. Khuri, and J.-X. Lu, *Phys. Rep.* **259**, 213 (1995); A. Giveon and D. Kutasov, 1998 preprint hep-th/9802067, to be published in *Rev. Mod. Phys.*

9. J. W. van Holten and A. Van Proeyen, *J. Phys. A: Math. Gen.* **15**, 3763 (1982).

10. J. A. de Azcárraga, J. P. Gauntlett, J. M. Izquierdo, and P. K. Townsend *Phys. Rev. Lett.* **63**, 2443 (1989).

11. D. Sorokin and P. K. Townsend, *Phys. Lett.* **B412**, 265 (1997); J. P. Gauntlett, J. Gomis, and P. K. Townsend, *J. High Energy Phys.* **9801**, 003 (1998).
12. The covariant field equations for these 2-branes were given by E. Bergshoeff, E. Sezgin, P. K. Townsend, Ref. 8 and *Ann. Phys. (NY)* **185**, 330 (1988). The demonstration that these 2-branes are solutions of the supergravity field equations that provide sources for the 3-form gauge field is due to M. J. Duff and K. Stelle, *Phys. Lett.* **B253**, 113 (1991).
13. The demonstration that these 5-branes are solutions of the supergravity field equations that provide sources for the 6-form gauge field was given by R. Gueven, *Phys. Lett.* **B276**, 49 (1992). The covariant field equations for these 5-branes were given by M. Aganagic, J. Park, C. Popescu, and J. Schwarz, *Nucl. Phys.* **B496**, 191 (1997); P. S. Howe and E. Sezgin, *Phys. Lett.* **B394**, 62 (1997); P. Pasti, D. Sorokin, and M. Tonin, *Phys. Lett.* **B398**, 41 (1997); P. S. Howe, E. Sezgin, and P. C. West, *Phys. Lett.* **B399**, 49 (1997); I. Bandos, K. Lechner, A. Nurmagambetov, P. Pasti, D. Sorokin, and M. Tonin, *Phys. Rev. Lett.* **78**, 4332 (1997).

Author Index

Where page numbers are given in *italics*, they refer to publications cited in lists of references.

Subject Index

Printed in the United States
By Bookmasters